Integrated Approaches to Water Pollution and Control

Integrated Approaches to Water Pollution and Control

Editor: Sheryl McMillan

Callisto Reference
www.callistoreference.com

Callisto Reference,
118-35 Queens Blvd., Suite 400,
Forest Hills, NY 11375, USA

Visit us on the World Wide Web at:
www.callistoreference.com

ISBN: 978-1-63239-941-0 (Hardback)

The publisher's policy is to use permanent paper from mills that operate a sustainable forestry policy. Furthermore, the publisher ensures that the text paper and cover boards used have met acceptable environmental accreditation standards.

Printed in China.

Cataloging-in-Publication Data

Integrated approaches to water pollution and control / edited by Sheryl McMillan.
p. cm.
Includes bibliographical references and index.
ISBN 978-1-63239-941-0
1. Water--Pollution. 2. Pollution prevention. I. McMillan, Sheryl.
TD420 .I58 2018
628.168--dc23

Table of Contents

Preface

Every book is a source of knowledge and this one is no exception. The idea that led to the conceptualization of this book was the fact that the world is advancing rapidly; which makes it crucial to document the progress in every field. I am aware that a lot of data is already available, yet, there is a lot more to learn. Hence, I accepted the responsibility of editing this book and contributing my knowledge to the community.

The introduction of harmful pollutants into water bodies is known as water pollution. Sources of water pollution include point sources such as factories, non-point sources such as pollution through runoff and groundwater pollution through aquifers, etc. Contaminants include organic substances like detergents, pesticides and insecticides, petroleum, etc. and inorganic pollutants like ammonia, heavy metals, fertilizers, etc. The topics included in this book on water pollution are of utmost significance and are bound to provide incredible insights to readers. It is a complete source of knowledge on the present status of this important field.

While editing this book, I had multiple visions for it. Then I finally narrowed down to make every chapter a sole standing text explaining a particular topic, so that they can be used independently. However, the umbrella subject sinews them into a common theme. This makes the book a unique platform of knowledge.

I would like to give the major credit of this book to the experts from every corner of the world, who took the time to share their expertise with us. Also, I owe the completion of this book to the never-ending support of my family, who supported me throughout the project.

Editor

1

Do Contaminants Originating from State-of-the-Art Treated Wastewater Impact the Ecological Quality of Surface Waters?

Daniel Stalter[1*¤], Axel Magdeburg[1], Kristin Quednow[2], Alexandra Botzat[3], Jörg Oehlmann[1]

1 Department Aquatic Ecotoxicology, Biological Sciences Division, Goethe University Frankfurt am Main, Frankfurt am Main, Germany, **2** Analytical Environmental Chemistry, Institute for Atmospheric and Environmental Sciences, Goethe University Frankfurt am Main, Frankfurt am Main, Germany, **3** Department of Ecology – Conservation Ecology, Faculty of Biology, Philipps University of Marburg, Marburg, Germany

Abstract

Since the 1980s, advances in wastewater treatment technology have led to considerably improved surface water quality in the urban areas of many high income countries. However, trace concentrations of organic wastewater-associated contaminants may still pose a key environmental hazard impairing the ecological quality of surface waters. To identify key impact factors, we analyzed the effects of a wide range of anthropogenic and environmental variables on the aquatic macroinvertebrate community. We assessed ecological water quality at 26 sampling sites in four urban German lowland river systems with a 0–100% load of state-of-the-art biological activated sludge treated wastewater. The chemical analysis suite comprised 12 organic contaminants (five phosphor organic flame retardants, two musk fragrances, bisphenol A, nonylphenol, octylphenol, diethyltoluamide, terbutryn), 16 polycyclic aromatic hydrocarbons, and 12 heavy metals. Non-metric multidimensional scaling identified organic contaminants that are mainly wastewater-associated (*i.e.*, phosphor organic flame retardants, musk fragrances, and diethyltoluamide) as a major impact variable on macroinvertebrate species composition. The structural degradation of streams was also identified as a significant factor. Multiple linear regression models revealed a significant impact of organic contaminants on invertebrate populations, in particular on Ephemeroptera, Plecoptera, and Trichoptera species. Spearman rank correlation analyses confirmed wastewater-associated organic contaminants as the most significant variable negatively impacting the biodiversity of sensitive macroinvertebrate species. In addition to increased aquatic pollution with organic contaminants, a greater wastewater fraction was accompanied by a slight decrease in oxygen concentration and an increase in salinity. This study highlights the importance of reducing the wastewater-associated impact on surface waters. For aquatic ecosystems in urban areas this would lead to: (i) improvement of the ecological integrity, (ii) reduction of biodiversity loss, and (iii) faster achievement of objectives of legislative requirements, *e.g.*, the European Water Framework Directive.

Editor: Silvana Allodi, Federal University of Rio de Janeiro, Brazil

Funding: This work was supported by grants of the Hessian Ministry for Science and the Arts (Germany). The funders had no role in study design, data collection and analysis, decision to publish, or preparation of the manuscript.

Competing Interests: The authors have declared that no competing interests exist.

* E-mail: d.stalter@uq.edu.au

¤ Current address: National Research Centre for Environmental Toxicology, The University of Queensland, Brisbane, Australia

Introduction

In urban areas, the water quality of small streams is predominantly impacted by structural degradation of stream morphology (*e.g.*, channelization and straightening), agricultural land use in the catchment, and a high load of treated wastewater (WW). These are also the most apparent drivers of ecological conditions in anthropogenically disturbed surface waters. Together, these stressors have led to significant declines in aquatic fauna populations and biodiversity, in particular in freshwater ecosystems, which in turn has had a profound impact on the ecological integrity of many aquatic ecosystems [1–5].

The ecological quality of surface water is commonly assessed by analyzing benthic macroinvertebrate species assemblages [6]. The highly diverse group of aquatic invertebrate organisms encompasses many euryoecious and stenoecious species [7]. The latter have narrow ecological requirements in terms of physico-chemical water parameters and structural characteristics of the water body. Consequently, assessing the invertebrate community can provide a picture of the 'health' of the water body and can be used to identify factors affecting its ecological quality (*e.g.*, structural deterioration, acidification, oxygen deficiency). The technique involves comparing the composition and abundance of the present invertebrate taxa to established reference communities from anthropogenically least-disturbed reference sites.

Evaluating and improving the ecological status of surface waters is increasingly perceived to be important, which is reflected by legislation changes in this area [6,8]. For this purpose, worldwide many different macroinvertebrate-based rating systems for ecological status evaluation with hundreds of indices, metrics, and evaluation tools have been developed [6,8]. In the European Union, the Water Framework Directive (WFD) [9]–enacted in 2000–set an objective that all coastal and inland waters of Europe be of a 'good status' by 2015. One tool for evaluating the quality of these waters is the ASTERICS software (AQEM/STAR Ecological River Classification System), which assists the assessment of ecological status. If the outcomes of ASTERICS analysis indicate a poor ecological status, key causal factors and mitigation and remediation options must be identified and explored in order to achieve the 'good status' benchmark.

Until the early 1980s in Germany, the primary driver of the deterioration of aquatic ecosystems was the large input of nutrients via WW discharge [10]. However, as progressive WW treatment and other restoration measures increased throughout the 1980s, nutrient inputs were diminished, eutrophication reduced, and water quality impacts were ameliorated, hence allowing aquatic ecosystems to recover [11]. Generally, in high-income countries, eutrophication is largely under control at present [12,13] because of the advances in, and the prevalence of WW treatment [14].

Despite the effective removal of nutrients and other contaminants, treated effluent often contains trace organic contaminants that have been demonstrated to have negative impacts on aquatic ecosystems [13]. These anthropogenic water pollutants and their transformation products are often present at low to very low concentrations (*e.g.*, ng/L). Despite such minimal trace level concentrations, the large spectrum of organic contaminants occurring in surface waters [15] may pose a potential threat to aquatic wildlife, particularly with regard to mixture toxicity. Pharmaceuticals and personal care products in particular often exhibit high biological activity–some of which are also persistent in the environment [16], thus giving them the potential to affect aquatic species [17–23]. The presence of these contaminants in urban surface water has been primarily attributed to environmental releases of municipal and industrial WW [13]. Accordingly, we anticipate there to be a clear negative impact on macroinvertebrate assemblages and the general ecological status of surface waters in proximity to WW treatment plants.

Many studies have addressed the impact of water pollution on invertebrate communities, but frequently they consider only severely contaminated sites or individual pollutants (*e.g.*, [24]). In this study, we have instead focused on aquatic invertebrate exposure to the 'subtle pollution' (*i.e.*, mixtures of organic micropollutants) occurring in German lowland streams. All treatment plants currently operating in the study area are equipped with biological activated sludge treatment–the most widely used WW treatment system in high-income countries [25]. As well as the elimination of nutrients, this technology has also been shown to effectively reduce the contaminant load and toxicity of WW [26,27]. However, a considerable fraction of toxic contaminants remains prevalent in WW treated by biological activated sludge [27].

This study aimed to enhance the understanding of factors that disturb the ecological integrity of surface waters in order to assist decision-making around remediation and mitigation of factors affecting the quality of surface waters. The primary objective was to evaluate whether lowland stream macroinvertebrate assemblages are affected by their loads of WW-associated trace organic contaminants (OC) in the water phase. Additionally, the potential impacts of heavy metals (HM) and polycyclic aromatic hydrocarbons (PAH) in sediment were investigated.

Materials and Methods

Study Area

The study was conducted in the Hessian Ried (HR) close to Frankfurt (Main), Germany. The region extends about 1200 km^2 and is bound by the river Main in the South, the lower mountain range Odenwald in the East, and the river Rhine in the West. The accumulation of sand and gravel, as a result of aggradational deposits, forms a large aquifer that serves as drinking water reservoir for the Frankfurt/Rhine-Main conurbation [28]. The population of the HR is approximately 800,000. Types of land use include intensive agriculture (predominantly cultivation of vegetables, fruit, wheat, and forage crops) and industry (*e.g.*, chemical and manufacturing industries), and high-density traffic highways occur in the region [28]. Due to this high level of anthropogenic disturbance, the importance of the aquifer in the region, and the homogenous distribution of urbanization, the HR was selected as a representative urban study area to investigate the impacts of water pollution, physico-chemical factors, and structural characteristics on the ecological quality of urbanized aquatic ecosystems.

Study Design

To identify key impact factors on the ecological quality, we assessed the benthic macroinvertebrate community of the streams, recorded a range of environmental and anthropogenic variables, and evaluated their potential impact on the species composition with the major focus on wastewater-associated contaminants. We conducted this study from 2005 to 2006 at 26 sampling sites in four river systems that run through the study area in an east-west direction and discharge into the Rhine River (Table 1). Most sampling sites belong to the water body type 19 (small streams in floodplains) according to the WFD [9,29]. Exceptions are sampling sites 1 to 6 in the Weschnitz stream (type 9: mid-sized siliceous rivers in the lower mountain ranges with fine to crude sediment) as well as Sw1 (type 6: small calcareous sediment rich streams in the lower mountain ranges) and Mo1 (type 5: small siliceous cobble-bottom streams in the lower mountain ranges [30]). 28 sewage treatment plants are currently operating in the catchment. The location of the sampling sites relative to these plants, as well as the quantity of effluent discharged to streams, resulted in different WW loads at each of the sampling sites. This factor was defined on a scale from 0–100% (Figure S1–4). The only sampling site without WW was Wi1.

Structural Quality of the Streams

To classify the streams with respect to their structural modification, we recorded the morphology or water structure quality following Zumbroich *et al.* [31] from class 1 (near-natural) to class 7 (completely modified). Additionally, the stream bed structure was characterized in detail (relative coverage of 10 different sediment types, Table S1) according to the macroinvertebrate sampling procedure described in Hering *et al.* [7].

Table 1. River systems with respective streams, abbreviations (abbr.), and number of sampling sites (no.).

river systems	streams	abbr.	no.
Schwarzbach-Landgraben-system	Schwarzbach, Apfelbach	Sw	6
Schwarzbach-Landgraben-system	Landgraben, Darmbach	La	2
Modau-Sandbach-system	Sandbach	Sa	3
Modau-Sandbach-system	Modau	Mo	4
Winkelbach	Winkelbach	Wi	5
Weschnitz	Weschnitz	We	6

Benthic Macroinvertebrate Community

We sampled the invertebrates in autumn and spring (September 2005 and March 2006) following the multi-habitat sampling procedure described in Hering *et al.* [7]. No specific permits were required for the described field studies as none of the sampling sites are privately-owned or protected in any way. The responsible Hessian State Office for Environment and Geology was informed prior to the study. Before sampling, we assessed the substrate composition of the sampling sites of a 100-m transect of each water body. The different substrate types were sampled according to their relative coverage with a total of 20 sub-samples per sampling site. We used a Surber-sampler (manufactured by precision engineers at the Goethe University of Frankfurt, Germany) with 500-µm mesh size and 20×25-cm frame size for macroinvertebrate sampling, resulting in 1-m^2 sampling plots per sampling site. Endangered or protected species were identified in the field and immediately released back to the streams. The remaining sample material was stored in 70% ethanol (denatured, Carl Roth GmbH, Darmstadt, Germany). Macroinvertebrates were separated as described in Haase & Sundermann [32] and determined to species level, if possible, but at least to the minimum taxonomic level recommended by Haase *et al.* [33] (Table S2–3). For separation and identification of the species we used stereo microscopes (Stemi 2000, Carl Zeiss AG, Oberkochen, Germany) and a microscope (BX 50, Olympus, Tokio, Japan) if a higher magnification was required, *e.g.*, for the identification of Ephemeroptera species. We could not sample Mo4 and Sw6 during the autumn sampling campaign due to high water levels.

Contaminants

As WW-associated pollutants commonly correlate in their occurrence, distinguishing their individual impacts is hardly feasible. Accordingly, a number of representative pollutants were selected to account for the burden of WW-associated OCs in general. At 11 sampling campaigns, water samples from each sampling site were analyzed for 12 common organic pollutants: five organophosphates; two musk fragrances; bisphenol A; the alkylphenols, nonylphenol and octylphenol; and the insect repellent diethyltoluamide (Table 2). Pollutants were selected due to their ubiquitous occurrence in municipal WW. Unexpectedly, terbutryn–in Europe formerly used as agricultural herbicide, which is still authorized as a biocide in antifouling paints and coatings–was detected in all water samples from the beginning of the sampling campaigns, and therefore was also included in the analysis suite. For more detailed information on analytical procedures and pollutant concentrations, see Quednow & Puettmann [34–37]. To take into account a possible influence of sediment-bound pollutants, the sediment load of 12 HMs and metalloids as well as 16 PAHs (Table 2)–the latter selected as proposed by the US EPA [38]–was determined according to standard methods [39,40]. Average contaminant concentrations are given in Table S4–6.

Water and Sediment Quality

We determined the organic carbon content and average grain size of sediments according to standard methods [41,42]. We recorded conductivity, temperature, O_2-concentration and pH (using multiparameter instrument Multi 350i, WTW, Weilheim, Germany) as well as flow velocity (using a hydrometric vane from OTT Hydromet, Kempten, Germany) during sampling events. Colorimetric on-site tests (Aquaquant®, Merckoquant®, Merck, Darmstadt, Germany) were used to determine ammonium, chloride, phosphate, total hardness, and carbonate hardness. Biological oxygen demand (BOD_5) was analyzed according to a standard method [43].

Data Analysis

We used the ASTERICS software [44] to calculate the ecological status (from 1–5: bad, poor, moderate, good, high) and biotic metrics (*e.g.*, saprobic index, Shannon index, Simpson index, number of EPT taxa) for each sampling site. To reduce the number of contaminant variables, separate principal component analyses (PCAs) were performed using SPSS (version 14, SPSS Inc., Chicago, USA, 2005) with the concentrations of OCs in the water phase and concentrations of PAHs and HMs in the sediments (Table S7–9). We only considered components with an eigenvalue >1 to limit the number of resulting components. The component scores were determined by regression and were standardized with a mean of 0 and standard deviation of 1 [45]. The resulting components (OC1–4, HM1–3, and PAH1–2) with the corresponding component scores were used to replace the contaminant concentrations in all further analyses.

Using the statistic software R (R Development Core Team 2011) we applied non-metric multidimensional scaling (NMDS) using Bray-Curtis dissimilarity as implemented in R package vegan 2.0–2 [46] based on abundance data to detect differences of macroinvertebrate taxa composition among the 26 sample sites. NMDS displays dissimilarities in community composition nonlinearly onto ordination space, can cope with nonlinear species responses, and is not constrained by predictors [46]. We excluded rare species that occurred less than three times and that were present in less than three sampling sites. We used taxa scores in the biplot to illustrate differences between sampling sites. Non-correlated variables were selected (structural quality, average discharge, organic carbon content of the sediments, and following contaminant factors: OC1, OC2, HM1, HM3, PAH1, PAH2; Table S1) and were fitted post-hoc to the ordination and their significance was tested via random permutations (1000 iterations).

We used multiple linear regression models to test the impact of WW-associated contaminants and environmental variables on the following dependent variables: total abundance of macroinvertebrates, number of macroinvertebrate taxa, Simpson and Shannon diversity of macroinvertebrates, number of EPT taxa, and saprobic index. We started with models comprising the same set of non-correlated variables as used for the NMDS. We used the step function implemented in R for model selection to reach minimum adequate models. This selection process is based on minimizing information loss according to Akaike's Information Criterion values [47].

We performed Spearman rank correlation analyses using species abundances, biotic metrics, the PCA-derived contaminant factors, and all environmental variables (Table S1). To take into account the toxicity of the analyzed WW-associated contaminants, toxic units (TUs) were calculated as described in Liess & Von Der Ohe [48] and included in the correlation analysis. The TU values are based on effect concentrations (LC_{50} and 'no observed effect concentration') from acute tests with *Chironomus riparius* and chronic tests with *Daphnia magna*). In addition to the median pollutant concentrations, the 90th percentile of the concentrations was used for TU calculations to estimate whether the peak loads influenced the taxa composition more than the mean concentrations. TUs were summed for each sampling site and included in the correlation analysis.

Table 2. Analyzed contaminants.

phase analyzed	contaminant group	contaminants
water	organophosphates	TBP (tributyl phosphate), TBEP (tris(2-butoxyethyl)phosphate), TCEP (tris(2-chloroethyl)phosphate),TCPP (tris(2-chloro-, 1-methyl-ethyl)-phosphate), TDCPP (tris(1,3-dichloro-2-propyl) phosphate)
	biphenols	BPA (bisphenol A)
	musk fragrances	HHCB (1,3,4,6,7,8-hexahydro-4,6,6,7,8,8-hexamethylcyclopenta-γ-2-benzopyran), AHTN (6-Acetyl-1,1,2,4,4,7- hexamethyltetraline)
	alkylphenols	octylphenol, nonylphenol
	triazines	terbutryn
	amides	DEET (diethyltoluamide)
sediment	heavy metals	aluminium, arsenic, barium, cadmium, cobalt, chromium, copper, iron, manganese, nickel, lead, zinc
	polycyclic aromatic hydrocarbons	naphthalene, acenaphtylene, acenaphthene, fluorene, phenanthrene, anthracene, fluoranthene, pyrene, benzo(a)anthracene, chrysene, benzo(b)fluoranthene, benzo(k)fluoranthene, benzo(a)pyrene, dibenz(ah)anthracene, benzo(ghi)perylene, indeno(1,2,3-cd)pyrene

Results

Sampling Sites and Species Data

The running waters of the study area were mostly structurally distorted with channel-like character, usually with trapezoidal profiles, low depth variance, large profile depth, and almost a lack of curvature erosion. Accordingly, the degree of anthropogenic deformation varied mostly between class 4 (significantly modified) to 7 (completely modified) with only La2 and Mo1 as slightly modified (class 2) and Sw1 as single near-natural stream (class 1; Figure S5). We identified a total of 141 different taxa across all sampling sites (Table S2–3). The number of taxa exceeded 50 at Wi1 and Wi2 only, while falling below 20 at La1 and Sw6 (Figure S5). The lowest Shannon diversity was determined for La1 (0.8) and the highest for Sw1 (3.0). At La1 and Sw2, the invertebrate abundance was dominated by Oligochaeta and *Gammarus roeselii*, contributing more than 50% to the total number of individuals (dominance structure in Figure S5 given as contribution of individual taxa to the total number of individuals per m^2 in decreasing order). Plecoptera were found at Sw1 and We1 only. Invasive species like *Corbicula fluminea*, *Potamopyrgus antipodarum*, and *Dikerogammarus villosus* were commonly found in the sampling area. At Sa2 *P. antipodarum* dominated species assemblage in terms of individuals per m^2. The ecological status was evaluated as 'good' for two sampling sites only (Wi2 and Wi4), while 50% of the sites were evaluated as insufficient or poor (Figure S5).

PCA-derived Variable Reduction of Contaminants

Four components of the PCA of OCs explained *ca.* 88% of the total variance while component loadings (*i.e.*, correlation coefficients between the variables and components) ranged from 0.55–0.97. The first component (OC1) encompassed the organophosphates as well as the musk fragrances and DEET (Table S7). OC1 is the contaminant factor that correlated to the highest degree with the WW load (Spearman's $\rho = 0.932$, Table S1) and is therefore regarded as predominantly WW-associated. Octylphenol and terbutryn were mainly represented by the second component (OC2), bisphenol A by the third (OC3), and nonylphenol by the fourth (OC4) component. The three components of HMs explained *ca.* 78% of the total variance. The component loadings ranged from 0.64–0.92. The first component (HM1) represented primarily Co, Cr, Cu, Fe, Mn, Ni, Pb, and Zn (Table S8) while the second (HM2) largely encompassed Al, As, and Ba. The third component (HM3) represented almost exclusively Cd. The two components calculated for PAHs explained more than 92% of the total variance, whereas component loadings were mainly >0.9 (except for Acyl and Fl, Table S9). The first component (PAH1) embodied 14 of the 16 PAHs while the second (PAH2) represented acenaphtylene and fluorine. PCA reduced the total number of contaminant variables from 40 variables to 9 variables, which were then used for subsequent analyses.

Dissimilarities in Macroinvertebrate Assemblages Related to Structural Degradation and Contaminants

The NMDS revealed the water body structure as one of the major impact variables in both sampling campaigns ($R^2 = 0.65$ and 0.56; $p = 0.005$ and 0.002; Figure 1A, B). Other significant impact factors for the taxa composition in spring included OC1 ($R^2 = 0.43$, $p = 0.006$) and HM3 ($R^2 = 0.72$, $p = 0.005$). In the autumn sampling event, HM1 was identified as a significant impact factor ($R^2 = 0.33$, $p = 0.014$), while OC1 and HM3 showed no significant effect (OC1: $R^2 = 0.18$, $p = 0.11$; HM3: $R^2 = 0.04$, $p = 0.57$).

Structural degradation was associated with increased abundances of Diptera and Turbellaria, demonstrating the euryoecious biology of respective species (*e.g.*, Chironomidae; Figure 1). On the other hand, abundances of more stenoecious Plecoptera species decreased with increased structural modification (Figure 1).

For the spring sampling event, the multiple linear regression model identified OC1 as the most significant impact variable on all biotic metrics apart from the number of individuals per m^2, which was more affected by PAH1 and PAH2 (Table 3). In the autumn sampling event, water body structure was the most significant impact variable on individuals per m^2 and number of taxa, while for metrics covering more sensitive species (EPT taxa and saprobic index; *e.g.*, [49]), the WW-associated contaminant factor, OC1, was the most prominent impact variable (Table 3). This might indicate a higher susceptibility of respective taxa to corresponding organic contaminants. HM1, HM3, PAH1 and average discharge are all related to Shannon and Simpson diversity. The affecting variables identified by the regression model (water body structure, OCl, HM1 and HM3) were similar to the results of the NMDS

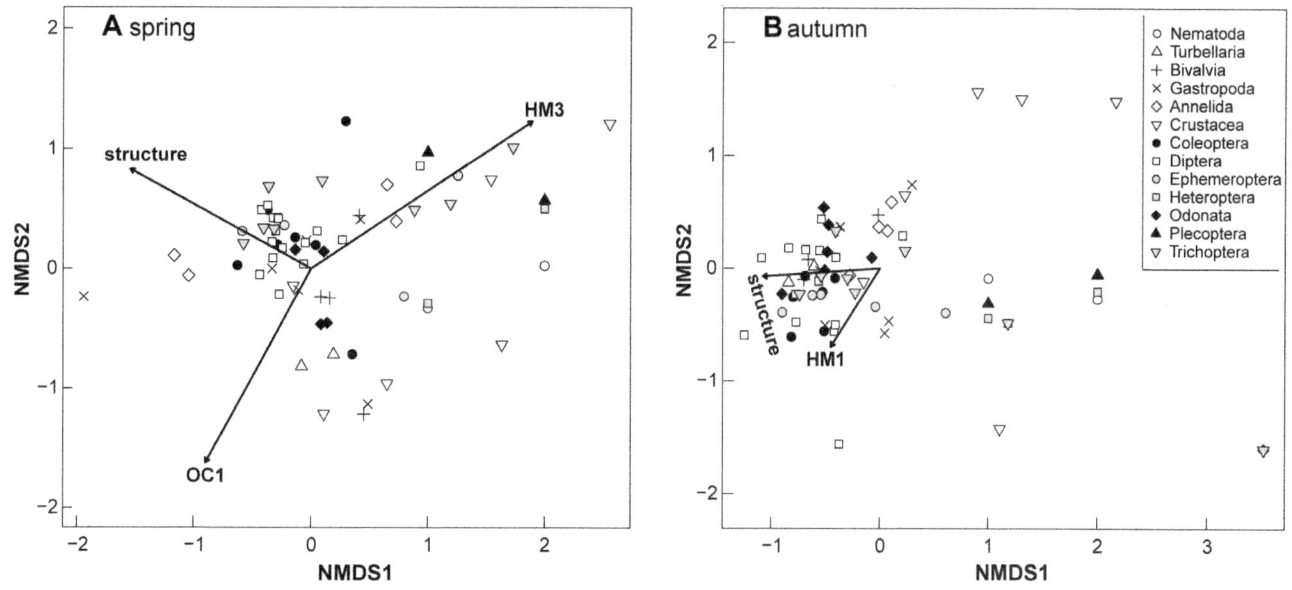

Figure 1. NMDS biplot of taxa and environmental variables. Displayed are variables with a significant impact ($p<0.05$) for sampling campaign in spring (A) and autumn (B). HM, components of the principal component analysis (PCA) with heavy metals; OC, components of the PCA with organic contaminants; structure, structural degradation. Spring: two convergent solutions, two dimensions, stress = 0.17; autumn: two convergent solutions, two dimensions, stress = 0.21).

although with the NMDS we considered the whole set of recorded species. Seasonal differences in species composition between both sampling campaigns might explain differing results between spring and autumn (*e.g.*, [50]).

The correlation analyses–including macroinvertebrate taxa, biotic indices, and environmental variables–revealed consistent and highly significant correlations between organic pollutant concentration (OC1) and indices encompassing sensitive taxa. Biotic metrics like the saprobic index (Figure 2), number of EPT taxa (Figure 3), and EPTCOB taxa (Ephemeroptera, Plecoptera, Trichoptera, Coleoptera, Odonata, Bivalvia) significantly correlated with OC1 ($p<0.001$) revealing correlation coefficients (Spearman's ρ) ranging from −0.63 to −0.81 (ρ = 0.73 and ρ = 0.77 in case of the saprobic index). Moreover, single genera of the orders Ephemeroptera and Coleoptera correlated significantly with OC1 (*Ephemera danica*: ρ = −0.70; *Elmis/Limnius/Haliplus*: ρ ranging from −0.45 to −0.53). Accordingly, numbers of Ephemeroptera and Coleoptera taxa and their abundances also highly negatively correlated with OC1 ($p<0.001$; $\rho<-0.7$).

Discussion

Key Factors Impacting Species Composition

The results of NMDS and the multiple linear regressions suggest that WW-associated organic contaminants (OC1)–alongside with structural degradation–play a pivotal role in shaping macroinvertebrate species composition in anthropogenically disturbed surface waters. The correlation analyses confirmed a deleterious relationship and indicated a reduced biodiversity of EPT taxa and Coleoptera with increasing concentrations of WW-associated trace organic contaminants. The number of EPT taxa as well as the saprobic index did not correlate with other contaminant factors that were identified as significantly affecting the macroinvertebrates in NMDS and multiple linear regression models (OC2, PAH1, HM1, HM3). This might indicate that OC1 and/or co-correlated variables were among the main negative impact factors for the macroinvertebrate communities.

Significant impact variables in the NMDS and the regression model did not necessarily have a deleterious effect. For instance, HM3 was identified as significantly affecting the species composition in the NMDS and regression analyses. HM3 represents the presence of Cd in the sediments (Table S8) suggesting a deleterious relationship due to toxic effects of Cd. However, a positive correlation of HM3 with the distance to the next WW treatment plant and a negative correlation with the WW load ($p<0.01$) indicated a non-causal relationship (Table S1). This is supported by predominantly low Cd concentrations of <0.9 mg/kg sediment (dry weight; apart from Mo1 with 2.4 mg/kg). Regarding factor PAH1, concentrations of PAHs exceed 'probable effect levels'–as proposed by Macdonald and colleagues [51]–at La1 and Mo3 by up to 20 times. At these sampling sites macroinvertebrate communities might be adversely affected. However, neither NMDS nor correlation analyses were able to support a significant impact of PAHs.

Furthermore, macroinvertebrate taxa and biotic metrics were not correlated with the water body structure. This might, however, mainly be a result of the rather homogenous sampling sites in terms of structural degradation (predominantly very highly modified, Figure S5) and does not indicate that water body structure is less relevant.

The ecological quality class determined via ASTERICS did not correlate with the contaminant factors. However, ASTERICS is designed to detect general degradation that is caused by multiple factors. The pervasive structural deterioration of the sampling sites (Figure S5) might have a predominant impact on the quality class determination masking any potential relationship between quality class and pollution.

In general, the correlation analyses revealed a principle problem with identifying environmental variables affecting species assemblages. The saprobic index and the number of EPT taxa, for example, were significantly correlated ($p<0.01$) because many

Table 3. Multiple linear regression models testing the effect of environmental parameters and contaminants on biotic response variables: the total number of individuals and taxa, Simpson and Shannon diversity, number of EPT taxa and the saprobic index.

response variable	significant impact variables									
	spring					autumn				
		p	df	F	R^2		p	df	F	R^2
individuals/m²	PAH2	★★	1	18.77	n.a.	water structure	★	1	5.14	n.a.
	PAH1	★★	1	8.63	n.a.	HM1	★	1	4.53	n.a.
	full model	★★	5,20	7.09	0.55	*full model*	n.s.	3,20	2.76	0.19
number of taxa	OC1	★★	1	22.02	n.a.	water structure	★★	1	9.09	n.a.
	PAH1	★	1	5.02	n.a.	average discharge	★★	1	8.47	n.a.
	HM3	★	1	4.37	n.a.	PAH1	★	1	7.39	n.a.
	full model	★★	4,21	10.62	0.61	*full model*	★★	3,20	5.69	0.38
Simpson diversity	OC1	★★	1	8.79	n.a.	HM3	★	1	5.35	n.a.
	OC2	★	1	6.05	n.a.	PAH1	★	1	4.89	n.a.
	HM1	★	1	6.19	n.a.	*full model*	n.s.	8,15	1.88	0.23
	full model	★★	5,20	10.48	0.65					
Shannon diversity	OC1	★★	1	12.30	n.a.	PAH1	★★	1	12.90	n.a.
	full model	★★	5,20	7.78	0.58	HM3	★★	1	8.97	n.a.
						average discharge	★	1	5.93	n.a.
						HM1	★	1	5.32	n.a.
						full model	★	5,18	3.37	0.34
EPT taxa	OC1	★★	1	10.50	n.a.	OC1	★★	1	15.82	n.a.
	full model	★★	4,21	4.99	0.39	water structure	★★	1	10.15	n.a.
						HM3	★★	1	9.18	n.a.
						full model	★★	5,18	4.68	0.44
saprobic index	OC1	★★	1	17.18	n.a.	OC1	★★	1	24.27	n.a.
	HM1	★	1	4.40	n.a.	HM3	★	1	5.70	n.a.
	full model	★★	4,21	9.07	0.56	HM1	★	1	4.86	n.a.
						full model	★★	6,17	7.48	0.63

Given are df-, R^2-, F- and p-values for full models after stepwise deletion of non-significant terms (n.s.) and of significant model parameters.
★, $p<0.05$;
★★, $p<0.01$;
★★★, $p<0.001$; n.a., not available.

EPT taxa exhibit low tolerance towards low oxygen conditions and hence have a low saprobic rate resulting in a considerable impact on the saprobic index. As the saprobic index is a measure of O_2 deficiencies due to decaying organic matter, impacts by insufficient O_2 concentrations and pollutant toxicities are hardly distinguishable. In our case, OC1 was significantly correlated with O_2 concentrations and biological oxygen demand (Table S1). In many studies low oxygen concentrations have been revealed to have a significant impact on benthic invertebrates, in particular on Ephemeroptera, Plecoptera, and Trichoptera (*e.g.*, [52]). However, our field measurements suggested good dissolved oxygen concentrations, with the bulk of measurements falling between 9.9 and 13.2 mg/L, with one exception: La1 with an average of 7.1 mg/L. Pronounced changes in aquatic species composition are not expected at concentrations >8 mg/L [53]. However, the presence and potential impact of an oxygen deficiency may have been underestimated because we did not record oxygen concentrations at night and increased nocturnal oxygen consumption by algae or submerged aquatic macrophytes can reduce oxygen saturation considerably [54].

OC1 was also associated with increased salinity ($\rho = 0.93$, Table S1). Salinity has been suspected to affect species composition elsewhere [53] and it is possible that the relationship observed between OCl and species assemblages was more related to salinity and not the WW-associated contaminants [55]. Again, Ephemeroptera, Plecoptera, and Trichoptera have been shown to react more sensitively to salinity than other benthic invertebrate species [55]. Therefore, a negative correlation of EPT taxa with salinity might be expected. Pond [56] documented a significant loss of mayfly taxa in Appalachian mining regions when conductivity exceeded 175 µS/cm. The average conductivity at our sampling sites ranged from 419 to 1350 µS/cm indicating a potential threat towards salinity-sensitive species. However, salinity in lowland streams is generally higher compared to streams in higher mountain ranges. Hence, lowland species should be more adapted to higher salinity levels and therefore salinity effects were expected to be less pronounced.

In the present study, salinity increase was mainly caused by the WW burden as indicated by the high correlation coefficient ($\rho = 0.93$) and was therefore regarded as anthropogenic disturbance. Depending on the causal factors influencing environmental

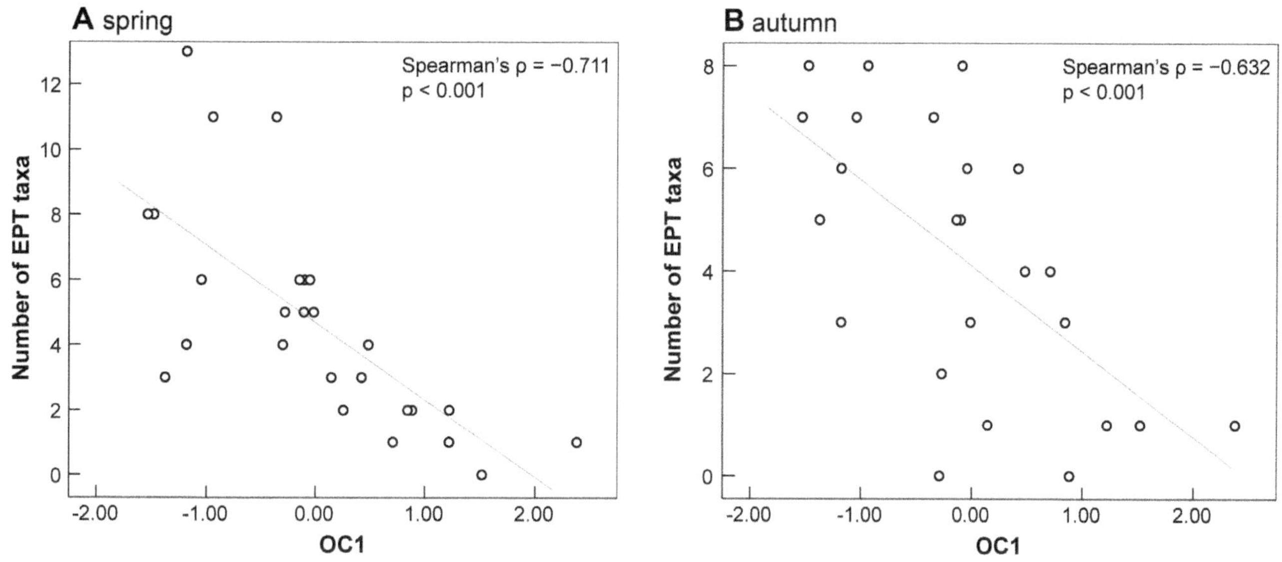

Figure 2. Number of EPT taxa (Ephemeroptera, Plecoptera, Trichoptera) correlating with the first component of the PCA with organic contaminants (OC1). Displayed are results for sampling campaign in spring (A) and autumn (B). Please note different scaling of y-axes in A and B.

salinity levels, high salt concentrations might often be accompanied by high contaminant levels (*e.g.*, in mining regions [49]). Accordingly, impacts by contaminants and salinity can hardly be distinguished. However, an indicator for WW-associated contaminants having an impact is the significant correlation of OC1 with Coleoptera genera (*Elmis*/*Limnius*/*Haliplus*: ρ ranging from −0.45 to −0.53) because they are much less salinity-sensitive than other groups of macroinvertebrate species [57].

In General, EPT taxa are more susceptible to toxicants than other groups of macroinvertebrates [58] and in particular to OCs [59]. Accordingly, a toxicity-related impact can be assumed. However, the contaminants represented by OC1 are apparently of minor ecotoxicological relevance to the EPT taxa at the quantified concentrations, as the lowest observed effect concentrations are 3–4 orders of magnitudes higher [60–67]. This indicates that organophosphates, synthetic musk fragrances, and DEET were not responsible for the relationship between OC1 and species composition. This is further supported by weak correlations of the TU-based factors (to account for toxic characteristics of the considered contaminants) with biota (data not shown). Additionally, the inclusion of peak concentrations (90% percentiles of measurements) did not enhance correlations. Nevertheless, our results demonstrate that compounds represented by OC1 are a good surrogate for contaminants associated with municipal WW and they may be a suitable set of markers for the identification of WW-contamination as their main entry pathway to surface waters

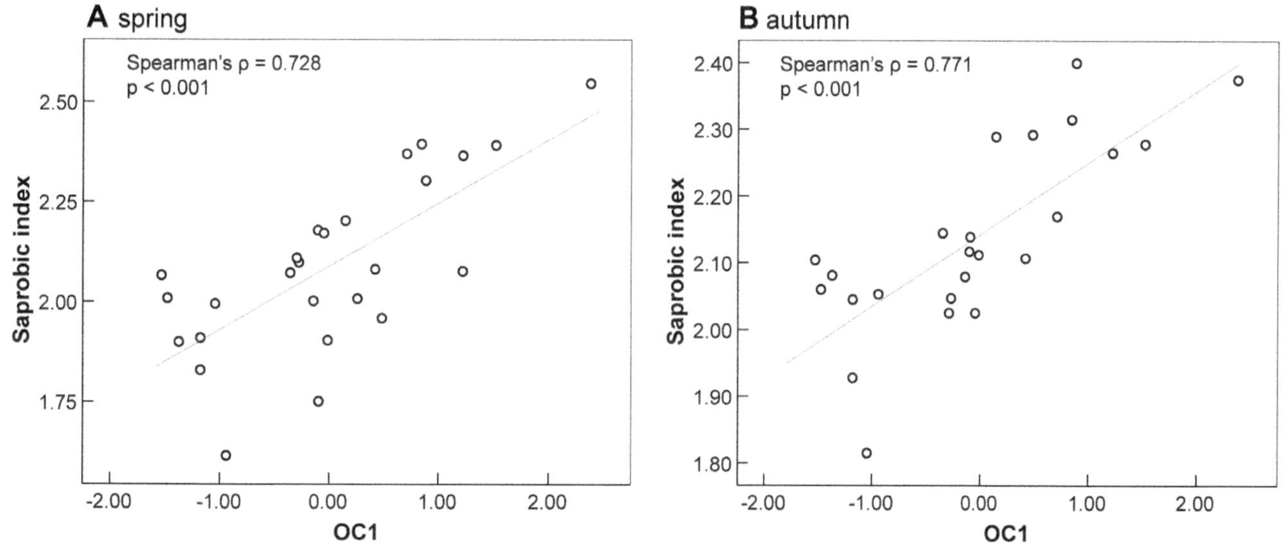

Figure 3. Saprobic index correlating with the first component of the PCA with organic contaminants (OC1). Displayed are results for sampling campaign in spring (A) and autumn (B). Please note different scaling of y-axes in A and B.

is through the sewage system. Among OC1 pollutants, synthetic musk fragrances have previously been proposed as a marker for municipal WW by Buerge *et al.* [68]. Therefore, we assume that the basic cause for the relationship between species composition and OC1 was due to alterations of physico-chemical water parameters and/or other contaminants present in sewage effluents that were not considered in our study.

The WW-associated factor responsible for the deterioration of aquatic fauna populations–oxygen concentration, salinity or contaminants–could not be unequivocally identified with the available data. It is possible that all factors contributed to the observed impact on macroinvertebrate assemblages. Long-term observations of wastewater impacted streams may help to tease apart these factors and identify the primary factor; particularly at sites upgraded with advanced WW treatment technologies for enhanced pollutant removal [27,69]. If an upgrade with enhanced pollutant removal technologies resulted in an increase in biodiversity, it could be unequivocally determined that the pollutants were primarily responsible for the decline in invertebrate populations.

Advanced WW treatment is supposed to be an appropriate measure to improve the ecological status of WW-impacted surface waters via contaminant degradation/removal. However, the potential benefits and adverse impacts have to be carefully evaluated to ensure that upgrading WW treatment facilities is a worthwhile exercise. For example, a potential adverse impact associated with enhanced pollutant removal technologies, like advanced oxidation processes, is the formation of hazardous transformation products. Environmental releases of these by-products may outweigh the benefits of treatment plant upgrades to include enhanced pollutant removal [70–73].

Conclusion

Our study emphasizes a clear wastewater-associated impact on the ecological quality, *e.g.*, decline in biodiversity, of surface waters, despite state-of-the-art wastewater treatment with biological activated sludge. As main factor threatening the ecological quality we identified wastewater-associated contaminants– despite their minute concentrations in the ng/L range–alongside with the structural modification of the streams. Consequently, we suggest that restoration measures should aim at improving both water as well as structural quality of surface waters in order to increase their ecological quality and help to conserve freshwater biodiversity.

Supporting Information

Figure S1 Sampling sites characteristics of the Schwarzbach river system.

Figure S2 Sampling sites characteristics of the Modau/ Sandbach river system.

Figure S3 Sampling sites characteristics of the Winkelbach river system.

Figure S4 Sampling sites characteristics of the Weschnitz river system.

Figure S5 Sampling sites characteristics.

Table S1 Environmental variables considered in the analyses and their correlations.

Table S2 List of taxa of aquatic invertebrates found during the sampling campaign in September 2005 at 24 sampling points.

Table S3 List of taxa of aquatic invertebrates found during the sampling campaign in March 2006 at 26 sampling points.

Table S4 Average concentrations of 12 organic contaminants in the water phase.

Table S5 Average concentrations of 12 heavy metals/ metalloids in the sediments.

Table S6 Average concentrations of 16 polycyclic aromatic hydrocarbons in the sediments.

Table S7 Loading matrix of principle components calculated by a principle component analysis of organic contaminants listed in Table 2.

Table S8 Loading matrix of principle components calculated by a principle component analysis of heavy metals listed in table 2.

Table S9 Loading matrix of principle components calculated by a principle component analysis of polycyclic aromatic hydrocarbons listed in table 2.

Acknowledgments

The authors would like to thank all participants of the INTAFERE project (http://www.intafere.de/english/). Madeleine Payne (Smart Water Research Centre, Griffith University, Gold Coast, Australia) is greatly acknowledged for reviewing and editing the manuscript. Matthias Oetken and Maren Heß are acknowledged for fruitful discussions and helpful advices. Furthermore, the authors thank Heike Heidenreich and Gerlinde Liepelt from the International Institute of Higher Education (Zittau, Germany) for their analyses of sediment-bound pollutants.

Author Contributions

Conceived and designed the experiments: DS AM KQ AB JO. Performed the experiments: DS AM KQ AB JO. Analyzed the data: DS AM KQ AB JO. Contributed reagents/materials/analysis tools: DS AM KQ AB JO. Wrote the paper: DS AM KQ AB JO.

References

1. Ricciardi A, Rasmussen JB (1999) Extinction rates of North American freshwater fauna. Conservation Biology 13: 1220–1222.
2. Jenkins M (2003) Prospects for biodiversity. Science 302: 1175–1177.
3. Spanhoff B, Arle J (2007) Setting attainable goals of stream habitat restoration from a macroinvertebrate view. Restoration Ecology 15: 317–320.
4. Brack W, Apitz SE, Borchardt D, Brils J, Cardoso AC, et al. (2009) Toward a holistic and risk-based management of european river basins. Integrated Environmental Assessment and Management 5: 5–10.
5. Ginebreda A, Munoz I, de Alda ML, Brix R, Lopez-Doval J, et al. (2010) Environmental risk assessment of pharmaceuticals in rivers: Relationships

between hazard indexes and aquatic macroinvertebrate diversity indexes in the Llobregat River (NE Spain). Environment International 36: 153–162.
6. Borja A, Dauer DM (2008) Assessing the environmental quality status in estuarine and coastal systems: Comparing methodologies and indices. Ecological Indicators 8: 331–337.
7. Hering D, Moog O, Sandin L, Verdonschot PFM (2004) Overview and application of the AQEM assessment system. Hydrobiologia 516: 1–20.
8. Diaz RJ, Solan M, Valente RM (2004) A review of approaches for classifying benthic habitats and evaluating habitat quality. Journal of Environmental Management 73: 165–181.
9. Commission E (2000) Establishing a framework for community action in the field of water policy. Directive 2000/60/EC of the European parliament and of the council.
10. Karlson K, Rosenberg R, Bonsdorff E (2002) Temporal and spatial large-scale effects of eutrophication and oxygen deficiency on benthic fauna in Scandinavian and Baltic waters - A review. Oceanography and Marine Biology 40: 427–489.
11. Vaughan IP, Ormerod SJ (2012) Large-scale, long-term trends in British river macroinvertebrates. Global Change Biology 18: 2184–2194.
12. BMU (2007) Bundesministerium fuer Umwelt, Naturschutz und Reaktorsicherheit [German Federal Ministry for the Environment, Nature Conservation and Nuclear Safety]. Wastewater flow processed in German municipal wastewater treatment plants. Bundesministerium fuer Umwelt website. Available: http://www.bmu.de/binnengewaesser/abwasser/doc/3142.php. Accessed 2013 Mar 5.
13. Schwarzenbach RP, Escher BI, Fenner K, Hofstetter TB, Johnson CA, et al. (2006) The challenge of micropollutants in aquatic systems. Science 313: 1072–1077.
14. UNSD (2010) United Nations Statistic Division. Environmental Indicators: Inland Waters Resources – Wastewater (UNSD, New York).
15. Daughton CG, Ternes TA (1999) Pharmaceuticals and personal care products in the environment: Agents of subtle change? Environmental Health Perspectives 107: 907–938.
16. Daughton CG (2003) Cradle-to-cradle stewardship of drugs for minimizing their environmental disposition while promoting human health. I. Rationale for and avenues toward a green pharmacy. Environmental Health Perspectives 111: 757–774.
17. Jobling S, Casey D, Rodgers-Gray T, Oehlmann J, Schulte-Oehlmann U, et al. (2003) Comparative responses of molluscs and fish to environmental estrogens and an estrogenic effluent. Aquatic Toxicology 65: 205–220.
18. Oetken M, Nentwig G, Loffler D, Ternes T, Oehlmann J (2005) Effects of pharmaceuticals on aquatic invertebrates. Part I. The antiepileptic drug carbamazepine. Archives of Environmental Contamination and Toxicology 49: 353–361.
19. Parrott JL, Blunt BR (2005) Life-cycle exposure of fathead minnows (*Pimephales promelas*) to an ethinylestradiol concentration below 1 ng/L reduces egg fertilization success and demasculinizes males. Environmental Toxicology 20: 131–141.
20. Oehlmann J, Schulte-Oehlmann U, Bachmann J, Oetken M, Lutz I, et al. (2006) Bisphenol A induces superfeminization in the ramshorn snail *Marisa cornuarietis* (Gastropoda : Prosobranchia) at environmentally relevant concentrations. Environmental Health Perspectives 114: 127–133.
21. Nentwig G (2007) Effects of pharmaceuticals on aquatic invertebrates. Part II: The antidepressant drug fluoxetine. Archives of Environmental Contamination and Toxicology 52: 163–170.
22. Triebskorn R, Casper H, Scheil V, Schwaiger J (2007) Ultrastructural effects of pharmaceuticals (carbamazepine, clofibric acid, metoprolol, diclofenac) in rainbow trout (*Oncorhynchus mykiss*) and common carp (*Cyprinus carpio*). Analytical and Bioanalytical Chemistry 387: 1405–1416.
23. Bundschuh M, Zubrod JP, Schulz R (2011) The functional and physiological status of *Gammarus fossarum* (Crustacea; Amphipoda) exposed to secondary treated wastewater. Environmental Pollution 159: 244–249.
24. Beasley G, Kneale P (2002) Reviewing the impact of metals and PAHs on macro invertebrates in urban watercourses. Progress in Physical Geography 26: 236–270.
25. Liu Z, Kanjo Y, Mizutani S (2009) Removal mechanisms for endocrine disrupting compounds (EDCs) in wastewater treatment - physical means, biodegradation, and chemical advanced oxidation: A review. Science of The Total Environment 407: 731–748.
26. Koh YKK, Chiu TY, Boobis AR, Scrimshaw MD, Bagnall JP, et al. (2009) Influence of operating parameters on the biodegradation of steroid estrogens and nonylphenolic compounds during biological wastewater treatment processes. Environmental Science & Technology 43: 6646–6654.
27. Stalter D, Magdeburg A, Wagner M, Oehlmann J (2011) Ozonation and activated carbon treatment of sewage effluents: Removal of endocrine activity and cytotoxicity. Water Research 45: 1015–1024.
28. Fahrenberger G (1986) Hessisches Ried und Bergstraße. Hessen.: Staatliche Landesbildstelle.
29. Ehlert T, Hering D, Koenzen U, Pottgiesser T, Schuhmacher H, et al. (2002) Typology and type specific reference conditions for medium-sized and large rivers in North Rhine-Westphalia: Methodical and biological aspects. International Review of Hydrobiology 87: 151–163.
30. Pottgiesser T, Sommerhäuser M (2003) Map of German stream types December 2003. Fließgewaesserbewertung website. Available: http://www.fliessgewaesserbewertung.de/download/typologie/. Accessed 2013 Mar 5.
31. Zumbroich T, Müller A, Friedrich G (1999) Strukturguete von Fließgewaessern - Grundlagen und Kartierung. Springer, Berlin.
32. Haase P, Sundermann A (2004) Standardisierung der Erfassungs- und Auswertungsmethoden von Makrozoobenthosuntersuchungen in Fließgewässern. Forschungsinstitut Senckenberg, Gelnhausen. Fließgewaesserbewertung website. Available: http://www.fliessgewaesserbewertung.de/download/probenahme-sortierung/. Accessed 2013 Mar 5.
33. Haase P, Sundermann A, Schindehütte K (2006) Informationstext zur Operationellen Taxaliste als Mindestanforderung an die Bestimmung von Makrozoobenthosproben aus Fließgewässern zur Umsetzung der EU-Wasserrahmenrichtlinie in Deutschland. Forschungsinstitut Senckenberg, Gelnhausen. Fließgewaesserbewertung website. Available: http://www.fliessgewaesserbewertung.de/download/bestimmung/. Accessed 2013 Mar 5.
34. Quednow K, Püttmann W (2007) Monitoring terbutryn pollution in small rivers of Hesse, Germany. Journal of Environmental Monitoring 9: 1337–1343.
35. Quednow K, Puttmann W (2008) Endocrine disruptors in freshwater streams of Hesse, Germany: Changes in concentration levels in the time span from 2003 to 2005. Environmental Pollution 152: 476–483.
36. Quednow K, Puttmann W (2008) Organophosphates and synthetic musk fragrances in freshwater streams in Hessen/Germany. Clean-Soil Air Water 36: 70–77.
37. Quednow K, Puttmann W (2009) Temporal concentration changes of DEET, TCEP, terbutryn, and nonylphenols in freshwater streams of Hesse, Germany: possible influence of mandatory regulations and voluntary environmental agreements. Environmental Science and Pollution Research 16: 630–640.
38. US-EPA (1982) The U.S. Environmental Protection Agency. List of Priority Pollutants. Office of Water. Water Quality Standards Database. Office of the Federal Registration (OFR) Appendix A: priority pollutants. Fed Reg. 47: 52309.
39. DIN (1996) Bestimmung von 61 Elementen durch Massenspektrometrie mit induktiv gekoppeltem Plasma (ICP-MS). Fachgruppe Wasserchemie der GDCh und Normenausschuss Wasserwesen (NAW) im DIN Deutsches Institut für Normung e.V. (Hrsg.): *Deutsche Verfahren zur Wasser-, Abwasser- und Schlammuntersuchung.* Band 1. VCH Weinheim, Beuth-Verlag, Berlin. ed.
40. DIN (1996) Bestimmung von 6 polyzyklischen aromatischen Kohlenwasserstoffen (PAK) mittels Hochleistungs-Flüssigkeitschromatographie. Fachgruppe Wasserchemie der GDCh und Normenausschuss Wasserwesen (NAW) im DIN Deutsches Institut für Normung e.V. (Hrsg.): *Deutsche Verfahren zur Wasser-, Abwasser- und Schlammuntersuchung.* Band 1. VCH Weinheim, Beuth-Verlag, Berlin. ed.
41. Hakanson L, Jansson M (1983) Principles of Lake Sedimentology. Springer, Berlin.
42. DIN (1987) Bestimmung der Partikelgrößenverteilung eines dispersen Gutes durch Siebanalyse. Fachgruppe Wasserchemie der GDCh und Normenausschuss Wasserwesen (NAW) im DIN Deutsches Institut für Normung e.V. (Hrsg.): *Deutsche Verfahren zur Wasser-, Abwasser- und Schlammuntersuchung.* Band 1. VCH Weinheim, Beuth-Verlag, Berlin. ed.
43. DIN (1987) Bestimmung der Sauerstoffzehrung in *n* Tagen. Fachgruppe Wasserchemie der GDCh und Normenausschuss Wasserwesen (NAW) im DIN Deutsches Institut für Normung e.V. (Hrsg.): *Deutsche Verfahren zur Wasser-, Abwasser- und Schlammuntersuchung.* Band 1. VCH Weinheim, Beuth-Verlag, Berlin. ed.
44. AQEM (2006) ASTERICS - Software manual - Version 3.0. Fließgewaesserbewertung website. Available: http://www.fliessgewaesserbewertung.de/download/berechnung/. Accessed 2013 Mar 5.
45. Backhaus K, Erichson B, Plinke W, Weiber R (2006) Multivariate Analysemethoden - Eine anwendungsorientierte Einführung. Springer, Berlin.
46. Oksanen J, Blanchet FG, Kindt R, Legendre P, Minchin PR, et al. (2011) vegan: Community Ecology Package. R package version 2.0–2. R-project website. Available: http://CRAN.R-project.org/package = vegan. Accessed 2013 Mar 5.
47. Crawley MJ (2007) The R Book. Wiley; 1 edition (June 19, 2007).
48. Liess M, Von der Ohe PC (2005) Analyzing effects of pesticides on invertebrate communities in streams Environmental Toxicology and Chemistry 24: 954–965.
49. Clements WH, Carlisle DM, Lazorchak JM, Johnson PC (2000) Heavy metals structure benthic communities in colorado mountain streams. Ecological Applications 10: 626–638.
50. Linke S, Bailey RC, Schwindt J (1999) Temporal variability of stream bioassessments using benthic macroinvertebrates. Freshwater Biology 42: 575–584.
51. Macdonald DD, Carr RS, Calder FD, Long ER, Ingersoll CG (1996) Development and evaluation of sediment quality guidelines for Florida coastal waters. Ecotoxicology 5: 253–278.
52. Coimbra CN, Graca MAS, Cortes RM (1996) The effects of a basic effluent on macroinvertebrate community structure in a temporary Mediterranean river. Environmental Pollution 94: 301–307.
53. Jacobsen D, Rostgaard S, Vasconez JJ (2003) Are macroinvertebrates in high altitude streams affected by oxygen deficiency? Freshwater Biology 48: 2025–2032.
54. Laursen AE, Seitzinger SP (2004) Diurnal patterns of denitrification, oxygen consumption and nitrous oxide production in rivers measured at the whole-reach scale. Freshwater Biology 49: 1448–1458.
55. Kefford BJ, Marchant R, Schafer RB, Metzeling L, Dunlop JE, et al. (2011) The definition of species richness used by species sensitivity distributions approximates

observed effects of salinity on stream macroinvertebrates. Environmental Pollution 159: 302–310.
56. Pond GJ (2010) Patterns of Ephemeroptera taxa loss in Appalachian headwater streams (Kentucky, USA). Hydrobiologia 641: 185–201.
57. Kefford BJ, Hickey GL, Gasith A, Ben-David E, Dunlop JE, et al. (2012) Global scale variation in the salinity sensitivity of riverine macroinvertebrates: Eastern Australia, France, Israel and South Africa. Plos One 7: e35224.
58. Beketov MA (2004) Comparative sensitivity to the insecticides deltamethrin and esfenvalerate of some aquatic insect larvae (Ephemeroptera and Odonata) and Daphnia magna. Russian Journal of Ecology 35: 200–204.
59. Von Der Ohe PC, Liess M (2004) Relative sensitivity distribution of aquatic invertebrates to organic and metal compounds. Environmental Toxicology and Chemistry 23: 150–156.
60. WHO (1998) Flame retardants: Tris(chloropropyl) phosphate and tris(2-chloroethyl) phosphate - Environmental Health Criteria 209, Genf.
61. WHO (2000) Flame retardants: Tris(2-butoxyethyl) phosphate, tris(2 ethylhexyl) phosphate and tetrakis(hydroxymethyl) phosphonium salts - Environmental Health Criteria 218, Genf.
62. EC (2000) European Commission - IUCLID Dataset: Tributyl Phosphate. European Chemicals Bureau Luxembourg.
63. EC (2000) European Commission - IUCLID Dataset: 4-(1,1,3,3-Tetramethylbuthyl) Phenol. European Chemicals Bureau Luxembourg.
64. EC (2000) European Commission - IUCLID Dataset: Tris(2-Chloroethyl) Phosphate. European Chemicals Bureau Luxembourg.
65. EC (2002) European Commission - 4-Nonylphenol (branched) and Nonylphenol. European Union Risk Assessment Report 10. European Chemicals Bureau Luxembourg. EUR 20387 EN.
66. EC (2003) European Commission - 4,4′-Isopropylidenephenol (Bisphenol-A). European Union Risk Assessment Report 37. European Chemicals Bureau Luxembourg. EUR 20843 EN.
67. Artola-Garicano E, Sinnige TL, van Holsteijn I, Vaes WHJ, Hermens JLM (2003) Bioconcentration and acute toxicity of polycyclic musks in two benthic organisms (*Chironomus riparius* and *Lumbriculus variegatus*). Environmental Toxicology and Chemistry 22: 1086–1092.
68. Buerge IJ, Buser H-R, Mueller MD, Poiger T (2003) Behavior of the polycyclic musks HHCB and AHTN in lakes, two potential anthropogenic markers for domestic wastewater in surface waters. Environmental Science and Technology 37: 5636–5644.
69. Hollender J, Zimmermann SG, Koepke S, Krauss M, McArdell CS, et al. (2009) Elimination of organic micropollutants in a municipal wastewater treatment plant upgraded with a full-scale post-ozonation followed by sand filtration. Environmental Science & Technology 43: 7862–7869.
70. Magdeburg A, Stalter D, Oehlmann J (2012) Whole effluent toxicity assessment at a wastewater treatment plant upgraded with a full-scale post-ozonation using aquatic key species. Chemosphere 88: 1008–1014.
71. Stalter D, Magdeburg A, Oehlmann J (2010) Comparative toxicity assessment of ozone and activated carbon treated sewage effluents using an *in vivo* test battery. Water Research 44: 2610–2620.
72. Stalter D, Magdeburg A, Weil M, Knacker T, Oehlmann J (2010) Toxication or detoxication? *In vivo* toxicity assessment of ozonation as advanced wastewater treatment with the rainbow trout. Water Research 44: 439–448.
73. Weemaes M, Fink G, Lachmund C, Magdeburg A, Stalter D, et al. (2011) Removal of micropollutants in WWTP effluent by biological assisted membrane carbon filtration (BIOMAC). Water Science and Technology 63: 72–79.

2

Polycyclic Aromatic Hydrocarbons in Coastal Sediment of Klang Strait, Malaysia: Distribution Pattern, Risk Assessment and Sources

Seyedeh Belin Tavakoly Sany[1,2]*, Rosli Hashim[1], Aishah Salleh[1], Majid Rezayi[2,3], Ali Mehdinia[4], Omid Safari[5]

1 Institute of Biological Sciences University of Malaya, Kuala Lumpur, Malaysia, **2** Food Science and Technology Research Institute, ACECR Mashhad Branch, Mashhad, Iran, **3** Chemistry Department, Faculty of Science, University Malaya, Kuala Lumpur, Malaysia, **4** Department of Marine Science, Marine Living Group, Iranian National Institute for Oceanography, Tehran, Iran, **5** Faculty of Natural Resources and Environment, Ferdowsi University of Mashhad, Mashhad, Iran

Abstract

Concentration, source, and ecological risk of polycyclic aromatic hydrocarbons (PAHs) were investigated in 22 stations from surface sediments in the areas of anthropogenic pollution in the Klang Strait (Malaysia). The total PAH level in the Klang Strait sediment was 994.02±918.1 μg/kg dw. The highest concentration was observed in stations near the coastline and mouth of the Klang River. These locations were dominated by high molecular weight PAHs. The results showed both pyrogenic and petrogenic sources are main sources of PAHs. Further analyses indicated that PAHs primarily originated from pyrogenic sources (coal combustion and vehicular emissions), with significant contribution from petroleum inputs. Regarding ecological risk estimation, only station 13 was moderately polluted, the rest of the stations suffered rare or slight adverse biological effects with PAH exposure in surface sediment, suggesting that PAHs are not considered as contaminants of concern in the Klang Strait.

Editor: Karl Rockne, University of Illinois at Chicago, United States of America

Funding: This work and survey components were supported by the High Impact Research Grant (UM.C/625/1/HIR/162) from the Ministry of Higher Education (Malaysia) and University Malaya Research Grant (RP004A-SUS). Study designs and data collections were done by the academic staff at the Institute of Biological Science, University Malaya, Kuala Lumpur, Malaysia. Analyses undertaken and the decision to prepare and publish this manuscript were made by Institute staff in conjunction with the Chemistry Department of University Malaya. The funders had no role in study design, data collection and analysis, decision to publish, or preparation of the manuscript.

Competing Interests: The authors have declared that no competing interests exist.

* E-mail: belintavakoli332@gmail.com

Introduction

Polycyclic aromatic hydrocarbon (PAH) contamination is a major hazard that is a concern for aquatic life in marine sediments, particularly in areas close to anthropogenic sources [1,2]. Many PAHs are at the same time persistent, bioaccumulative, and toxic for humans and aquatic organism [3,4,5,6,7].

In environmental research, the aromatic fraction of C_{11}–C_{22} was selected as being representative of aromatic hydrocarbon compounds for the purpose of assessment of ecological and human risks. This fraction is associated with the release of petroleum products to the environment, and is potentially toxic due to its mobility and stability in sediments [8,9]. Physicochemical properties of these fractions are provided in the Supplementary data (Table S1).

Similar to other pollutants, the sources of PAHs are divided into major groups; anthropogenic and lithogenic. Anthropogenic sources of PAHs originated from pyrogenic and petrogenic sources. Pyrogenic PAHs are usually made up of high molecular weight PAHs with 4–6 rings that includes, fluoranthene (Fla), pyrene (Py), benzo(a)anthracene (BaA), chrysene (Chy), benzo(b)-fluoranthene (BbF), benzo(k)fluoranthene (BkF), benzo(a)pyrene (BaP), dibenzo(a,h)anthracene (DibA), benzo(g,h,i)perylen (BghiP) and Indeno[1,2,3,(c,d)]pyrene; (InP) [2,10]. The pyrogenic are mainly detected in incomplete combustion of organic compounds, such as fossil fuels (heating oil, cooking, coal burning, vehicle emissions, waste tire), and biomass burning (fireplace, controlled burning) [2,11,12]. Pyrogenic sources are thought to be more thermodynamically stable and toxic than petrogenic sources due to their high concentrations of non-alkylated PAHs [11,13,14].

Petrogenic PAHs involve naphthalene, acenaphthylene, acenaphthene, and fluorine, and belong to the alkyl-substituted PAHs or low molecular weight PAHs, possessing 2–3 rings. These hydrocarbons are thought to have originated from oil spills from fresh or used crankcase oil, crude and fuel oil, chronic or accidental leakages of marine and land pipelines, and domestic and industrial wastes [2,10,12,14].

The Klang Strait is surrounded by the west coast of Malaysia and the Straits of Malacca in Southeast Asia. The strategic location of this strait has made it into one of the busiest shipping routes in the world, corresponding to huge economic demands from the Middle East and the Far East [15]. This strait also experienced the rapid development of industrialization, urbanization, and motorization over the past few decades. Thus, this area is under constant threat from multiple sources of energy, such as petroleum [16,17]. In this research, a hypothesis is defined based on the serious threat posed by PAHs contamination to the Klang Strait.

Hence, this study tries to estimate the concentration and distribution of PAHs in coastal sediments from the Klang Strait to identify the possible sources of PAHs in the Port, and to conduct an ecological risk assessment to recognize the possible adverse ecological effects on the biological community due to the exposure to PAHs concentrated in Klang Strait sediments.

Ecological assessment of PAHs compounds in this study has been highlighted via two main problems: the first problem was related to the scarcity of background and updated databases on the PAHs concentrations in Klang Strait sediments. This problem is due to the limitations in collecting samples such as strong currents, high traffic density of shipping activity. The second problem was related to the lack of SQGs (Sediment Quality Guidelines) for coastal waters of Malaysia, since no SQGs were available to assess biological effects. Therefore, the results from this study will be applied in the form of the managerial tools in order to control the pollution occurrence and protect living organisms to assure the safe future of the marine environment of Peninsular Malaysia. Moreover it can be practical as background data for future studies.

Materials and Methods

Ethics Statement

The Ethical Review Committee of University Malaya approved this study. All necessary permits were obtained for the described field study from University Malaya and Port Klang Authorities (Permit Number UM.G/KB4/6/1). No specific permission was required for water and sediment surveys as they were conducted at these stations. Likewise, the field studies did not involve endangered or protected species, and no lethal sampling was conducted.

Study area and sample collection

The Klang Strait is divided into three main Ports, most of which are well-sheltered by several mangrove islands and mudflats, forming natural enclosures. Experimental samples were collected from three sites (North Port, South Port and West Port) in the Klang Strait, which is located on the western coastal region (03° 00′N to 101° 24′E) of Peninsula Malaysia at the north end of the Malacca Strait, and over 573 km^2 [18].

The sites were chosen based on in their unique activities. The North Port is a terminal of container to transship the coal, oil, and other chemical products, while the South Port is a terminal jetty for fishing boats, ferries, and yachts, and is linked to the Klang river, which is known as the most polluted river in Malaysia [15,19].

The West Port is a main container terminal for massive cargo transport by ocean ships. Similarly, an industrial complex, including cement industries and a food, and palm oil factories have been constructed at the West Port [15].

In this research, 22 stations were selected from three transects parallel to the coastline, with three different distances (Figure 1), and a station was selected as a control point 21 km from the Klang Strait at a remote location. The sampling stations were adjacent coastal industrial outlets and busy shipping lanes. The sampling stations were identified a progressive number: six up-gradient stations in North Port (Station 1–6), six stations in South Port close to Klang River (station 7–15), and nine down-gradient stations in West Port (station 16–21) (Table 1).

The study area lies within the humid tropical part with rainy (North monsoon, November to March) and dry seasons (south monsoon, April to October). In the Klang Strait, the average salinity is 30.25‰ (±1.36), the average temperature is 30.04°C (±0.62), and the average surface dissolved oxygen (DO) is 5.38 mg/l (±0.17), while the monthly average surface and bottom pH values were between 7.85 to 8.25. The sediment samples were taken with a Peterson grab sampler (0.07 m^2) at two dates per season during November 2011 until October 2012. The top 10 cm of the sediment sample was removed with a stainless steel spoon for subsequent analysis. All of the samples were packed into aluminum boxes and transported on ice to the laboratory. The samples were air-dried, crushed, and sieved (<2 mm) to remove residual roots and stones, and immediately stored at −20°C in pre-cleaned amber glass bottles until further treatments. For PAHs analysis, three replicates of grab samples were used to investigate the spatial and temporal distributions.

Experimental method

Chemical and reagents. A standard solution of 16 USEPA priority PAHs (Table 2) and a five-surrogate standard (acenaphthene-d10, chrysene-d12, perylened12, naphthalene-d8 and phenanthrene-d10) were purchased from Ultra Scientific, Inc. (North Kingstown, RI, USA). A standard reference material (SRM) was obtained from the National Institute of Standards and Technology (NIST, Gaithersburg, MD, USA). Neat (99%) hexamethylbenzene was purchased from Aldrich Chemical Company (Milwaukee, WI, USA). All solvents (acetone, dichloromethane, methanol and hexane) applied in the analyses were of analytical grade and redistilled twice.

The alumina (120–200 mesh) and Silica gel (80–100 mesh) were extracted for a period 72 hours in a Soxhlet apparatus, activated in the oven at 150°C and 180°C for a period 12 hours, respectively, and then deactivated by distilled water at a ratio of 3% (m/m). The deionised water was obtained from a Milli-Q system.

Extraction and fractionation. For the extraction procedure, about 5 g of dried sediment samples were Soxhlet-extracted by dichloromethane (150 ml) for a period of 72 hours. A mixture of surrogate standard was added to all of the samples as a recovery surrogate standards. Elemental sulfur was removed from the extracts using activated copper granules. After extraction, the extract was concentrated up using a rotary vacuum evaporator. The volume of the extract was adjusted to 2–3 ml, and solvent-exchanged into 10 ml n-hexane, which caused it to further decrease to approximately 1–2 ml.

Clean up and fraction procedures were performed with a 1:2 alumina/silica gel column. PAHs were obtained by eluting it with 60 ml of hexane/dichloromethane (1:1). PAHs fraction were re-concentrated using a rotary evaporator, while the volume was concentrated to up to 1 ml. The fraction was further reduced to 0.2 ml under a stream of filtered purified nitrogen gas. An aliquot of 0.2 ml of each extract was applied to gas chromatography–mass spectrometer (GC–MS) analysis. Likewise, hexamethylbenzene was added as an internal standard prior to the GC–MS analysis.

Instrumental analysi. The PAHs were estimated using a Hewlett–Packard 5890 series gas chromatograph interfaced with 5972 mass-selective detector (MSD) in the selective ion-monitoring (SIM) mode. A fused silica capillary column (50 m, 0.32 mm, 0.17 μm) coated with HP-5MS (film thickness 0.25 lm) was used for separation, with helium as a carrier gas at a flow rate of 2 ml/min, a head pressure of 12.5 psi, and a linear velocity of 39.2 cm/s at 290°C.

The temperature was programmed to vary from 80 to 290°C at 4°C/min, and was held at the final temperature for 30 minutes. The injector temperature and transfer line temperature were 250°C and 180°C, respectively. The mass spectrometer operated at electron energy of 70 eV, with an ion source temperature of 250°C.

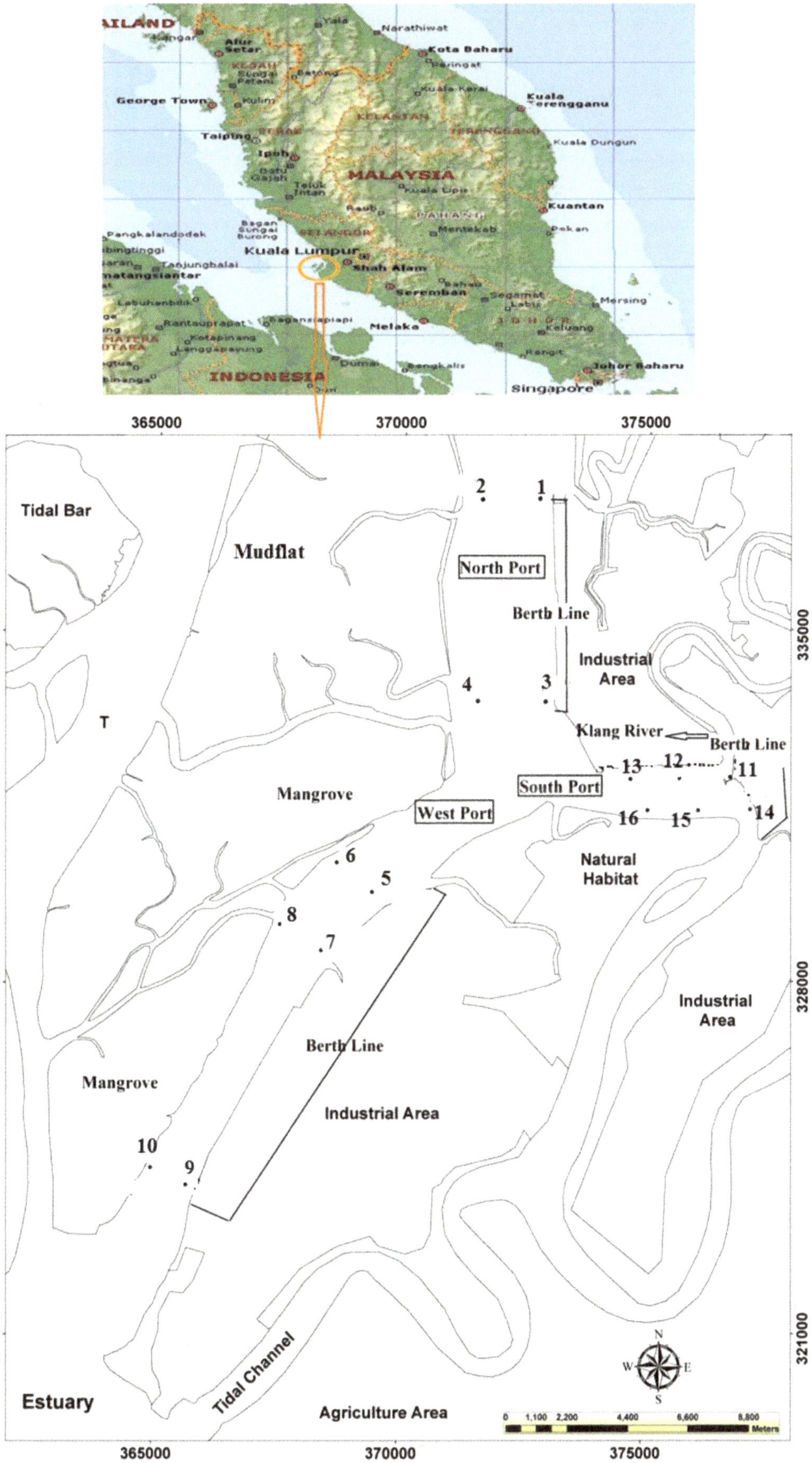

Figure 1. Location of the sampling stations in Klang Strait.

The interface temperature and injection were programmed at 290°C. The oven temperature was initially isothermal at 80°C for 5 minutes, programmed from 80 to 290°C at 3°C/min, and held at the final temperature (290°C) for 30 minutes. A 1 ml sample was manually injected in the split-less injector with a 1-minute solvent delay. The mass spectrometer was operated at electron impact (EI) of 70 eV, with an ion source temperature of 250°C. The ranges of mass scanning were between m/z 50 and m/z 500.

Quality assurance. For every set of samples, a procedural method blanks (solvent), sample duplicates, spiked matrices (standards spiked into solvent), and standard reference material (SRM1941) were used to assess quality control and assurances. Individual PAHs were quantified according to the retention time and m/z ratio of PAHs mixed standard (Sigma), while a standard calibration curve was used to calibrate the concentrations of each PAH.

The method detection limits (MDLs) for the investigated organic pollutants were determined according to USEPA [20]. The MDL of PAHs were estimated based on six analyses of the minimal solutions (Σ16PAH) of the standard PAH mixture, which was subjected to the full analytical procedure in the range of 0.0641 ng/g to 1.0183 ng/g. These values were within the acceptable range of the EPA method [20]. The percentage of relative standard deviation (RSD) for all of the investigated PAHs were lower than 20.0% in the fortification experiment and in the replicates used to determine the MDL. The reported results were corrected with the recoveries of the surrogate standards. The average recoveries of surrogate standards varied from 63.28–96.75. The recoveries of all the PAHs and RSD falls within the range of 78.4–95.2% and 2.6–13.5%, respectively, which met the acceptance criteria of the EPA method [20]. A detailed recovery and RSD values were provided in the Supplementary data (Table S1).

Measurement of TOC and sediment grain size. The sediment grain size was determined using a multi-wavelength particle size analyzer (model LS 13 320), and the results were divided into sand (>64 µm), silt (2 µm<size<64 µm), and clay (<2 µm) fractions for the determination of PAHs in the contaminated sediments [21].

Freeze-dried sediment samples were grounded, and was treated with 10% (v/v) HCl to remove carbonates from the sediment samples. After that, the samples were dried at 60°C in an oven. Total organic carbon (TOC) was determined in the surface sediments using a carbon analyzer (Horbia Model 8210).

PAHs source identification

Ratio values such as an Ant/Ant+Phn and a Fla/Fla+Pyr had been used to distinguish between pyrogenic and petrogenic sources [22,23]. Sediments with Ant/Ant+Phn ratio <0.100 were mainly contaminated by petrogenic inputs, while a ratio >0.100 indicates

Table 1. Physicochemical description of sampling stations.

Sites	Station number	Latitude	Longitude	Silt and clay (<64 µm)%	TOC (%)	Depth (m)	Salinity (‰)	Description of stations
North Port	1	3° 3′1.49″N	101°21′18.70″E	58.20	41.79	14.3	30.15	Liquid and dry berth line
	2	3° 3′1.33″N	101°20′56.04″E	49.63	50.36	20.5	30.18	Remote
	3	3° 3′1.47″N	101°20′33.11″E	73.77	26.22	10.3	31.24	Mangrove
	4	3° 0′53.11″N	101°21′20.25″E	59.78	40.21	13.5	30.81	Container berths
	5	3° 0′52.64″N	101°20′58.04″E	50.89	49.10	21.6	31.02	Remote
	6	3° 0′52.49″N	101°20′34.54″E	65.19	34.80	11.2	31.36	Mangrove
West Port	7	2°58′44.00″N	101°19′21.02″E	53.57	46.42	12.5	30.86	Dry berth and Cement factory outlets
	8	2°58′54.12″N	101°19′9.06″E	45.96	54.03	19.5	30.98	Remote
	9	2°59′3.12″N	101°18′58.38″E	63.42	36.57	7.8	30.86	Mangrove
	10	2°58′6.34″N	101°18′48.14″E	56.33	43.66	13.3	30.44	Liquid berth and palm oil factory oulets
	11	2°58′14.90″N	101°18′34.56″E	41.10	58.89	20.3	30.58	Remote
	12	2°58′23.07″N	101°18′20.99″E	70.81	29.18	8.8	30.79	Mangrove
	13	2°55′34.43″N	101°17′18.76″E	52.31	47.68	15.5	30.51	Container berths
	14	2°55′39.38″N	101°17′7.57″E	50.69	49.30	21.1	30.63	Remote
	15	2°55′45.02″N	101°16′55.55″E	70.36	29.63	6.8	30.77	Mangrove
South Port	16	2°59′59.08″N	101°23′18.88″E	95.39	4.60	7.5	26.10	Dry berths, Klang river
	17	2°59′58.17″N	101°22′45.35″E	93.16	6.83	10.5	26.12	Klang river
	18	2°59′57.66″N	101°22′12.45″E	64.69	35.30	12.4	30.11	Semi-urban
	19	2°59′38.25″N	101°23′32.32″E	69.50	30.49	10.3	29.45	Liquid berth
	20	2°59′37.70″N	101°22′57.93″E	69.72	30.27	11.3	29.54	Mangrove
	21	2°59′37.23″N	101°22′23.78″E	57.73	42.26	10.4	30.50	Mangrove
Control Point	22	3° 6′55.95″N	101°12′44.70″E	51.60	48.39	17.5	31.54	Remote

Table 2. Concentration of PAHs (μg/kg dw) in surface sediments of Klang Strait.

Station	Nap	Acy	Ace	Flr	Phn	Ant	Fla	Pyr	BaA	Chy	BbF	BkF	BaP	DibA	Bghip	Inp	∑PAHs
1	ND	ND	2.1	34.2	47.5	100.4	ND	482.8	613.0	ND	ND	ND	ND	ND	ND	ND	**1280.4**
2	1.3	ND	ND	9.7	13.3	12.7	ND	131.3	31.5	ND	ND	ND	ND	ND	ND	ND	**199**
3	1.7	ND	8.1	118.5	89.0	58.8	42.7	62.9	ND	ND	ND	ND	ND	ND	ND	ND	**381.7**
4	0.4	16.32	ND	1.1	10	59.8	12.5	12	66.7	50.0	852.8	ND	1495.8	190.2	ND	99.8	**2852.2**
5	0.8	ND	0.3	16.8	12.6	12	10	195.6	ND	ND	11.0	ND	ND	ND	ND	ND	**259.1**
6	4.0	ND	0.4	57.3	36.2	37.9	26.0	31.8	179.4	ND	ND	ND	ND	ND	ND	42.3	**415.2**
7	1.3	ND	ND	27.6	127.6	267.8	10.3	12.7	204.0	ND	ND	47.6	ND	ND	ND	ND	**698.9**
8	5.7	ND	1.0	39.0	6.8	71.4	13.5	15.8	50.5	ND	ND	ND	122.1	ND	64.2	77.4	**467.4**
9	1.2	ND	ND	7.7	5.1	8.2	ND	28.1	191.3	198.0	ND	ND	ND	67.8	ND	ND	**507.4**
10	0.8	ND	ND	23.7	5.0	10.1	ND	236.5	278.8	ND	ND	ND	ND	ND	ND	68.7	**623.7**
11	0.7	ND	0.7	48.8	15.6	37.8	17.6	25.1	29.5	ND	ND	367.6	ND	ND	ND	ND	**543.4**
12	2.0	ND	1.0	39.8	24.8	233.9	21.5	341.5	774.7	ND	ND	1.3	1.0	7.3	ND	ND	**1448.8**
13	16.1	645.3	14.8	520.8	225.6	308.0	166.5	243.8	390.9	ND	11.6	28.6	825.2	ND	49.4	ND	**3446.9**
14	3.0	ND	0.5	33.8	75.8	181.3	15.8	14.0	146.3	ND	11.3	ND	59.9	ND	ND	31.8	**573.5**
15	2.2	ND	ND	8.7	9.6	3.8	ND	ND	78.3	ND	12.8	ND	ND	130.2	ND	348.4	**594.0**
16	109.0	ND	1.2	120.6	66.3	100.5	92.2	99.7	974.8	116.0	ND	ND	716.8	ND	ND	ND	**2397.1**
17	9.6	ND	4.0	225.6	94.3	501.2	ND	226.2	1292.0	549.2	ND	ND	ND	ND	ND	ND	**2902.1**
18	1.3	ND	ND	20.8	8.5	11.6	ND	119.3	191.8	ND	378.6	ND	ND	ND	ND	ND	**731.9**
19	2.2	ND	ND	35.6	39.7	33.3	16.5	17.1	267.3	ND	10.2	ND	103.9	ND	ND	ND	**525.8**
20	2.1	ND	ND	37.1	38.6	30.2	16.3	17.0	226.8	ND	9.3	ND	89.9	ND	ND	ND	**467.5**
21	2.4	ND	0.6	37.2	16.7	16.5	16.0	11.2	194.6	ND	22.0	ND	ND	129.6	ND	ND	**483.5**
22	1.8	ND	ND	12.5	1.7	4.1	ND	ND	56.8	ND	23.4	ND	ND	0.0	ND	ND	**100.3**
Mean	7.7	30.04	1.6	67.13	43.6	95.5	21.2	105.7	282.7	41.5	61.0	20.2	155.8	23.9	5.2	32.0	**994.02**
SD	22.6	137.39	3.453	131.0	53.01	142.3	38.0	144.0	363.8	121.7	191.9	77.5	377.7	53.7	16.7	76.1	**918.1**
Min	ND	ND	ND	1.1	1.7	3.8	10	11.2	ND	ND	ND	ND	ND	ND	ND	ND	**100.3**
Max	109.0	645.3	14.8	520.8	225.6	501.2	169	599	1292.0	549.2	852.8	367.6	1495.8	190.2	64.2	348.4	3446.9
***TEL**	34.6	5.87	6.71	21.2	86.7	46.9	113	153	74.8	108	320	280	88.8	6.22	430	*	1684
***PEL**	391	128	88.9	144	544	245	1494	1398	693	846	1880	1620	763	135	1600	*	16770

TELs (threshold effects level) and PELs (probable effects level), ** <MDL, below the method detection limit; ND, not detect.

a dominance of combustion [10,22,23]. A Fla/Fla+Pyr ratio of 0.500 is usually defined as the petroleum/combustion transition point. This boundary appears to be less definitive than 0.100 for Ant/Ant+Phn.

The Fla/Fla+Pyr ratio is below 0.4 for most petroleum samples, between 0.400–0.500 for liquid fossil fuel (vehicle and crude oil) combustion, and exceeds 0.500 in kerosene, grass, most coals, and wood combustion [22,23].

Ecological risk assessment

A Screening Level Ecological Risk Assessment (SLERA) was performed to evaluate the possibility of adverse ecological effects according to the framework developed by the United States Environmental Protection Agency (USEPA) [11,24,25]. The average concentration of individual PAHs in Klang Strait was used as the exposure concentration in a conservative assumption.

The sediment quality guideline, which was previously developed, was applied in the risk characterization step. In the present study, a consensus approach based on MacDonald et al. was used according to the following specific values of TEL (threshold effects level) and PEL (probable effects level) [26,27]. In the risk characterization step, the hazard quotient (HQ) was calculated based on the maximum concentration of each individual PAHs (MCPAHs) and the estimated consensus based sediment quality guideline (CCPEL) according to the following equation [11,27]:

$$HQ = \frac{MCPAHs}{CCPEL} \tag{1}$$

The PEL quotient (PELq) is a unique factor to describe the contamination effect of PAHs on biological organisms according to the analyses of chemical data and matching toxicity from 1,068 sediment samples from coastal and estuaries water in the USA [28]. The PELq factor is the average of the ratios between the PAHs concentration in the sediment sample and the related PEL value [11,29]. PELq factors were divided into four categories, which can be used to describe the sediment as non-adverse effect (PELq<0.1), slightly adverse effect (0.1<PELq>0.5), moderately effect (0.5<PELq>1.5), and heavily effect (PELq>1.5) [11,30]. This factor is practical for comparing different historical episodes or other study areas, and to facilitate the decision maker's work in sediment quality assessment [31].

The Nemerow composite index (Nemerow Index) for all of the sediments samples was estimated based on [32]. This method used the individual contamination index, which directly reflect the sediments' contamination degree. The equation is as follows:

$$P=\sqrt{\frac{(P_{iave})^2+(P_{iave})^2}{2}} \quad (2)$$

Where $P_{i\ max}$ is the maximum concentration of the individual contamination indices, while $P_{i\ ave}$ is the average concentration of the individual contamination indices. When $P_i = C_i/S_i$, P_i is the individual contamination index of PAHs contaminates i, C_i is the determined concentration of PAHs from at least 5 sampling points, and S_i is the required standard based on the evaluation criteria. In the present study, the adopted evaluation criteria were specific values of the sediment quality guideline, as Malaysian coastal waters have no stipulated guideline on PAHs in sediments. The Nemerow pollution index (P) is classified as the quality of sediment into 5 pollution grades safety domain ($P<0.7$), precaution domain ($0.7<P\leq1.0$), slightly polluted domain ($1.0<P\leq2.0$), moderately polluted domain ($2.0<P\leq3.0$), and seriously polluted domain ($P>3.0$) [32,33].

Statistical analysis

Statistical analyses were performed using Microsoft Excel and SPSS 17 software (SPSS, Chicago, IL) for statistical data evaluation. Multivariate techniques are sensitive to the distribution of non-parametric geochemical data. Thus, evaluating normal distribution and data transformation are essential to appropriately pretreat data. The data were arranged based on the samples and estimated values, and converted into a single matrix formed by the data points (cases) and the concentration values of the 16 PAHs analyzed (variables), forming a [22×16] data matrix. The methods were selected based on the results of the Shapiro-Wilk Normality test, the levene test for homogeneity of variances, and the Bartlett's test of equal variances. Normally, the distributed data were evaluated via one-way ANOVA, with an alpha level of 0.05. Data that did not pass the tests of normality and homogeneity was evaluated with the Kruskal-Wallis one-way nonparametric ANOVA. The differences were determined to be significant, where $p<0.05$. The data transformation was also done using the Box–Cox method with an optimal transformation for data that were not normally distributed. This method was performed based on the Kannel et al. 2007 to reduce the mean squared error [71]. Statistical tests, multivariate analyses, diagnostic ratios, and risk calculations include non-detect (or left-censored) data with substitution by zero. The nonparametric correlation method (Kendall's tau-b) was used to obtain the correlation coefficient and the significance of the correlation among physicochemical parameters in the sediments. Multivariate techniques, such as cluster analysis (CA) and principal components analysis (PCA) were used to estimate structure and relationships in a multivariate data. PCA is also known as a dimensional reduction because this method is able to decrease the dimensionality of the primary set of data (measured PAH contents in sediment samples) and compress data into a lower dimensional matrix (principal components) [34]. Specifically, the KMO and Bartlett's test of sphericity were applied to extract all factors with eigenvalues that exceed 1, and were rotated by the Varimax method. CA analysis was used to classify a set of data into different groups based on similarity. The geostatistical analysis was performed using Surfer 8 software (GPS value of stations) based on the geospatial methods to better understand the contaminant pathways and to provide a comprehensive contour map of the spatial distribution of contaminants over a large range.

Results and Discussion

Spatial distribution of PAHs in sediments

In the Klang Strait, fine-grained sediments are predominated in almost all stations, with the highest amount of fine sediment in the vicinity of stations near the mangrove edge and the mouth of the Klang River (Table 1). TOCs' distribution was synchronous with that of fine-grain-sized sediment in most parts of the study areas (with the notable exception of stations 9 and 12) (Table 1). The details related to the distribution and concentration of TOC in the Klang strait was recorded by Tavakoly Sany et al. in 2013 [35]. Generally, the PAH concentration showed an insignificant correlation with the percentage of TOC and fine-grained sediments ($p<0.05$, $r=0.15$ and $r=0.11$) in the surface sediments of the Klang Strait. Thus, it might be probable that fine-grained sediments and TOC are not effective parameters in controlling the distribution of PAHs in the Klang Strait.

In the Klang Strait, the concentration of total PAHs ranged from 100.3 to 3446.9 µg/kg dw, with an average concentration of 994.02±918.1 µg/kg dw (Table 2). The highest concentration of PAHs was estimated at station 13, while the lowest concentration was observed at the control point. The PAH contamination was estimated for surface sediments all over the strait, with higher concentrations observed at the stations close to the berth line (except station 12), especially in front of the container terminal in the West Port (station 13: 3446.9 µg/kg dw), in front of the dry and liquid terminal in the North Port (station 1: 1280.4 µg/kg dw and station 4: 2851.2 µg/kg dw), and in stations located adjacent to the mouth of the Klang River in the South Port (16: 2397.1 µg/kg dw and 17: 2902.1 µg/kg dw). PAH concentrations were generally lower in more remote stations and mangrove side, except for station 12, which had an elevated concentration of PAHs (1448.8 µg/kg dw) (Figure 2A). A Kruskal–Wallis test showed that there are significant differences ($p<0.05$, df=21, sig =0.003) among the concentrations of PAHs at all stations.

The concentrations of high-molecular-weight PAHs (HMWPAHs) are usually estimated in order to assess the combustion values [2,10,11]. In our study, combustion values ranged from 80.3 µg/kg dw to 2767.8 µg/kg dw, with an average concentration of 767.2 µg/kg dw (Table 3). High-molecular-weight PAHs represented between 27.66% and 97.08% of the total concentration of PAHs, with a mean value of 77.18%. Moreover, the combustible PAHs constituted a significant portion of the total PAHs (more than 50%) at all stations, with the exception of stations 13 (in front of the container terminal) and 7 (close to cement outlets) in the West Port, and stations 3 in the North Port (Figure 2b).

Some high-molecular-weight PAHs, such as BaA, Chy, BbF, BkF, BaP, InP and DibA are regarded as known as toxic PAHs, due to their mutagenic and carcinogenic effects on humans and other organisms [11,36]. Their concentration made up an average of 61.25% of the total concentration of PAHs, ranging from 0–97.3%. These concentrations were highest in stations 1, 4, 13, 15, 16, and 17, whereas at certain remote stations (2, 3, 5 and 22), the toxics' PAH concentrations were significantly lower.

Distribution maps were used to better visualize of the distribution of contamination at the spatial scales (Figure 2C). At the South Port, the distributions of PAH compound ($PAHs_{combust}$ and PAH_{toxic}) sediments exhibited a homogeneous pattern of decreasing concentration in the north-to-south direction, and

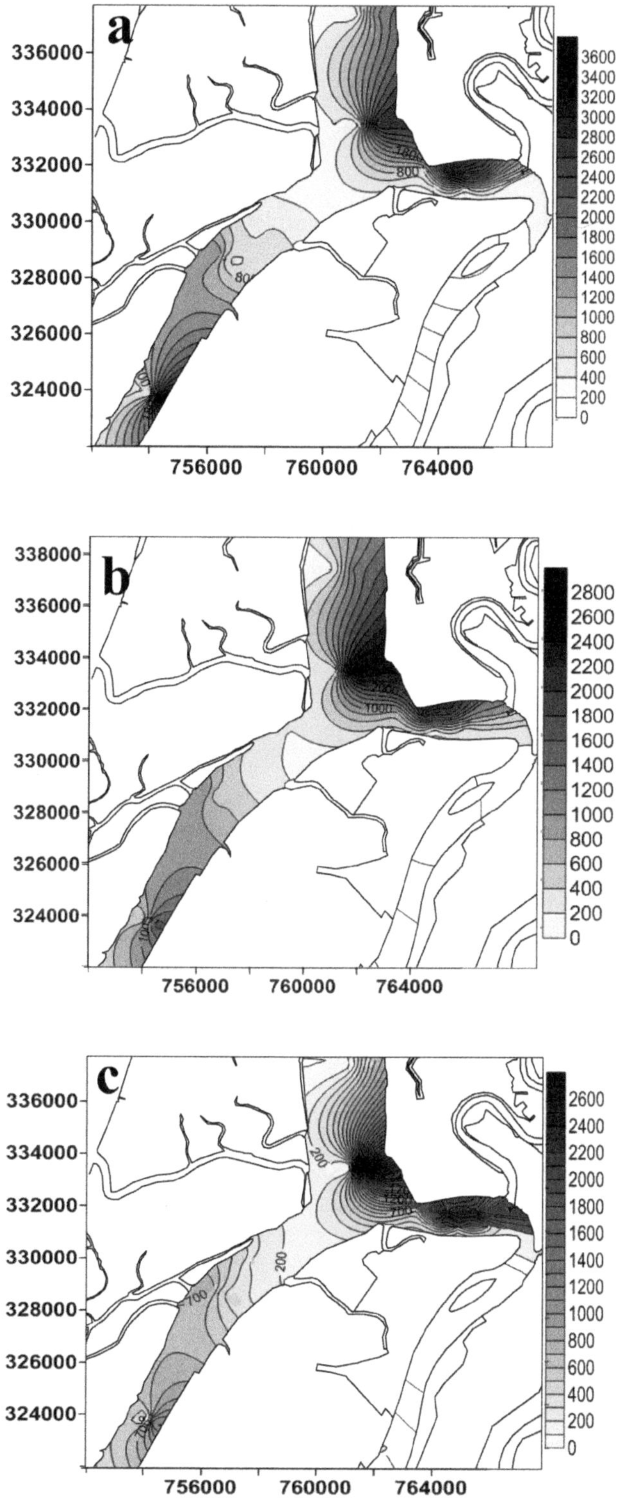

Figure 2. Spatial variations of PAHs compounds in surface sediment of the Klang Strait (a: Total PAHs, b: PAHs $_{combust}$, c: PAHs $_{toxic}$).

concentrations were high at the stations16 and 17, which are parallel to the mouth of the Klang River (Figure 2 and Table 3). This pattern supports the view that the Klang River may be the primary source of contamination in the Klang Strait, and influences the concentration and distribution of PAHs as water and suspended solids were easily exchanged between the South Port and the polluted Klang River, which contains industrial effluents and untreated municipal waste. Water currents in the vicinity of the South Port are weak; therefore, there is enough time for the deposit of organic components by suspended solids to enter surface sediments. Organic compounds are not easily deposited onto bottom sediments where there are strong water currents [15].

At both the West Port and North Port, the spatial distribution of PAHs exhibited a decreasing east–west gradient. These stations were located close to the near shore area, and are thus strongly influenced by port activities, especially at stations 4, 13 and 14, which were close to the terminal containers in the North and West Port. Additionally, land-based run-offs directly release organic compounds in the vicinity of these stations. Further research revealed significant differences in the sources and concentration of PAHs in sediment samples of near-shore vs. offshore areas. PAH concentrations in sediment samples collected from near-shore sites (city hinterland) were greater than the PAH concentrations in offshore areas [11,30,37,38,39]. This implies that PAH concentrations in near-shore areas are influenced by lateral transport, such as run-off and the transportation of water due to daily rainfall [12,40]. In 2005, Ikaneka indicated that heavy rainfall and floods significantly contributed to the pollution of marine sediments. Moreover, the results of their work indicated that near-shore areas received both burnt material (pyrogenic) and oil products (petrogenic), whereas offshore stations were primarily influenced by the burnt materials [9,24,40].

Overall, the distinct pattern in the distributions of measured parameters revealed that multiple sources contributed significantly to the PAH loads in the Klang Strait. These sources include the large-scale inflow from industries, such as the palm oil, cement, and food manufacturers that are located along the coastline of the North and West Ports, vessel-based discharges, Klang River outflow, land-based run-off, sedimentation, and siltation.

Source identification

Several methods were used to identify the source of PAHs according to the correlation analysis, diagnostic PAHs ratios, PCA analysis, and significant differences in the concentration of the individual PAHs.

A correlation analysis (Table S2) revealed a significant correlation between some individual PAHs related to pyrogenic and petrogenic compounds. This implied that the sediment samples are polluted by anthropogenic sources of PAHs because unpolluted sediment samples do not exhibit an oil fingerprint, and only some PAH compounds may be found in this sediment [12,40].

Mahyar et al. in 2010 studied the characteristics and possible origins of PAHs in the developed and developing areas around the coastal waters of Peninsular Malaysia [12,40]. The collected sediment core samples were used to assess the historical profile of PAHs from 1875 to 2007. Their research agrees well with our results, as because they revealed that older sediment samples (1875–1899) did not exhibit an oil fingerprint, and that only some PAH compounds were found. Sediment samples that are polluted by oil products generally exhibit high concentrations of phenanthrene and its methylated derivative. Thus, there is a high correlation between crude oil and petrogenic and pyrogenic compounds, whereas this correlation in old sediment samples with natural inputs (with no oil fingerprints) ranges from low to negative values [12,40].

Ratios of PAHs, such as Ant/Ant+Phn and Fla/(Fla+Pyr) had been applied to provide an accurate estimation of PAH sources

Table 3. Molecular indices of PAHs, content in surface sediments of Klang Strait.

Station	$\sum$ PAH $_{combust}$	$\sum$ PAH $_{toxic}$	Ant/Ant+Phn	Fla/Fla+Pyr	L/H-PAH
1	1195.8	613.0	0.68	0.00	0.15
2	162.8	31.5	0.49	0.00	0.23
3	105.6	0.0	0.40	0.40	2.61
4	2767.8	2755.3	0.85	0.51	0.03
5	206.6	11.0	0.49	0.05	0.20
6	279.4	221.7	0.51	0.45	0.49
7	274.6	251.6	0.68	0.45	1.55
8	343.5	250.0	0.91	0.46	0.36
9	485.2	457.1	0.61	0.00	0.05
10	584.1	347.6	0.67	0.00	0.07
11	439.8	397.1	0.71	0.41	0.24
12	1046.8	784.3	0.90	0.08	0.29
13	1715.9	1256.3	0.58	0.41	1.07
14	279.2	249.3	0.71	0.53	1.05
15	569.7	569.7	0.28	0.00	0.04
16	2099.5	1907.6	0.60	0.48	0.19
17	2267.4	2041.3	0.86	0.00	0.41
18	689.7	570.4	0.58	0.00	0.06
19	415.0	381.5	0.46	0.49	0.27
20	359.4	326.1	0.44	0.49	0.30
21	410.1	382.8	0.50	0.59	0.18
22	80.3	80.3	0.71	0.00	0.25
Mean	767.2	631.1	0.70	0.16	0.33
Standard deviation	233.2	213.8	0.22	0.28	0.32
Minimum	80.3	11	0.28	0.4	0.04
Maximum	2767.8	2755.3	1	1	2.61

$\sum$ **PAH** $_{combust}$: Combustion value of PAHs calculated as sum of concentration of high-molecular-weight PAHs include Fla, Pyr, BaA, Chy, BbF, BkF, BaP, InP, DibA, BghiP; $\sum$ **PAH** $_{toxic}$: Toxic value of PAHs calculated as sum of concentration of BaA, Chy, BbF, BkF, BaP, InP and DibA; L/H: ratio of low molecular weight hydrocarbon (**n**-C10–**n**-C14) to high molecular weight hydrocarbon (**n**-C16–**n**-C22).

(Table 3) [41]. The ratio of Ant/Ant+Phn was greater than 0.1 for all stations, implying that the origin of the PAHs in the sediments of the Klang Strait is primarily combustion. However, the ratio of Fla/(Fla+Pyr) showed the mixed pattern of pyrogenic and petrogenic sources in most stations. In the present study, ratio analyses showed the contradictions in certain stations.

In several researches, PAH ratios have been considered for "ideal" samples dominated by a single source, such as smoke, vehicle emission, and wood. However, some authors indicated that the limits between the diagnostic PAHs ratios are imprecise, as most environmental samples contain PAHs from mixed sources, none of which dominates the PAH profile [2,40,41]. Likewise, the validity of many diagnostic PAHs ratios has been examined for source distinction based on the source material in tropical Asian waters [12,14,42].

Although these ratios are usable for the differentiation of pyrogenic and petrogenic sources, they are not definitive, since several exceptions have been found. For example, Saha et al. in 2009 reported that, although all of the crude oil samples have Ant/(Ant+Phn)<0.1 and Phn/Ant>10, implying a petrogenic source, one crude oil sample (Miri) has Anh/(Anh+Phn)>0.1 and Phn/Ant<10, showing a pyrogenic source, and one wood sample has Anth/(Anh+Phn)<0.1; implying a petrogenic sources of PAHs [12,14,41]. Hence, these PAHs ratios are not authoritative enough to be exact; they only provide a rough idea for sources-distinction. These contradictions are most likely due to inconsistencies associated with the conventional ratios, large range of thermodynamic stability (discrimination ability) of different parent PAHs, and different homolog distributions between pyrogenic and petrogenic sources [2,14,42]. Despite these exceptions, the diagnostic PAHs ratios are now routinely analyzed. Several researchers suggest using more indices and cross-plot of diagnostic PAHs ratios to provide a more robust interpretation [10,41,43].

In the present study, the PAH ratio of Fla/(Fla+Pyr) was plotted against Ant/Ant+Phn for all of the sampling stations in Klang Strait (Figure 3). Figure 3 showed a mixed pattern of contamination from pyrogenic and petrogenic sources (petroleum combust) at some of the stations (3, 6, 7, 8, 11, 13, 16, 19 and 20). Simultaneously, wood combustion fingerprints (more frequent PAHs representing wood, coal and grass combustion) were strongly present in stations 4, 14, and 21. Other stations (1, 2, 5, 9, 10, 12, 15, 17, 18 and 22) have primarily originated from combustion sources with a significant contribution of petroleum inputs. Furthermore, An Ant/(Ant+Phn) ratio >0.10 almost always corresponds to a Fla/(Fla+Pyr) ratio >0.40, supporting the conclusion that combustion products are a main component of the contamination [10,14,41].

The purpose of principal component analysis (PCA) is to represent the total variation of the PAHs data with a minimum number of the factor loadings. By critically quantifying the percentage of contribution of individual PAH sources and identifying the occurrence of the sources responsible, each factor can be realized [2,23,44]. As shown in Table 4, PCA analysis classified the dataset of PAHs into five principal components (PCs) that control 81.89% of the variability in the data.

The first factors (PC1) explained a total variance of 26.73% in the data. These factors were strongly weighted the context of Acy, Ace, Flr, Phn and Fla, and were believed to be from petrogenic sources of PAHs, and were also been identified as tracers for volatilization or spill of petroleum products [2,12,14,23,41,42]. Phn and Fla are the thermodynamically stable triaromatic isomer, and their high concentration are regarded as characteristic of petrogenic sources [14,23]. As indicated by a previous researcher, Phn and Fla were more abundant in crude oil, crankcase oil, and unburnt fuel, and originated from soot collected from diesel engines [12,41,45].

The second factor (PC2) responsible for 21.08% of the total variance was strongly related to BbF, BaP, DibA, Bghip, with moderate loading of InP and BkF. This factor are predominately composed of high molecular weight PAHs with 4–6 rings, and are basically known to be derived from the incomplete combustion and pyrolysis of fuel [44,46]. The source of this individual PAHs have been found in both vehicular emission sources (diesel and gas engine), road dust, and gasoline combusted ships [2,44]. InP and BkF have been found in gasoline vehicle soot, and both gas and diesel engine emissions [2,12,47,48]. Bghip and BaP have been identified as a tracer of auto emissions because there were determined to be rich in road dust from urban-areas and traffic tunnels. The abundance of these compounds was considered as an index for the determination of autoexhaust contribution of gasoline-engines. Therefore, this factor is selected to represent the combustion source of PAHs [2,12,14,41,48].

The third factor is responsible for 17.25% of the total variance, and is predominately composed of Ant, Pyr, Chy, and BaA (4-ring PAHs), with moderate loadings of Fla. More researchers have used Ant as the marker of wood combustion source [2] (Harrison et al., 1996). Also, several literature have identified Pyr, BaA, Chy, Fla (4-rings) and Bap as a tracer of auto emissions, as it was found to be enriched in road dust collected from Kuala Lumpur, Bangkok, Shanghai [2,12,14,47,48], and traffic tunnels [10,44,47,48]. The PAH profiles of road dust in these cities showed that the ratio of PAHs (4 rings) to PAHs (5–6 rings) exceeds 1 [12,47]. In the present paper, we showed that the ratio of PAHs (4 rings) to PAHs (5–6 rings) in the surface sediment, 0–126.44 with a mean of 7.805 is significantly greater than that in the road dust from Kuala Lumpur (1.58). This result indicates that there might be an additional PAH source (not traffic-related), leading to a higher ratio of these PAHs [2]. Several researchers have identified Pyr, BaA, Chy, and Fla as markers of coal combustion because these individual PAHs are dominant in coal-combustion profiles [2,12,14,47,48]. In Malaysia, coal is the main energy source, and is used widely for industrial and domestic and industrial purposes, such as cement, palm oil, tire, and power industry [2,12,14]. It is reasonable to assign these PAHs to coal combustion and traffic-related source of PAHs.

Table 4. Rotated component matrix of 16 PAHs from the Klang Strait sediment.

PAHs	PC1	PC2	PC3	PC4	PC5
Nap	.095	0.000	0.138	**0.954**	−0.034
Acy	**.957**	0.037	0.009	0.006	0.012
Ace	**.918**	0.061	.185	−0.013	0.031
Flr	**.910**	−0.042	0.332	0.114	0.069
Phn	**.824**	−0.106	0.322	0.091	.124
Ant	.429	0.063	**.814**	−0.064	.429
Fla	**.866**	0.017	.466	0.465	0.07
Pyr	.201	−0.335	**.779**	−0.021	−0.236
BaA	.058	−0.142	**.861**	0.392	−0.112
Chy	−0.087	0.094	**.865**	0.029	0.043
BbF	−0.092	**0.871**	−0.016	−0.016	0.036
BkF	.012	0.521	−0.139	0.108	**0.852**
BaP	0.288	**0.777**	.026	.395	0.013
DibA	−0.152	**0.847**	−0.096	−0.137	−0.217
BghiP	0.185	0.584	−0.185	−0.058	0.173
InP	0.081	0.609	0.139	−0.221	−0.492
Variance (%)	26.73	21.08	17.25	9.66	7.17

*Rotation method: Varimax with Kaiser normalization.
*Bold loadings >0.70.

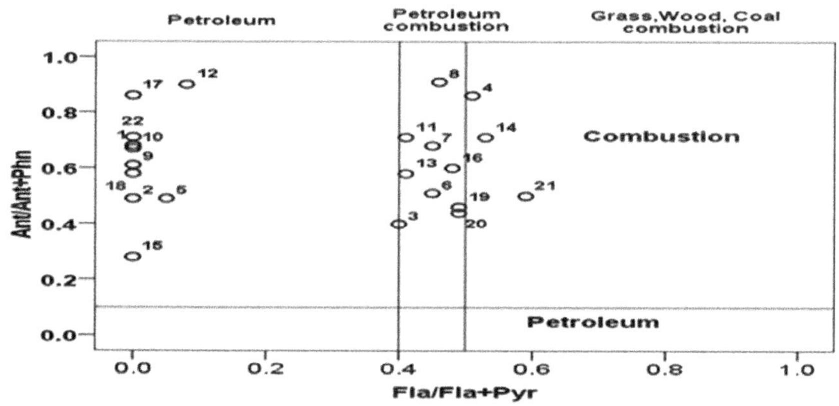

Figure 3. Plots of PAH isomer pair ratios for source identification (Ant/(Ant+Phy) vs Flu/(Flu+Pyr).

PC4 explained 9.66% of the total variance, with a high correlation with Nap. Nap is used as a tracer of oil spills, such as Acy and Ace. It can also originate from termite activities on vascular land plants and woody material in tropical areas and the Amazonian region [49,50]. The fifth factor (PC5) is responsible for 7.17% of the total variance and was strongly related to BkF, which are known to be derived from the incomplete combustion and pyrolysis of fuel [44,46].

The PCA analysis described 45.5% of the total variance (PC2, PC3 and PC5) in the PAH load, including individual PAHs that are known to originate from coal combustion and vehicular emissions or traffic-related sources.

Similarly, 36.39% of the total variance (PC1 and PC4) in PAHs load related to petrogenic sources and the variance in PAHs was unrelated to unknown and biogenic sources.

There are two origins for vehicular sources in the Klang Strait, one is the traffic-related PAHs in road dust entering the surface sediment by land-based runoff, others due to the exhaust emission from cargo vessels, ships, ferries, and fishing boats [2,12,14,47]. Furthermore, the cargo vessels play the main role as PAHs contributor in the Klang Strait compared to traffic related PAHs, as cargo vessels are responsible in transferring most coal to coal–fired power plants located along the West coastal water of Malaysia, such as Connaught bridge power, Kapar and Tanjung Bin, Manjung, and Jimah. Most fishing boats and ferries are still unequipped with catalytic converters [51,52].

Coal and wood combustions mainly originated from agricultural, industrial, and habitats activities. Recently, coal usage near the Klang Strait is concentrated in the Kapar Power Plant in the Klang Valley region with high consumption of coal (more than16 million tons). Additionally, It is also the only power plant in Malaysia with triple fuel firing capability (natural gas, oil, and coal). It started operating in in 1987 and was the first coal–fired power plant in Malaysia [51,52]. Moreover, Mahyar et al. in 2010 indicated that, in the near-shore and offshore areas near Klang, high levels of PAHs originated from pyrogenic sources, such as the combustion of fossil fuel and the burning of biomass and wood. The dominance of pyrogenic sources in this area is due to the products of combusted petroleum from industries and automobiles emissions [12,40].

Petrogenic sources include petroleum products and crude oil, which are identified as major fossil fuels being used in Malaysia (around 49% of total energy consumption) [14]. Zakaria et al. in 2002 reported that several sources lead to the input of petrogenic PAHs into Malaysian coastal water, namely spillage of unburnt fuel and waste crankcase oil, leakage of crankcase oils from vehicles, with subsequent land-based runoff, wastewater of industry outlets, and vessel maintenance [12,14,47].

In addition, other studies have reported that contamination from various seas contribute to increase petrogenic pollution in the coastal waters of western Malaysia [16,53]. For example, in 1999, Abdullah revealed that land-based run-off and contamination from an offshore oil field near Sumatra Island contributed to the petroleum contamination in the Malacca Strait, which is a narrow channel of marine water located between Sumatra Island, Indonesia, and Peninsular Malaysia. This strait is vulnerable to contamination caused by tanker operation and oil spills [16,53,54]. The Strait of Malacca also plays an important role in transporting pollution into the Klang Strait. Pauzi Zakaria in 2001 used a biomarker compound to identify the source of tar-balls off the coast of Malaysia.

They identified several petrogenic sources of contamination in the western coastal waters of Malaysia, such as accidental oil spills and tanker-derived sources in the Strait of Malacca and crude oils originating from land-based run-off from human activities. Additionally, they clearly showed that the western coastal waters of Malaysia have received approximately 30% of their petroleum pollution from Middle East crude oil (MECO) and South-East Asian crude oil (SEACO), which was probably transported to Malaysia via marine currents, tankers and shipping discharges, including ballast water and tank-washing water [16].

Generally, the results of the PCA analyses are in concordance with the evidence from the pair isomer ratio of PAHs, which revealed a mixture of pyrogenic-and petrogenic-derived PAHs. PAHs contamination in the sediment in the Klang Strait are derived primarily from vehicular sources, coal and wood combustions, and crude oil and coal (petroleum combustion sources), which can originate from land-based run-off, oil spill, tanker operation, shipping activities, and industrial discharges such as effluents from cement, food and oil factories.

Ecological risk assessment

The results obtained from the ecological risk assessment of PAHs in the surface sediments of Klang Strait are summarized in Table 5. These results were arranged based on the concentration of PAHs measured at 22 stations. Generally, the risk assessment revealed that total PAHs are likely to cause slightly adverse effects on the biological communities at stations 4, 7, 12, 16, and 17 as these stations showed a value of $0.1<PELq<0.5$ and $1.0<P\leq2.0$. Additionally, moderately adverse biological effects were shown for station 13 ($0.5<PELq<1.5$, $2.0<P\leq3.0$), but the rest of the stations showed rare, adverse ecological effects due to the PAH exposure in surface sediments ($PELq<0.1$, $P<3$). The cluster analysis allows for a better understanding of the similarity between sampling stations and PAHs contamination (Figure 4). The dendrograms clearly classified the sampling stations into three main clusters based on adverse biological effects. Station 13 (cluster C) was completely separated from the other stations as it demonstrated the moderate adverse biological effect by PAHs compound, while stations 4, 16, and 17 are in the same group (cluster B), with slightly adverse effects. The other stations are grouped into cluster A, and exhibited non-adverse effects ($PELq<0.1$, $P<3$), with the exception of stations 7 and 12. However, the average toxicity effects of PAHs in stations 7 and 12 are in the slight range, but these stations were arranged in cluster A, because their values are closer to 0.1 compared to the stations in cluster B. Based on a comparison with the sediment quality guidelines, only certain individual PAHs, such as Acy, Flr, Ant, BaA, BaP and DibA were associated with adverse biological effects, and only at stations 4 (close to the container terminal in the North Port), 7 (cement outlet), 13 (container terminal in the West Port), 12 (around the mangrove forest in the West Port) 16, and 17 (vicinity of the mouth of the Klang River) (Table 5).

Moreover, the first three components of PCA analysis are strongly related to these individual PAHs that controls 63.36% of changes in PAHs contamination in surface sediment of the Klang Strait. According to the cluster and PCA classification, stations 4, 7, 13, 16 and 17 can be regarded as the main entrances for loading individual PAHs, especially Acy, Flr, Ant, BaA, BaP and DibA.

The present study shows that generally, the sediment is slightly polluted with PAHs ($PELq = 0.16$ and $P = 1.01$) in the Klang Strait. According to the results of ecological risk, it can be concluded that PAHs are not a primary pollution concern in the Klang Strait, and that only station 13 can be considered as a vulnerable station in the Klang area.

In this research, our hypothesis was defined based on the serious threat posed by petroleum contamination to the Klang Strait due

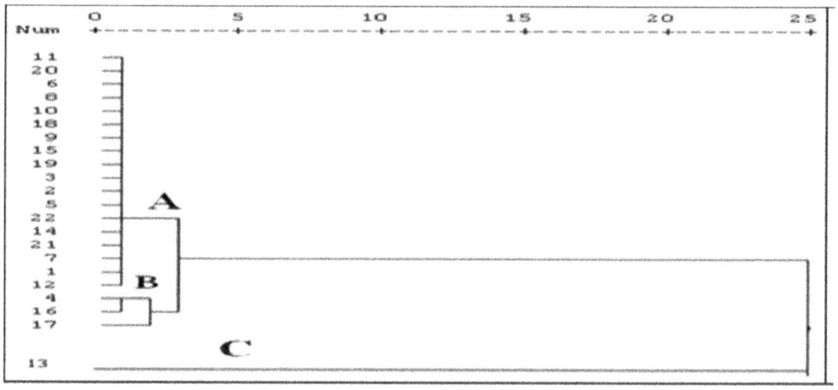

Figure 4. Cluster analyses to classify stations based on adverse biological effect of PAHs in surface sediment.

to several evidences, which was reported in previous literature [16,40,53,55].

The highest value for PAH contamination was estimated to be 1700 ng/g d.w. in near-shore Klang sediment samples from 1976–1999, with a predominant input of burnt materials and combusted petroleum discharges from industries, oil spill events and ship ballasting or bilge pumping. From 1976–1999, rapid development of the West Port and maritime-related industries that caused greater numbers of ocean going ships to enter the strait from the South China Sea and to the Far East to transport goods. Similarly, in 1997, a collision between two tankers in the straits of Singapore caused 25000 tons of heavy fuels oil to be released into the marine environment, of which 700 tons of oil were released into the Malacca strait, negatively affecting the coastal area of the Klang Strait [12,40].

According to the present study, the hypothesis was rejected due to the fact that in most of the stations, the average PAHs concentration was significantly lower than the threshold effects

Table 5. Ecological risk calculated for individual PAHs in the surface sediment of different station.

Stations	Nap	Acy	Ace	Flr	Phn	Ant	Fla	Pyr	BeA	Chy	BbF	BkF	BaP	DibA	BghiP	PELq	P
1	0.00	0.00	0.02	0.23	0.08	0.36	0.00	0.41	0.8	0.00	0.00	0.00	0.00	0.00	0.00	0.09	0.92
2	0.00	0.00	0.00	0.07	0.02	0.05	0.00	0.09	0.05	0.00	0.00	0.00	0.00	0.00	0.00	0.02	0,77
3	0.00	0.00	0.06	0.82	0.16	0.24	0.03	0.04	0.00	0.00	0.00	0.00	0.00	0.00	0.00	0.09	0,83
4	0.00	0.00	0.00	0.01	0.02	0.24	0.01	0.01	0.10	0.06	0.45	0.00	**1.96**	**1.41**	0.00	**0.28**	**1,56**
5	0.00	0.00	0.00	0.12	0.02	0.05	0.01	0.14	0.00	0.00	0.01	0.00	0.00	0.00	0.00	0.02	0,79
6	0.01	0.00	0.00	0.40	0.07	0.15	0.02	0.02	0.26	0.00	0.00	0.00	0.00	0.00	0.00	0.06	0,83
7	0.00	0.00	0.00	0.19	0.23	**1.09**	0.01	0.01	0.29	0.00	0.00	0.03	0.00	0.00	0.00	**0.12**	**1.02**
8	0.01	0.00	0.01	0.27	0.01	0.29	0.01	0.01	0.07	0.00	0.00	0.00	0.16	0.00	0.04	0.06	0,85
9	0.00	0.00	0.00	0.05	0.01	0.03	0.00	0.02	0.28	0.23	0.00	0.00	0.00	0.50	0.00	0.08	0,86
10	0.00	0.00	0.00	0.16	0.01	0.04	0.00	0.17	0.40	0.00	0.00	0.00	0.00	0.00	0.00	0.05	0,90
11	0.00	0.00	0.01	0.34	0.03	0.15	0.01	0.02	0.04	0.00	0.00	0.23	0.00	0.00	0.00	0.06	0,87
12	0.01	0.00	0.01	0.28	0.05	0.95	0.01	0.24	**1.12**	0.00	0.00	0.00	0.00	0.05	0.00	**0.18**	**1,14**
13	0.04	**7.26**	0.12	**4.31**	0.41	**1.26**	0.11	0.17	0.56	0.00	0.01	0.02	**1.08**	0.00	0.03	**1.03**	**2.03**
14	0.01	0.00	0.00	0.23	0.14	0.74	0.01	0.01	0.21	0.00	0.01	0.00	0.08	0.00	0.00	0.10	0,88
15	0.01	0.00	0.00	0.06	0.02	0.02	0.00	0.00	0.11	0.00	0.01	0.00	0.00	0.96	0.00	0.08	0,89
16	0.28	0.00	0.01	0.84	0.12	0.41	0.06	0.07	**1.41**	0.14	0.00	0.00	**1.07**	0.00	0.00	**0.30**	**1,42**
17	0.02	0.00	0.03	**1.57**	0.17	**2.45**	0.00	0.16	**2.15**	0.65	0.00	0.00	0.00	0.00	0.00	**0.48**	**1,57**
18	0.00	0.00	0.00	0.14	0.02	0.05	0.00	0.09	0.28	0.00	0.20	0.00	0.00	0.00	0.00	0.05	0,93
19	0.01	0.00	0.00	0.25	0.07	0.14	0.01	0.01	0.39	0.00	0.01	0.00	0.14	0.00	0.00	0.07	0,87
20	0.01	0.00	0.00	0.26	0.07	0.12	0.01	0.01	0.33	0.00	0.00	0.00	0.12	0.00	0.00	0.06	0,85
21	0.01	0.00	0.00	0.25	0.03	0.07	0.01	0.01	0.26	0.00	0.01	0.00	0.00	0.96	0.00	0.09	0.86
22	0.00	0.00	0.00	0.09	0.00	0.02	0.00	0.00	0.08	0.00	0.01	0.00	0.00	0.00	0.00	0.01	0,74
Mean	0.02	0.33	0.01	0.50	0.08	0.41	0.01	0.08	0.42	0.05	0.03	0.01	0.21	0.18	0.00	0.16	1,01

*Nemerow pollution index (P).
*The PEL quotient (PELq).

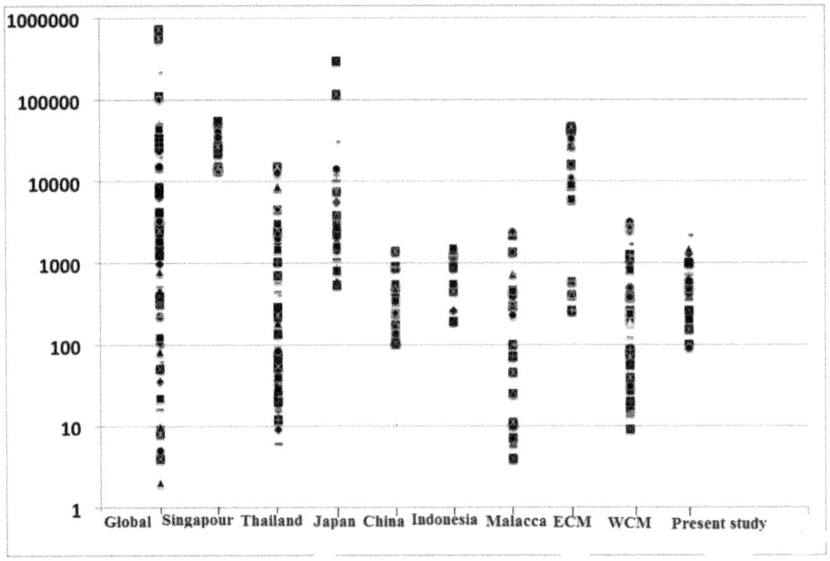

Figure 5. Total PAHs concentration (ng/g dw) in Klang Strait sediments in comparison to those of the reported concentrations for global, Southeast Asian countries, and Malaysian costal zones. Data for global sediments and others areas are derived from several literatures [12,14,40,44,45,52,55,56,58,59].

level (TELs and PELs), and rarely reached a level likely to cause adverse biological effects.

Several studies recorded a significant decline in PAH discharges around the west coastal water of Malaysia after 2000 due to the establishment of an integrated management programme, weathering, and meteorological conditions [16,40,53,55].

An integrated management programme was established in 1996, where several responsible organizations contributed to solving the environmental problems in Malaysia's marine environments. These organizations ratified specific regulations related to marine pollution to control petroleum and chemical contaminations in Malaysian coastal waters. These regulations focused on strategies that were based on international agreements to prevent and control pollution from ships, platform draining, and industrial inputs [16].

The weathering process can greatly deplete the concentration of PAH in a marine environment. The L/H-PAH ratio is accepted as a practical method for assessing weathering based on the differences between the low and high molecular weights of PAH compounds [16,40,53,55]. These results show that the L/H-PAH ratio was below 1 at most of the stations, and exceeds 1 only at stations 3, 7, 13, and 14 (Table 3). Generally, a low ratio of L/H-PAHs is attributed to the high resistances of the high-molecular-weight PAHs to microbial degradation, which is consistent with other studies. Such a low ratio could also be due to the high solubility and volatility of the low-molecular-weight PAHs, which would lead to their depletion [16,53,56].

In addition, there is also evidence of a greater depletion of low-molecular-weight PAHs by the weathering process, which is clearly indicated by Pauzi Zakaria (2001) [16]. He revealed that the tar-ball samples from western Malaysian coastal waters had undergone significant weathering, as their L/H-PAH ratios were much lower (0.23–1.48) than the ratios of crude oil samples (8–44 for SEACO crude oil and 12.13–20.3 for MECO crude oil) [16]. In the present study, this ratio ranged from 0.02–2.61, which was significantly lower than that of the crude oil in this region. Several possibilities have been described to explain why the PAH compositions had a lower ratio of L/H in the western coastal waters of Malaysia, such as wandering due to long-haul transportation, ballast water, and tank washings [16,53,56].

Additional research has indicated that meteorological conditions play a major role in controlling PAH concentrations over spatial and temporal scales, and significantly negative linear relationships were found between wind speeds, temperatures, and PAH concentrations, because atmospheric turbulence causes a dilution of contaminant concentrations and hastens the weathering process, especially if the wind speed is greater than 5.8 km h^{-1} and the temperature is exceeds than 20°C [30,57]. Thus, the higher temperature (approximately 30°C) of the Klang Strait as a tropical area could potentially increase the depletion of PAHs.

Figure 5 shows the concentration of PAH in comparison with global, Southeast Asian countries, and Malaysian costal zones. Most values of PAHs concentrations in the present study, west coastal water of Malaysia, Malacca, Indonesia, China and Thailand ranged from tens to thousands of ng/g. Thus, PAHs values in Klang Strait sediments are comparable to these areas. On the other hand, the current PAHs concentration are 1–2 orders of magnitude lower than those in Japan, Singapore, and the eastern coastal water of Malaysia, as the value of PAHs in these countries ranged over thousands of ng/g. Thus, PAHs concentrations in Klang Strait can be categorized as low to moderate on a global scale.

Conclusion

Exploratory source identification of PAHs in Klang Strait sediment was based on the distribution pattern of the individual PAHs, principal component analysis, and paired isomer ratios. The results of the distribution pattern showed that the pyrogenic sources are the major source of PAHs in most stations. Petrogenic input appears be a main source of PAHs in station 13 (in front of the container terminal) and stations 7 (close to cement outlets) in the West Port, and station 3 in the North Port because the LMWPAHs constituted a significant portion of the total PAHs in these stations.

The results of diagnostic ratios showed that PAHs are derived from a mixture of sources (both petrogenic and pyrogenic), with high frequencies of pyrogenic sources in most stations. The PCA analysis method showed that the contributions of spill of oil products, traffic-related pollution, and coal and wood combustion are dominant in Klang Strait sediments.

The pyrogenic sources (traffic-related pollution and coal combustion) are responsible for 45.5% of the total variance in the Klang Strait sediment sediments. As well as pyrolytic input as a main source, petrogenic input was also regarded as a source for PAHs in the Klang Strait due to direct petrogenic sources from oil leaks (oil spill, urban run-off, tanker operation and shipping activities) in agreement with their location.

Despite the inherent limitations of this research, the adopted approach in the present study highlighted that no adverse biological effects are associated with the exposure to PAHs levels in the Klang Strait, and only the areas around terminal container (station 13) are moderately polluted. Several factors caused the decline of these contaminants, such as an integrated management programme and regulation, meteorological conditions, and weathering. This paper represents a baseline study for future studies, which may include the whole Malaysian estuary and coastal waters for the purpose of promoting our knowledge of the distribution of PAHs in coastal and estuary ecosystems.

Acknowledgments

We would like to acknowledge Professor A. Sasekumar, Professor Chong Ving Ching and Dr Wong Chee Seng and others involved in field collection of samples, for their support and inspiration. Special thanks to Chemistry Department of Marine Science, Iranian National Institute for Oceanography, for valuable contribution to laboratory analysis.

Author Contributions

Conceived and designed the experiments: SBTS RH AS AM. Performed the experiments: SBTS MR AM. Analyzed the data: SBTS OS. Contributed reagents/materials/analysis tools: AS RH. Wrote the paper: SBTS MR OS.

References

1. Veltman K, Huijbregts MAJ, Rye H, Hertwich EG (2012) Including impacts of particulate emissions on marine ecosystems in life cycle assessment: The case of offshore oil and gas production. Integr Environ Assess Manag 7: 678–686.
2. Liu Y, Chen L, Huang Q-h, Li W-y, Tang Y-j, et al. (2009) Source apportionment of polycyclic aromatic hydrocarbons (PAHs) in surface sediments of the Huangpu River, Shanghai, China. Sci Total Environ 407: 2931–2938.
3. Barkhordarian A (2012) Investigating the Influence of Anthropogenic Forcing on Observed Mean and Extreme Sea Level Pressure Trends over the Mediterranean Region. The Scientific World Journal 2012.
4. Zhou Q, Jiang G, Liu J (2002) Organotin pollution in China. The Scientific World Journal 2: 655–659.
5. Tavakoly Sany SB, Salleh A, Sulaiman AH, Tehrani GM (2012) Ecological risk assessment of poly aromatic hydrocarbons in the North Port, Malaysia. World Academy of Science, Engineering and Technology 69: 43–46.
6. Ahmadzadeh S, Kassim A, Rezayi M, Rounaghi GH (2011) Thermodynamic Study of the Complexation of p-Isopropylcalix [6] arene with Cs+ Cation in Dimethylsulfoxide-Acetonitrile Binary Media. Molecules 16: 8130–8142.
7. Rezayi M, Kassim A, Ahmadzadeh S, Yusof NA, Naji A, et al. (2011) Conductometric Determination of Formation Constants of tris (2-pyridyl) methylamine and Titanium (III) in Water-Acetonitryl Mixture. Int J Electrochem Sci 6: 4378–4387.
8. MADEP (2007) Sediment Toxicity of Petroleum Hydrocarbon Fractions,Prepared for Massachusetts Department of Environmental Protection Office of Research and Standards. 397 Washington Street, Duxbury, MA 02332: Battelle.
9. Tavakoly Sany SB, Hashim R, Salleh A, Safari O, Mehdinia A, et al. (2013) Risk assessment of polycyclic aromatic hydrocarbons in the West Port semi-enclosed basin (Malaysia). Environmental Earth Sciences: 1–14.
10. Riccardi C, Di Filippo P, Pomata D, Di Basilio M, Spicaglia S, et al. (2013) Identification of hydrocarbon sources in contaminated soils of three industrial areas. Sci Total Environ 450: 13–21.
11. Khairy MA, Kolb M, Mostafa AR, EL-Fiky A, Bahadir M (2009) Risk assessment of polycyclic aromatic hydrocarbons in a Mediterranean semi-enclosed basin affected by human activities (Abu Qir Bay, Egypt). J hazard mater 170: 389–397.
12. Zakaria MP, Takada H, Tsutsumi S, Ohno K, Yamada J, et al. (2002) Distribution of polycyclic aromatic hydrocarbons (PAHs) in rivers and estuaries in Malaysia: a widespread input of petrogenic PAHs. Environ Sci Technol 36: 1907–1918.
13. Beyer J, Jonsson G, Porte C, Krahn MM, Ariese F (2010) Analytical methods for determining metabolites of polycyclic aromatic hydrocarbon (PAH) pollutants in fish bile: A review. Environmental Toxicology and Pharmacology 30: 224–244.
14. Saha M, Togo A, Mizukawa K, Murakami M, Takada H, et al. (2009) Sources of sedimentary PAHs in tropical Asian waters: differentiation between pyrogenic and petrogenic sources by alkyl homolog abundance. Mar Pollut Bull 58: 189–200.
15. Tavakoly Sany SB, Salleh A, Suliman AH, Sasekumar A, Rezayi M, et al. (2012) Heavy metal contamination in water and sediment of the Port Klang coastal area, Selangor, Malaysia. Environmental Earth Sciences: 1–13.
16. Pauzi Zakaria M, Okuda T, Takada H (2001) Polycyclic aromatic hydrocarbon (PAHs) and hopanes in stranded tar-balls on the coasts of Peninsular Malaysia: applications of biomarkers for identifying sources of oil pollution. Mar Pollut Bull 42: 1357–1366.
17. Tehrani GM, Hshim R, Sulaiman AH, Tavakoly Sany SB, Salleh A, et al. (2012) Distribution Of Total Petroleum Hydrocarbons And Polycyclic Aromatic Hydrocarbons In Musa Bay Sediments (Northwest Of The Persian Gulf). Environment Protection Engineering 15: 115–128.
18. Yap (2005) Ecology of Klang Strait. In (eds: Sasekumar, A. and V.C. . Chong) Pollution. Kuala Lumpur: University of Malaya. pp. 225–235.
19. Naji A, Ismail A (2012) Sediment quality assessment of Klang Estuary, Malaysia. Aquat Ecosyst Health Manage 15: 287–293.
20. US-EPA (2010) United States Environmental Protection Agency, Method 610, Polynuclear Aromatic Hydrocarbons.
21. Rauret G (1998) Extraction procedures for the determination of heavy metals in contaminated soil and sediment. Talanta 46: 449–455.
22. Budzinski H, Mazeas O, Tronczynski J, Desaunay Y, Bocquene G, et al. (2004) Link between exposure of fish (Solea solea) to PAHs and metabolites: Application to the "Erika" oil spill. Aquat Living Resour 17: 329–334.
23. Wang C, Wang W, He S, Du J, Sun Z (2011) Sources and distribution of aliphatic and polycyclic aromatic hydrocarbons in Yellow River Delta Nature Reserve, China. Appl Geochem.
24. Dsikowitzky L, Nordhaus I, Jennerjahn TC, Khrycheva P, Sivatharshan Y, et al. (2011) Anthropogenic organic contaminants in water, sediments and benthic organisms of the mangrove-fringed Segara Anakan Lagoon, Java, Indonesia. Mar Pollut Bull 26: 1330–1336.
25. EPA (1991) Sediment quality guidelines. Chicago: United States Environmental Protection Agency, Draft report.Region V.
26. Macdonald DD, Carr RS, Calder FD, Long ER, Ingersoll CG (1996) Development and evaluation of sediment quality guidelines for Florida coastal waters. Ecotoxicology 5: 253–278.
27. United States Environmental Protection Agency, National Oceanic and Atmospheric Administration (2005) Predicting toxicity to Amphipods from sediment chemistry. Washington, DC: National Center for Environmental Assessment, Office of Research and Development.
28. Long ER, MacDonald DD, Smith SL, Calder FD (1995) Incidence of adverse biological rtile sulfide. Environ. Toxicol. effects within ranges of chemical concentrations in marine and estuarine sediments. Environ Manag 19: 81–97.
29. Fdez-Ortiz de Vallejuelo S, Arana G, De Diego A, Madariaga JM (2010) Risk assessment of trace elements in sediments: The case of the estuary of the Nerbioi–Ibaizabal River (Basque Country). J hazard mater 181: 565–573.
30. Montuori P, Triassi M (2012) Polycyclic aromatic hydrocarbons loads into the Mediterranean Sea: Estimate of Sarno River inputs. Mar Pollut Bull 64: 512–520.
31. McCready S, Birch G.F, Long E.R (2006) Metallic and organic contaminants in sediments of Sydney Harbour, Australia and vicinity–a chemical dataset for evaluating sediment quality guidelines. Environ Inte 32: 455–465.

32. Liang J, Chen C, Song X, Han Y, Liang Z (2011) Assessment of heavy metal pollution in soil and plants from Dunhua sewage irrigation area. Int J Electrochem Sci 6: 5314–5324.
33. Ogunkunle CO, Fatoba PO (2013) Pollution Loads and the Ecological Risk Assessment of Soil Heavy Metals around a Mega Cement Factory in Southwest Nigeria. Polish Journal of Environmental Studies 22: 487–49.
34. Primpas I, Karydis M (2010) Scaling the trophic index (TRIX) in oligotrophic marine environments. Environ Monit Assess: 1–13.
35. Tavakoly Sany SB, Salleh A, Rezayi M, Saadati N, Narimany L, et al. (2013) Distribution and Contamination of Heavy Metal in the Coastal Sediments of Port Klang, Selangor, Malaysia. Water Air Soil Pollut 224: 1–18.
36. Hale SE, Lehmann J, Rutherford D, Zimmerman A, Bachmann RT, et al. (2012) Quantifying the total and bioavailable PAHs and dioxins in biochars. Environ Sci Technol 46: 2830–2838.
37. Pan K, Wang WX (2011) Trace metal contamination in estuarine and coastal environments in China. Sci total environ 421–422: 4–16.
38. Saadati N, Abdullah MP, Zakaria Z, Tavakoly Sany SB, Rezayi M, et al. (2013) Limit of detection and limit of quantification development procedures for organochlorine pesticides analysis in water and sediment matrices. Chem Central J 7: 1–10.
39. Tavakoly Sany SB, Hashim R, Rezayi M, Salleh A, Safari O (2013) A review of strategies to monitor water and sediment quality for a sustainability assessment of marine environment. Environ Sci Pollut Res 21: 813–833.
40. Safari M, Zakaria MP, Mohamed CAR, Lajis NH, Chandru K, et al. (2010) The history of petroleum pollution in Malaysia; urgent need for integrated prevention approach. Environ Asia 3: 131–142.
41. Yunker MB, Macdonald RW, Vingarzan R, Mitchell RH, Goyette D, et al. (2002) PAHs in the Fraser River basin: a critical appraisal of PAH ratios as indicators of PAH source and composition. Organic Geochem 33: 489–515.
42. Guo Z, Lin T, Zhang G, Yang Z, Fang M (2006) High-resolution depositional records of polycyclic aromatic hydrocarbons in the central continental shelf mud of the East China Sea. Environ SciTechnol 40: 5304–5311.
43. Wagener AdL, Meniconi MdFtG, Hamacher C, Farias CO, da Silva GC, et al. (2012) Hydrocarbons in sediments of a chronically contaminated bay: The challenge of source assignment. Mar Pollut Bull 64: 284–294.
44. Zhang Y, Guo C-S, Xu J, Tian Y-Z, Shi G-L, et al. (2012) Potential source contributions and risk assessment of PAHs in sediments from Taihu Lake, China: Comparison of three receptor models. Water Research 46: 3065–3073.
45. Boonyatumanond R, Wattayakorn G, Togo A, Takada H (2006) Distribution and origins of polycyclic aromatic hydrocarbons (PAHs) in riverine, estuarine, and marine sediments in Thailand. Mar Pollut Bull 52: 942–956.
46. Chen H-y, Teng Y-g, Wang J-s (2012) Source apportionment of polycyclic aromatic hydrocarbons (PAHs) in surface sediments of the Rizhao coastal area (China) using diagnostic ratios and factor analysis with nonnegative constraints. Sci Total Environ 414: 293–300.
47. Boonyatumanond R, Murakami M, Wattayakorn G, Togo A, H T (2007) Sources of polycyclic aromatic hydrocarbons (PAHs) in street dust in a tropical Asian mega-city, Bangkok, Thailand. Sci Total Environ 384: 420–432.
48. Larsen IR, Baker J (2003) Source apportionment of polycyclic aromatic hydrocarbons in the urban atmosphere: a comparison of three methods. Environ Sci Technol 37: 1873–1881.
49. Okere U, Semple K (2012) Biodegradation of PAHs in 'Pristine'Soils from Different Climatic Regions. J Bioremed Biodegrad S 1: 2.
50. Ekpo BO, Oyo-Ita OE, Oros DR, Simoneit BRT (2011) Distributions and sources of polycyclic aromatic hydrocarbons in surface sediments from the Cross River estuary, SE Niger Delta, Nigeria. Environmonit assess: 1–11.
51. Alam L, Mohamed CAR (2011) Natural radionuclide of Po210 in the edible seafood affected by coal-fired power plant industry in Kapar coastal area of Malaysia. Environ Health 10: 43.
52. Koike T, Koike H, Kurumisawa R, Ito M, Sakurai S, et al. (2012) Distribution, source identification, and historical trends of organic micropollutants in coastal sediment in Jakarta Bay, Indonesia. J Hazard mater 217: 208–216.
53. Elias M, Wood AK, Hashim Z, Hamzah MS, Rahman SA, et al. (2009) Identification sources of polycyclic aromatic hydrocarbons (PAHs) in sediments from the straits of Malacca. Fresenius Environmental Bulletin 18: 843–847.
54. Mirsadeghi SA, Zakaria MP, Yap CK, Shahbazi A (2011) Risk assessment for the daily intake of polycyclic aromatic hydrocarbons from the ingestion of cockle (Anadara granosa) and exposure to contaminated water and sediments along the west coast of Peninsular Malaysia. J Environ Sci 23: 336–345.
55. Nasher E, Heng LY, Zakaria Z, Surif S (2013) Assessing the Ecological Risk of Polycyclic Aromatic Hydrocarbons in Sediments at Langkawi Island, Malaysia. The Scientific World Journal 2013.
56. Chandru K, Zakaria MP, Anita S, Shahbazi A, Sakari M, et al. (2008) Characterization of alkanes, hopanes, and polycyclic aromatic hydrocarbons (PAHs) in tar-balls collected from the East Coast of Peninsular Malaysia. Mar Pollut Bull 56: 950–962.
57. Neff JM, Stout SA, Gunster DG (2005) Ecological risk assessment of PAHs in sediments. Identifying sources and toxicity.Integr. Environ Assess Manage 1: 22–33.
58. Dsikowitzky L, Nordhaus I, Jennerjahn TC, Khrycheva P, Sivatharshan Y, et al. (2011) Anthropogenic organic contaminants in water, sediments and benthic organisms of the mangrove-fringed Segara Anakan Lagoon, Java, Indonesia. Mar Pollut Bullet 62: 851–862.
59. Isobe T, Takada H, Kanai M, Tsutsumi S, Isobe KO, et al. (2007) Distribution of polycyclic aromatic hydrocarbons (PAHs) and phenolic endocrine disrupting chemicals in South and Southeast Asian mussels. Environ Monit and Assess 135: 423–440.

3

Variation in Patterns of Metal Accumulation in Thallus Parts of *Lessonia trabeculata* (Laminariales; Phaeophyceae): Implications for Biomonitoring

Claudio A. Sáez[1,2,3*], M. Gabriela Lobos[2], Erasmo C. Macaya[4], Doris Oliva[3], Waldo Quiroz[5], Murray T. Brown[1]

1 Plymouth University, School of Marine Science & Engineering, Faculty of Science and Technology, Drake Circus, Plymouth, United Kingdom, 2 Universidad de Valparaíso, Departamento de Química y Bioquímica, Facultad de Ciencias. Valparaíso, Chile, 3 Universidad de Valparaíso, Departamento de Biología y Ciencias Ambientales, Facultad de Ciencias. Valparaíso, Chile, 4 Universidad de Concepción, Departamento de Oceanografía, Casilla 160-C, Facultad de Ciencias Naturales y Oceanográficas, Concepción, Chile, 5 Pontificia Universidad Católica de Valparaíso, Laboratorio de Química Analítica y Ambiental, Instituto de Química, Curauma, Valparaíso, Chile

Abstract

Seaweeds are well known to concentrate metals from seawater and have been employed as monitors of metal pollution in coastal waters and estuaries. However, research showing that various intrinsic and extrinsic factors can influence metal accumulation, raises doubts about the basis for using seaweeds in biomonitoring programmes. The thallus of brown seaweeds of the order Laminariales (kelps) is morphologically complex but there is limited information about the variation in metal accumulation between the different parts, which might result in erroneous conclusions being drawn if not accounted for in the biomonitoring protocol. To assess patterns of individual metals in the differentiated parts of the thallus (blade, stipe, holdfast), concentrations of a wide range of essential and non-essential metals (Fe, Cr, Cu, Zn, Mn, Pb, Cd, Ni and Al) were measured in the kelp *Lessonia trabeculata*. Seaweeds were collected from three sampling stations located at 5, 30 and 60 m from an illegal sewage outfall close to Ventanas, Chile and from a pristine location at Faro Curaumilla. For the majority of metals the highest concentrations in bottom sediment and seaweed samples were found at the site closest to the outfall, with concentrations decreasing with distance from the outfall and at control stations; the exception was Cd, concentrations of which were higher at control stations. The patterns of metal concentrations in different thallus parts were metal specific and independent of sampling station. These results and the available literature suggest that biomonitoring of metals using seaweeds must take account of differences in the accumulation of metals in thallus parts of complex seaweeds.

Editor: Vishal Shah, Dowling College, United States of America

Funding: Source of funding were FONDECYT (Project 11080235) and CONICYT Becas Chile Scholarship (72110557)www.conicyt.cl. The funders had no role in study design, data collection and analysis, decision to publish, or preparation of the manuscript.

Competing Interests: The authors have declared that no competing interests exist.

* E-mail: claudio.saez.a@gmail.com

Introduction

Metals entering estuaries and coastal waters from land-based mining, industrial, agricultural and domestic activities can be readily accumulated by the resident biota, which may display symptoms of toxicity beyond certain threshold concentrations [1,2]. Quantification of metal concentrations in sediments and water provides limited information about bioavailability and therefore analyses of organisms are routinely used to assess the quantity and availability of metals in the environment. The rationale for using organisms to assess levels of environmental pollution and the criteria by which they are selected as biomonitors are well documented [3,4]. Their wide-spread distribution, local abundance, longevity, benthic nature and capacity to accumulate metals to concentrations several thousand times higher than those dissolved in seawater make seaweeds choice candidates. However, because extrinsic and intrinsic factors (e.g. environmental variables affecting growth rates, allometric parameters, interactions between metals) can influence the bioaccumulation of metals, the interpretation of the toxicological significance of thallus burdens may not necessarily be straightforward [5–8].

Brown algae belonging to the orders Fucales and Laminariales have been widely employed in studies on metal pollution in estuarine and coastal ecosystems from temperate latitudes [9–12]. Fucoids and kelps are dominant members of intertidal and sublittoral communities and have vital roles as primary producers and ecosystem bio-engineers [13–16]. They also have the greatest morphological complexity of all seaweeds and there is evidence that the accumulation capacity can differ between thallus parts [10,12,17–19]. This has implications for using seaweeds as biomonitors, as the choice of thallus part sampled could, inadvertently, influence how the data is interpreted [12].

Lessonia trabeculata is the most abundant kelp species along the south-east coast of South America, between 18° and 42° S, and constitutes one of the most important components of shallow subtidal ecosystems [20]. Structurally, the thallus can be separated into three main parts: annually renewable blades that are responsible for most nutrient absorption and photosynthetic activity, perennial stipes, providing structural support, and holdfast that anchor the thallus to the substratum [20].

The aim of this study was to assess the patterns of metal concentrations in different metals in the three main components, blades, stipes and holdfasts, of thalli sampled along a pollution gradient emanating from a sewage outfall located in Ventanas, Central Chile, and from a pristine location with similar oceanographic characteristics that acted as a control. The concentrations of the essential metals Fe, Cr, Cu, Zn and Mn and non-essential metals Pb, Cd, Ni and Al were measured. The implications of our findings for using morphologically complex seaweeds in biomonitoring programmes are discussed.

Materials and Methods

Study stations

At Ventanas, Central Chile (32°44′32,4′′S; 71°29′56,3′′W), an illegal sewage outfall was identified that constantly discharged wastes within the intertidal zone. The plume from the pipe (30 cm diameter) flowed in a north easterly direction as a consequence of the prevailing wind and water currents from the south west [21]. Three sampling stations, 1V, 2V and 3V, were established along the plume at distances of 5, 30 and 60 m from the outfall, respectively. For comparison, control stations were located at Faro Curaumilla (33°05′38,03′′S; 71°44′08,25′′W), 45 km south of Ventanas (fig. 1), as this area has similar oceanographic and topographic features [22]. Three sampling stations, 1F, 2F and 3F, were established in a northly direction at similar depths to those of the outfall stations and at 5, 30 and 60 m from the shore (low tide mark or chart datum point), respectively. Turbidity of the water at each station was estimated using secchi disks. Three measurements, separated by an interval of 60 min, were taken at each of the six stations between 1100 and 1300 hours. From a boat, the compensation depths (1% of surface incident light) were estimated by the recorded depth of a 25 cm diameter secchi disk multiplied by the constant 2.7 [23].

No permits were required for the described field studies, as both locations are not privately-owned or protected by the Republic of Chile. Locations were public and land-reachable. This study did not involve endangered or protected species.

Sampling and laboratory pre-treatment

Sampling from Ventanas and Faro Curaumilla was carried out in austral summer, on the 8th and 10th January 2009, respectively. Three independent sediment samples were collected at each sampling station. Sediments were collected with a small plastic shovel, stored in plastic containers and transported to the laboratory where they were dried in an oven (Memmert, model UL30790986) at 60°C to constant weight. Samples were homogenized through a 63 μm nylon sieve. Concurrently with sampling in the outfall zone, 3 samples of effluent were taken at 2 hour intervals for chemical analyses.

At each of the six sampling stations three individual thalli of *Lessonia trabeculata*, holdfasts ranging between 25 to 30 cm diameter with healthy appearance and free from excessive grazing, were collected from the centre of the kelp forest by SCUBA divers. Each thallus was separated into three parts: middle thallus blades, main stipes and central portion of holdfast (see fig. 2). Material was transported to the laboratory in clean plastic bags; seaweeds were rinsed with milli-Q water and a plastic brush was used to remove remainders of sediment and epibiota, dried in an oven at 60°C until constant weight and then pulverized with a titanium blender to powder. Gledhill et al. [24] found that the most effective method for cleaning seaweed material was a solution of 10% ethanol in seawater as evidences from the significant decrease in Fe concentration (compared with washing in milli-Q water). However, the presence of some epibiota on thr surface of *L. trabeculata* cannot be ruled out, as these cleaning methods have been assessed mainly for fucoids. Metal concentrations in three sub-samples of each of the different thallus parts were determined, giving a total of 1458 measurements (9 metals ×3 sub-samples ×3 parts ×3 individual thalli ×3 stations ×2 zones).

Analytical methods

A protocol of the Environmental Protection Agency was adapted for metal analysis of sediments (EPA, Method 3050B); 0.5 g DW were weighed in an analytical balance and placed in Erlenmeyer flasks. Subsequently, 3 ml of fuming hydrochloric acid (HCl, Merck, 37% G.R. for analyisis), 2 ml hydrofluoric acid (HF, Merck, 48%) and 9 mL HNO_3 (Merck, 65% G. R. for analysis) were added for 24 h for pre-digestion. When complete, 6 ml HCl, 2 ml HNO_3 and 8 ml of hydrogen peroxide (H_2O_2, Merck, 30% G.R. for analysis) were added and digested at 150°C for ~5 h until transparent. The volume was made up to 25 ml with milli-Q water in a volumetric flask. For effluent samples, wastewaters were mixed and 200 ml were concentrated by evaporation to c.5 ml at 95°C. Then, 10 ml of HNO_3 were added and samples were digested for 3 hours at 150°C until the digested turned transparent. The final volume was made up to 25 ml with milli-Q water.

The protocol for algal digestion was modified from Burger et al. [12]. Briefly, 1 g dry biomass was placed in an Erlenmeyer flask and pre-digested overnight at room temperature with 20 ml concentrated HNO_3 and 4 ml of supra H_2SO_4 (Aldrich, supra 96%). Then, 4 ml of H_2O_2 were added and digested for 4 to 5 h at 150°C. The solution was cooled to room temperature, 2 ml $HClO_4$ (Riedel- de Haen, 70% G. R. for analysis) then added and the solution re-heated to 150°C for 20–30 min until full digestion. The solution was then filtered (Whatman N°5) and made-up to 25 ml in a volumetric flask with milli-Q water.

The concentrations of Fe, Cr, Cu, Zn, Mn, Pb, Cd, Ni and Al in sediment, wastewaters, and seaweed samples were measured using inductively coupled plasma optical emission spectrometry (ICP-OES, Perkin-Elmer ICP Optima 2000 DV). To validate the analytical procedures, digestion and protocols for metals, measurements were also applied to marine sediment (BCSS-1) and algal (sea lettuce, *Ulva lactuca*, BCR-279) reference materials, the results of which are presented in Table 1.

Statistical analyses

To assess the requirements of normality and homogeneity of variances, Kolmogorov-Smirnov and Bartlett Test analyses were performed, respectively. One-way ANOVA and post-hoc Tukey test at 95% confidence was estimated for metal concentrations in sediments among sampling stations to identify significant differences. A multivariate analysis of variance (MANOVA) and post-hoc Tukey tests at 95% confidence were performed in order to estimate significant differences between the main factors (zone, stations and tissues) with 9 dependent variables (metals). The error intervals in figures and tables are calculated at 95% confidence ($n = 3$).

Multivariate principal component analysis (PCA), of metal concentrations in blades, stipes and holdfasts from the two locations was also undertaken. This analysis provides a small number of linear combinations of the 9 metal concentrations that describe most of the variability in the data. In addition to the quantitative analysis by the MANOVA, the PCA provides a visual

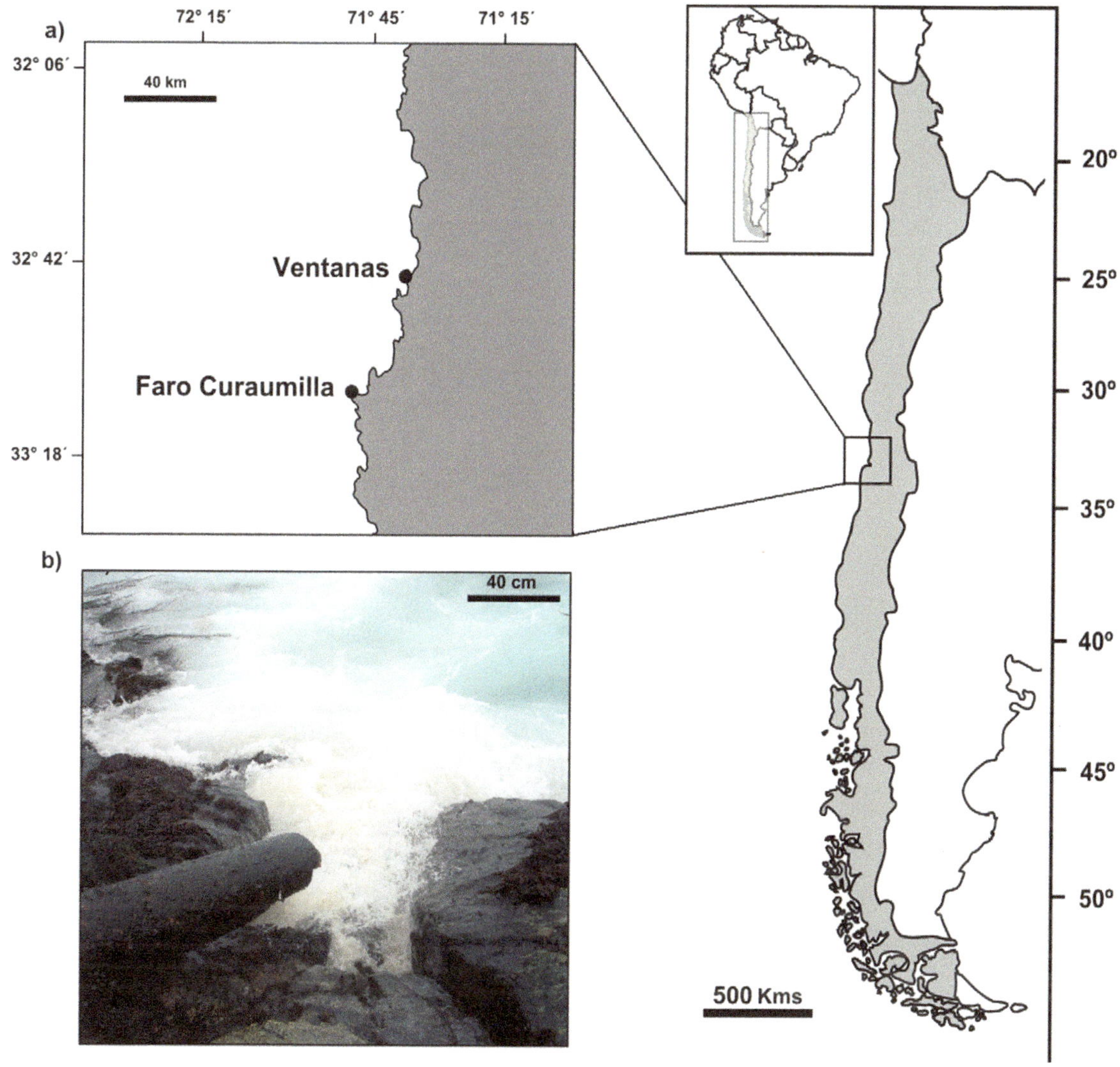

Figure 1. Map of study zones in Central Chile. a) Enlargement of study zones. b) Photo of the outfall.

qualitative representation of the data for ease of discerning relationships between metals and sampling stations.

Results

Physical measures

Table 2 provides information on the bathymetry and compensation depths of the sampling stations at the two locations (Ventanas and Faro Curaumilla). Despite the same bathymetry at corresponding stations in the two zones, the compensation depths differed significantly ($p<0.001$). The mean compensation depth at the outfall stations was c. 20 m compared with c. 45 m at the control stations.

Metals in sediments and wastewaters

The metal concentrations in sediments are presented in Table 3. The highest concentrations of Al, Fe, Cr, Cu, Mn, Ni, Pb and Zn were found at the station closest to the outfall (1V), showing significant differences with all the rest stations assessed ($p<0.001$). The concentrations of all these metals decreased significantly with distance from the outflow. For Al, Fe, Mn, Pb and Zn concentrations at 3V were significantly higher than control stations ($p<0.001$) whereas there were no significant differences ($p>0.05$) between 3V and the controls for Cr, Cu and Ni. Due to the detection limit of Cd, this metal was only measurable at stations 2F and 3F.

Metal concentrations measured in the wastewaters and comparisons with the legal limits of permitted concentrations of metals in wastewaters entering the sub-tidal [25], are presented in Table 3. With the exception of Cd (which was below the detection limit of the methodology applied), the concentrations of all other metals measured were higher than the legal permissible limits.

Metal concentrations in *Lessonia trabeculata*

The results from MANOVA performed using the different metals, sampling stations and algal parts are shown in Table 4. Significant differences ($p<0.00001$) were found for all main effects and 2-way and 3-way interactions. The results from a post-hoc Tukey test allowed for more detailed analysis of these results.

The metal concentrations in *L. trabeculata* are presented in Figure 3, and overall tendencies per algal part assessed in the PCA in Figure 4. For all metals analysed with the exception of Cd, the highest concentrations were found in seaweeds sampled from

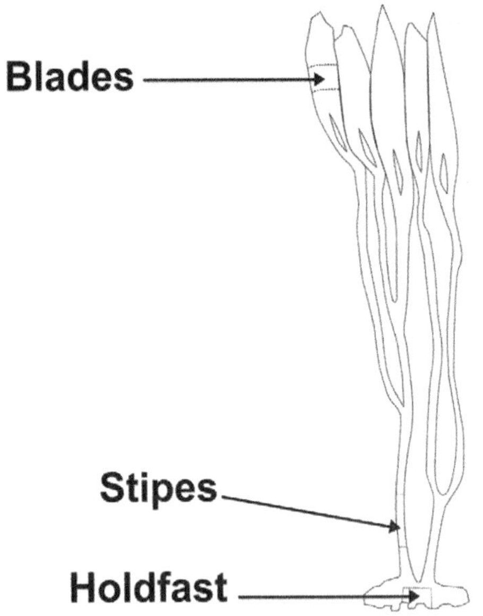

Figure 2. Diagram of algal thallus parts of *Lessonia trabeculata* sampled for metal analyses.

station 1V with a significant decrease with distance from the outflow pipe. Typically, concentrations at station 3V were significantly higher than those at the control sites. By contrast, Cd concentrations were higher in samples collected in control stations; moreover, in addition to Cd, Zn was out of tendency for blades and stipes while Ni for holdfasts. Some departures from this general trend become apparent when different parts of the seaweed are analysed separately. Concentrations of Fe, Cr, Cu, Pb and Al in blades of seaweeds from station 3V were significantly higher ($p<0.05$) than those sampled from control stations. For Mn there was a trend of decreasing concentrations with distance from the shore at both locations and Cd concentrations were highest at station 3F. There was no obvious pattern for Zn and Ni concentrations were very low at all sampling stations.

For stipes, concentrations of Fe, Cr, Ni, and Al were significantly higher at station 3V than control sites whereas Cu concentrations did not differ significantly between stations. Except for stations 1V and 2V, there were no measurable concentrations of Pb in stipes. The patterns of Mn and Cd concentration in stipes were the same as those of blades and, as with blades, there was no discernible pattern for Zn. The concentrations of Fe, Cr, Mn, Cu, Zn and Al in holdfasts declined with increasing distance from the outfall (1V) and, with the exception of Al, concentrations at station 3V were significantly higher than those at control stations. Concentrations of Pb were detectable in samples from 1V and 2V only, with the highest concentrations at the former station. The trend in Cd was the same as that reported for blades and stipes (Fig. 4).

Table 5 provides a summary of the differences in the concentrations of each metal between the three parts of the seaweed thallus. The highest concentrations of Cu, Pb and Al were present in blades, whereas for Ni it was stipes and for Cd the holdfasts. The lowest concentrations of Cr and Ni were found in blades and for Cd and Al in stipes. For Fe, Mn and Zn there was either no difference or no discernible pattern. With the exception of Zn (blades and stipes) and Ni (holdfasts), there was a significant positive correlation between metal concentrations in the sediment and seaweeds samples (Table 6). Due to the lack of data rather in

Table 1. Metal measures in certified material BCSS-1 (marine sediment) and BCR-279 (sea lettuce, *Ulva lactuca*). Confidence intervals at 95% (n = 3).

	Al_2O_3	Fe_2O_3	Cu	Cr	Cd	Mn	Ni	Pb	Zn
found value (mg Kg^{-1})	1.0±1.0[a]	5.00±0.40[a]	17.0±0.8	120±8	0.25±0.04	214±15	56.0±3.0	24.0±2.0	120±8
certified value (mg Kg^{-1})	1.8±0.4[a]	4.70±0.14[a]	18.5±2.7	123±14	0.32±0.06	229±15	55.3±3.6	22.7±3.4	119±12
	Al	**Fe**	**Cu**	**Cr**	**Cd**	**Mn**	**Ni**	**Pb**	**Zn**
found value (mg Kg^{-1})	2.0±0.2 (×103)	2.5±0.2 (×103)	12.8±0.4	11.0±0.7	0.28±0.03	1.98±0.12 (×103)	15.6±0.5	13.1±0.5	52±2
certified value (mg Kg^{-1})	.90±0.05[b] (×103)	2.4±0.1[b] (×103)	13.1±0.4	10.7±0.9[b]	0.27±0.02	2.09±0.08[b] (×103)	15.9±0.4[b]	13.5±0.4	51.3±1.2

[a]concentration in percentage w/w.
[b]Indicative Values.

Table 2. Compensation depths and bathymetry in sampling stations from the outfall and control zones.

Sampling stations	Compensation depth (m)	Bathymetry (m)
1V	19.4±0.2	14
2V	20.2±0.2	17
3V	21.2±0.2	20
1F	43.2±2.7	14
2F	44.1±4.1	17
3F	45.9±2.7	20

Sampling stations located at 5,30 and 60 m from the sewage pipe at the outfall zone, named 1V, 2V, and 3V, respectively; Same sampling design at control zone, sediment samples taken at 5, 30, and 60 m from the shore (reference point, no outfall), named 1F, 2F, and 3F, respectively. Compensation depths correspond to bathymetry multiplied by the constant 2.7 [23]. Confidence intervals at 95% (n = 3).

sediments or algae material, for Cd in all the tallus parts assessed and for Pb in stipes and holdfasts, correlation analyses were not possible.

Discussion

Concentrations of Fe, Al, Cu, Cr, Mn, Ni, Pb and Zn in most thallus parts of *L. trabeculata* collected from the metal impacted stations of Ventanas were significantly higher than in material collected from the pristine stations of Faro Curaumilla, with concentrations of metals associated with thalli decreasing with distance from the sewage outflow pipe. The exception to this general pattern was Cd, the concentrations of which were higher in algal parts from sampling stations of the control zone. There are several possible interpretations for these findings. For example, the presence of high concentrations of metals in the outfall zone might have impeded the accumulation of Cd. Simultaneous exposure of several metals can modify sorption and accumulation of certain metals, including Cd. For instance, Zhou et al. [26] found that the sorption capacity of *Laminaria japonica* for Cd decreased from 97 mg L^{-1} when exposed to 400 mg L^{-1} to 39 mg L^{-1} in the presence of Cu cations. However, considering that highest concentrations of Cd found in sediments were from stations of the control zone and were undetectable at the outfall it is most likely that the upwelling characteristics of Faro Curaumilla, together with the lack of competition from other metals for binding sites, are responsible for the higher than expected levels of Cd in *L. trabeculata*. Faro Curaumilla is located within a recognized upwelling system [22] and it is known that Cd has a close affinity with phosphates and nitrates in the marine environment [27–31]. Complexed Cd is removed from the upper layer of the water column by planktonic organisms and is incorporated into their biomass. After sedimentation of the biogenic material, organic oxidation returns nutrients and Cd to the water column which can then be transported to coastal zones by upwelling processes [32] and becomes bioavailable to seaweeds.

The most interesting results of this study is the lack of consistency in the patterns of accumulation for the different metals between thallus parts of *L. trabeculata*. Brown algae are known to accumulate metals to several orders of magnitude higher than concentrations dissolved in seawater, and they integrate short-term temporal fluctuation in concentrations. For these reasons they have been often used as monitors of metal pollution

Table 3. Metal concentrations in sediments and measured in the outfall wastewaters.

Sediments in stations	Al	Fe	Cd	Cr	Cu	Mn	Ni	Pb	Zn
1V	131±15[a]	464±18	<DL	23±3	68±3	262±13	4,0±0,4	6,8±0,3	51±6
2V	84±3[a]	251±38	<DL	9,6±0,5	58±4	201±10	3,1±0,1	3,5±0,4	28±4
3V	65±5[a]	111±7	<DL	5,6±0,3	18±2	109±6	2,0±0,1	1,9±0,1	20±1
1F	45±4[a]	51±4	<DL	4,8±0,2	16±1	67±8	1,8±0,3	1,0±0,1	14±2
2F	48±5[a]	54±3	0,45±0,01	5,0±0,2	14±1	70±10	1,9±0,1	1,1±0,1	12±2
3F	43±4[a]	50±1	0,45±0,04	4,7±0,3	14±1	70±2	1,9±0,2	1,1±0,1	15±2
Wastewaters	**Al**	**Fe**	**Cd**	**Cr**	**Cu**	**Mn**	**Ni**	**Pb**	**Zn**
Sewage	2269±221	242±11	<DL	4,9±0,1	240±7	66±4	5,1±0,2	3,6±0,3	132±15
Permitted concentrationb	1.0	10.0	0.02	2.5	1.0	2.0	2.0	0.2	5.0

[a]Concentrations in mg g^{-1}.
[b]According to SEGEGOB (2001), local law in regard to the permitted limits of metal concentrations in wastewaters discarded to the subtidal.
<DL: Below the detection limit.
Last row describes permitted metal concentrations in sewage by local law. Sediment samples taken at 5,30 and 60 m from the sewage pipe at the outfall zone, named 1V, 2V, and 3V, respectively; Same sampling design at control zone, sediment samples taken at 5, 30, and 60 m from the shore (reference point, no outfall), named 1F, 2F, and 3F, respectively. Metal concentrations in sediments and water in µg g^{-1} and µg L^{-1}, respectively. Confidence intervals at 95% (n = 3).

Table 4. Multivariate analysis of variance (MANOVA) at 95% confidence showing interactions between main factors (zones, sampling stations, and algal parts) and 9 dependent variables assessed (metals).

Effect	Value	F	Hypothesis df	Error df	Signifficance
Intercept	0.000	32150.869	9.000	28.000	0.000
Zone	1.000	a	0.000	32.000	
Sation	0.000	94.725	36.000	106.666	0.000
Algal part	0.000	1471.306	18.000	56.000	0.000
Zone*Station	1.000	a	0.000	32.000	
Zone*Algal part	1.000	a	0.000	32.000	
Sation*Algal part	0.000	63.557	72.000	177.897	0.000
Zone*Station*Algal part	1.000	a	0.000	32.000	

a. Exact statistic due to factor Zone has 2 levels, the outfall and control zones.
Wilks' lambda distribution is shown for multivariate hypothesis testing.

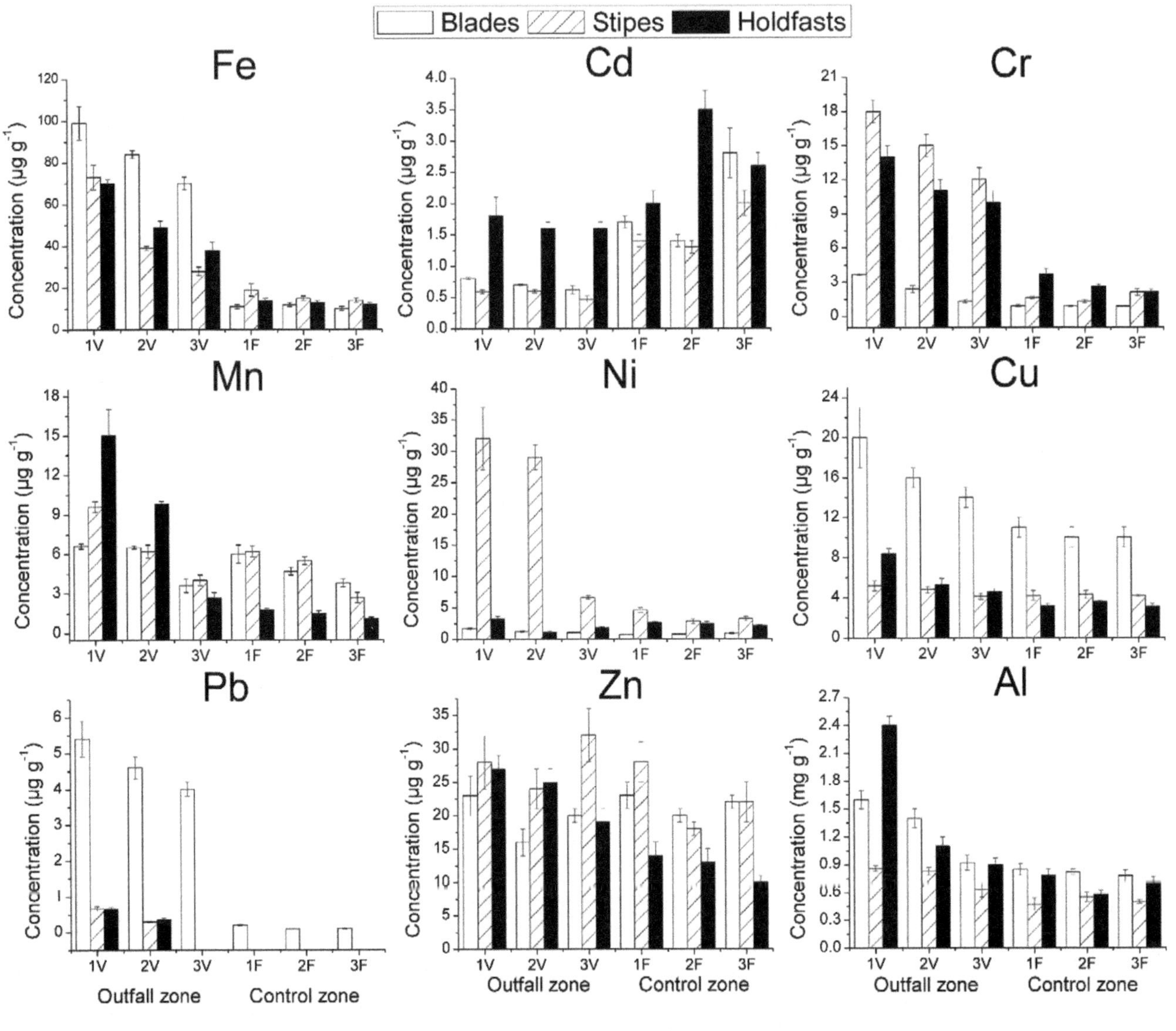

Figure 3. Metal concentrations in different thallus parts of *L. trabeculata* collected from different sampling stations. Sampling stations located at 5,30 and 60 m from the sewage pipe at the outfall zone, named 1V, 2V, and 3V, respectively; Same sampling design at control zone, sediment samples taken at 5, 30, and 60 m from the shore (reference point, no outfall), named 1F, 2F, and 3F, respectively. Sampling carried out during austral summer (January) of 2009. Intervals at 95% confidence (n = 3).

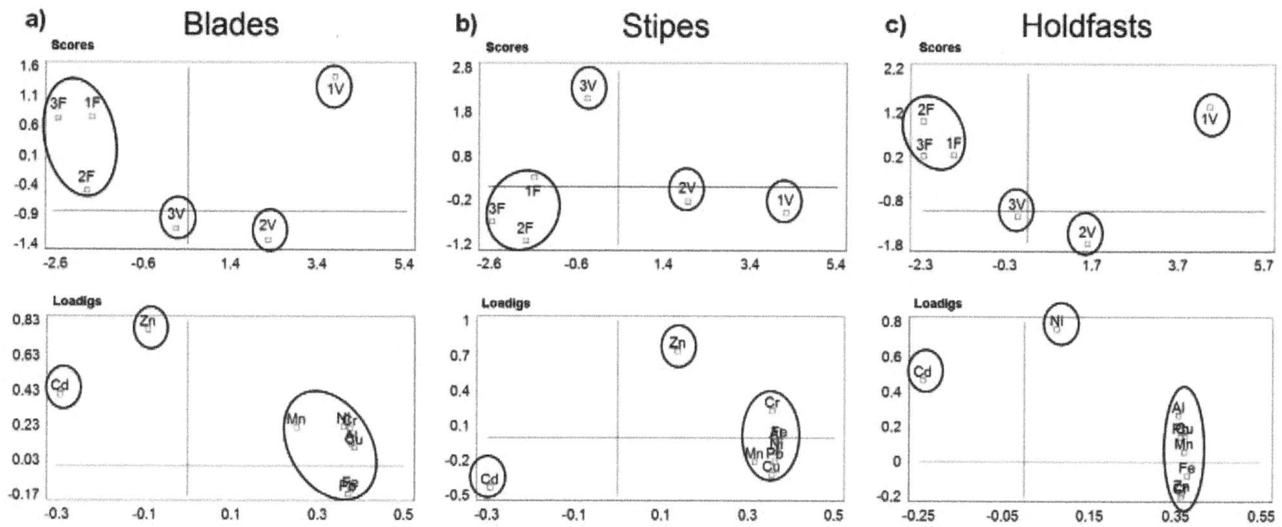

Figure 4. Principal component analysis (PCA) of metals measured in blades, stipes and holdfasts of *L. trabeculata* from the different sampling stations. Sampling stations located at 5, 30 and 60 m from the sewage pipe at the outfall zone, named 1V, 2V, and 3V, respectively; Same sampling design at control zone, sediment samples taken at 5, 30, and 60 m from the shore (reference point, no outfall), named 1F, 2F, and 3F, respectively.

[33]. The accumulation and uptake of metals by brown algae are affected by both environmental factors that influence metabolism and growth (e.g. light, temperature) and alter metal speciation, and also by intrinsic factors (e.g. age, allometry, tissue type) [33]. For example, it has been observed that both essential and non-essential metals can have an influence on accumulation and uptake, for example by increasing Cu accumulation during growth in *Ecklonia radiata* [18] or by not having influence on growing tissue, in case of the non- essential Cd in *Laminaria saccharina* [34]. Several studies have identified the holdfast as the structure that accumulates the highest concentration of metals [10,12,35–41], while others found no or different patterns within thalli [10,42,43]. For example, Stengel et al. [10] reported higher concentrations of Cu, Fe and Mn in holdfasts of *Laminaria digitata*, compared to other parts of the thallus at both polluted and pristine locations. They attributed to time dependent accumulation in perennial holdfasts and to their closer proximity to leached ions that might be released from bottom sediments. However, in *Fucus vesiculosus* no such differences in metal concentrations along the thallus were observed. In this study on *L. trabeculata*, concentrations of Cu and Pb were highest in blades, Ni in stipes, Cd in holdfasts and Cr in holdfasts and stipes. We could not observe signs of saturation, even in the point where concentrations of the majority of metals in sediments were highest (sampling site closest to the outfall). Thus, the reported intra-individual variation in metal burdens between tissue structures appears inconsistent and seemingly dependent on the metal and species of seaweed. The reasons for the different patterns of accumulation are not understood but are likely to be related to a number of physico-chemical and biological factors.

It is generally agreed that the first line of defence against metal toxicity is the cell wall. [33,44]. The main components of cell walls in brown algae (Phaeophyta) are cellulose, alginate, fucoidan and mucilage [45]. The alginate content of different brown algae can vary between 10 and 40% of the dry weight [46], and is most important macromolecule involved in metal chelation. The carboxylic groups between segments of polymers forming alginate

Table 5. Tendencies from MANOVA and *posteriori* Tukey test.

Metal	Alga parts distribution
Al	blades > holdfasts > stipes
Cu	blades > stipes ~ holdfasts
Pb	blades > stipes ~ holdfasts
Ni	stipes > holdfasts > blades
Cd	holdfasts > blades > stipes
Cr	holdfasts ~ stipes > blades
Fe	blades ~ stipes ~ holdfasts
Mn	blades ~ stipes ~ holdfasts
Zn	blades ~ stipes ~ holdfasts

In algal parts distribution signs > and < represent higher and lower than, respectively, when significant differences were found ($p<0.05$), while ~ means no significant differences ($p>0.05$) or no pattern recognized between metal concentrations among algal parts.

Table 6. Correlations (R values) of metal concentrations in sediments and thallus parts of *L. trabeculata*.

Metals	Blades	Stipes	Holdfasts
Al	0.9587	0.9146	0.9639
Fe	0.8681	0.9907	0.9553
Cd	ID	ID	ID
Cr	0.9618	0.7721	0.7846
Cu	0.9272	0.9662	0.8975
Mn	0.7843	0.7692	0.9919
Ni	0.9616	0.9646	0.2164
Pb	0.8302	ID	ID
Zn	0.1005	0.3564	0.8682

ID: Insufficient data to perform representative correlations.

are key targets for metal ions such as Pb^{2+}, Cu^{2+}, Cd^{2+}, and Zn^{2+} [47,48]. Alginic acid is composed of linear polysaccharides containing mannuronic (M) and guluronic (G) acid residues, organized as (–M–)n, (–G–)n, and (–MG–)n [49]. Changes in the molecular conformation of alginic acid are responsible for differences in their affinity for metals. For example, Haug [48] found that the lower M/G ratio the binding capacity of alginates for the divalent metal ions Cu, Zn, Pb, and Cd, increased. As well as different algal species varying in quantity and chemical composition of alginate, different structural components of the brown seaweed thallus can have different forms and quantities of alginate, which are influenced by the environmental conditions in which the seaweeds are growing [50]. For example, Chandia et al. [51] observed no systematic differences in M/G ratios between holdfasts, stipes and blades in *L. trabeculata*, from specific sampling sites; however, when comparing individuals from different locations, significant differences in M/G ratios were observed between thallus parts. In addition, Venegas et al. [52] found that M/G ratios were lower in blades of *L. trabeculata* collected from a beach exposed to strong currents, in comparison to blades collected from a protected bay. The information suggests that M/G ratios can vary between *L. trabeculata* populations exposed to different environmental factors and the development of divergent ecotypes are likely to occur under different environmental conditions which could influence responses to metal exposure. Further research on the influence of environmental change on alginate content and composition of thallus parts would contribute to our understanding of why metal bioaccumulation differs between thallus parts of *L. trabeculata*.

Seaweeds exude dissolved organic compounds from their cells that chelate metals in the water column, thereby altering their bioavailability in the surrounding water and reducing accumulation and uptake into cells [53]; Of these phenolic compounds are released in large quantities by brown algae and are important chelators of metals in the environment [54–59]. Phenolic compounds are associated with physodes in cells of brown seaweeds. Their content varies between brown algae species and also in different thallus parts of an individual [60,61]. They can be variously induced by, for example, herbivory, nitrogen availability as well as exposure to metals, all of which could influence the metal accumulation capacity of specific algal tissues [57–59].

Metals transported across the cell membrane and entering cells are potentially toxic to cell functions. In order to prevent this occurring metals must be effectively sequestered by, for example, polyphenols, glutathione, phytochelatins, metallothioneins and then either stored in organelles such as vacuoles and physodes or subsequently released to the environment [62]. The method we used to measure concentrations of metals associated with seaweed tissues does not allow for discrimination between externally bound metals to cell surfaces and the internalised fractions. However, previous studies have shown that the proportion of extra-cellular to intra-cellular bound metal differs for different metals, between seaweed species, and can vary seasonally [63].

The behaviour of Zn was another interesting point of this investigation, while holdfasts showed clear influence of the outfall pipe and therefore environmental representativeness, structures such as blades and stipes did not evidence differences that could be attributed to surroundings levels of Zn. Concentration of metals Fe, Cd, Cr, Cu and Al showed environmental patterns in all the algal parts assessed. Tendencies were observed in the rest of the metal measured, however, they were clear only for certain thallus parts; for Mn and Pb blades were more representative, while for Ni and Zn, stipes and holdfasts, respectively, were the most representative thallus parts. Even though it might be argued that total metal concentrations in sediments do not represent the bioavailable fraction for uptake and accumulation, the strong relationship between total metal concentrations in sediments and in the different algal parts of *L. trabeculata* suggests that also the bioavailable fraction of these metals followed a decay from the point of sewage discharge, and lower concentrations in stations from the control zone (except Cd). In our case, the information gathered on metal burdens in *L. trabeculata* suggests that the outfall zone in Ventanas is polluted by action of the sewage pipe. Taking into account that the wastewaters released from the outfall far exceeded the permitted concentrations of metals by the local legislation [25] and, moreover, the records available on the negative effects of this outfall on surrounding biological communities [16], indicate that *L. trabeculata* metal concentrations can reliably constitute a pollution discriminatory tool for biomonitoring the coastal rocky shore.

Our investigation can be considered to be a more complex example of 'passive' biomonitoring (PBM), in which metal measurements have been taken from native individuals that had recruited and grown in the ecosystems assessed. Several authors [49,64–66] have suggested that 'active' biomonitoring (ABM) is a better option to study environmental pollution. This approach employs specific and responsive organisms that have been cultured in the laboratory under controlled conditions and then introduced into locations of environmental interest for a defined period of time. ABM provides several advantages including: using individuals of similar age and state of development and physiology, responses can be observed between introduced and native individuals, experimental locations can be selected despite the absence of the biomonitoring species. However, not all species are necessarily appropriate for this approach. In the case of *L. trabeculata*, and in the majority of complex seaweeds, as well as the variability in metal accumulation between different thallus parts, their culture and growth in the laboratory is demanding of time and logistics. While it would be possible to culture and deploy young kelp sporophytes, it may be easier to use morphologically less complex, perennial, species of seaweeds for ABM, albeit only for short-term evaluations.

Conclusions

Metal concentrations varied between blades, stipes and holdfasts of *L. trabeculata*. In most cases patterns of accumulation were independent of the sampling station assessed (pristine or polluted). The results indicate that the concentrations of metals associated with the thalli of *L. trabeculata* can discriminate between metal-polluted and pristine locations. However, the information must be treated carefully, as metal accumulation can vary significantly among thallus parts, and that this variation can be metal specific. Our results suggest that different thallus parts of *L. trabeculata* could be sampled for the assessment of particular metals, but no one part is suitable for all metals. Given the widespread use of seaweeds as biomonitors of metal pollution, it is important to consider the outcome of our investigation when other morphologically complex seaweeds are included in evaluating levels of environmental pollution. The known innumerable environmental and species specific intrinsic factors that can influence metal uptake and accumulation bring uncertainties that need to be addressed prior to the selection of candidate species for biomonitoring programmes in order to prevent false diagnosis.

Acknowledgments

The cooperation of René Sáez, Danilo Gutierrez, Claudio Galarce, Felipe Santander and Patricia Díaz during the different stages of this research is greatly appreciated. We are grateful of Dr. Miguel Franco of Plymouth University for his valuable statistical advice. We also thank the comments of two anonymous reviewers that helped improve this manuscript.

Author Contributions

Conceived and designed the experiments: CAS MGL DO. Performed the experiments: CAS MGL. Analyzed the data: CAS MGL LECM DO WQ MTB. Contributed reagents/materials/analysis tools: CAS MGL DO WQ. Wrote the paper: CAS MGL LECM DO WQ MTB.

References

1. Beltrame MO, De Marco SG, Marcovecchio JE (2010) Effects of zinc on molting and body weight of the estuarine crab *Neohelice granulata* (Brachyura: Varunidae). Science of the Total Environment 408: 531–536.
2. Ferrer L, Andrade S, Asteasuain R, Marcovecchio J (2006) Acute toxicities of four metals on the early life stages of the crab *Chasmagnathus granulata* from Bahia Blanca estuary, Argentina. Ecotoxicology and Environmental Safety 65: 209–217.
3. Markert BA, Breure AM, Zechmeister HG, editors (2003) Bioindicators & Biomonitors: Principles, Concepts, and Applications. Amsterdam: Elsevier.
4. Butterworth FM, Gunatilaka A, Gonsebatt ME, editors (2001) Biomonitors and biomarkers as indicators of environmental change 2: a handbook. New York: Plenum Publishers. 508 p.
5. Pereira P, de Pablo H, Pacheco M, Vale C (2010) The relevance of temporal and organ specific factors on metals accumulation and biochemical effects in feral fish (*Liza aurata*) under a moderate contamination scenario. Ecotoxicology and Environmental Safety 73: 805–816.
6. Hedouin L, Metian M, Teyssie JL, Fowler SW, Fichez R, et al. (2006) Allometric relationships in the bioconcentration of heavy metals by the edible tropical clam *Gafrarium tumidum*. Science of the Total Environment 366: 154–163.
7. Amiard JC, Perrein-Ettajani H, Gerard A, Baud JP, Amiard-Triquet C (2004) Influence of ploidy and metal-metal interactions on the accumulation of Ag, Cd, and Cu in oysters *Crassostrea gigas* thunberg. Archives of Environmental Contamination and Toxicology 48: 68–74.
8. Bunluesin S, Pokethitiyook P, Lanza GR, Tyson JF, Kruatrachue M, et al. (2007) Influences of cadmium and zinc interaction and humic acid on metal accumulation in *Ceratophyllum demersum*. Water Air and Soil Pollution 180: 225–235.
9. Phillips DJH (1977) The use of biological indicator organisms to monitor trace metal pollution in marine and estuarine environments – a review. Environmental Pollution 13: 281–317.
10. Stengel DB, McGrath H, Morrison LJ (2005) Tissue Cu, Fe and Mn concentrations in different-aged and different functional thallus regions of three brown algae from western Ireland. Estuarine Coastal and Shelf Science 65: 687–696.
11. Barreiro R, Picado L, Real C (2002) Biomonitoring heavy metal in estuaries: A field comparison of two brown algae species inhabiting upper estuarine reaches. Environmental Monitoring and Assessment 75: 121–134.
12. Burger J, Gochfeld M, Jeitner C, Gray M, Shukla T, et al. (2007) Kelp as a bioindicator: Does it matter which part of 5 m long plant is used for metal analysis? Environmental Monitoring and Assessment 128: 311–321.
13. Dayton PK (1985) Ecology of kelp communities. Annual Review of Ecology and Systematics 16: 215–245.
14. Holbrook SJ, Carr MH, Schmitt RJ, Coyer JA (1990) Effect of Giant-Kelp on Local Abundance of Reef Fishes – the Importance of Ontogenic Resource Requirements. Bulletin of Marine Science 47: 104–114.
15. Davenport AC, Anderson TW (2007) Positive indirect effects of reef fishes on kelp performance: The importance of mesograzers. Ecology 88: 1548–1561.
16. Sáez CA, Pérez-Matus A, Lobos MG, Oliva D, Vásquez JA, et al. (2012) Environmental assessment in a shallow subtidal rocky habitat: Approach coupling chemical and ecological tools. Chemistry and Ecology 28: 1–15.
17. Munda IM, Hudnik V (1988) The effects of Zn, Mn, and Co accumulation on growth and chemical composition of *Fucus vesiculosus* L. under different temperature and salinity conditions. Marine Ecology Progress Series 9: 213–225.
18. Higgins HW, Mackey DJ (1987) Role of *Ecklonia radiata* (C. Ag.) J. Agardh in determining trace metal availability in coastal waters. II. Trace metal speciation. Australian Journal of Marine and Freshwater Research 38: 317–328.
19. Munda IM (1986) Differences in heavy metal accumulation between vegetative parts of the thalli and receptacles in *Fucus spiralis* L. Botanica Marina 29: 341–349.
20. Villouta E, Santelices B (1986) *Lessonia trabeculata* Sp-nov (laminariales, phaeophyta), a new kelp from Chile. Phycologia 25: 81–86.
21. Strub P, Mesías J, Montecinos V, Rutllant J, Marchant S (1998) Coastal ocean circulation of western South America. In: Robinson A, Brink K, editors. The Sea, vol 14A: The Global Coastal Ocean: Interdisciplinary Regional Studies and Syntheses. New York: John Wiley and Sons. 273–314.
22. Silva N, Valdenegro A (2003) Evolución de un evento de surgencia frente a punta Curaumilla, Valparaíso Investigaciones marinas 31: 73–89.
23. Margalef R (1983) Limnologia. Barcelona: Omega.
24. Gledhill M, Brown MT, Nimmo M, Moate R, Hill SJ (1998) Comparison of techniques for the removal of particulate material from seaweed tissue. Marine Environmental Research 45: 295–307.
25. SEGEGOB (2001) Decreto Supremo N°90: Establece norma de emisión para la regulación de contaminantes asociados a las descargas de residuos líquidos a aguas marinas y continentales superfificales. In: MINSEGPRES DS/2000. 51 p.
26. Zhou JL, Huang PL, Lin RG (1998) Sorption and desorption of Cu and Cd by macroalgae and microalgae. Environmental Pollution 101: 67–75.
27. Abe K, Matsumaga K (1988) Mechanism controlling Cd and PO4 concentration in Funka Bay, Japan. Marine Chemistry 23: 145–152.
28. Bruland K, Orinas K, Cowen J (1994) Reactive trace metals in the stratified central north Pacific. Geochimica et Cosmochimica Acta 58: 3171–3182.
29. Elbaz-Poulichet F, Martin J (1987) Dissolved Cd behavior in some selected French and Chinese estuaries. Consequences on Cd supply to the ocean. Marine Chemistry 22: 125–136.
30. de Baar H, Saager P, Nolting R, Vandermeer J (1994) Cadmium versus phosphate in the world ocean. Marine Chemistry 46: 261–281.
31. Abe K (2001) Cd in the western equatorial Pacific Marine Chemistry 74: 197–211.
32. Bruland K (1980) Oceanographic distribution of cadmium, zinc, nickel and copper in the north Pacific. Earth and Planetary Sciences Letter 47: 176–198.
33. Brown MT, Depledge MH (1998) Determinants of trace metal concentrations in marine organisms. In: Bebianno MJ, Langston WJ, editors. Metabolism of Trace Metals in Aquatic Organisms. London: Chapman & Hall. 185–217.
34. Markham JW, Kremer BP, Sperling KR (1980) Effects of cadmium on *Laminaria saccharina* in culture. Marine Ecology Progress Series 3: 31–39.
35. Black WAP, Mitchell RL (1952) Trace elements in the common brown algae and sea water. Journal of the Marine Biological Association of the United Kingdom 30: 575–584.
36. Bryan GW (1969) The absorption of zinc and other metals by the brown seaweed *Laminaria digitata*. Journal of the Marine Biological Association of the United Kingdom 49: 225–243.
37. Bryan GW, Hummerstone LG (1973) Brown seaweed as an indicator of heavy metals in estuaries in South-west England. Journal of the Marine Biological Association of the United Kingdom 53: 705–720.
38. Fuge R, James KH (1973) Trace metal concentrations in brown seaweeds, Cardigan Bay, Wales. Marine Chemistry 1: 281–293.
39. Fuge R, James KH (1974) Trace metal concentrations in *Fucus* from the Bristol Channel. Marine Pollution Bulletin 5: 9–12.
40. Young ML (1975) The transfer of Zn65 and Fe59 along a *Fucus serratus* (L.) to *Littorina obtusata* (L.) food chain. Journal of the Marine Biological Association of the United Kingdom 55: 583–610.
41. Phillips DJH (1990) Use of macroalgae and invertebrates as monitors of metal levels in estuarine and coastal waters. In: Furness RW, Rainbow PS, editors. Heavy Metals in the Marine Environment. Boca Raton: CRC. 81–100.
42. Ray S, McLeese DW, Metcalfe CF, Burridge LE, Waiwood BA (1980) Distribution of cadmium marine biota in the vicinity of Belledune. Canadian Technical Report of Fisheries and Aquatic Sciences 96: 12–27.
43. Sanchiz G, Garcia-Carrascosa AM, Pastor A (1999) Bioaccumulation of Hg, Cd. Pb, and Zn in four marine phanerogams and the alga *Caulerpa prolilfera* (Foersskal) Lamouroux from the east coast of Spain. Botanica Marina 42: 157–164.
44. Salgado LT, Andrade LR, Amado GM (2005) Localization of specific monosaccharides in cells of the brown alga *Padina gymnospora* and relation to heavy metal accumulation. Protoplasma 225: 123–128.
45. Lee R (2008) Phycology. Cambridge: Cambridge University Press. 547 p.
46. Percival E, McDowell RH (1967) Chemistry and Enzymology of Marine Algal Polysaccharides. London: Academic Press.
47. Puranik PR, Modak JM, Paknikar KM (1999) A comparative study of the mass transfer kinetics of metal biosorption by microbial biomass. Hydrometallurgy 52: 189–198.
48. Haug A (1967) The affinity of some divalent metals to different types of alginates. Acta Chemica Scandinavica 15: 1794–1795.
49. Collen J, Pinto E, Pedersen M, Colepicolo P (2003) Induction of oxidative stress in the red macroalga *Gracilaria tenuistipitata* by pollutant metals. Archives of Environmental Contamination and Toxicology 45: 337–342.
50. Davis TA, Volesky B, Mucci A (2003) A review of the biochemistry of heavy metal biosorption by brown algae. Water Research 37: 4311–4330.
51. Chandia NP, Matsuhiro B, Vasquez AE (2001) Alginic acids in *Lessonia trabeculata*: characterization by formic acid hydrolysis and FT-IR spectroscopy. Carbohydrate Polymers 46: 81–87.
52. Venegas M, Matsuhiro B, Edding ME (1993) Alginate composition of *Lessonia trabeculata* (phaeophyta, laminariales) growing in exposed and sheltered habitats. Botanica Marina 36: 47–51.
53. Connan S, Stengel DB (2011) Impacts of ambient salinity and copper on brown algae: 2. Interactive effects on phenolic pool and assessment of metal binding capacity of phlorotannin. Aquatic Toxicology 104: 1–13.

54. Gledhill M, Nimmo M, Hill SJ, Brown MT (1999) The release of coppercomplexing ligands by the brown alga *Fucus vesiculosus* (Phaeophyceae) in response to increasing total copper levels. Journal of Phycology 35: 501–509.
55. Ragan MA, Ragan CM, Jensen A (1980) Natural chelators in sea water: detoxification of Zn^{2+} by brown algal polyphenols. Journal of Experimental Marine Biology and Ecology 44: 261–267.
56. Sueur S, Van den Berg CMG, Riley JP (1982) Measurement of the metal complexing ability of exudates of marine macroalgae. Limnology and Oceanography 27: 536–543.
57. Ragan MA, Jensen A (1979) Quantitative studies on brown algal phenols. III. Lightmediated exudation of polyphenols from *Ascophyllum nodosum* (L.) Le Jol. Journal of Experimental Marine Biology and Ecology 36: 91–101.
58. Shibata T, Hama Y, Miyasaki T, Ito M, Nakamura T (2006) Extra-cellular secretion of phenolic substances from living brown algae. Journal of Applied Phycology 18: 787–794.
59. Swanson AK, Druehl LD (2002) Induction, exudation and the UV protective role of kelp phlorotannins. Aquatic Botany 73: 241–253.
60. Pedersen A (1984) Studies on phenol content and heavy metal uptake in fucoids. Hydrobiologia 116–117: 498–504.
61. Haug A, Larsen B (1958) Phenolic compounds in brown algae. I. The presence of reducing compounds in *Ascophyllum nodosum*. Acta Chemica Scandinavica 12: 650–657.
62. Pawlik-Skowronska B, Pirszel J, Brown MT (2007) Concentrations of phytochelatins and glutathione found in natural assemblages of seaweeds depend on species and metal concentrations of the habitat. Aquatic Toxicology 83: 190–199.
63. Mehta SK, Gaur JP (2005) Use of algae for removing heavy metal ions from wastewater: Progress and prospects. Critical Reviews in Biotechnology 25: 113–152.
64. Brown MT, Newman JE, Han T (2012) Inter-population comparisons of copper resistance and accumulation in the red seaweed, *Gracilariopsis longissima*. Ecotoxicology 21: 591–600.
65. Piotrowska-Niczyporuk A, Bajguz A, Zambrzycka E, Godlewska-Zylkiewicz B (2012) Phytohormones as regulators of heavy metal biosorption and toxicity in green alga *Chlorella vulgaris* (Chlorophyceae). Plant Physiology and Biochemistry 52: 52–65.
66. Elbaz A, Wei YY, Meng QA, Zheng Q, Yang ZM (2010) Mercury-induced oxidative stress and impact on antioxidant enzymes in *Chlamydomonas reinhardtii*. Ecotoxicology 19: 1285–1293.

4

Water Quality Conditions Associated with Cattle Grazing and Recreation on National Forest Lands

Leslie M. Roche[1]*, Lea Kromschroeder[1], Edward R. Atwill[2], Randy A. Dahlgren[3], Kenneth W. Tate[1]

1 Department of Plant Sciences, University of California, Davis, California, United States of America, **2** School of Veterinary Medicine, University of California University of California, Davis, California, United States of America, **3** Department of Land, Air, and Water Resources, University of California, Davis, California, United States of America

Abstract

There is substantial concern that microbial and nutrient pollution by cattle on public lands degrades water quality, threatening human and ecological health. Given the importance of clean water on multiple-use landscapes, additional research is required to document and examine potential water quality issues across common resource use activities. During the 2011 grazing-recreation season, we conducted a cross sectional survey of water quality conditions associated with cattle grazing and/or recreation on 12 public lands grazing allotments in California. Our specific study objectives were to 1) quantify fecal indicator bacteria (FIB; fecal coliform and *E. coli*), total nitrogen, nitrate, ammonium, total phosphorus, and soluble-reactive phosphorus concentrations in surface waters; 2) compare results to a) water quality regulatory benchmarks, b) recommended maximum nutrient concentrations, and c) estimates of nutrient background concentrations; and 3) examine relationships between water quality, environmental conditions, cattle grazing, and recreation. Nutrient concentrations observed throughout the grazing-recreation season were at least one order of magnitude below levels of ecological concern, and were similar to U.S. Environmental Protection Agency (USEPA) estimates for background water quality conditions in the region. The relative percentage of FIB regulatory benchmark exceedances widely varied under individual regional and national water quality standards. Relative to USEPA's national *E. coli* FIB benchmarks–the most contemporary and relevant standards for this study–over 90% of the 743 samples collected were below recommended criteria values. FIB concentrations were significantly greater when stream flow was low or stagnant, water was turbid, and when cattle were actively observed at sampling. Recreation sites had the lowest mean FIB, total nitrogen, and soluble-reactive phosphorus concentrations, and there were no significant differences in FIB and nutrient concentrations between key grazing areas and non-concentrated use areas. Our results suggest cattle grazing, recreation, and provisioning of clean water can be compatible goals across these national forest lands.

Editor: A. Mark Ibekwe, U. S. Salinity Lab, United States of America

Funding: This research was funded by USDA Forest Service, Pacific Southwest Region. The funders did provide field data collection assistance. The funders had no role in data analysis, decision to publish, or preparation of the manuscript.

Competing Interests: The authors have declared that no competing interests exist.

* E-mail: lmroche@ucdavis.edu

Introduction

Livestock grazing allotments on public lands managed by the United States Forest Service (USFS) provide critical forage supporting ranching enterprises and local economies [1–3]. Surface waters on public lands are used for human recreation and consumption, and serve as critical aquatic habitat. Concerns have been raised that microbial and nutrient pollution by livestock grazing on public lands degrades water quality, threatening human and ecological health [4–7]. Some of the contaminants of concern include fecal indicator bacteria (FIB), fecal coliform (FC) and *Escherichia coli* (*E. coli*), as well as nitrogen (N) and phosphorus (P). FIB are regulated in an attempt to safeguard public health from waterborne pathogens such as *Cryptosporidium parvum* and *E. coli* O157:H7 and human enteroviruses including adenoviruses and coliphages [8]. Concerns about elevated N and P concentrations in surface water stem from the potential for eutrophication of aquatic systems [9].

The USFS must balance the many resource use activities occurring on national forests (e.g., livestock grazing, recreation). National forests in the western United States support 1.8 million livestock annually, provisioning 6.1 million animal unit months (AUM) of forage supply allocated through 5,220 grazing permits held by private ranching enterprises [10]. In California (USFS Region 5), 500 active grazing allotments annually supply 408,000 AUM of forage to support 97,000 livestock across 3.2 million ha on 17 national forests. With an annual recreating population of over 26 million [11], California's national forests are at the crossroad of a growing debate about the compatibility of livestock grazing with other activities (e.g., recreation) dependent upon clean, safe water.

There is a paucity of original research on water quality conditions on public grazing lands, and the conclusions of these reports are often inconsistent. For example, in California's Sierra Nevada, Derlet and Carlson [6] found surface water samples collected below horse and cattle grazing areas on USFS-administered lands were more likely to have detectable *E. coli* than non-grazed sites in national parks. Derlet *et al.* [12] reported algal coverage, algal-*E. coli* associations, and detection of waterborne *E. coli* to be greatest at sites below cattle grazing and lowest below sites experiencing little to no human or cattle activity, with human recreation sites being intermediate. Also in the central

Sierra Nevada, Myers and Whited [13] found FIB increased in surface waters below key grazing areas on USFS allotments following the arrival of cattle. However, Roche *et al.* [14] found no evidence of degradation of Yosemite toad breeding pool water quality in key grazing areas on three allotments in the Sierra National Forest of central California. Examining land-use and water quality associations in watersheds throughout the Cosumnes River Basin, Ahearn *et al.* [15] also reported water quality conditions in upper forested watersheds, which include USFS grazing allotments, to be well below levels of ecological concern.

The purpose of this study was to quantify microbial pollutant and nutrient concentrations during the summer cattle grazing and recreation season on 12 representative allotments across 5 national forests in northern California. Specific objectives were to 1) quantify FC, *E. coli*, total nitrogen, nitrate, ammonium, total phosphorus, and soluble-reactive phosphate concentrations in surface waters; 2) compare these results to a) water quality regulatory benchmarks, b) maximum nutrient concentrations recommended to avoid eutrophication, and c) estimates of nutrient background concentrations for this region; and 3) examine relationships between water quality, environmental conditions, and cattle grazing and recreation (i.e., resource uses).

Methods

Ethics Statement

Permission for site access was granted by the US Forest Service, and no permits were required.

Study Area

This cross sectional, longitudinal water quality survey was completed across 12 grazing allotments on USFS-managed public

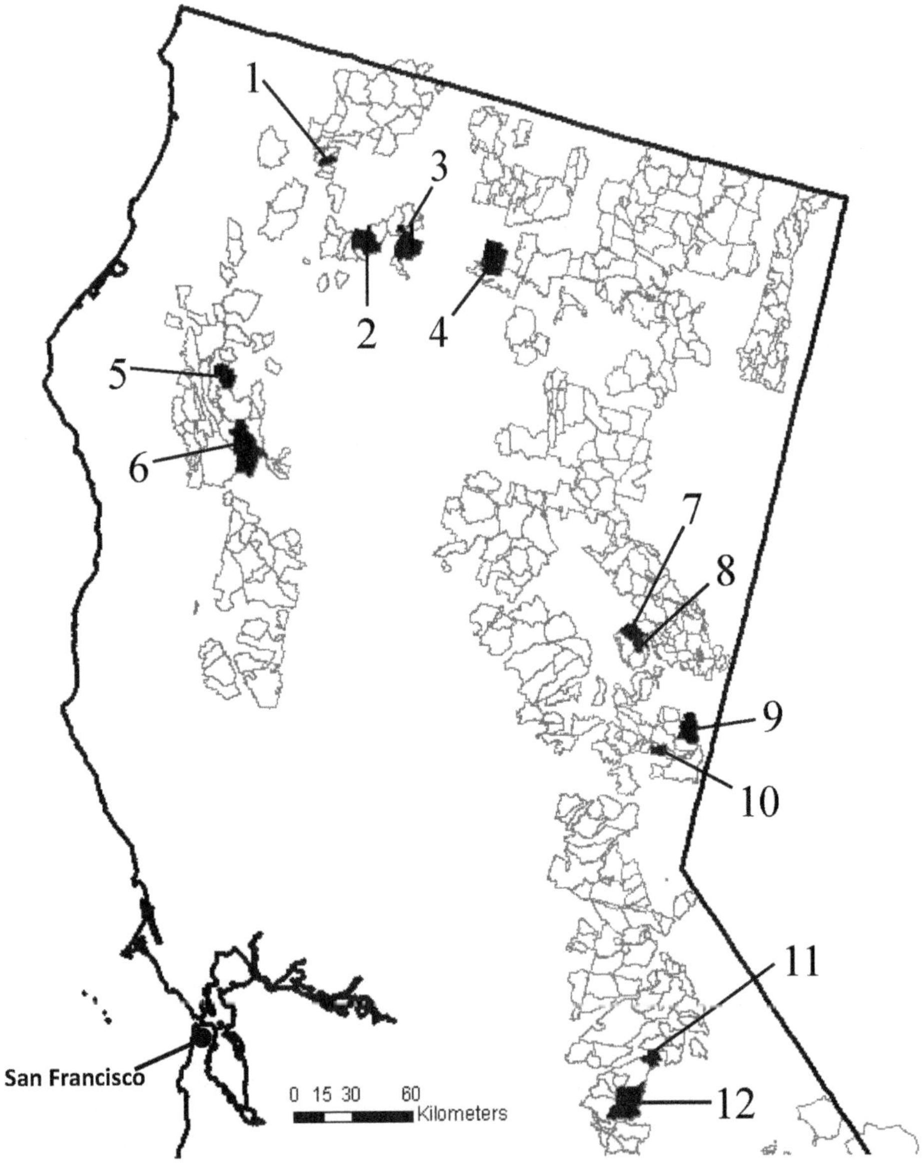

Figure 1. The 12 U.S. Forest Service grazing allotments (shaded polygons) in northern California enrolled in this cross-sectional longitudinal study of stream water quality between June and November 2011. Unshaded polygons are other U.S. Forest Service grazing allotments in the study area.

Table 1. Concentrations of total nitrogen (TN), nitrate (NO_3-N), ammonium (NH_4-N), total phosphorus (TP), and phosphate (PO_4-P) for 743 stream water samples collected across 155 sample sites on 12 U.S. Forest Service grazing allotments in northern California.

Nutrient	Mean[a] (μg L^{-1})	Median (μg L^{-1})	Maximum (μg L^{-1})	Below Detection[b] (%)	Eutrophication[c] (μg L^{-1})	Background[d] (μg L^{-1})
TN	58±2.7	33	675	5	-	60–530
NO_3-N	19±0.9	5	221	51	300	5–40
NH_4-N	11±0.4	5	146	61	-	-
TP	21±2.8	9	1321	32	100	9–32
PO_4-P	7±0.3	5	83	40	50	-

Published estimates of concentrations of general concern for eutrophication of stream water, and estimates of background concentrations for the study area are provided for context.
[a]The '±' indicates 1 standard error of the mean.
[b]Percentage of samples below minimum analytical detection limit. Limits were 10 μg L^{-1} for nitrogen and 5 μg L^{-1} for phosphorous. Observations below detection limit were set to one half detection limit (5 μg L^{-1} for nitrogen and 2.5 μg L^{-1} for phosphorus) for calculation of mean and median concentrations.
[c]Concentrations if exceeded indicate potential for eutrophication of streams [38–42].
[d]Estimated range of background concentrations for the three U.S. Environmental Protection Agency Level III sub-ecoregions (5, 9, 78) included in the study [43].

lands in northern California, USA (Fig. 1). Allotments were selected to represent the diversity of climate, soil, vegetation, water quality regulatory agencies, and resource use activities found across this landscape. The study area ranged from 41°40′ to 37°55′ N latitude and 123°30′ to 120°10′ W longitude, and included national forests in the Klamath, Coast, Cascade, and Sierra Nevada Mountain Ranges. Allotments were located on the Klamath (Allotments 1, 2), Shasta-Trinity (Allotments 3–6), Plumas (Allotments 7, 8), Tahoe (Allotments 9, 10), and Stanislaus (Allotments 11, 12) National Forests (Fig. 1). The study area totaled approximately1,300 km^2 and elevation ranged from 207 to 3,016 m (Table S1). The prevailing climate is Mediterranean with cool, wet winters and warm, dry summers. The majority of precipitation falls as snow between December and April, with snow melt generally occurring between May and June. Soils in Allotments 1–2, 5–7, and 11 are dominated by Inceptisols; Allotments 3, 10, and 12 are dominated by Alfisols; Allotment 8 and 9 are dominated by Mollisols; and Allotment 4 is dominated by Andisols [16] (Table S1).

All allotments were located in mountainous watersheds with canopy cover of mesic and xeric forests ranging from 9 to 89 and 2 to 93% cover, respectively [17]. Cooler mesic conifer forests were dominated by white fir (*Abies concolor*), red fir (*Abies magnifica*), and Douglas fir (*Pseudotsuga menziesii*). The relatively drier xeric conifer forests were dominated by ponderosa pine (*Pinus ponderosa*) and Jeffrey pine (*Pinus jeffreyi*). Montane hardwood and shrub cover ranged from 0 to 20%, and grass and forb cover from 1 to 9%. Wet meadows and other riparian plant communities covered 1 to 5% of allotment areas, and were the primary forage source for cattle grazing in these allotments.

Grazing Management

Cattle grazing management strategies on the study allotments reflect those widely found on western public grazing lands, such as those reviewed in Delcurto et al. [18] and George et al. [19]. Study allotments were grazed with commercial beef cow-calf pairs during the June to November grazing-growing season, following allotment-specific management plans designed to achieve annual herbaceous forage use standards (Table S1). Herbaceous use standards are set as an annual management target to protect ecological condition and function of meadow and riparian sites [20], and vary by national forest, allotment, and meadow ecological conditions [21–27].

Cattle stocking densities ranged from 1 animal unit (~450 kg cow with or without calf) per 18 ha to 1 animal unit per 447 ha (Table S1). Timing of grazing (turn on and turn off dates for cattle), duration of grazing season, and number of cattle are permitted by the USFS on an allotment-specific basis. Animal unit month (AUM) is the mass of forage required to sustain a single animal unit for a 30-day period, and is the standard metric of grazing pressure on USFS allotments.

Foraging, and thus spatial distribution of cattle feces and urine, is non-uniform across these allotments. Areas receiving relatively concentrated use by cattle are referred to as key grazing areas. Key grazing areas are often relatively small, stream-associated meadows and riparian areas that are preferentially grazed by cattle due to high forage quantity and quality and drinking water availability. For the most part, allotments are not cross-fenced to create pastures, which would improve grazing distribution. Where cross-fences exist, resulting pasture sizes are large (>2000 ha) with few pastures per allotment (<3).

Sample Site Selection

Key grazing areas and concentrated recreation areas within 200 m of streams in each allotment were identified and enrolled in the study in collaboration with local USFS managers and forest stakeholders. Water sample collection sites were established in streams immediately above, beside, and/or below sites with each activity to characterize water quality associated with these activities. Recreational activities included developed and undeveloped campgrounds, swimming-bathing areas, and trailheads used by hikers and recreational horse riders (i.e., pack stock). Key grazing areas were meadows and riparian areas that cattle were known to graze and occupy frequently and/or for extended periods throughout the grazing season. Additional sites were established at perennial flow tributary confluences with no concentrated use activities, enabling us to objectively include comparison sites across allotments with no concentrated grazing and/or recreation. While cattle use was concentrated primarily in key grazing areas, cattle grazing could occur throughout each allotment; therefore, it was not possible to determine water quality conditions in the complete absence of cattle.

A total of 155 stream water sample collection sites were identified and sampled monthly throughout the 2011 summer grazing-recreation period. Sample collection sites per allotment ranged from 7 to 18, depending upon the number of key grazing

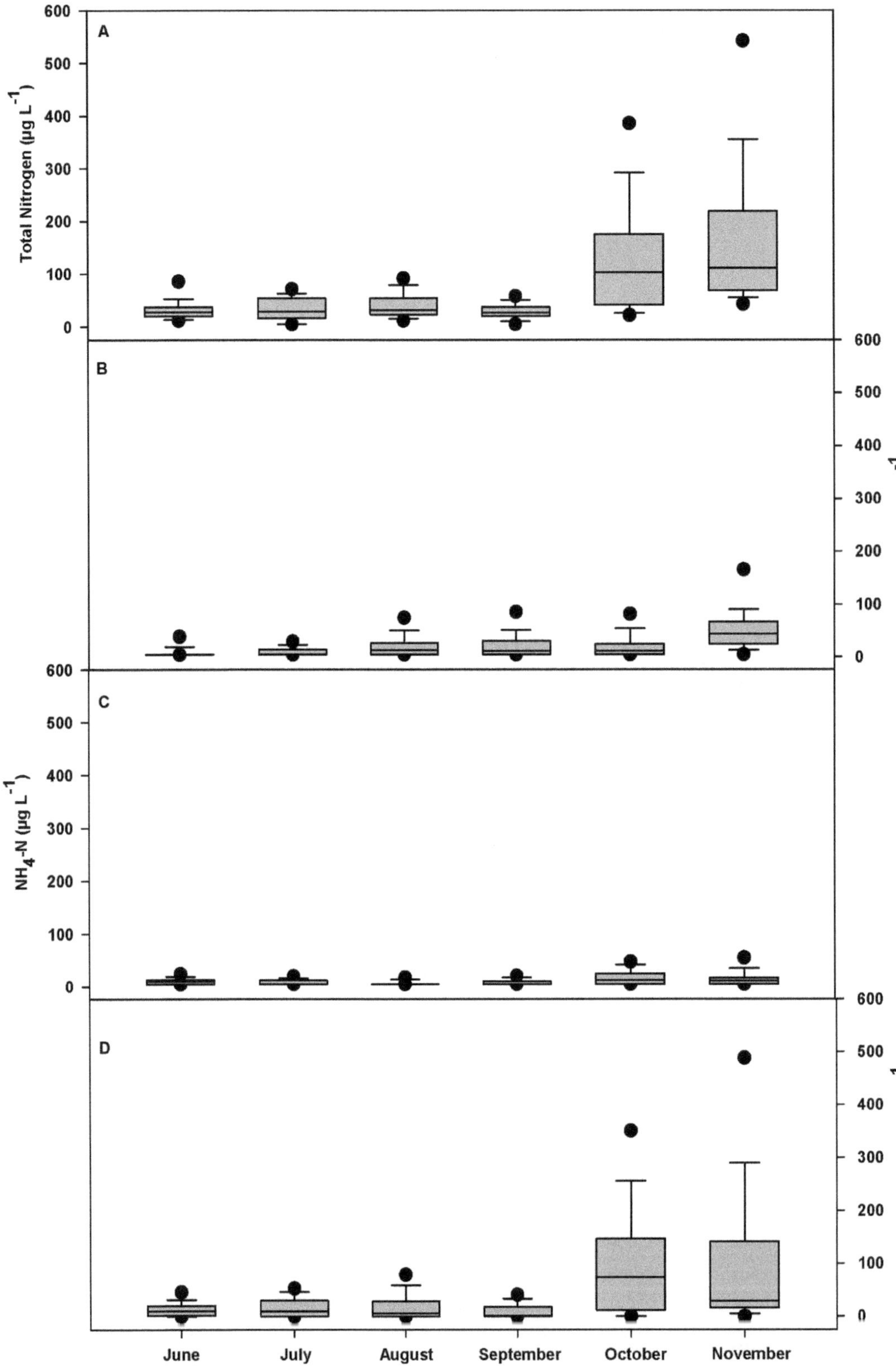

Figure 2. Overall monthly nitrogen concentrations for 743 stream water samples collected from 155 sample sites across 12 U.S. Forest Service grazing allotments in northern California enrolled in this cross-sectional longitudinal study between June and November 2011. (A) Total nitrogen, (B) nitrate (NO_3-N), and (C) ammonium (NH_4-N) were measured directly. (D) Organic nitrogen represents the difference between total nitrogen and NO_3-N plus NH_4-N. Bottom and top of shaded box are the 25th and 75th percentile of data, horizontal line within shaded box is median value, ends of vertical lines are 10th and 90th percentiles of data, and black dots are 5th and 95th percentiles of data. June $n = 135$; July $n = 150$; August $n = 178$; September $n = 120$; October $n = 127$; November $n = 33$.

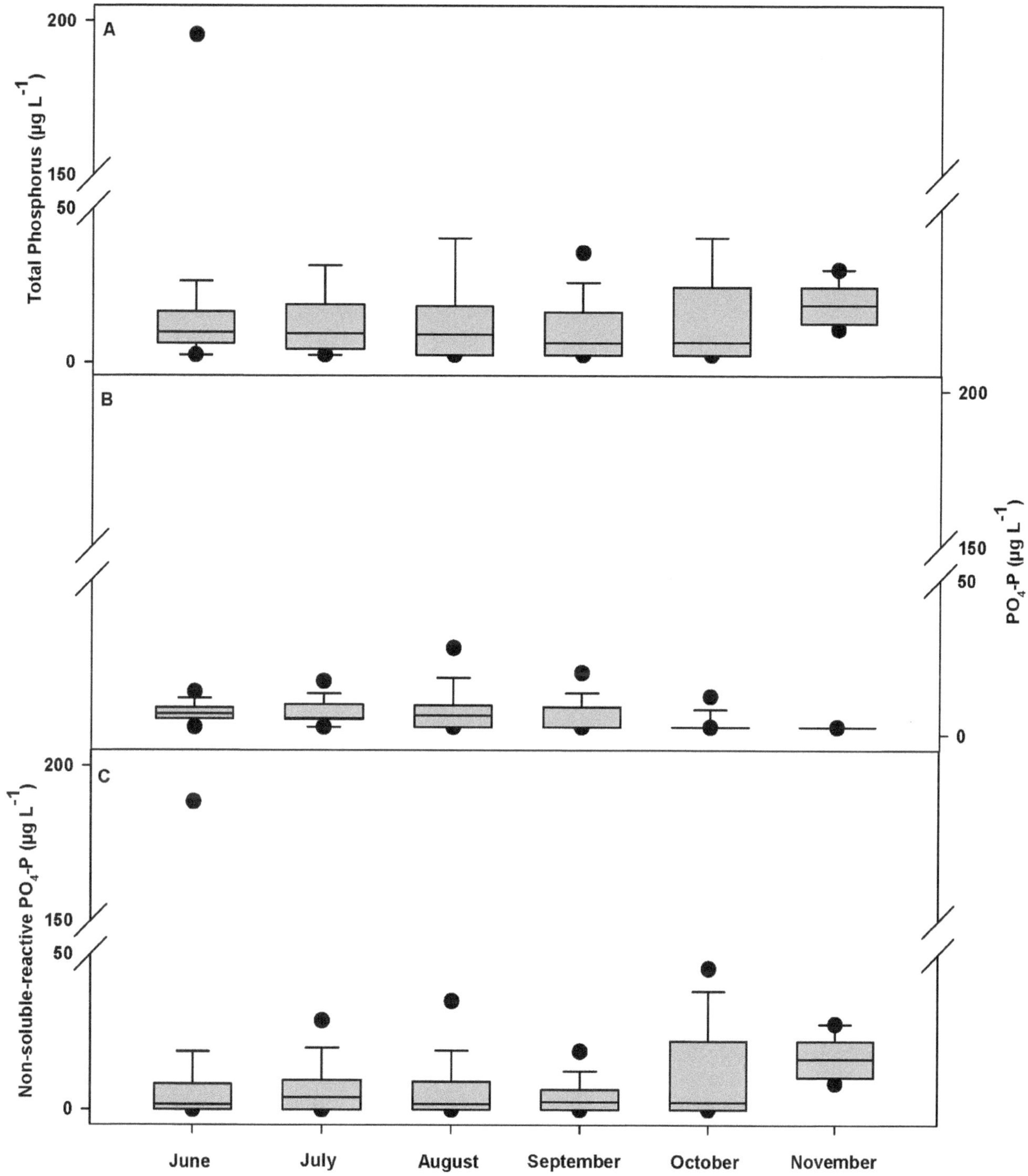

Figure 3. Overall monthly phosphorus concentrations for 743 stream water samples collected from 155 sample sites across 12 U.S. Forest Service grazing allotments in California enrolled in this cross-sectional longitudinal study between June and November 2011. (A) Total phosphorus (B) and soluble-reactive phosphorus (PO_4-P) were measured directly. (C) Non-soluble-reactive phosphorus represents the difference between total phosphorus (measured on unfiltered sample and treated with digesting agent) and soluble-reactive phosphorus. Bottom and top of shaded box are the 25th and 75th percentile of data, horizontal line within shaded box is median value, ends of vertical lines are 10th and 90th percentiles of data, and black dots are 5th and 95th percentiles of data. June $n=135$; July $n=150$; August $n=178$; September $n=120$; October $n=127$; November $n=33$.

and recreation areas identified, and number of tributary confluences (Table S1). Sixty-three percent of sample sites were associated with key grazing areas, 17% were associated with recreation activities, and 20% were tributary confluences with no concentrated use activities.

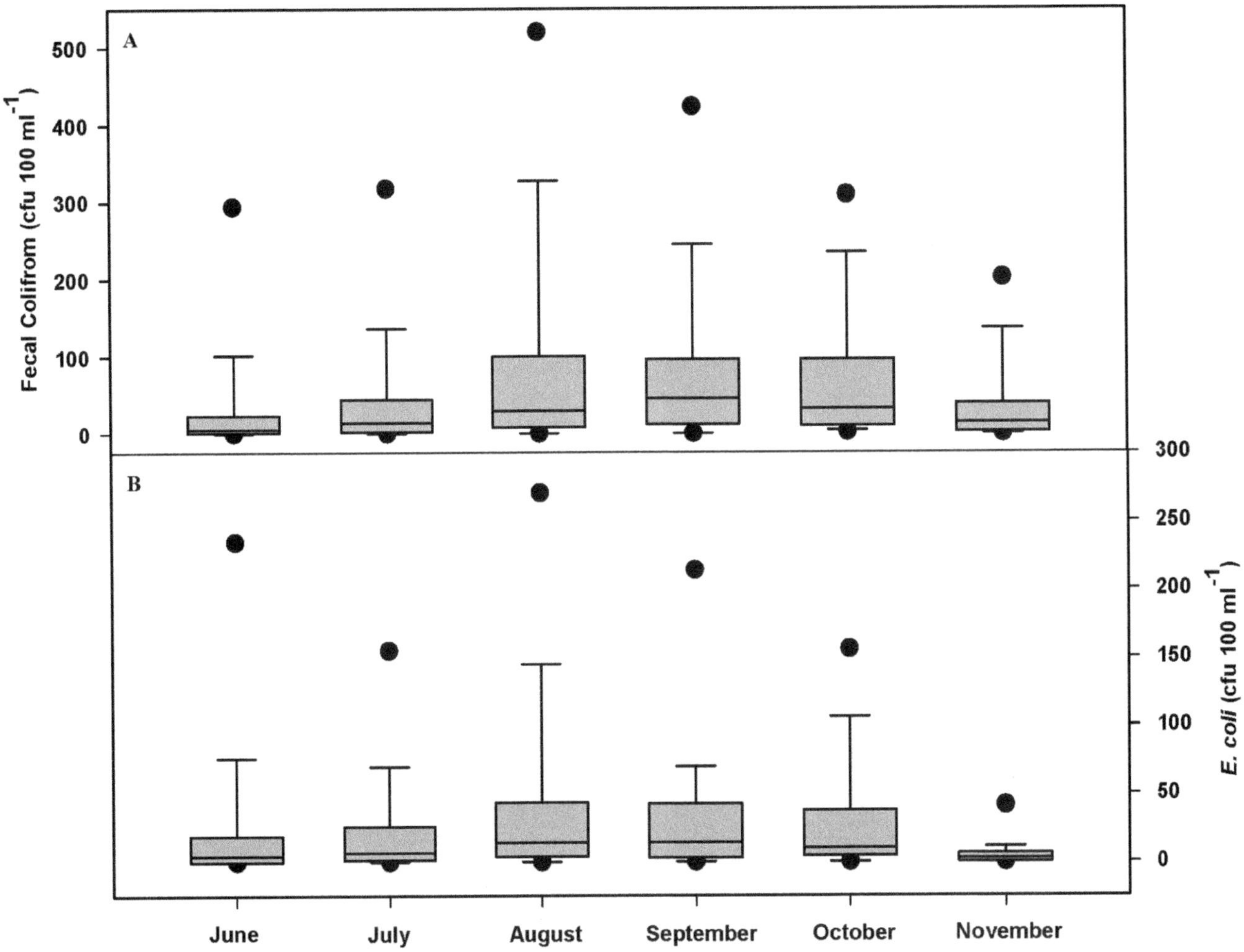

Figure 4. Overall monthly (A) fecal coliform and (B) *E. coli* concentrations for 743 stream water samples collected from 155 sample sites across 12 U.S. Forest Service grazing allotments in northern California enrolled in this cross-sectional longitudinal study between June and November 2011. Bottom and top of shaded box are the 25th and 75th percentile of data, horizontal line within shaded box is median value, ends of vertical lines are 10th and 90th percentiles of data, and black dots are 5th and 95th percentiles of data. June $n = 135$; July $n = 150$; August $n = 178$; September $n = 120$; October $n = 127$; November $n = 33$.

Sample Collection and Analysis

In 2011, a total of 743 water samples were collected and analyzed during the June 1 through November 9 study period, which captured the period of overlapping cattle grazing and recreation activities across these allotments. On each allotment, sampling occurred monthly throughout the grazing-recreation season. All sites in an allotment were sampled on the same day. Total sample numbers per allotment ranged from 40 to 88 (Table S1).

At the time of sample collection, environmental conditions and/or resource use activities that may have affected water quality were recorded. Specifically, the following conditions were noted (yes/no): 1) stagnant-low stream flow (<2 liters per second); 2) turbid stream water; 3) recreation (i.e., swimming-bathing, camping, hiking, fishing, horse riding); 4) cattle; and 5) any activities (i.e., low stream flow, turbid water, precipitation, cattle, recreation users) observed that may affect water quality. If algae, periphyton, or other aquatic autotrophic organisms were present at high to moderate levels (>20% of substrate cover) at time of sampling, then these conditions were recorded.

A vertical, depth-integrated stream water collection was made at the stream channel thalweg [28]. Water was collected in sterilized, acid-washed one liter sample containers, which were immediately stored on ice. All samples were analyzed for FC and *E. coli* within 8 hours of field collection. A 250 ml subsample was taken from each sample, frozen within 24 hours of collection, and processed for nutrient concentrations within 28 days of field collection. FC and *E. coli* concentrations as colony forming units (cfu) per 100 ml of water sample were determined by direct one step membrane filtration (0.45 μm nominal porosity filter) and incubation (44.5°C, 22–24 hours) on selective agar following standard method SM9222D [29]. Difco mFC Agar (Becton, Dickinson and Company, Spars, MD, USA) and CHROMagar *E. coli* (ChromAgar, Paris, France) were used for FC and *E. coli*, respectively. Total N (TN) and total phosphorus (TP) were measured after persulfate digestion of non-filtered subsamples following Yu *et al.* [30] and standard method SM4500-P.D [29], respectively. Concentrations of nitrate (NO_3-N), ammonium (NH_4-N), and soluble-reactive phosphorus (PO_4-P) were determined from filtered (0.45 μm nominal porosity filter) subsamples following Doane and Horwath [31], Verdouw *et al.* [32], and Eaton *et al.* [29],

Table 2. Percentage of 743 stream water samples collected across 155 sample sites on 12 U.S. Forest Service grazing allotments in northern California which exceeded water quality benchmarks relevant to the study area, specifically, and the nation, broadly.

Benchmark	Overall (% of 743)	Key Grazing Area (% of 462)	Recreation Area (% of 125)	No Concentrated Use Activities (% of 156)
FC >20 cfu 100 ml^{-1} [a]	50	48	46	58
FC >50 cfu 100 ml^{-1} [b]	31	28	27	42
FC >200 cfu 100 ml^{-1} [c]	10	10	6	13
FC >400 cfu 100 ml^{-1} [d]	4	5	2	4
E. coli >100 cfu 100 ml^{-1} [e]	9	8	7	11
E. coli >126 cfu 100 ml^{-1} [f]	7	7	6	8
**E. coli* >190 cfu 100 ml^{-1} [g]	5	4	4	6
**E. coli* >235 cfu 100 ml^{-1} [h]	3	3	3	4
E. coli >320 cfu 100 ml^{-1} [i]	2	2	2	2
E. coli >410 cfu 100 ml^{-1} [j]	1	2	2	1
NO_3-N >300 µg L^{-1} [k]	0	0	0	0
TP>100 µg L^{-1} [l]	2	2	2	<1
PO_4-P>50 µg L^{-1} [m]	<1	1	0	0

Results are reported for samples collected across all sample sites (overall) as well as for samples collected at sample sites monitored to characterize specific resource use activities across the allotments.
*Indicates the most relevant and contemporary standards for this study.
[a]Fecal coliform (FC) benchmark designated by Lahontan Regional Water Quality Control Board (LRWQCB) (based on geometric mean (GM) of samples collected over a 30-day interval) [36].
[b]FC benchmark designated by North Coast Regional Water Quality Control Board (NCRWQCB) (based on a median of samples collected over a 30-day interval) [37].
[c]FC benchmark designated by Central Valley Regional Water Quality Control Board (CVRWQCB) (based on GM of samples collected over a 30-day interval) [35].
[d]FC benchmark designated by CVRWQCB and NCRWQCB (maximum threshold value not to be exceeded by more than 10% of samples over a 30-day interval) [35].
[e]*E. coli* benchmark designated by U.S. Environmental Protection Agency (USEPA) [34] for an estimated illness rate of 32 per 1,000 primary contact recreators (based on GM of samples collected over a 30-day interval).
[f]*E. coli* benchmark designated by USEPA [34] for an estimated illness rate of 36 per 1,000 primary contact recreators (based on GM of samples collected over a 30-day interval).
[g]*E. coli* benchmark designated by USEPA [34] for an estimated illness rate of 32 per 1,000 primary contact recreators (for a single grab sample, approximates the 75th percentile of a water quality distribution based on desired GM).
[h]*E. coli* benchmark designated by USEPA [34] for an estimated illness rate of 36 per 1,000 primary contact recreators (for a single grab sample, approximates the 75th percentile of a water quality distribution based on desired GM).[i] *E. coli* benchmark designated by USEPA [34] for an estimated illness rate of 32 per 1,000 primary contact recreators (approximates the 90th percentile of a water quality distribution based on desired GM).
[j]*E. coli* benchmark designated by USEPA [34] for an estimated illness rate of 36 per 1,000 primary contact recreators (approximates the 90th percentile of a water quality distribution based on desired GM).[k] Maximum concentrations of nitrate as nitrogen (NO_3-N) recommended by USEPA [38,39].
[l]Maximum concentrations of total phosphorus (TP) recommended by USEPA [39,40].
[m]Maximum concentrations of phosphate as phosphorus (PO_4-P) recommended by USEPA [39,41].

respectively. Minimum detection limits were ~10 µg L^{-1} for TN, NH_4-N, and NO_3-N and ~5 µg L^{-1} for TP and PO_4-P. Organic nitrogen (ON) was calculated as *TN – [NO_3-N+NH_4-N]*, and non-soluble-reactive PO_4-P was calculated as *TP – PO_4-P*. Laboratory quality control included replicates, spikes, reference materials, control limits, criteria for rejection, and data validation methods [33].

Data Analysis and Interpretation

Descriptive statistics were calculated for the overall dataset as well as by 1) key grazing areas, recreation areas, and sample sites with no concentrated resource use; 2) activity observed at time of sample collection; 3) and month. Results were compared to numerous FIB benchmark concentrations used in the formulation of contemporary microbial water quality standards, maximum nutrient concentrations recommended to avoid eutrophication, and background nutrient concentration estimates for surface waters across the study area. The United States Environmental Protection Agency (USEPA) nationally recommends and has provided guidance on *E. coli* FIB-based standards ranging from 100 to 410 cfu 100 ml^{-1}, dependent upon selected illness rate benchmarks and frequency of sample collection over a 30 day period [34]. The study area falls within the jurisdiction of three semi-autonomous California Regional Water Quality Control Boards (RWQCBs), each of which has established enforceable standards based on FC benchmarks [35–37] ranging from 20 to 400 cfu 100 ml^{-1}. We report study results relative to each of these benchmarks to allow for comparisons to the various national and regional policies. For our study, which is based on monthly monitoring of multiple land-use activity types and environmental conditions across a broad regional scale (spanning approximately1,300 km^2), the most relevant and contemporary comparisons are the national U.S. Environmental Protection Agency (USEPA) *E. coli* single sample-based [8,34] standards of 190 cfu 100 ml^{-1} (estimated illness rate of 32 per 1,000 primary contact recreators) and 235 cfu 100 ml^{-1} (estimated illness rate of 36 per 1,000 primary contact recreators).

General recommendations for maximum concentrations to prevent eutrophication of streams and rivers are 300, 100, and 50 µg L^{-1} for NO_3-N, TP, and PO_4-P, respectively [38–42]. The study area is within three USEPA Level III Sub-Ecoregions (5, 9, and 78), and estimated background concentrations for TN, NO_3-N, and TP in these sub-regions range from 60 to 530, 5 to 40, and 9 to 32 µg L^{-1}, respectively [43].

Table 3. Percentage of 155 stream water sample sites on 12 U.S. Forest Service grazing allotments in northern California which had at least one exceedance of water quality benchmarks relevant to the study area, specifically, and the nation, broadly.

Benchmark	Overall (% of 155)	Key Grazing Area (% of 97)	Recreation Area (% of 27)	No Concentrated Use Activities (% of 31)
FC >20 cfu 100 ml^{-1} [a]	83	82	81	87
FC >50 cfu 100 ml^{-1} [b]	65	61	63	81
FC >200 cfu 100 ml^{-1} [c]	34	36	22	39
FC >400 cfu 100 ml^{-1} [d]	18	20	11	19
E. coli >100 cfu 100 ml^{-1} [e]	29	31	22	29
E. coli >126 cfu 100 ml^{-1} [f]	25	28	19	23
**E. coli* >190 cfu 100 ml^{-1} [g]	17	16	15	19
**E. coli* >235 cfu 100 ml^{-1} [h]	14	13	11	16
E. coli >320 cfu 100 ml^{-1} [i]	8	6	11	10
E. coli >410 cfu 100 ml^{-1} [j]	6	6	7	3
NO_3-N >300 μg L^{-1} [k]	0	0	0	0
TP>100 μg L^{-1} [l]	8	10	7	3
PO_4-P>50 μg L^{-1} [m]	2	3	0	0

Results are reported for all sample sites (overall) as well as for sample sites monitored to characterize specific resource use activities across the allotments. *Indicates the most relevant and contemporary standards for this study.

[a]Fecal coliform (FC) benchmark designated by Lahontan Regional Water Quality Control Board (LRWQCB) (based on geometric mean (GM) of samples collected over a 30-day interval) [36].

[b]FC benchmark designated by North Coast Regional Water Quality Control Board (NCRWQCB) (based on a median of samples collected over a 30-day interval) [37].

[c]FC benchmark designated by Central Valley Regional Water Quality Control Board (CVRWQCB) (based on GM of samples collected over a 30-day interval) [35].

[d]FC benchmark designated by CVRWQCB and NCRWQCB (maximum threshold value not to be exceeded by more than 10% of samples over a 30-day interval) [35].

[e]*E. coli* benchmark designated by U.S. Environmental Protection Agency (USEPA) [34] for an estimated illness rate of 32 per 1,000 primary contact recreators (based on GM of samples collected over a 30-day interval).

[f]*E. coli* benchmark designated by USEPA [34] for an estimated illness rate of 36 per 1,000 primary contact recreators (based on GM of samples collected over a 30-day interval).

[g]*E. coli* benchmark designated by USEPA [34] for an estimated illness rate of 32 per 1,000 primary contact recreators (for a single grab sample, approximates the 75th percentile of a water quality distribution based on desired GM).

[h]*E. coli* benchmark designated by USEPA [34] for an estimated illness rate of 36 per 1,000 primary contact recreators (for a single grab sample, approximates the 75th percentile of a water quality distribution based on desired GM).[i] *E. coli* benchmark designated by USEPA [34] for an estimated illness rate of 32 per 1,000 primary contact recreators (approximates the 90th percentile of a water quality distribution based on desired GM).

[j]*E. coli* benchmark designated by USEPA [34] for an estimated illness rate of 36 per 1,000 primary contact recreators (approximates the 90th percentile of a water quality distribution based on desired GM).[k] Maximum concentrations of nitrate as nitrogen (NO_3-N) recommended by USEPA [38,39].

[l]Maximum concentrations of total phosphorus (TP) recommended by USEPA [39,40].

[m]Maximum concentrations of phosphate as phosphorus (PO_4-P) recommended by USEPA [39,41].

At the sample site-scale, we used bivariate generalized linear mixed effects models (GLMMs) and zero-inflated count models to test for mean FIB and nutrient concentration (dependent variables were fecal coliform, *E. coli*, TN, NO_3-N, NH_4-N, TP, and PO_4-P)

Table 4. Mean concentrations for fecal coliform (FC) and *E. coli*, total nitrogen (TN), nitrate as nitrogen (NO_3-N), ammonium as nitrogen (NH_4-N), total phosphorus (TP), and phosphate as phosphorus (PO_4-P) for 743 total stream water samples collected across 155 sample locations on 12 U.S. Forest Service grazing allotments in northern California.

	Key Grazing Area (462 samples)	Recreation Area (125 samples)	No Concentrated Use Activities (156 samples)
FC (cfu 100 ml^{-1})	87±12 a	55±9 b	90±12 a
E. coli (cfu 100 ml^{-1})	42±6 a	29±7 b	43±8 a
Total N (μg L^{-1})	61±4 a	38±3 b	64±6 a
NO3-N (μg L^{-1})	17±1 ab	16±1 a	25±2 b
NH4-N (μg L^{-1})	11±0.6 a	10±1 a	10±0.7 a
Total P (μg L^{-1})	24±4 a	14±4 a	17±2 a
PO4-P (μg L^{-1})	7±0.3 a	5±0.2 b	8±0.6 a

Results reported are mean concentration for each resource use activity category. The '±' indicates 1 standard error of the mean. Different lower case letters indicate significant ($P<0.05$ with Bonferroni-correction for multiple comparisons) differences between resource use activity categories.

Table 5. Mean concentrations for fecal coliform (FC) and *E. coli*, total nitrogen (TN), nitrate as nitrogen (NO_3-N), ammonium as nitrogen (NH_4-N), total phosphorus (TP), and phosphate as phosphorus (PO_4-P) for 743 total stream water samples collected across 155 sample locations on 12 U.S. Forest Service grazing allotments in northern California.

	Low Stream Flow[a]		Turbid Water[b]		Cattle Present[c]		Recreation[d]		Activities Observed[e]	
	Yes	No	Yes	No	Yes	No	Yes	No	Yes	No
No. Occurrences	51	692	37	706	130	613	28	715	341	402
FC (cfu 100 ml^{-1})	216±67**	72±7	212±64**	76±8	205±39**	56±5	36±13	84±8	115±16**	54±6
E. coli (cfu 100 ml^{-1})	114±45*	35±3	142±56**	35±3	115±21**	24±3	14±5*	41±4	61±9*	23±3
Total N (µg L^{-1})	87±16	55±3	95±12	56±3	44±4	60±3	27±3**	59±3	48±3	65±4
NO_3-N (µg L^{-1})	17±3	19±1	19±1	16±3	19±2	18±1	16±3	19±1	17±1	20±1
NH_4-N (µg L^{-1})	15±3	10±0.4	10±0.4	13±2	9±1	11±0.5	7±0.7**	11±0.4	10±0.6	11±0.5
Total P (µg L^{-1})	30±5	20±3	107±37**	16±2	20±3	21±3	10±2	21±3	27±6*	15±1
PO_4-P (µg L^{-1})	13±2**	7±0.2	11±2**	7±0.2	10±1*	6±0.2	6±0.5**	7±0.3	7±0.5	5±0.3

Results are reported by category of field observation of resource use activities and environmental conditions observed at the time of sample collection. The '±' indicates 1 standard error of the mean, * indicates different at $P<0.05$ (Bonferroni-adjusted), and ** indicates different at $P<0.01$ (Bonferroni-adjusted).
[a]Stagnant or low stream flow (<2 liters per second).
[b]Stream water turbid.
[c]Cattle observed.
[d]Recreational activities only (i.e., no cattle present) observed.
[e]Any activities (low stream flow, turbid water, precipitation, cattle, or recreation) observed that potentially impact water quality.

differences between 1) key grazing areas, recreation areas, and sample sites with no concentrated resource use; and 2) occurrence of stagnant-low stream flow, turbid stream water, cattle, and recreation at the time of sample collection. We used GLMMs to analyze dependent variables with overdisperison (i.e., greater variance than expected) (fecal coliform, *E. coli*, TN) using the Poisson probability distribution function with robust standard errors [44]. For the GLMMs, we specified allotment identity and sample site identity as sequential random effects to account for hierarchical nesting and repeated measures [44,45]. Data with evidence of both overdispersion and zero-inflation can be produced by either unobserved heterogeneity or by processes that involve different mechanisms generating zero and nonzero counts [46–48]. For dependent variables with apparent overdispersion and zero-inflation (>25% zeros; NO_3-N, NH_4-N, TP, and PO_4-P), we used likelihood ratio tests to evaluate relative fits of zero-inflated negative binomial versus zero-inflated Poisson models [46–48]; we used simple Vuong tests [49] to evaluate relative fits of zero-inflated versus standard count models; and we used either likelihood ratio tests or Akaike Information Criterion (AIC), as appropriate, to compare relative fits between negative binomial and Poisson models. To account for the within-cluster correlation due to repeated measures, we specified sample site identity as a clustering variable in the final models to obtain robust variance estimates [50].

We also examined allotment-scale relationships of FIB and nutrient concentrations with environmental conditions and grazing management. We used bivariate zero-truncated count models to test associations between mean allotment values of response variables (fecal coliform, *E. coli*, TN, NO_3-N, NH_4-N, TP, and PO_4-P; mean of all samples collected for each allotment) and cattle grazing duration, animal unit months (AUM) of grazing, cattle density as cow-calf pairs 100 ha^{-1}, mean allotment elevation, and 2011–2012 water year precipitation [42] (independent variables). We used likelihood ratio tests to compare Poisson and negative binomial models [48]. For all analyses, when multiple response variables were predicted with the same independent variables, we interpreted significance levels using Bonferroni corrections to safeguard against Type I errors. Bonferroni adjusted p-values were considered significant at 0.0071 (dividing $P=0.05$ by the 7 water quality indicators tested) and 0.0014 (dividing $P=0.01$ by the 7 water quality indicators tested). All statistical analyses were conducted in Stata/SE 11.1 [48].

Results

Surface Water Quality and Weather Conditions Observed during Study

Precipitation during the 2010–11 water year ranged from 88 to 173% of the 30-year mean annual precipitation for each allotment, with 11 of 12 allotments receiving over 100% of mean annual precipitation (Table S1). Overall, nutrient concentrations were low across the study area (Table 1). With the exception of TN, over 32% of samples were below minimum detection limits for all nutrients (<10 µg N L^{-1} and <5 µg P L^{-1}). Nitrogen concentrations increased in October and November with the onset of fall rains (Fig. 2), and phosphorus concentrations showed no seasonal patterns (data not shown). The sum of NO_3-N and NH_4-N concentrations was lower than organic N (*TN* – [*NO_3-N+NH_4-N*]) concentrations throughout the sampling season (Fig. 2), suggesting that the majority of nitrogen was in organic forms. Additionally, PO_4-P concentrations were much lower than TP (Table 1; Fig. 3), suggesting that the majority of phosphorus was either organic or inorganic P adsorbed to suspended sediments. Mean and maximum FC and *E. coli* concentrations per allotment ranged from 30 to 255 and 17 to 151 CFU 100 ml^{-1}, and from 248 to 3,460 and 74 to 1,920, respectively (Table S2). FIB concentrations were highest from August through October (Fig. 4).

Nutrient and FIB Concentrations Relative to Water Quality Benchmarks

Mean and median NO_3-N, TP, and PO_4-P concentrations were at least one order of magnitude below nutrient concentrations recommended to avoid eutrophication (Table 1). No samples exceeded the NO_3-N maximum recommendation (Table 1). Overall, less than 2% of samples exceeded eutrophication

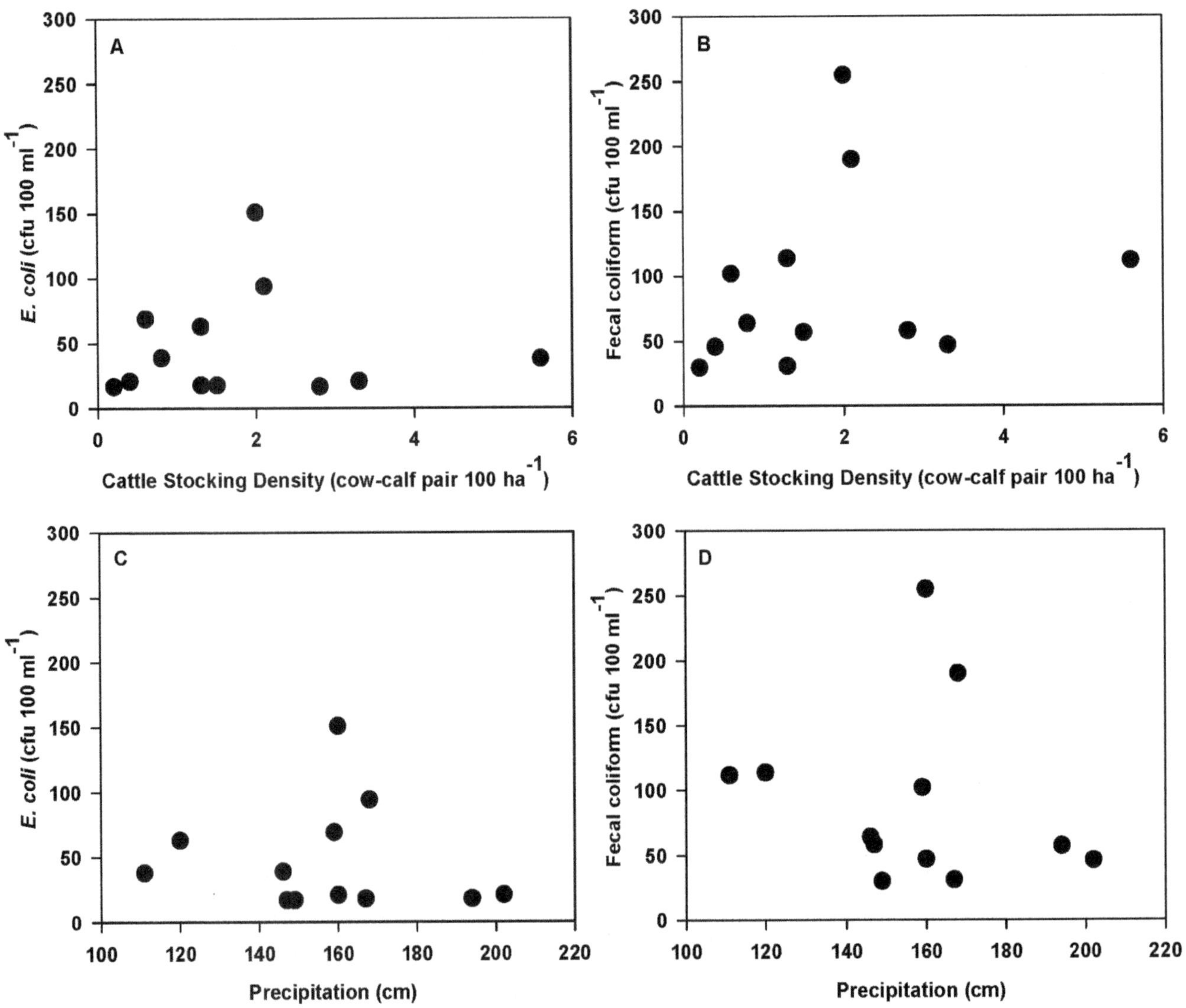

Figure 5. Trends in overall mean fecal indicator bacteria concentrations across sample sites during the June through November 2011sample period on 12 U.S. Forest Service grazing allotments in northern California enrolled in this cross-sectional longitudinal study. There were no significant relationships between allotment cattle stocking density and mean allotment concentrations of (A) *E. coli* ($P>0.9$) and (B) fecal coliform ($P>0.3$). During the study period, there were also no significant relationships between 2010–2011 water year precipitation and mean allotment concentrations of (C) *E. coli* ($P>0.6$) and (D) fecal coliform ($P>0.5$).

benchmarks (Table 2), and less than 8% of sites exceeded these benchmarks at least once (Table 3). Mean and median TN, NO_3-N, and TP concentrations were at or below estimated background concentrations for the study area (Table 1). The percentage of all samples (Table 2) exceeding FIB benchmarks ranged from 50% (benchmark FC = 20 cfu 100 ml^{-1}) to 1% (benchmark *E. coli* = 410 cfu 100 ml^{-1}), while the percentage of sites (Table 3) that exceeded a FIB benchmark at least once ranged from 83% (benchmark FC = 20 cfu 100 ml^{-1}) to 6% (benchmark *E. coli* = 410 cfu 100 ml^{-1}).

Nutrient and FIB Concentrations Relative to Grazing, Recreation, and Field Observations

Nutrient concentrations were at or below background levels, and only 0–10% of sites within each resource use activity category (i.e., key grazing areas, recreation areas, and non-concentrated use activities) had at least one nutrient benchmark exceedance (Table 3). The relative percentage of samples and sites exceeding FIB benchmarks for key grazing areas, recreation areas, and non-concentrated use areas varied by the individual benchmarks (Tables 2 and 3).

We found significantly ($P<0.002$) lower FC, *E. coli*, TN and PO_4-P concentrations at recreation areas than at key grazing areas and areas with no concentrated use activities (Table 4). Mean NO_3-N concentrations were also significantly lower ($P<0.001$) at recreation sites than at areas with no concentrated use activity; however, it is important to note that all nutrient concentrations were at or below background levels (Table 1), and none of the sites sampled ever exceeded the maximum recommended NO_3-N concentrations during the study (Tables 3).

Relative to conditions at time of sample collection, FC, *E. coli*, and PO_4-P concentrations were significantly ($P<0.0071$) higher when stream flow was low or stagnant, stream water was turbid, and when cattle were actively observed (Table 5). TP concentrations were also significantly higher ($P<0.001$) under turbid water conditions. *E. coli*, TN, NH_4-N, and PO_4-P concentrations were significantly lower ($P<0.006$) when recreation activities were observed at time of sampling, compared to sample events when recreation was not occurring (Table 5). Occurrence of high to moderate cover (>20% of substrate cover) of algae, periphyton, and other aquatic organisms at time of sampling was low (<2% of samples).

Allotment-scale Nutrient and FIB Concentrations Relative to Grazing Management and Environmental Conditions

Mean allotment-scale nutrient concentrations were not significantly related (at Bonferroni adjusted $P<0.0071$) to cattle density (TN: $P=0.3$; NO_3-N: $P=0.2$; NH_4-N: $P=0.2$; TP: $P=0.3$; PO_4-P: $P=0.1$), precipitation (TN: $P=0.09$; NO_3-N: $P=0.07$; NH_4-N: $P=0.73$; TP: $P=0.3$; PO_4-P: $P=0.04$), mean allotment elevation (TN: $P=0.02$; NO_3-N: $P=0.4$; NH_4-N: $P=0.07$; TP: $P=0.5$; PO_4-P: $P=0.2$), AUM (TN: $P=0.6$; NO_3-N: $P=0.5$; NH_4-N: $P=0.9$; TP: $P=0.1$; PO_4-P: $P=0.6$), or grazing duration (TN: $P=0.02$; NO_3-N: $P=0.5$; NH_4-N: $P=0.03$; TP: $P=0.6$; PO_4-P: $P=0.6$).

Mean allotment *E. coli* and FC concentrations showed increasing trends with increasing cattle densities and AUMs, and decreasing trends with increasing precipitation; however, these relationships were not statistically significant ($P>0.2$; Fig. 5). Mean allotment elevation ($P>0.8$), and cattle grazing duration ($P>0.7$) were also not correlated to mean allotment FIB concentrations (data not shown).

Discussion

Nutrient Conditions Relative to Water Quality Benchmarks

Mean and median nutrient concentrations observed across this grazed landscape were well below eutrophication benchmarks and background estimates (Table 1) [38–43]. Observed peak values in nitrogen and phosphorus concentrations were largely organic (or inorganic P adsorbed to suspended sediments) (Figs. 2 and 3), which are not considered readily available to stimulate primary production and eutrophication [39,51]. These results do not support concerns that excessive nutrient pollution is degrading surface waters on these USFS grazing allotments [4,12]. Our nutrient results are consistent with other examinations of surface water quality in similarly grazed landscapes. In the Sierra Nevada, Roche *et al.* [14] found nutrient concentrations of surface waters within key cattle grazing areas (mountain meadows) to be at least an order of magnitude below levels of ecological or biological concern for sensitive amphibians. On the Wallowa-Whitman National Forest in northeastern Oregon, Adams *et al.* [52] also reported nutrient levels to be at or below minimum detection levels in surface waters at key grazing areas.

Our results also agree with other studies of nutrient dynamics in the study area [53,54]. Headwater streams, such as those draining the study allotments, typically make up 85% of total basin scale drainage network length, have high morphological complexity, and high surface to volume ratios–which make them particularly effective at nutrient processing and retention [55]. Leonard *et al* [54] found that drainages in the western Tahoe Basin recovering from past disturbances and undergoing secondary succession tend to act as sinks for nutrients. Several studies have reported nutrient limitations across montane and subalpine systems resulting in low riverine nutrient export [56].

FIB Concentrations Relative to Water Quality Benchmarks

Overall mean and median *E. coli* were 40 and 8 cfu 100 ml^{-1}, and mean and median FC were 82 and 21 cfu 100 ml^{-1} (Table S2)– indicating that the nationally recommended *E. coli* FIB-based benchmarks would be broadly met, and that the more restrictive, FC FIB-based regional water quality benchmarks would be commonly exceeded across the study region. Clearly, assessments of microbial water quality and human health risks are dependent upon which FIB benchmarks are used for evaluation (Tables 2 and 3).

The scientific and policy communities are currently evaluating the utility of, and guidance for, FIB-based water quality objective effectiveness for safe-guarding recreational waters. As reviewed in Field and Samadpour [8], *E. coli* and FC are not always ideal indicators of fecal contamination and risk to human health from microbial pathogens. Poor correlations between bacterial indicators and pathogens such as *Salmonella* spp., *Giardia* spp., *Cryptosporidium* spp., and human viruses undermine the utility of these bacteria as *indicators* of pathogen occurrence and human health risk [8]. The ability of FIB to establish extra-intestinal, non-animal, non-human associated environmental strains and to grow and reproduce in water, soil sediments, algal wrack, and plant cavities also erodes their utility as *indicators* of animal or human fecal contamination [8]. Citing scientific advancements in the past two decades, the USEPA now recommends adoption of an indicator *E. coli* water quality objective as an improvement over previously used general indicators, including FC [34]. This guidance is based, in part, on *E. coli* exhibiting relatively fewer of the fecal indicator bacteria utility issues listed above, and on evidence that *E. coli* is a better predictor of gastro-intestinal illness than FC. Therefore, comparing our results to the most relevant and scientifically defensible *E. coli* FIB-based recommendations, 17% of all sites exceeded the 190 cfu 100 ml^{-1} benchmark, and 14% of all sites exceeded the 235 cfu 100 ml^{-1} benchmark [34]. This analysis, based on the best available science and USEPA guidance, clearly contrasts with the FC FIB-based interpretations currently in use by several regional regulatory programs, which suggest that as many as 83% of all sites in our study present potential human health risks.

Temporal Patterns in Water Quality

We observed a marked increase in total nitrogen concentrations in October and November, driven primarily by increased organic nitrogen, and to a lesser extent NO_3-N (Fig. 2). This coincided with the first rainfall-runoff events of fall that initiated flushing of solutes and particulates. The annual fall flush occurs subsequent to the summer drought and base flow period during which organic and inorganic nutrient compounds accumulate in soil and forest litter [54,57–60]. The disparity between TN and inorganic nitrogen (NO_3-N+NH_4-N) indicates the majority of flushed nitrogen was either particulate or dissolved organic nitrogen (Fig. 2). Consequently, most of the nitrogen flushed was likely in a relatively biologically unavailable form [51], with limited risk (relative to inorganic forms) of stimulating primary production and eutrophication. However, in nitrogen limited systems, increased biological utilization of organic nitrogen can occur [61].

FIB concentrations were highest from August through October (Fig. 4), which coincides with the period of maximum number of cattle turned out (Table S1). There is clear evidence that FIB concentrations increase with the introduction of cattle into a landscape, and increase with increasing cattle numbers [62–65]. The observed seasonal pattern of peak FIB concentrations also tracks the progression of stream flow from high, cold spring snowmelt to low, warm late-summer base flow conditions. Warm, low-flow conditions have been associated with elevated FIB [66–68]. Across this region, stream water temperatures are at their annual maximum in August and stream flows are at their annual minimum in September [69,70]. We observed stagnant-low flow conditions to be significantly associated with increased FIB concentrations (Table 5). It is likely that the seasonal peak of FIB concentrations is driven by timing of maximum annual cattle numbers, as well as optimal environmental conditions for growth and in-stream retention of both animal-derived and environmental bacteria (e.g., wildlife sources) [71–73]. Similar temporal trends in FIB concentrations have been observed in surface waters of Oregon, Wyoming, and Alaska [65,74,75].

Water Quality, Grazing, Recreation, and Environmental Conditions

Mean FIB concentrations at key grazing and non-concentrated use areas were higher than recreation sites, but did not exceed USEPA *E. coli* FIB-based benchmarks (Table 4). Mean FIB concentrations for all resource use activity categories exceeded the most restrictive regional FC FIB-based benchmarks of 20 and 50 cfu 100 ml^{-1}. *E. coli* FIB-based benchmark comparisons were generally comparable across sites, with recreation sites exhibiting overall lower numbers of exceedances; however, the different FC FIB-based benchmark comparisons indicated inconsistent results for water quality conditions across sites (Table 3). Similar to other surveys in the region [6,12,13], FIB concentrations were significantly greater when cattle were present at time of sample collection (Table 5). Tiedemann *et al.* [65] observed the same trend, with higher stream water FC concentrations on forested watersheds experiencing relatively intensive cattle grazing compared to ungrazed watersheds. Gary *et al.* [63] found grazing to have relatively minor impacts on water quality, though a statistically significant increase in stream water FC concentrations was induced at a relatively high stocking rate.

Mean allotment FIB concentrations showed apparent increasing trends with greater cattle densities (Fig. 5A and 5B); however, these allotment-level relationships were not statistically significant. Decreasing cattle density lowers fecal-microbial pollutant loading [76], which has been shown to reduce FIB concentrations in runoff from grazed landscapes [77]. Decreasing cattle density may also reduce stream bed disturbance and re-suspension of FIB-sediment associations by cattle [78–82]. Attracted to streams for shade, water, and riparian forage, cattle have been shown to spend approximately 5% of their day within or adjacent to a stream [63], depositing about 1.5% of their total fecal matter within one meter of a stream [83]. In a comprehensive review, George et al. [19] found that management practices that reduce livestock densities, residence time, and fecal and urine deposition in streams and riparian areas can reduce nutrient and microbial pollutant loading of surface water.

Samples associated with turbid stream water at the time of sample collection had significantly higher mean FIB concentrations than samples associated with non-turbid conditions (Table 5). It has been well documented that stream sediments contain higher concentrations of FIB than overlying waters [78–80,82], and that re-suspension of sediments in the water column by factors such as cattle disturbance or elevated stream flow is associated with elevated water column FIB concentrations [81]. FIB concentrations were also significantly higher under stagnant-low flow conditions (Table 5). Schnabel *et al.* [75] found a negative correlation between stream discharge and FIB concentrations at some sites, possibly due to the absence of a dilution effect under low flow conditions.

Although not statistically significant, we observed decreasing mean allotment FIB concentrations with greater precipitation during the 2010–2011 water-year (October 1 to September 30) (Fig. 5C and 5D). It is likely that precipitation during the 2010–2011 water-year is primarily reflecting snowpack, which supported higher than historical stream flow volumes during the study period. This potential relationship possibly reflects capacity of higher base flow volumes to dilute FIB concentrations. Lewis *et al.* [84] observed a similar negative correlation between surface runoff FC concentrations and annual cumulative precipitation on California coastal dairy pastures. Our observation that maximum FIB concentrations occurred under stagnant-low flow conditions (Table 5) also supports the potential for a negative relationship between FIB concentrations and annual precipitation.

Our results do not support previous concerns of widespread microbial water quality pollution across these grazed landscapes, as concluded in other surveys [6,12,13]. Although we did find apparent trends between cattle density and FIB concentrations (Figs. 5A and 5B) and significantly greater FIB concentrations when cattle were actively present, only 16% and 13% (Table 3) of key grazing areas (n = 97) exceeded the *E. coli* FIB-based benchmarks of 190 cfu 100 m^{-1} and 235 cfu 100 m^{-1}, respectively. Only 5 and 3% of total samples collected exceeded the *E. coli* FIB-based benchmarks of 190 cfu 100 m^{-1} and 235 cfu 100 m^{-1}, respectively (Table 2). In contrast, Derlet et al. [6] reported 60% and 53% of cattle grazing sites (n = 15) exceeded the 190 cfu 100 m^{-1} and 235 cfu 100 m^{-1} benchmarks, respectively. We also found no significant differences in FIB concentrations among key grazing areas and areas of no concentrated use activities (Table 4), which contrasts with previous work in the Sierra Nevada [6,12]. Finally, in this landscape of mixed livestock grazing and recreational uses, we found FIB concentrations to be lowest at recreation sites, indicating that water recreation objectives can be broadly attained within these grazing allotments. There are three important distinctions that separate our study from previous work: 1) in reaching our conclusions, we compared our study results to regulatory and background water quality benchmarks, which are based on current and best available science and policy; 2) these co-occurring land-use activities were directly compared on the same land units managed by a single agency (USFS), as opposed to previous comparisons between these land-uses occurring on different management units administered by different agencies with very different land-use histories and policies (e.g., USFS and U.S. National Park Service); and 3) to date, this study is the most comprehensive water quality survey in existence for National Forest public grazing lands, including an assessment of seven water quality indicators at 155 sites across five National Forests.

Conclusions

Nutrient concentrations observed across this extensively grazed landscape were at least one order of magnitude below levels of ecological concern, and were similar to USEPA estimates for background conditions in the region. Late season total nitrogen concentrations increased across all study allotments due to a first flush of organic nitrogen associated with onset of fall rainfall-runoff events, as is commonly observed in California's Mediterranean climate. Similar to previous work, we found greater FIB concentrations when cattle were present; however, we did not find overall significant differences in FIB concentrations between key grazing areas and non-concentrated use areas, and all but the most restrictive, FC FIB-based regional water quality benchmarks were broadly met across the study region. Although many regional regulatory programs utilize the FC FIB-based standards, the

USEPA clearly states–citing the best available science–*E. coli* are better indicators of fecal contamination and therefore provide a more accurate assessment of water quality conditions and human health risks. Throughout the study period, the USEPA recommended *E. coli* benchmarks of 190 and 235 cfu 100 ml^{-1} were met at over 83% of sites. These results suggest cattle grazing, recreation, and clean water can be compatible goals across these national forest lands.

Supporting Information

Table S1 Geographic characteristics, study year precipitation, cattle grazing management, and water quality sample collection sites and sample numbers for 12 U.S. Forest Service grazing allotments in northern California enrolled in this cross-sectional longitudinal study of stream water quality between June and November 2011.

Table S2 Mean, median, and maximum fecal coliform (FC) and *E. coli* concentrations for 743 stream water samples collected across 155 sample sites on 12 U.S. Forest Service grazing allotments in northern California. All concentrations are reported as colony forming units per 100 ml of sample water (cfu 100 ml^{-1}).

Acknowledgments

We thank Anne Yost, Barry Hill, and staff from the Klamath, Shasta-Trinity, Plumas, Tahoe, and Stanislaus National Forests for their assistance with study plan development and field data collection; Tom Lushinsky, D.J. Eastburn, Natalie Wegner, Donna Dutra, Mark Noyes, and Xien Wang for lab sample processing; University of California Cooperative Extension and USFS District Rangers and Forest Supervisors who provided lab space; and Anne Yost, Barry Hill, and three anonymous reviewers for their valuable and constructive comments on this manuscript.

Author Contributions

Conceived and designed the experiments: KWT ERA RAD. Performed the experiments: KWT. Analyzed the data: LMR KWT. Contributed reagents/materials/analysis tools: KWT RAD. Wrote the paper: LMR LK ERA RAD KWT.

References

1. Gentner BJ, Tanaka JA (2002) Classifying federal public land grazing permittees. J Range Manage 55: 2–11.
2. Huntsinger L, Forero LC, Sulak A (2010) Transhumance and pastoralist resilience in the Western United States. Pastoralism 1: 9–36.
3. Sulak A, Huntsinger L (2002) The importance of federal grazing allotments to central Sierran oak woodland permittees: A first approximation. In: Standiford RB, editor. Fifth symposium on oak woodlands: Oaks in California's changing landscape, PSW-GTR-184. Albany, California: USDA Forest Service Pacific Southwest Research Station. 43–51.
4. Belsky AJ, Matzke A, Uselman S (1999) Survey of livestock influences on stream and riparian ecosystems in the western United States. J Soil Water Conserv 54: 419–431.
5. Brunson MW, Steel BS (1996) Sources of variation in attitudes and beliefs about federal rangeland management. J Range Manage 49: 69–75.
6. Derlet RW, Carlson JR (2006) Coliform bacteria in Sierra Nevada wilderness lakes and streams: What is the impact of backpackers, pack animals, and cattle? J Wilderness Med 17: 15–20.
7. Taylor FR, Gillman LA, Pedretti JW (1989) Impact of cattle on 2 isolated fish populations in Pahranagat Valley, Nevada. Great Basin Nat 49: 491–495.
8. Field KG, Samadpour M (2007) Fecal source tracking, the indicator paradigm, and managing water quality. Water Res 41: 3517–3538.
9. Conley DJ, Paerl HW, Howarth RW, Boesch DF, Seitzinger SP, et al. (2009) Controlling eutrophication: Nitrogen and phosphorus. Science 323: 1014–1015.
10. USFS (2011) Grazing statistical summary fiscal year 2009. Washington, D.C.: USDA Forest Service. 108 p.
11. USFS (2012) National visitor use monitoring results - USDA Forest Service national summary report. Washington, D.C.: USDA Forest Service. 31 p.
12. Derlet RW, Richards JR, Tanaka LL, Hayden C, Ger KA, et al. (2012) Impact of summer cattle grazing on the Sierra Nevada watershed: Aquatic algae and bacteria. J Env Pub Health 2012: 1–7.
13. Myers L, Whited B (2012) The impact of cattle grazing in high elevation Sierra Nevada mountain meadows over widely variable annual climatic conditions. J Env Protection 3: 823–837.
14. Roche LM, Allen-Diaz B, Eastburn DJ, Tate KW (2012) Cattle grazing and Yosemite toad (*Bufo canorus* Camp) breeding habitat in Sierra Nevada meadows. Rangeland Ecol Manag 65: 56–65.
15. Ahearn DS, Sheibley RW, Dahlgren RA, Anderson M, Johnson J, et al. (2005) Land use and land cover influence on water quality in the last free-flowing river draining the western Sierra Nevada, California. J Hydrol 313: 234–247.
16. Soil Survey Staff, Natural Resources Conservation Staff (2008) Soil Survey Geographic Database (SSURGO). Available: http://soils.usda.gov/survey/geography/ssurgo/. Accessed 2013 May 31.
17. USFS (2009) CALVEG 2000, USDA Forest Service Pacific Southwest Region Remote Sensing Lab website. Available: http://www.fs.fed.us/r5/rsl/projects/gis/data/vegcovs/cvalley/EvegTile33A_00_07_100k_v2.html. Accessed 2013 May 31.
18. Delcurto T, Porath M, Parsons CT, Morrison JA (2005) Management strategies for sustainable beef cattle grazing on forested rangelands in the Pacific Northwest. Rangeland Ecol Manag 58: 119–127.
19. George MR, Jackson RD, Boyd CS, Tate KW (2011) A scientific assessment of the effectiveness of riparian management practices. In: Bridke DD, editor. Conservation benefits of rangeland practices: Assessment, recommendations, and knowledge gaps. Lawrence: Allen Press. 213–252.
20. Clary WP, Leininger WC (2000) Stubble height as a tool for management of riparian areas. J Range Manage 53: 562–573.
21. BLM (1999) Utilization studies and residual measurements. Denver: USDI Bureau of Land Management. 174 p.
22. USFS (1988) Plumas National Forest land and resource management plan website. Available: http://www.fs.usda.gov/land/plumas/landmanagement. Accessed 2013 May 23.
23. USFS (1990) Tahoe National Forest land and resource management plan website. Available: http://www.fs.usda.gov/detail/tahoe/landmanagement. Accessed 2013 May 23.
24. USFS (1995) Shasta-Trinity National Forest land and resource management plan website. Available: http://www.fs.usda.gov/main/stnf/landmanagement/planning. Accessed 2013 May 23.
25. USFS (2004) 2004 Sierra Nevada Forest plan amendment record of decision. Vallejo, California: USDA Forest Service. 72 p.
26. USFS (2010) Klamath National Forest land and resource management plan website. Available: http://www.fs.usda.gov/main/klamath/landmanagement/planning. Accessed 2013 May 23.
27. USFS (2010) Stanislaus National Forest land and resource management plan website. Available: http://www.fs.usda.gov/main/stanislaus/landmanagement/planning. Accessed 2013 May 23.
28. Dahlgren RA, Tate KW, Ahearn DS (2004) Watershed scale, water quality monitoring – water sample collection. In: Down RD, Lehr JH, editors. Environmental instrumentation and analysis handbook. New York: John Wiley and Sons. 547–564.
29. Eaton AD, Clesceri LS, Rice EW, Greenberg AE, Franson MAH (2005) Standard methods for the examination of water and wastewater: Centennial edition. Washington, D.C.: American Public Health Association. 1368 p.
30. Yu ZS, Northup RR, Dahlgren RA (1994) Determination of dissolved organic nitrogen using persulfate oxidation and conductimetric quantification of nitrate-nitrogen. Commun Soil Sci Plan 25: 3161–3169.
31. Doane TA, Horwath WR (2003) Spectrophotometric determination of nitrate with a single reagent. Anal Lett 36: 2713–2722.
32. Verdouw H, Vanechteld CJA, Dekkers EMJ (1978) Ammonia determination based on indophenol formation with sodium salicylate. Water Res 12: 399–402.
33. Puckett M (2002) Quality assurance management plan for the state of California's surface water ambient monitoring program. Surface Water Ambient Monitoring Program website. Available: http://www.swrcb.ca.gov/water_issues/programs/swamp/tools.shtml. Accessed 2013 May 23.
34. USEPA (2012) Recreational water quality criteria. Washington D.C.: US Environmental Protection Agency Office of Water Regulations and Standards. 73 p.
35. CVRWQCB (2011) The water quality control plan (basin plan) for the California Regional Water Quality Control Board, Central Valley region. Sacramento: Central Valley Regional Water Quality Control Board. 148 p.
36. LRWQCB (2005) Water quality control plan for the Lahontan region. South Lake Tahoe: Lahontan Regional Water Quality Control Board. 55 p.
37. NCRWQCB (2011) Water quality control plan for the North Coast region. Santa Rosa: North Coast Regional Water Quality Control Board. 13 p.
38. Cline C (1973) The effects of forest fertilization on the Tahuya River, Kitsap Peninsula, Washington. Olympia: Washington State Department of Ecology. 55 p.

39. MacDonald LH, Smart AW, Wissmar RC (1991) Monitoring guidelines to evaluate effects of forestry activities on streams in the Pacific Northwest and Alaska, EPA 910/9-91-001. Seattle: US Environmental Protection Agency. 180 p.
40. Mackenthun KM (1973) Toward a cleaner aquatic environment. Washington, D.C.: US Environmental Protection Agency. 289 p.
41. USEPA (1986) Quality criteria for water: 1986. Washington, D.C.: US Environmental Protection Agency Office of Water Regulations and Standards. 477 p.
42. USEPA (1987) Surface water monitoring: A framework for change. Washington, D.C.: US Environmental Protection Agency Office of Water Regulations and Standards. 41 p.
43. USEPA (2000) Ambient water quality criteria recommendations: Rivers and streams in Ecoregion II. Washington, D.C.: US Environmental Protection Agency Office of Water Regulations and Standards. 120 p.
44. Rabe-Hesketh S, Skrondal A (2008) Multilevel and longitudinal modeling using stata. College Station: Stata Press. 562 p.
45. Pinheiro JC, Bates DM (2000) Mixed-effects models in S and S-PLUS. New York: Springer-Verlag. 528 p.
46. Cameron AC, Trivedi PK (1998) Regression analysis of count data. Cambridge: Cambridge University Press.
47. Long JS (1997) Regression models for categorical and limited dependent variables. Thousand Oaks: Sage.
48. StataCorp (2009) Stata Statistical Software: Release 11. College Station: StataCorp LP.
49. Vuong QH (1989) Likelihood ratio tests for model selection and non-nested hypotheses. Econometrica 57: 307–333.
50. Williams RL (2000) A note on robust variance estimation for cluster-correlated data. Biometrics 56: 645–646.
51. Campbell JL, Hornbeck JW, McDowell WH, Buso DC, Shanley JB, et al. (2000) Dissolved organic nitrogen budgets for upland, forested ecosystems in New England. Biogeochemistry 49: 123–142.
52. Adams MJ, Pearl CA, McCreary B, Galvan SK, Wessell SJ, et al. (2009) Short-term effect of cattle exclosures on Columbia spotted frog (*Rana luteiventris*) populations and habitat in northeastern Oregon. J Herpetol 43: 132–138.
53. Green MB, Fritsen CH (2006) Spatial variation of nutrient balance in the Truckee River, California-Nevada. J Am Water Resour As 42: 659–674.
54. Leonard RL, Kaplan LA, Elder JF, Coats RN, Goldman CR (1979) Nutrient transport in surface runoff from a subalpine watershed, Lake Tahoe Basin, California. Ecol Monogr 49: 281–310.
55. Peterson BJ, Wollheim WM, Mulholland PJ, Webster JR, Meyer JL, et al. (2001) Control of nitrogen export from watersheds by headwater streams. Science 292: 86–90.
56. Hill BH, McCormick FH, Harvey BC, Johnson SL, Warren ML, et al. (2010) Microbial enzyme activity, nutrient uptake and nutrient limitation in forested streams. Freshwater Biol 55: 1005–1019.
57. Ahearn DS, Sheibley RW, Dahlgren RA, Keller KE (2004) Temporal dynamics of stream water chemistry in the last free-flowing river draining the western Sierra Nevada, California. J Hydrol 295: 47–63.
58. Bernal S, Butturini A, Sabater F (2005) Seasonal variations of dissolved nitrogen and DOC : DON ratios in an intermittent Mediterranean stream. Biogeochemistry 75: 351–372.
59. Miller WW, Johnson DW, Denton C, Verburg PSJ, Dana GL, et al. (2005) Inconspicuous nutrient laden surface runoff from mature forest Sierran watersheds. Water Air Soil Poll 163: 3–17.
60. van Verseveld WJ, McDonnell JJ, Lajtha K (2008) A mechanistic assessment of nutrient flushing at the catchment scale. J Hydrol 358: 268–287.
61. Kaushal SS, Lewis WM (2005) Fate and transport of organic nitrogen in minimally disturbed montane streams of Colorado, USA. Biogeochemistry 74: 303–321.
62. Doran JW, Schepers JS, Swanson NP (1981) Chemical and bacteriological quality of pasture runoff. J Soil Water Conserv 36: 166–171.
63. Gary HL, Johnson SR, Ponce SL (1983) Cattle grazing impact on surface-water quality in a Colorado Front Range stream. J Soil Water Conserv 38: 124–128.
64. Schepers JS, Francis DD (1982) Chemical water-quality of runoff from grazing land in Nebraska. 1. Influence of grazing livestock. J Environ Qual 11: 351–354.
65. Tiedemann AR, Higgins DA, Quigley TM, Sanderson HR, Marx DB (1987) Responses of fecal-coliform in streamwater to 4 grazing strategies. J Range Manage 40: 322–329.
66. Allan JD (1995) Stream Ecology: Structure and function of running waters. London: Chapman & Hall. 388 p.
67. Edwards DR, Coyne MS, Daniel TC, Vendrell PF, Murdoch JF, et al. (1997) Indicator bacteria concentrations of two northwest Arkansas streams in relation to flow and season. T ASAE 40: 103–109.
68. Lau YL, Liu D (1993) Effect of flow-rate on biofilm accumulation in open channels. Water Res 27: 355–360.
69. Krupa M, Tate KW, Van Kessel C, Sarwar N, Linquist B (2011) Water quality in rice-growing watersheds in a Mediterranean climate. Agr Ecosyst Environ 144: 290–301.
70. Tate KW, Lancaster DL, Lile DF (2007) Assessment of thermal stratification within stream pools as a mechanism to provide refugia for native trout in hot, arid rangelands. Environ Monit Assess 124: 289–300.
71. Alm EW, Burke J, Hagan E (2006) Persistence and potential growth of the fecal indicator bacteria, *Escherichia coli*, in shoreline sand at Lake Huron. J Great Lakes Res 32: 401–405.
72. Byappanahalli MN, Whitman RL, Shively DA, Ting WTE, Tseng CC, et al. (2006) Seasonal persistence and population characteristics of *Escherichia coli* and enterococci in deep backshore sand of two freshwater beaches. J Water Health 4: 313–320.
73. Roszak DB, Colwell RR (1987) Survival strategies of bacteria in the natural-environment. Microbiol Rev 51: 365–379.
74. Clark ML, Norris JR (2000) Occurrence of fecal coliform bacteria in selected streams in Wyoming,1990–99: Water resources investigations report 00–4198. Cheyenne: US Geological Survey. 8 p.
75. Schnabel WE, Wilson T, Edwards R, Stahnke G, Maselko M, et al. (2010) Variability, seasonality, and persistence of fecal coliform bacteria in a cold-region, urban stream. J Cold Reg Eng 24: 54–75.
76. Roche LM, Latimer AM, Eastburn DJ, Tate KW (2012) Cattle grazing and conservation of a meadow-dependent amphibian species in the Sierra Nevada. PloS One 7: 1–11.
77. Knox AK, Tate KW, Dahgren RA, Atwill ER (2007) Management reduces *E. coli* in irrigated pasture runoff. Calif Agr 61: 159–165.
78. Jamieson R, Joy DM, Lee H, Kostaschuk R, Gordon R (2005) Transport and deposition of sediment-associated *Escherichia coli* in natural streams. Water Res 39: 2665–2675.
79. Jamieson RC, Gordon RJ, Tattrie SC, Stratton GW (2003) Sources and persistence of fecal coliform bacteria in a rural watershed. Water Qual Res J Can 38: 33–47.
80. Sherer BM, Miner JR, Moore JA, Buckhouse JC (1992) Indicator bacterial survival in stream sediments. J Environ Qual 21: 591–595.
81. Stephenson GR, Rychert RC (1982) Bottom sediment - a reservoir of *Escherichia-coli* in rangeland streams. J Range Manage 35: 119–123.
82. Van Donsel DJ, Geldreich EE (1971) Relationships of *Salmonellae* to fecal coliforms in bottom sediments. Water Res 5: 1079–1087.
83. Porath ML, Momont PA, DelCurto T, Rimbey NR, Tanaka JA, et al. (2002) Offstream water and trace mineral salt as management strategies for improved cattle distribution. J Anim Sci 80: 346–356.
84. Lewis DJ, Atwill ER, Lennox MS, Pereira MDG, Miller WA, et al. (2009) Reducing microbial contamination in storm runoff from high use areas on California coastal dairies. Water Sci Technol 60: 1731–1743.

5

Reduction in Acute Ecotoxicity of Paper Mill Effluent by Sequential Application of Xylanase and Laccase

Saurabh Sudha Dhiman[1,2]⁹, Gaurav Garg[3,4]⁹, Jitender Sharma[3], Vipin C. Kalia[5]*, Yun Chan Kang[1]*, Jung-Kul Lee[1,2]*

1 Department of Chemical Engineering, Konkuk University, Gwangjin–Gu, Seoul, Republic of Korea, **2** Institute of SK–KU Biomaterials, Konkuk University, Gwangjin–Gu, Seoul, Republic of Korea, **3** Department of Biotechnology, Kurukshetra University, Kurukshetra, Haryana, India, **4** Department of Biotechnology, Maharishi Markandeshwar University, Mullana-Ambala, Haryana, India, **5** Microbial Biotechnology and Genomics, CSIR-Institute of Genomics and Integrative Biology, Delhi University Campus, Delhi, India

Abstract

In order to reduce the ecotoxicity of paper mill, four different enzymatic pretreatment strategies were investigated in comparison to conventional chemical based processes. In strategy I, xylanase-aided pretreatment of pulp was carried out, and in strategy II, xylanase and laccase-mediator systems were used sequentially. Moreover, to compare the efficiency of *Bacillus stearothermophilus* xylanase and *Ceriporiopsis subvermispora* laccase in the reduction of ecotoxicity and pollution, parallel strategies (III and IV) were implemented using commercial enzymes. Conventional $C_DE_{OP}D_1D_2$ (C_D, Cl_2 with ClO_2; E_{OP}, H_2O_2 extraction; D_1 and D_2, ClO_2) and $X/XLC_DE_{OP}D_1D_2$ (X, xylanase; L, laccase) sequences were employed with non-enzymatic and enzymatic strategies, respectively. Acute toxicity was determined by the extent of inhibition of bioluminescence of *Vibrio fischeri* with different dilutions of the effluent. Two-fold increase was observed in EC_{50} values for strategy I compared to the control process. On the other hand, sequential application of commercial enzymes resulted in higher acute toxicity compared to lab enzymes. In comparison to the control process, strategy II was the most efficient and successfully reduced 60.1 and 25.8% of biological oxygen demand (BOD) and color of effluents, respectively. We report for the first time the comparative analysis of the ecotoxicity of industrial effluents.

Editor: Ligia O. Martins, Universidade Nova de Lisboa, Portugal

Funding: This work was supported by the Energy Efficiency & Resources Core Technology Program of the Korea Institute of Energy Technology Evaluation and Planning (KETEP), granted financial resource from the Ministry of Trade, Industry & Energy, Republic of Korea (201320200000420). This research was also supported by a grant from the Intelligent Synthetic Biology Center of Global Frontier Project (2011-0031955) funded by the Ministry of Science, ICT and Future Planning, Republic of Korea. This work was supported by 2013 KU Brain Pool fellowship of Konkuk University. The authors also acknowledge the support provided by Kurukshetra University, BILT and RCDP, India. VCK wish to thank CSIR-WUM (ESC0108) Government of India for providing necessary funds and facilities. The funders had no role in study design, data collection and analysis, decision to publish, or preparation of the manuscript.

Competing Interests: The authors have declared that no competing interests exist.

* Email: vckalia@igib.res.in (VCK); yckang@konkuk.ac.kr (YCK); jkrhee@konkuk.ac.kr (JKL)

⁹ These authors contributed equally to this work.

Introduction

In the last two decades, use of enzymes, especially hemicellulases, has revolutionized the pulp and paper industry and provided a glimpse of hope that application of enzymes at various levels can reduce the industrial pollution and effluent's toxicity. However, the current scenario continues to be challenging because of the high pollution load released by the pulp and paper industries, which are still using chlorine-based bleaching sequences [1]. The conventional bleaching methodology is chemical dependent and energy intensive; therefore, a novel enzymatic treatment with safe level of discharge needs to be developed [2]. Hydrolytic enzymes along with a laccase mediator system (LMS) have been more beneficial in reducing the pollution load of industries compared to other strategies [3]. An N-hydroxy-based synthetic mediator was predominantly used for this purpose [4]. However, toxicity and cost are two of the major hurdles, which hamper the industrial applications of these synthetic mediators. Therefore, the application of natural mediators in LMS is one of the alternatives to overcome these disadvantages, even though their application may cause grafting onto the pulp, an increase in kappa number, and a reduction in brightness of the pulp. Since chemical-intensive conventional strategies and enzymatic processes utilizing synthetic mediators release high levels of toxic compounds into water bodies, the entire processing of the effluents should be characterized to analyze their ecotoxicity and other hazardous properties [5].

It has been observed that pretreatment with xylanase alone cannot reduce the pollution load of pulp and paper industry significantly. Hence, it is believed that a cocktail of two or more enzymes could reduce the release of hazardous materials to safer levels [4]. Although enzymes are effective at the pretreatment level, effluents from the entire process should be analyzed to study the enzymatic after-effects. Few reports deal with the characterization of effluents from enzyme-aided bleaching processes, but without evaluating the interaction of technical parameters at the pretreatment level [2]. Therefore, in the present investigation, the individual and cumulative effects of physical parameters on the efficiency of both xylanase and laccase were optimized and were analyzed using a statistical model.

This is the first ever attempt where ecotoxicity of mixed effluents from the entire bleaching process was characterized using Microtox 81.9% basic toxicity assay method along with the evaluation of the reduction in pollution load in terms of biological oxygen demand (BOD) and color. A schematic study was done with four different strategies by supplementing the conventional bleaching sequence with xylanase from *Bacillus stearothermophilus* SDX and Pulpzyme VLBL (Novozyme, Denmark). In sequential strategies, *Ceriporiopsis subvermispora* laccase and the commercial laccase TM L603P (UK) were used with a natural mediator (syringaldehyde, SA) for the extraction of cellulosic fibers from agro-residual material for paper processing.

Materials and Methods

Microbial strains and their maintenance

The bacterial culture for xylanase production was isolated from a compost sample and it was identified by the Institute of Microbial Technology (IMTECH), Chandigarh, India, as *B. stearothermophilus* SDX and strain has been deposited at IMTECH and the accession number is 8508 [6]. A lignin-degrading fungal culture of *C. subvermispora* L-14807 SS-3 was used for laccase production [2].

Enzyme production, extraction and assay

Xylanase production was studied through two-step statistical (Table S1, S2 in File S1) modeling [7] under solid-state fermentation (SSF) conditions. An Erlenmeyer flask (250 mL) containing 5 g of wheat bran as sole carbon source was moistened with 25 mL of modified Horikoshi medium (pH 8.0) [8]. Sterilized substrate, cooled to room temperature, was inoculated with 15% (v/w) inoculum (18 h culture, $\sim 3.6 \times 10^6$ cells/mL) and incubated at 37°C in a humidified chamber (relative humidity 60–65%) for 96 h. The flasks were gently tapped intermittently to mix the contents. Colonized solid culture (1 g) was extracted twice with 10 mL of phosphate buffer (10 mM, pH 8.0) through a wet muslin cloth. Xylanase activity was determined through modified Bailey's method [9] by using 3,5-dinitrosalicylic acid (DNS) reagent [10]. All the parameters related to laccase production were optimized through the conventional "one variable at a time" method under submerged fermentation (SmF) conditions. The crude extract of *C. subvermispora* culture was partially purified using ammonium sulfate fractionation (50–70% cut) and was concentrated with a 30-kDa cutoff membrane filter (Amicon). The observed molecular weight of the laccase was approximately 50 kDa. To obtain the concentrated partially purified enzyme, a membrane with a 30-kDa cutoff was used. The laccase activity was determined according to a previously reported method [11]. Crude extracts of both the strains were assayed for other enzymatic activities viz. cellulase and amylase. Similar to xylanase from *B. stearothermophilus* SDX and laccase from *C. subvermispora*, commercial xylanase and laccase were also characterized.

Estimation of the protein content in all the enzyme samples was determined by Bradford's method [12]. All the experiments were carried out independently in triplicates and the results are presented as the mean of the three values.

Screening of natural mediators

Different phenolic compounds such as acetosyringone (AS), coniferyl alcohol (CA), *p*-coumaric acid (PCA), 2,6-dimethoxyphenol (DMP), 2,6-dimethylphenol (DP), guaiacol, 4-hydroxybenzoic acid (HBA), 4-hydroxybenzaldehyde (4-HA), and syringaldehyde (SA) were screened to determine their potential as natural mediators that facilitate decolorization. The potential of natural phenols (50 μM each) was determined using the decolorization assay of 25 μM Azure-B [13]. The catalytic efficiency of natural phenols was compared with that of the synthetic mediator 1-hydroxybenzatriazole (HBT) as a control.

Enzyme, chemicals and software

Commercial xylanase (Pulpzyme VLBL) and laccase (TM L603P) were kindly provided by Novozyme (Denmark) and Treforest Industrial Estate (Pontypridd, CF375 UD, Wales, United Kingdom), respectively. Agro-residual pulp composed of (w/w) wheat straw (*Triticum aestivum*) (78.8%), sarkanda (*Saccharum spontaneum*) (10.6%), and candy (*Eragrostis* sp.) (10.6%) cooked at 165–175°C for 30 min at a pressure of 7.0–7.5 kg m^{-3}, was obtained from an agro-based industry (Trident Paper, Barnala, Punjab) in India. The permanganate number, brightness (%ISO), yellowness, and viscosity (poise) of the unbleached pulp were 7.11±0.96, 42.5±5.6, 15.8±2.6, and 5.18±0.96, respectively. Experimental design was generated using the statistical software "Design-Expert 6.0" developed by Stat-Ease Inc. (Minneapolis, MN, USA) to study the individual and cumulative effect of independent variables on enzyme production and biobleaching.

Parametric optimization for pretreatment of pulp

Pretreatment of pulp with xylanase and laccase was carried out separately. Different levels of all the independent variables designed through central composite design of response surface methodology (CCD-RSM) are given in Table S3 in File S1. For xylanase-aided pretreatment, 10% (w/v) pulp consistence was used and statistical modeling was applied to four independent variables viz. pH, temperature (T_X), enzyme dose (ED_X), and retention time (RT_X, Table 1).

After extensive washing with distilled water, xylanase-treated pulp was subjected to LMS. A broad range of reaction parameters (viz. enzyme dose [ED_L], mediator concentrations [MC_L], and retention times [RT_L]) were optimized through CCD-RSM for the LMS (Table 2). The pulp was again washed thoroughly with distilled water before determining its strength. Similar to xylanase pretreatment, regression equation and responses were determined for the pretreatment of agro-residual pulp with the LMS.

Residual laccase activity

Laccase activity was determined under controlled conditions using 50 g oven-dried (OD) pulp and SA as mediator. Residual laccase activity was determined in the presence and absence of pulp and mediators and was expressed as percentage of initial activity, which was measured at the beginning (zero time) of the incubation time. Experiments were performed using 20 U of laccase and 1.5 mM of mediator.

Scheme ($X/XLC_DE_{OP}D_1D_2$) for biobleaching

The most extensively used bleaching sequence $C_DE_{OP}D_1D_2$ (C_D, chlorine and chlorine dioxide; E_{OP}, extraction with oxygen and hydrogen peroxide; D_1 and D_2, chlorine dioxide) was used as the control. In enzyme-aided strategies, only xylanase-assisted stage (X) and pretreatment with both xylanase and laccase along with natural mediator step (XL) were included prior to the conventional sequence.

Table 1. Experimental design (coded variables) and results of CCD-RSM for xylanase-aided pretreatment of pulp.

pH	T_x[a]	ED_x[b]	RT_x[c]	PN_x[d]		B_x[e]		Y_x[f]	
				Actual	Predicted	Actual	Predicted	Actual	Predicted
0	0	0	−α	6.92±1.72	6.82±1.13	43.1±5.7	41.5±4.2	14.2±3.1	14.3±1.5
+1	+1	−1	+1	7.12±1.44	7.31±1.32	48.5±5.7	44.7±4.5	14.8±2.4	15.0±1.6
−α	0	0	0	6.74±1.31	6.85±1.15	44.7±6.4	42.8±4.3	14.2±2.1	14.1±1.5
+1	+1	+1	−1	7.21±1.32	7.09±1.24	39.5±4.5	40.5±4.2	14.5±2.9	14.6±1.5
0	0	0	0	6.58±1.35	6.50±0.92	45.8±2.9	45.8±4.1	13.5±1.2	13.5±1.3
−1	+1	+1	−1	6.74±1.23	6.69±0.82	42.1±5.4	42.7±4.3	13.5±1.9	14.1±1.4
0	0	0	0	6.58±1.24	6.50±0.91	45.8±5.7	45.8±4.5	13.5±3.1	13.5±1.3
0	+α	0	0	6.94±1.22	6.85±0.82	43.4±5.6	43.9±4.1	14.1±2.5	14.4±1.4
0	0	+α	0	6.66±1.35	6.87±0.92	44.5±1.5	41.4±4.4	13.8±3.4	14.3±1.5
0	0	−α	0	7.78±1.33	7.33±1.34	37.2±5.4	39.4±3.1	15.4±1.9	14.7±.1.4
−1	+1	−1	−1	6.73±1.24	6.78±1.25	42.1±6.4	41.0±4.2	13.5±2.3	14.0±1.4
+α	0	0	0	7.62±1.42	7.35±1.24	38.5±9.4	39.5±3.3	16.9±3.1	15.7±1.6
−1	−1	+1	−1	6.97±1.85	6.74±0.92	42.2±4.6	45.5±4.4	14.2±2.4	14.4±1.5
+1	−1	−1	+1	7.24±1.23	7.26±1.12	39.6±2.9	40.3±3.1	14.9±1.2	14.8±1.5
+1	+1	+1	+1	6.98±1.34	6.93±1.11	42.5±4.2	43.3±4.2	14.1±2.6	14.4±1.6
0	0	0	0	6.53±1.22	6.50±0.91	45.8±6.8	45.8±4.3	13.5±2.6	13.9±1.3
−1	−1	+1	+1	7.05±1.25	6.88±0.82	42.2±2.5	43.0±4.4	14.0±3.2	13.9±1.4
0	0	0	0	6.53±1.34	6.50±0.81	45.8±4.1	45.8±4.1	13.5±2.1	13.9±1.3
+1	+1	−1	−1	7.12±1.22	7.28±1.12	39.5±6.8	40.1±3.2	14.3±3.2	14.7±1.5
−1	−1	−1	+1	7.04±1.24	7.16±1.12	42.8±5.7	41.4±4.3	14.1±2.5	14.2±1.5
+1	−1	+1	+1	6.96±1.15	6.88±0.93	39.9±6.8	40.6±4.5	14.6±3.1	14.4±1.5
0	0	0	0	6.58±1.42	6.50±0.81	45.8±5.8	45.8±4.1	13.5±2.5	13.9±1.3
0	−α	0	0	6.99±1.34	6.85±0.81	43.7±5.4	42.3±4.4	14.2±1.3	14.1±1.4
−1	−1	−1	−1	6.82±1.32	6.83±0.82	41.5±4.6	42.1±4.2	14.7±2.5	14.1±1.4
+1	−1	−1	−1	7.01±1.55	7.18±1.22	38.7±3.4	38.1±3.1	14.6±2.2	14.7±1.5
−1	+1	+1	+1	6.94±1.51	6.78±0.91	42.5±4.6	42.7±4.5	14.1±2.3	13.8±1.4
0	0	0	0	6.54±1.63	6.50±0.82	45.8±4.5	45.8±4.1	13.5±2.1	13.9±1.4
−1	+1	−1	+1	7.16±1.24	7.06±1.12	41.2±5.8	42.8±4.2	14.9±3.2	14.3±1.5
+1	−1	+1	−1	6.93±1.62	6.99±1.02	40.5±4.6	40.3±4.2	14.2±3.2	14.8±1.5
0	0	0	+α	7.02±1.14	6.98±0.91	42.8±5.3	43.5±4.2	14.1±2.4	13.9±1.4

a – Temperature (°C); b – Enzyme Dose (U/mg); c – Retention time (min); d – Permanganate Number; e – Brightness (%ISO) of pulp after pretreatment; f – Yellowness (b*) of pulp after pretreatment.

Dry plastic bags filled with 50 g of extensively washed OD pulp were subjected to different combinations of reaction parameters as designed using CCD-RSM. During chlorination stage, the consistency of the pulp was maintained at 3% for effective penetration of the fiber by the reaction liquor. Experiments were designed with different doses of Cl_2 at the C_D stage, and a kappa factor of 0.25 was used for calculating the Cl_2 dose. The E_{OP} stage was carried out in a stainless steel vessel at 70±1°C after adding the required amounts of sodium hydroxide (25 kg/T) and hydrogen peroxide (H_2O_2; 5 kg/T). The oxygen pressure was maintained at 5.0 and 2.5 kg/cm^2 for the initial 30 min and final 90 min, respectively. Final pH in the reaction vessel was maintained by adding dilute H_2SO_4 before the addition of ClO_2.

FTIR and XRD analysis

Fourier Transformed Infrared Spectroscopy (FTIR) spectra of pulp samples were recorded with a resolution of 4 cm^{-1} over the waves ranging from 4,000 to 400 cm^{-1}, using 32 scans per sample. FTIR spectrum was analyzed using the FTIR database.

Crystallinity of cellulose was determined for both untreated and enzymatically treated samples by X-ray diffraction (XRD). After freeze-drying, the samples were analyzed using X-ray diffractometer with an X-ray generator (from 2 to 50 of 2θ scattering angle). The crystalline index of cellulose (*Xc*) was calculated from the X-ray diffraction patterns using the following equation:

$$Xc = \frac{(I_{002} - I_{am})}{I_{002}} \times 100$$

Where I_{002} is the peak intensity from the (0 0 2) lattice plane ($2\theta = 22.3°$) and I_{am} is the peak intensity of amorphous phase ($2\theta = 15.7°$).

Analytical studies

Compositional analysis of the agro-residual pulp after pretreatment with different enzyme samples was carried out according to the standard National Renewable Energy Laboratory (NREL) assay procedures [14]. The permanganate number and viscosity were determined according to Technical Association of Pulp and Paper (TAPPI) standards [15]. The brightness, whiteness, optical and color dimensions of hand sheets were also measured (L & W-070; Sweden). The strength of hand sheets was tested according to TAPPI test methods.

Mixed effluents from the different biobleaching strategies have been characterized by analyzing BOD and color following the American Society for Testing and Materials (ASTM) standards, ASTM D-1252-00, and ASTM-1209-00, respectively.

Toxicity analysis of mixed effluents

Mixed effluents (C_D, E_{OP}, D_1, and D_2 in the ratio of 3.25:1:1:1) from all the strategies were assessed using the 81.9% toxicity test protocol with a Microtox Model 500 Analyzer (Modern Water, USA). Different dilutions of each sample were prepared, and the osmolarity of each sample was adjusted with the osmolarity adjusting solution provided by the manufacturer. The endpoint measured by the Microtox assay detects the decrease in the intensity of light emitted by the luminescent marine bacterium *Vibrio fischeri* after 5 and 15 min of exposure. Hence, the $EC_{50\text{-}5min}$ and $EC_{50\text{-}15min}$ average values denote the effective concentration (EC) of effluent that caused a 50% reduction in the luminescence of the bacteria [16,17]. These EC values were calculated using the standard 81.9% toxicity test for each collected effluent sub-sample to determine the acute toxicity of each collected sub-sample. Test results were entered into an Excel worksheet and corrected to normalize the color differences in different treated samples. The unit of acute toxicity (TOU) after 15-min incubation was also calculated for each sample to determine the EC of effluent required to kill 50% of the model organism. The sensitivity of each freeze-dried bacterial sample was periodically checked in our laboratory using zinc sulfate ($ZnSO_4 \cdot 7H_2O$) as a reference substance.

Results

Optimization of enzyme production

All the nutritional and physical parameters affecting the xylanase production were screened through Plackett-Burman (PB) designing (Table S1 in File S1). Based on the regression analysis data of PB design, four significant factors [peptone (X_1), KNO3 (X_2), temperature (X_3) and incubation time (X_4)] were selected and studied at five different levels (Table S2 in File S1) to analyze their individual and cumulative effect on xylanase production through CCD-RSM (Table S3 in File S1). An enhanced (4,220 U/g-dry cell weight) xylanase yield with a specific activity indicative of 597 U/mg-protein after 4 days incubation at 37°C under SSF conditions was achieved for *B. stearothermophilus* SDX by using two-step statistical modeling. The crude extract was free from cellulase and laccase activities and showed very low (12 U/mg-protein) amylase activity.

During laccase production under SmF by *C. subvermispora*, all the parameters were optimized using the "one variable at a time" approach. The highest specific activity of laccase in the crude extract was indicative of 10 U/mg-protein after 7 days incubation at 35°C and pH 5.0 under shaking conditions (50±5 rotations per minute) using 2.5% (w/v) wheat bran (Fig. 1). Crude extract of *C. subvermispora* was free from any cellulase and amylase activity, and contained very less (1.8 U/mg-protein) xylanase activity (Fig. 1). The commercial samples of xylanase and laccase showed specific activities indicative of 498 and 7 U/mg-protein, respectively.

Selection of natural mediator

C. subvermispora and commercial laccase were analyzed with Azure-B dye to verify the mediator capability of all the phenolic compounds. Decolorization of Azure-B was used to verify the potential of natural phenols to behave like mediators in lignin degradation [18]. Among all the tested natural mediators (AS, CA, PCA, DMP, DP, HBA, 4-HA, and SA), only SA showed a significant potential relative to other mediators (Fig. S1 in File S1). In comparison with HBT, AS showed a weak potential, which was reflected by only 5% decolorization (Fig. S1 in File S1).

Parametric optimization of enzymatic pretreatment

The operational parameters for xylanase-aided pretreatment were optimized through CCD-RSM. A second-order polynomial equation was used to analyze the individual and cumulative effects of four independent variables on the pulp properties (Table 1). Five different levels of each independent variable (Table S4 in File S1) were used to analyze the regression of dependent variables in order to obtain the corresponding responses Y_1, Y_2, and Y_3 respectively and facilitate their statistical analysis (Equation S1 in File S1). On the basis of the coefficient of determination ($R^2 = 0.9029$) and an adequate precision (6.446) for the model, 7.5 U of *B. stearothermophilus* xylanase at 60°C and pH 9.0 for 150 min of retention time (i.e., when all the independent variables were at their central point "0") was found to be the most effective for the pretreatment

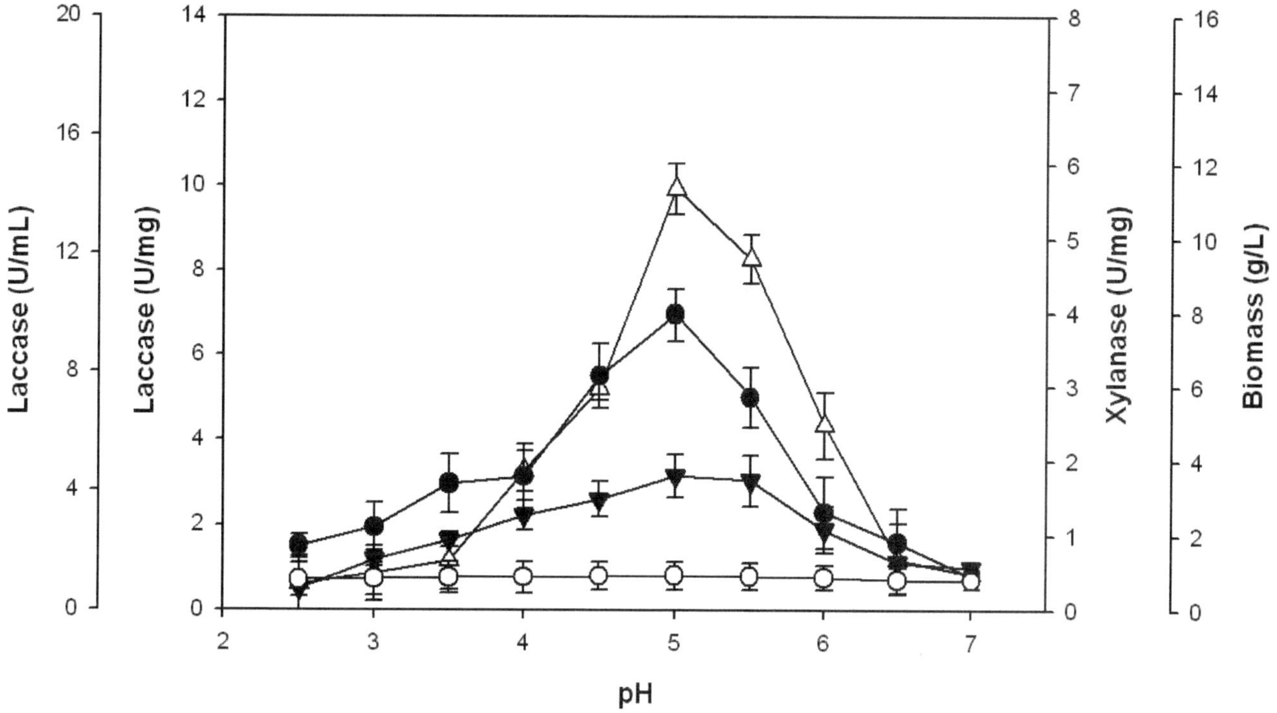

Figure 1. Effect of pH on laccase and xylanase production and growth of fungal mycelium of *C. subvermispora*. *Experiments were carried out at 35°C for 7 days using 2.5% (w/v) of wheat bran in a 250-mL Erlenmeyer flask containing 50 mL of production media. Laccase activity (●); specific activity (△); xylanase activity (▼); fungal biomass (○).

(Table 1). At optimal conditions, the viscosity of the xylanase produced in the lab and commercial xylanase pretreated pulps were 4.46±0.98 and 4.24±0.84 poise, respectively.

A similar strategy was applied to study the individual and cumulative effects of the laccase dose (ED_L), natural mediator concentration (MC_L), and retention time (RT_L) in the response of the dependent variables of the agro-residual pulp pretreatment (Table 2). Responses of the dependent variables were studied using the second-order polynomial equations (Equation S2 in File S1). The value of R^2 indicates that a variation of only 5.41% is possible in the values of the dependent variables. From the statistical modeling it was observed that 20 U (coded value "0") of *C. subvermispora* laccase with 1.5% of SA concentration (coded value "0") was the most effective for an incubation period of 225 min (coded value "+α", Table 2). Other physical parameters (viz. temperature and pH) were optimized by using the "one variable at a time" approach and it was observed that the most suitable conditions for the LMS corresponded to 50°C (Table S5 in File S1) and pH 5.0 (Table S6 in File S1). After sequential pretreatment of the pulp, the viscosity of the enzyme produced in the lab and commercial enzyme pretreated pulps were 4.17±0.76 and 3.89±0.45 poise, respectively. Enzyme supplementation significantly reduced the viscosity of the pretreated pulps compared to the untreated pulp (5.18±0.96 poise).

The optimized physical parameters at which *B. stearothermophilus* xylanase and *C. subvermispora* laccase are the most effective were used for strategies III and IV, involving commercial enzymes. Pretreated agro-residual pulp was further characterized for compositional analysis (Fig. S2 in File S1). Under optimized conditions, the combination of *B. stearothermophilus* xylanase and *C. subvermispora* laccase along with SA was found to be the most effective among all combinations for removal of interfering residual materials (Fig. S2 in File S1).

Residual laccase activity

Both enzyme samples were highly unstable in the presence of SA (natural mediator) as 94.5% and 82.5% of their activity was lost for *C. subvermispora* and commercial laccase, respectively (Fig. 2). However, the presence of pulp stabilized the enzyme samples leading to only 34.8% and 16.3% loss of activity for *C. subvermispora* and commercial laccase, respectively. The pulp in the reaction mixture efficiently prevented denaturation of the enzymes even after the addition of SA to the reaction mixture, resulting in 42.8% and 28.8% loss of activity for *C. subvermispora* and commercial laccase, respectively. A surfactant (Tween-20, Tween-80 and Triton X-100) and a preservative (glycine) were also used to improve the residual activity of *C. subvermispora* laccase. Addition of Tween-20 (1.5%; v/v) provided extra stability to the *C. subvermispora* laccase leading to only 34.4% loss of activity.

Development of the enzymatic process

With an optimized ED_X of 7.5 U per gram of OD pulp at pH 9.0, strategy I, which involved the use of *B. stearothermophilus* xylanase, resulted in a 20.6% reduction in the consumption of Cl_2 to give brightness identical to that observed for the conventional bleaching process. Strategy II (ED_L = 20 U at pH 5.0 and 50°C), which involves the sequential use of *C. subvermispora* laccase resulted in a 30.2% reduction in the consumption of Cl_2 (Table 3). In order to compare the biobleaching efficiency of *B. stearothermophilus* SDX xylanase (strategy I), commercial Novozyme xylanase (Pulpzyme VLBL) was also used in strategy III, and it resulted in only 17.8% reduction in consumption of Cl_2. On the other hand, when commercial xylanase and laccase were used sequentially in strategy IV, a reduction of 25.3% in Cl_2 consumption was observed (Table 3).

Table 2. Experimental design (coded variables) and results of CCD-RSM for laccase-aided pretreatment of pulp.

Run	RT_L[a]	ED_L[b]	MC[c]	PN_L[d]		B_L[e]		Y_L[f]	
				Actual	Predicted	Actual	Predicted	Actual	Predicted
1	0	0	0	5.69±0.32	5.23±0.53	47.6±4.1	47.8±4.5	12.8±1.2	12.7±1.1
2	+1	−1	−1	5.68±0.32	5.43±0.52	47.5±4.2	49.2±4.5	11.9±1.4	10.6±1.0
3	−1	+1	+1	5.41±0.51	4.76±0.42	48.2±4.2	47.0±4.5	11.5±2.1	11.5±1.3
4	0	−α	0	5.85±0.73	5.23±0.51	46.8±4.2	48.4±5.4	13.2±1.6	12.9±1.4
5	0	0	0	5.69±0.31	6.10±0.42	47.6±4.4	46.0±4.5	12.8±1.3	12.5±1.2
6	−1	+1	−1	5.89±0.72	5.46±0.51	47.6±4.1	46.7±4.5	13.2±1.1	12.3±1.2
7	−1	−1	−1	5.92±0.74	4.88±0.43	48.1±3.8	47.2±4.5	13.8±1.4	11.4±1.0
8	0	+α	0	5.48±0.72	5.46±0.43	48.6±3.9	46.4±4.6	11.3±1.3	11.7±1.7
9	+1	+1	−1	5.41±0.81	5.95±0.44	48.2±3.6	46.6±5.4	11.6±1.2	12.9±1.2
10	+1	−1	+1	6.52±0.83	4.96±0.61	47.6±4.1	47.4±5.5	11.6±1.4	11.5±1.2
11	+1	+1	+1	5.23±0.83	5.66±0.61	48.9±3.6	46.4±4.5	10.9±1.4	12.3±1.3
12	0	0	0	5.69±0.72	5.23±0.52	47.6±3.9	48.4±4.5	12.8±1.5	12.9±1.8
13	+α	0	0	5.36±0.72	5.22±0.52	48.5±4.1	47.0±4.5	11.5±1.4	11.5±1.3
14	0	0	0	5.67±0.72	6.02±0.52	47.6±3.6	47.2±4.5	12.8±1.5	12.6±1.5
15	−1	−1	+1	6.03±0.71	5.23±0.51	46.7±4.1	47.2±4.5	12.9±1.1	12.2±1.6
16	0	0	0	5.69±0.72	5.23±0.33	47.6±3.6	48.4±4.5	12.8±1.2	12.9±1.2
17	0	0	0	5.69±0.71	5.23±0.41	47.6±4.1	48.4±4.5	12.8±1.3	12.9±1.2
18	−α	0	0	6.24±0.83	5.23±0.52	45.9±3.7	47.1±4.4	13.7±1.2	12.1±1.1
19	0	0	+α	5.37±0.71	4.78±0.43	48.9±3.4	48.9±5.4	10.9±1.1	11.2±1.3
20	0	0	−α	5.15±0.72	5.29±0.62	45.4±3.2	47.6±5.4	13.3±1.2	13.0±1.3

a – Retention Time (min); b – Enzyme Dose (U/mg); c – Mediator Concentration (%); d – Permanganate Number; e – Brightness (%ISO) of pulp after *C. subvermispora* laccase mediated pretreatment; f – Yellowness (b*) of pulp after *C. subvermispora* laccase mediated pretreatment.

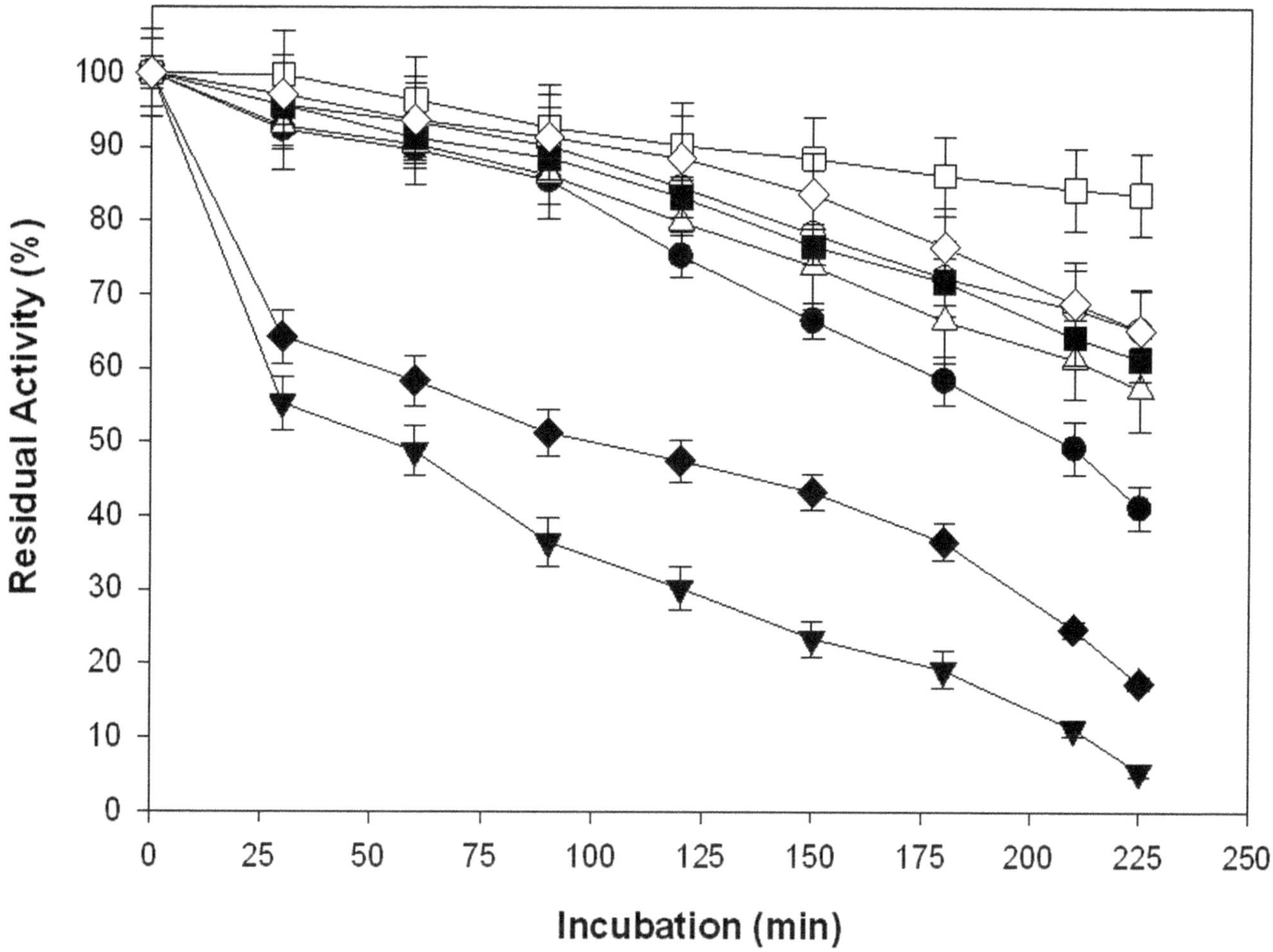

Figure 2. Effect of natural mediator and pulp on the residual activity of laccase during delignification. (●) *C. subvermispora* only; (○) *C. subvermispora* with pulp only; (▼) *C. subvermispora* with SA only; (△) *C. subvermispora* with pulp & SA; (■) Commercial laccase only; (□) Commercial laccase with pulp only; (◆) Commercial laccase with SA only; (◇) Commercial laccase with pulp & SA.

After the chlorination stage, an alkali has to be added to the reaction mixture before the addition of ClO_2. The main purpose of alkali addition is to eliminate the loosened and degraded lignin [19]. The addition of ClO_2 (0.9%) was performed in two steps. An excessive dose of ClO_2 or the addition of the entire volume of ClO_2 in a single step may reduce the strength of the pulp fiber [20]. Finally, sulfur dioxide (SO_2) was added to neutralize the remaining ClO_2 in the reaction mixture since residual ClO_2 might otherwise interfere with the final brightness [21]. By following the $XC_DE_{OP}D_1D_2$ bleaching sequence, an increase of 39.8% and 37.8% in the brightness was achieved after the D_1 stage for strategies I and III, respectively. On the other hand, the $XLC_DE_{OP}D_1D_2$ bleaching sequence resulted in a 48.9% and 44.2% increase in brightness after the same D_1 stage for strategy II and IV, respectively (Fig. S3 in File S1).

For the analysis of the strength of biobleached pulp, hand sheets (70 ± 1 gm^{-2}) were prepared and acclimatized overnight under standard temperature and humidity conditions [22]. Among all the processes, strategy II was the most efficient in improving the strength, while strategy IV was the next most efficient (Table 4). Increases of 13.6% and 12.8% were observed in the tear index ($mN\ m^2/g$) in strategies II and IV, respectively. The improvements in other physical properties such as breaking length and porosity were also observed (Fig. S4 in File S1).

X-ray diffraction (XRD) analysis

The X-ray diffraction analysis of control and enzymatically pretreated wheat straw pulp is shown in Figure S5 in File S1. In all the control and enzymatically pretreated samples, the peak intensity from (002) lattice plane ($2\theta = 22.3°$) represented the crystalline cellulose, while peak intensity of amorphous phase occurred at $2\theta = 15.7°$ [23]. The I_{002} intensities (in cm) of samples pretreated with xylanase produced in the lab and commercial xylanase at $2\theta = 22.3°$ were 5.83 and 5.80, respectively (Table S7 in File S1). For sequentially pretreated substrate at $2\theta = 22.3°$, the intensities were 9.90 and 9.16 for lab enzymes and commercial enzymes (Table S7 in File S1). Similarly I_{am} intensities (in cm) at $2\theta = 15.7°$ are given in Table S5 of File S1 for individual enzyme pretreated and sequentially pretreated biomasses. XRD analysis revealed that all of the strategies, i.e. individual enzyme pretreatment or sequential enzymatic pretreatment increased the crystallinity index of the pulp; however, the highest improvement in crystallinity was observed in the pretreatment strategy involving xylanase and laccase produced in the lab up to 68.6% (Table S7 in File S1).

FTIR spectral analysis

FTIR spectra for single enzyme-pretreated and sequentially pretreated pulp samples showed several characteristic and

Table 3. Different schemes for biobleaching agro-residual pulp with a reduced dose of chlorine*.

Parameters	Control	Strategy - I	Strategy - II	Strategy - III	Strategy - IV
	100%[a]	79.4%[b]	69.8%[c]	82.2%[d]	74.7%[e]
After enzymatic pretreatment stage (X/XL)					
Brightness (%ISO)	43.6	45.8	46.9	45.2	46.4
Yellowness	15.9	13.9	10.9	14.4	11.7
Whiteness	23.5	24.1	25.4	23.8	24.6
Permanganate no	7.1	6.5	5.3	6.9	5.6
Kappa no	11.5	10.8	9.9	11.2	10.4
Chlorine, chlorine dioxide treatment stage (C_D)					
Total Cl_2 added (%)	2.50	1.99	1.75	2.06	1.87
Cl_2:ClO_2	90:10	90:10	90:10	90:10	90:10
Cl_2 Consumed (%)	2.48	1.62	1.52	1.88	1.61
pH	2.5	2.5	2.5	2.5	2.5
Time (min.)	45	45	45	45	45
Brightness (%ISO)	53.6	53.6	57.2	52.7	55.9
Yellowness	11.6	10.6	9.95	11.2	10.4
Whiteness	8.06	8.41	8.82	8.19	8.27
Alkali stage (E_{OP})					
Alkali added (%)	1.8	1.8	1.8	1.8	1.8
Time (min)	120	120	120	120	120
Temp (°C)	70	70	70	70	70
pH	12	12	12	12	12
Brightness (%ISO)	58.4	58.4	59.6	58.2	59.7
Yellowness	9.15	8.85	6.95	9.07	7.92
Whiteness	17.0	18.6	19.5	17.8	18.8
D_1 stage					
ClO_2 added (%)	0.9	0.678	0.628	0.739	0.672
ClO_2 consumed (%)	0.806	0.552	0.504	0.627	0.517
pH	3.41	3.25	3.18	3.42	3.35
Time (min.)	180	180	180	180	180
Temp (°C)	70	70	70	70	70
Brightness (%ISO)	84.5	81.6	88.8	80.2	86.1
Yellowness	2.95	2.66	2.05	2.77	2.12
Whiteness	71.5	75.9	73.9	71.6	70.9
D_2 stage					
ClO_2 added (%)	0.4	0.301	0.279	0.328	0.298
ClO_2 consumed (%)	0.317	0.263	0.191	0.313	0.211
pH	3.85	3.56	3.37	3.54	3.65
Time (min)	180	180	180	180	180
Temp (°C)	70	70	70	70	70
Brightness (%ISO)	88.1	88.8	89.6	88.6	88.4
Yellowness	1.42	1.27	1.01	1.40	1.11
Whiteness	81.6	82.5	83.2	82.1	82.8
SO_2 stage					
Brightness (%ISO)	88.9	89.4	89.9	89.1	89.2

*Volume of elemental chlorine (Cl_2) used in control process was considered as 100%.
a: Pulp without pretreatment with 100% Cl_2; b: Pretreated (with *B. stearothermophilus* xylanase) pulp having 75.4% of Cl_2; c: Pretreated (with *B. stearothermophilus* xylanase and *C. subvermispora* laccase) pulp having 69.8% of Cl_2; d: Pretreated (with commercial xylanase) pulp having 82.2% of Cl_2; e: Pretreated (with commercial xylanase and laccase) pulp having 74.7% of Cl_2.

Table 4. Effect of enzymatic treatment on different physical properties of the agro-based pulp.

Properties	Control	Strategy-I	Strategy-II	Strategy-III	Strategy-IV
Burst factor	38.1±3.9	40.2±4.4	41.4±4.6	39.8±4.1	41.2±4.3
Burst index (kN/g)	3.72±0.42	3.94±0.41	4.06±0.42	3.90±0.41	4.04±0.47
Tear factor	47.1±4.3	51.8±5.3	52.6±5.5	51.3±4.9	52.2±4.7
Tear index (mNm^2/g)	4.62±0.61	5.07±0.62	5.15±0.72	5.03±0.32	5.11±0.72
Double fold (no.)	65.0±8.1	74.0±10.1	76.1±11.2	72.2±8.1	76.0±9.1
Tensile strength (Nm/g)	55.3±5.8	57.2±6.1	58.1±5.35	57.1±5.8	57.9±6.2
Breaking length (m)	5640±230	5830±380	5920±420	5820±380	5910±440
Porosity (sec/100 ml)	86.4±8.3	88.3±7.8	89.1±7.9	87.9±8.4	88.9±9.1

prominent changes (Fig. S6 in File S1). The sequentially pretreated pulp showed absorption structures similar to single enzyme pretreated strategy but with different intensities (Fig. S6 in File S1). The main bands around the region 3,420–3,460 cm^{-1} depict –OH stretching of hydrogen bonding. The increase in their relative intensity after enzymatic treatments is attributed to the increase in cellulosic content of the pulp [24]. The variation in the relative intensity of bands at 2,921–2,917 cm^{-1} and the decrease in the intensity of bands at 607–617 cm^{-1}, assigned to CH asymmetrical stretching vibration in CH_3, CH_2, and CH, in sequentially pretreated pulp indicated the degradation of aliphatic side chains. The relative intensity around 1,617–1,652 cm^{-1} increased in sequentially pretreated pulp, which is attributed to the release of free carbonyl groups (C = O) due to the action of the enzyme on lignin's aromatic ring. In single enzyme pretreated pulp, these carbonyl groups remain associated with aromatic rings resulting in lesser absorption. The relative intensity of sequentially pretreated pulp decreased at 1,436–1,416 cm^{-1}, a band assigned to aromatic skeletal vibrations combined with –CH_3 in-plane deformations, showing that some methoxy groups were removed during the enzymatic treatment. The band at 1,320–1,382 cm^{-1} was assigned to aliphatic C–H stretching in CH_3 (not in –OCH_3) and phenol-OH group. The decrease in its relative intensity illustrated that either the side chains or phen–OH of lignin decreased after the enzymatic treatment. The decrease in relative intensity at 1,266–1,252 cm^{-1} indicates the degradation of guaiacyl groups. A new band that appeared at 1,737 cm^{-1} in pulp pretreated with lab produced xylanase was assigned to C = O stretching vibration in β-C = O, COOH, and ester, indicating that residual lignin after an enzymatic pretreatment was enriched in these types of functional groups because of the removal of xylan. The changes in pulp crystallinity were derived from A1430/A897 ratio in accordance with the X-ray diffraction studies [24].

Characterization of effluents

Strategies II and IV resulted in 60.1% and 47.9% reduction in BOD, respectively (Table S8 in File S1), compared to the control process. These results were further supported by the color reduction values. The highest (25.8%) and lowest (12.1%) color reductions were observed with strategies II and III, respectively (Table S8 in File S1). The observed reduction in the color of the effluent was higher than that reported previously, where 11% reduction was reported with Pulpzyme HC [25]. Similarly, 60.1% reduction in BOD was observed with strategy II compared to 23.0% reduction with Kraft pulp in the previous report [26].

Acute toxicity assay

The chemical intensive control process significantly inhibited the bioluminescence of *Vibrio fischeri* (Fig. 3a, b), as shown by the lower $EC_{50\text{-}5min}$ and $EC_{50\text{-}15min}$ values (Table 5). However, strategy I using *B. stearothermophilus* SDX xylanase more significantly reduced the acute toxicity of the mixed effluent (Fig. 3c, d), as shown by EC_{50} values that were more than double the EC_{50} of the control process (Table 5). Biological processes using Pulpzyme VLBL also reduced the acute toxicity (Fig. 3e, f); however, it was less efficient than the enzyme system produced in the lab (Table 5). Similar TOU (%) results were observed with these enzymes. A 2-fold improvement in TOU (%) was observed for strategy I compared to the control process, whereas the commercial xylanase resulted in 53.6% improvement in TOU (%). A slight increase in acute toxicity was observed with strategies involving the sequential application of xylanase and laccase compared to the single enzyme systems. In fact, each of these strategies alone, i.e., strategy II and IV, was significantly better than the control process in terms of the EC_{50} and TOU (%) values (Table 5). Sequential application of xylanase and laccase produced in lab, reduced the acute toxicity by 43.4–50.0% (Fig. 3 g, h) compared to the control process at different sampling times. In contrast, strategies involving commercial enzymes reduced the acute toxicity by 32.7–38.4% (Fig. 3i, j) compared to the control process at different sampling times (Table 5).

Discussion

Model for pulp pretreatment

Owing to the differences in the optimal pH values for the two enzymes and the inhibition of the activity of the mediator system under alkaline conditions, the pulp was subjected to sequential application of xylanase and laccase enzymes. Pretreatment was carried out with xylanase first because the mediators used during laccase pretreatment might generate free radicals capable of hindering the hydrolytic activity of xylanase [18]. Three-dimensional contour plots were also drawn, and it was observed that in xylanase-aided pretreatment, significant decrease in permanganate number was observed (Fig. S7 in File S1). On the other hand, a significant increase in the brightness was observed after the sequential application of enzymes (Fig. S8 in File S1). This might be because all the inhibitory components, which were generated after the use of individual enzymatic pretreatment, were neutralized in the sequential treatment process. Although numerous reports are available on the sequential use of xylanase and laccase

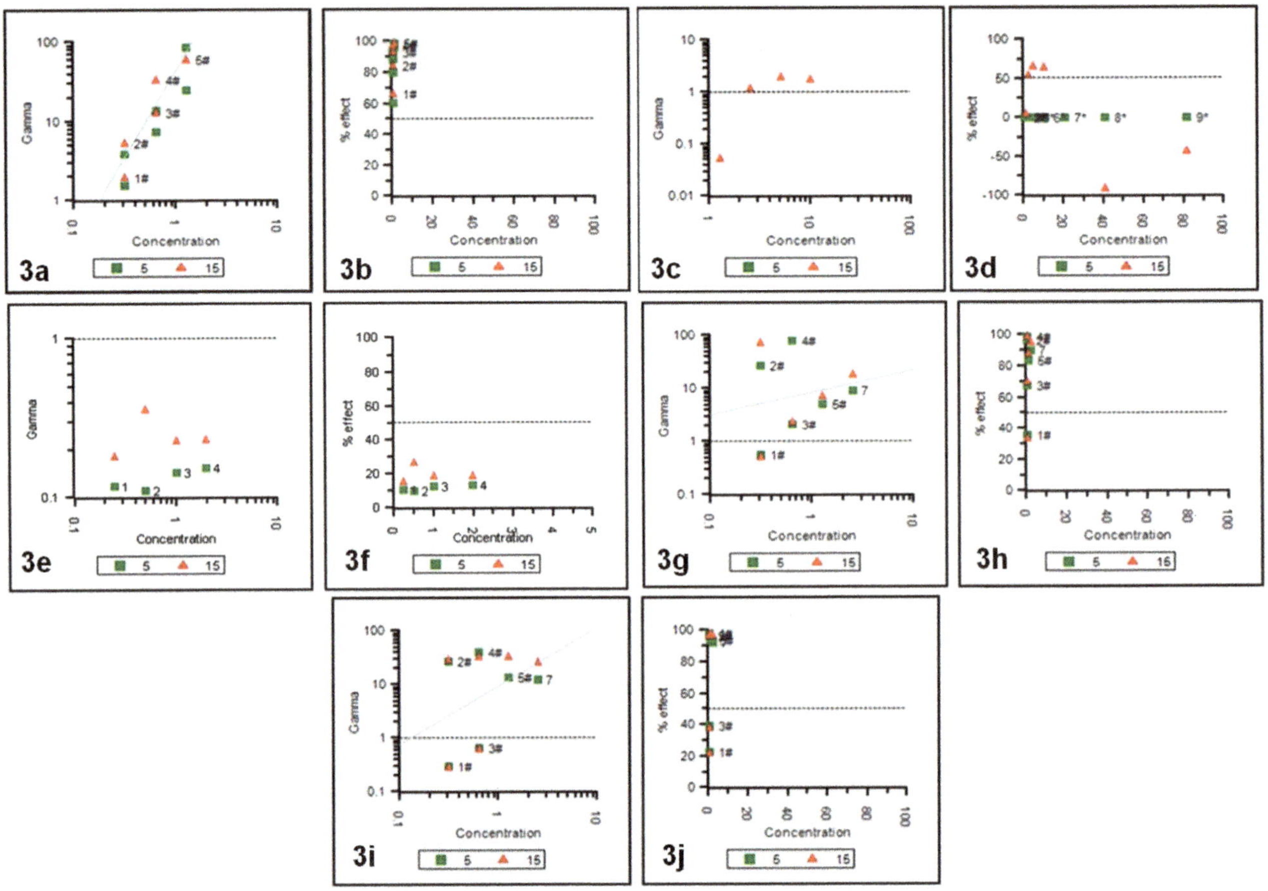

Figure 3. Gamma effect and EC_{50} values for (a, b) control; (c, d) strategy I; (e, f) strategy II; (g, h) strategy III; (i, j) strategy IV. The EC calculated by log linear plotting of Concentration (C) vs. % Light Decrease (%Δ), or more precisely by plotting Gamma (Γ) (which is the corrected ratio of the amount of light lost to the amount of light remaining) versus Concentration.

for pulp pretreatment [3,18], this is the first report on the sequential application of xylanase and laccase for biobleaching of the agro-residual pulp.

Natural mediator and laccase pretreatment

SA was selected as the mediator because of its plant origin and its higher efficiency to decolorize the Azure-B (Fig. S1 in File S1). Further, the improvement observed in the residual activity of the

Table 5. Reduction in acute toxicity of effluent samples determined through Microtox analyzer and represented by the average values of EC_{50} and unit of toxicity (TOU)*.

Sample	Parameter	EC_{50} (%)	TOU (%)
Control	$EC_{50\text{-}5min}$	39±4	221±26
	$EC_{50\text{-}15min}$	32±3	
Strategy I	$EC_{50\text{-}5min}$	84±9	119±14
	$EC_{50\text{-}15min}$	82±9	
Strategy II	$EC_{50\text{-}5min}$	69±7	152±19
	$EC_{50\text{-}15min}$	64±7	
Strategy III	$EC_{50\text{-}5min}$	72±7	134±16
	$EC_{50\text{-}15min}$	69±8	
Strategy IV	$EC_{50\text{-}5min}$	58±6	166±21
	$EC_{50\text{-}15min}$	52±6	

*$EC_{50\text{-}5min}$: determination of inhibition in bioluminescence after 5 min of incubation; $EC_{50\text{-}15min}$: determination of inhibition in bioluminescence after 15 min of incubation; TOU (%) determination of toxicity unit after 15 min of incubation.

enzyme in the presence of pulp could be due to the cellulosic content of the pulp. Pulp fibers might provide an additional substrate for free radicals generated by the action of the natural mediator. The rate at which these free radicals target the enzyme is decreased owing to the presence of the pulp fiber; therefore, the residual enzyme activity increased. A similar observation was also reported by Fillat et al. [27]. Use of natural mediators will be beneficial in pollution reduction because the higher concentration of free radicals generated due to the action of synthetic mediators that make the whole bleaching process highly toxic and raise environmental concerns [19,28]. Since commercial laccase has lower specific activity, it was less effective for delignification compared to *C. subvermispora* laccase, which has higher specific activity.

Properties of pulp obtained from enzymatic pretreatment strategies

Improvement in the strength after enzymatic pretreatment is directly related to the hydrolytic action of xylanase. Hydrolytic enzymes disrupt the surface of cellulosic fiber and generate microfibrils, which in turn lead to a cross-networked, condensed packing of pulp fibers and hence they give extra strength [22]. Since only xylanase has hydrolytic properties, even sequential application of oxidizing enzymes (strategies II and IV) did not improve the strength significantly (Fig. S4 in File S1). Although commercial xylanase alone (Ecopulp, Cartazyme NS-10, and Pulpzyme HC) has been used previously [20], no report is available on the sequential application of xylanase and laccase for processing wheat straw pulp.

Improvement in crystallinity means a decrease in amorphous cellulose and increase in crystalline cellulose of the pulp. The increase in crystallinity might be due to the removal of hemicelluloses and lignin and components adhered to lignin as a result of sequential pretreatment, thereby increasing the cellulose content of the pulp [24]. Sequential pretreatment strategies involve separate laccase supplementation, and this is probably the reason for higher increase in crystallinity in sequential approach than in single enzyme strategy.

Reduction in pollution load

Achieving less or non-toxic discharge from pulp and paper industry is the biggest challenge in the current scenario. Specific characteristics of the effluents such as BOD and color were determined, and strategy II was found to be the most effective in reducing the pollution load (color), resulting in 25.8% reduction. This might be because of the sequential use of *B. stearothermophilus* SDX xylanase and *C. subvermispora* laccase that removes a significant amount of lignin during pretreatment, which is further removed in the subsequent washes. Washing away the oxidized ligneous material and other hemicellulosic waste reduces the consumption of oxidizing chemicals during bleaching, thereby resulting in effluents with less BOD and colored units. Lesser pollution reduction efficiency of strategy III and IV could be attributed to the presence of surfactants and preservatives in commercial enzyme samples [20]. Media components present in crude enzymes produced in the lab do not interfere with the catalytic properties of the enzymes and hence do not alter their pollution reduction efficiencies [29]. The use of natural mediators in sequential enzymatic approach resulted in better performance in terms of reduction in pollution load. Therefore, the costs associated with the use of natural mediators can be compensated.

Reduction in acute toxicity

The main reason for the reduced acute toxicity observed with the biological processes was the reduced consumption of oxidizing materials during bleaching. However, a variation was observed in the efficiency of the enzymes produced in lab and commercial enzyme systems. This difference in the efficiency could be explained by the presence or absence of chelators and preservatives. Commercial enzymes contain unidentified preservatives and other compounds to preserve their activity for an extended period of time, and these compounds have been shown to be crucial during the determination of acute toxicity and units of toxicity (http://www.brenntagspecialties. com/en/downloads/Products/Food/Novozymes/Cellulast_1_5L.pdf). Application of commercial enzymes liberates these compounds in the effluent and subsequently increases the biological load and hence the acute toxicity (http://www.sfm.state.or.us/cr2k_subdb/MSDS/CELLUCLAST. PDF). This could explain the lower toxicity of effluents generated using enzymes produced in lab.

Although the sequential application of xylanase and laccase efficiently removes the ligneous material during pretreatment, augmentation of the reaction by mediators caused an increase in the biological load of the effluent generated from these biological processes. Although the mediator was from a natural source, its addition still caused the formation of free radical compounds [13]. Once they formed, they immediately interact with other organic compounds to form highly stable toxicants, which increased the acute toxicity. Because of their natural origin, the concentration of the generated free radical compounds was much less than the intermediates and toxicants generated by the application of synthetic organic mediators [30]. Therefore, application of natural mediator is preferred over synthetic mediator.

Conclusions

In the current study, sequential biobleaching of agro residual pulp has been proved to be a beneficial strategy compared to the conventional bleaching sequence. Significant reduction in the release of toxic organic compounds was achieved by incorporating sequential enzymatic pretreatment strategy with the conventional bleaching ($C_DE_{OP}D_1D_2$) process. Sequential pretreatment strategy combined with natural mediators further improved the eco-friendly nature of the process. Significant reduction in ecotoxicity also suggests the potential of this approach in the development of green technology for pulp and paper industry.

Supporting Information

File S1 Supporting equations, figures, and tables. Equation S1, Second order polynomial equation showing pretreatment of agro-residual pulp with *B. stearothermophillus* xylanase. A, B, C and D are independent variables corresponds to pH, temperature (°C), enzyme dose (U/mg) and retention time (min), respectively. Y_1, Y_2 and Y_3 are final response for permanganate number (PN_X), brightness (B_X) and yellowness (Y_X) after pretraetment of agro-residual pulp with *B. stearothermophillus* xylanase. **Equation S2,** Second order polynomial equation showing pretreatment of agro-residual pulp with *C. subvermispora* laccase. where E, F and G are independent variables corresponds to enzyme dose (U/mg), mediator conc. (%) and retention time (min), respectively Y_4, Y_5 and Y_6 are final response for permanganate number (PN_L), brightness (B_L) and yellowness (Y_L) after pretreatment of agro-residual pulp with *C.*

subvermispora laccase. **Figure S1,** Decolorization of Azure-B using natural mediators (syringaldehyde and acetosyringone) and synthetic mediator (1-hydroxybenzatriazole) with *C. subvermispora* and the commercial laccase. *C. subvermispora* laccase with HBT (•), SA (▼), AS (■); the commercial laccase with HBT (○), SA (△) and AS (□). **Figure S2,** Effect of different enzymatic treatments on lignin removal. BP: before pulping; AP: after pulping; X: after xylanase treatment; X_L: after sequential use of xylanase and laccase; X_C: after treatment with commercial xylanase; X_C+L_C: after sequential treatment with commercial xylanase and laccase. **Figure S3,** Brightness (%ISO) value of the agro-residual pulp at different bleaching stages for the different strategies. Enz: after enzymatic pretreatment; C_D: after chlorination stage; E_{OP}: after alkaline peroxide stage; D_1: after first ClO_2 stage; D_2: after second ClO_2 stage; SO_2: after final treatment with SO_2. **Figure S4,** Improvement in the different strength properties of hand sheets made from the agro-residual pulp for various enzymatic treatment strategies. BF: burst factor; BI: burst index (kN/g); TF: tear factor; TI: tear index (mNm^2/g); DF: double fold (number); TS: tensile strength (Nm/g); BL: breaking length (m); P: porosity (sec/100 ml). **Figure S5,** X-ray diffraction analysis of (a) control sample and enzymatically pretreated sample after (b) strategy I; (c) strategy II; (d) strategy III and (e) strategy IV. **Figure S6,** FTIR spectral analysis of (a) control sample and enzymatically pretreated sample after (b) strategy I; (c) strategy II; (d) strategy III and (e) strategy IV. **Figure S7,** Three dimensional contour plot for the reduction in the permanganate number. *pH and temperature were used as independent variables for xylanase-aided pretreatment. **Figure S8,** Three dimensional contour plot for the increase in the brightness. *enzyme dose and retention time were considered as independent variables for laccase-mediated pretreatment. **Table S1,** Study of interaction among the various nutritional and physical parameters on xylanase production through Plackett-Burman design. #xylanase activity in Unit per gram of dry bacterial bran. *Experiments were performed at pH 8.0. a – Beef extract (%); b – Galactose (%); c – KNO_3 (%); d – Temperature (°C); e – Incubation time (h); f – Substrate (wheat bran) concentration (%); g – Particle (wheat bran) size (mm); h – Peptone (%); i – Yeast extract (%); j – Dummy variable (blank). **Table S2,** Different levels of variables affecting the xylanase production through CCD-RSM modeling. **Table S3,** Study of interaction among different levels of variables and their effect on xylanase production under SSF. *Xylanase activity in Units per gram of dry wheat bran. a – Peptone (%); b – KNO_3 (%); c – Temperature (°C); d – Incubation time (h). **Table S4,** List of values of independent variables used for the pretreatment of the agro-residual pulp. **Table S5,** Effect of various temperatures on different properties of wheat straw pulp. *Experimentation was carried out with 15 U/mg of laccase from *C. subvermispora* using 2% of ABTS for 240 min of incubation at pH 5.5. a – temperature in °C; b – permanganate number of pulp; c – Brightness (%ISO) of pulp; d – yellowness (b*) of pulp. **Table S6,** Effect of various pH values on different properties of wheat straw pulp. *Experimentation was carried out with 15 U/mg of laccase from *C. subvermispora* using 2% of ABTS for 240 min of incubation at 50°C. a – permanganate number of pulp; b – Brightness (%ISO) of pulp; c – yellowness (b*) of pulp. **Table S7,** X-ray diffraction analysis of wheat straw pulp after different enzymatic pretreatments. *crystallinity index. X_C represents pretreatment with commercial xylanase; X_L represent pretreatment with lab-scale xylanase; X_C+L_C represents sequential pretreatment with commercial xylanase followed by commercial laccase; X_L+L_L represents sequential pretreatment with lab-scale xylanase followed by lab-scale laccase. **Table S8,** Characterization of effluents from the entire process for different enzymatic combinations.

Author Contributions

Conceived and designed the experiments: SSD JS JKL. Performed the experiments: SSD GG. Analyzed the data: SSD JKL. Contributed reagents/materials/analysis tools: YCK. Wrote the paper: SSD VCK YCK JKL.

References

1. Garg G, Dhiman SS, Mahajan R, Kaur A, Sharma J (2011) Bleach-boosting effect of crude xylanase from *Bacillus stearothermophilus* SDX on wheat straw pulp. New Biotechnol 28: 58–64.
2. Yadav R, Chaudhary S, Dhiman SS (2010) Biopulping and its potential to reduce effluent loads from bleaching of hardwood kraft pulp. BioRes 5: 159–171.
3. Valls C, Vidal T, Roncero MB (2010) Boosting the effect of a laccase mediator system by using a xylanase stage in pulp bleaching. J Hazard Mat 177: 586–592.
4. Fillat U, Roncero MB (2009) Biobleaching of high quality pulps with laccase mediator system: influence of treatment time and oxygen supply. Biochem Eng J 4: 193–198.
5. Johannes C, Majcherczyk A (2000) Natural mediators in the oxidation of polycyclic aromatic hydrocarbons by laccase mediator systems. Appl Environ Microbiol 66: 524–528.
6. Dhiman SS, Sharma J, Battan B (2008) Pretreatment processing of fabrics by alkalothermophilic xylanase from *Bacillus stearothermophilus* SDX. Enz Microb Technol 43: 262–269.
7. Plackett RL, Burman JP (1946) The design of optimum multifunctional experiments. Biometals 33: 305–32.
8. Horikoshi K (1979) Production and industrial application of beta-cyclodextrin. Process Biochem 14: 26.
9. Bailey MJ, Biely P, Poutanen K (1992) Interlaboratory testing of methods for assay of xylanase activity. J Biotechnol 23: 257–270.
10. Miller GL (1959) Use of Dinitrosalicylic Acid Reagent for Determination of Reducing Sugar. Anal Chem 31: 426–428.
11. Bourbonnais R, Paice MG (1990) Oxidation of non-phenolic substrates. An expanded role for laccase in lignin biodegradation. FEBS Lett 267: 99–102.
12. Bradford MM (1976) A rapid and sensitive method for the quantitation of microgram quantities of protein utilizing the principle of protein-dye binding. Anal Biochem 72: 248–254.
13. Camarero S, Ibarra D, Martinez DT, Romero J, Gutierrez A, et al. (2007) Paper pulp delignification using laccase and natural mediators. Enzyme Microb Technol 40: 1264–1271.
14. Sluiter A, Hames B, Hyman D, Payne C, Ruiz R, et al. (2008) Determination of structural carbohydrates and lignin in biomass. National Renewable Energy Laboratory (NREL), Golden Company.
15. TAPPI Test Methods (1996) Technical association of the pulp and paper industry. Atlanta, GA, TAPPI Press.
16. An J, Jeong S, Moon HS, Jho EH, Nam K (2012) Prediction of Cd and Pb toxicity to *Vibrio fischeri* using biotic ligand-based models in soil. J Hazard Mater 203–204: 69–76.
17. Valls C, Vidal T, Roncero MB (2010) Boosting the effect of a laccase-mediator system by using a xylanase stage in pulp bleaching. J Hazard Mater 177: 586–592.
18. Aracri E, Vidal T (2011) Xylanase- and laccase-aided hexenuronic acids and lignin removal from specialty sisal fibres. Carbo Poly 83: 1355–1362.
19. Ibarra D, Romero J, Martínez M, Martínez A, Camarero S (2006) Exploring the enzymatic parameters for optimal delignification of eucalypt pulp by laccase mediator. Enzyme Microb Technol 39: 319 1327.
20. Vicuna R, Escobar F, Osses M, Jara A (1997) Bleaching of eucalyptus Kraft pulp with commercial xylanases. Biotechnol Lett 19: 575–578.
21. Dhiman SS, Garg G, Mahajan R, Garg N, Sharma J (2009) 'Single lay out' and 'mixed lay out' enzymatic processes for bio-bleaching of kraft pulp. Bioresour Technol 100: 4736–4741.
22. Lundquist L, Willi F, Leterrier Y, Manson JAE (2004) Compression behavior of pulp fiber networks. Poly Eng Sci 44: 45–55.
23. Kaushik A, Singh M (2011) Isolation and characterization of cellulose nanofibrils from wheat straw using steam explosion coupled with high shear homogenization. Carbo Res 346: 76–85.
24. Virk AP, Puri M, Gupta V, Capalash N, Sharma P (2013) Combined enzymatic and physical deinking methodology for efficient eco-friendly recycling of old newsprint. PLOS ONE 8: 10.137/journal.pone.0072346.

25. Sudha B, Veeramani H, Sumathi S (2003) Bleaching of bagasse pulp with enzyme pre-treatment. Water Sci Technol 147: 163–168.
26. Yadav RD, Chaudhry S, Gupta S (2012) Novel application of fungal *Phanerochaete* sp. and xylanase for reduction in pollution load of paper mill effluent. J Environ Biol 33: 223–226.
27. Fillat A, Colom JF, Vidal T (2010) A new approach to the biobleaching of flax pulp with laccase using natural mediators. Bioresour Technol 101: 4104–4110.
28. Amann M (1997) The lignozym process coming closer to the mill. In: Nineth ISWFPC Protocol. Volume F4. pp. 1–5.
29. Lefebvre V, Vaes G (1989) The enzymatic evaluation of procollagenase and collagenase inhibitors in crude biological media. Biochem Biophys Acta 992(3): 355–361.
30. Camarero S, Ibarra D, Martinez MJ, Martinez AT (2005) Lignin-Derived compounds as efficient laccase mediators for decolorization of different types of recalcitrant dyes. Appl Environ Microbiol 71: 1775–1784.

6

Final Opportunity to Rehabilitate an Urban River as a Water Source for Mexico City

Marisa Mazari-Hiriart[1]*, Gustavo Pérez-Ortiz[1,2], María Teresa Orta-Ledesma[3], Felipe Armas-Vargas[4], Marco A. Tapia[2,5], Rosa Solano-Ortiz[5], Miguel A. Silva[6], Isaura Yañez-Noguez[3], Yolanda López-Vidal[7], Carlos Díaz-Ávalos[8]

1 Laboratorio Nacional de Ciencias de la Sostenibilidad, Departamento de Ecología de la Biodiversidad, Instituto de Ecología, Universidad Nacional Autónoma de México, México, D.F., México, **2** Departamento de Biología, Facultad de Ciencias, Universidad Nacional Autónoma de México, México, D.F., México, **3** Coordinación de Ingeniería Ambiental, Instituto de Ingeniería, Universidad Nacional Autónoma de México, México, D.F., México, **4** Posgrado en Ciencias de la Tierra, Instituto de Geología, Universidad Nacional Autónoma de México, México, D.F., México, **5** Posgrado en Ciencias Biológicas, Facultad de Ciencias-Instituto de Ecología, Universidad Nacional Autónoma de México, México, D.F., México, **6** Posgrado de Ciencias Bioquímicas, Departamento de Ingeniería Celular y Biocatálisis, Instituto de Biotecnología, Universidad Nacional Autónoma de México, Cuernavaca, Morelos, México, **7** Programa de Inmunología Molecular Microbiana, Departamento de Microbiología y Parasitología, Facultad de Medicina, Universidad Nacional Autónoma de México, México, D.F., México, **8** Departamento de Probabilidad y Estadística, Instituto de Investigaciones en Matemáticas Aplicadas y en Sistemas, Universidad Nacional Autónoma de México, México, D.F., México

Abstract

The aim of this study was to evaluate the amount and quality of water in the Magdalena-Eslava river system and to propose alternatives for sustainable water use. The system is the last urban river in the vicinity of Mexico City that supplies surface water to the urban area. Historical flow data were analyzed (1973–2010), along with the physicochemical and bacteriological attributes, documenting the evolution of these variables over the course of five years (2008–2012) in both dry and rainy seasons. The analyses show that the flow regime has been significantly altered. The physicochemical variables show significant differences between the natural area, where the river originates, and the urban area, where the river receives untreated wastewater. Nutrient and conductivity concentrations in the river were equivalent to domestic wastewater. Fecal pollution indicators and various pathogens were present in elevated densities, demonstrating a threat to the population living near the river. Estimates of the value of the water lost as a result of mixing clean and contaminated water are presented. This urban river should be rehabilitated as a sustainability practice, and if possible, these efforts should be replicated in other areas. Because of the public health issues and in view of the population exposure where the river flows through the city, the river should be improved aesthetically and should be treated to allow its ecosystem services to recover. This river represents an iconic case for Mexico City because it connects the natural and urban areas in a socio-ecological system that can potentially provide clean water for human consumption. Contaminated water could be treated and reused for irrigation in one of the green areas of the city. Wastewater treatment plants and the operation of the existing purification plants are urgent priorities that could lead to better, more sustainable water use practices in Mexico City.

Editor: Judi Hewitt, University of Waikato (National Institute of Water and Atmospheric Research), New Zealand

Funding: The authors would like to acknowledge the financial support of Gobierno del Distrito Federal, Secretaría de Medio Ambiente in conducting the project Plan Maestro de Manejo Integral y Aprovechamiento Sustentable de la Cuenca del Río Magdalena del Distrito Federal, through the Programa Universitario de Estudios sobre la Ciudad, Universidad Nacional Autónoma de México (2008–2009), and the project Sistema de Indicadores para el Rescate de los Ríos Magdalena y Eslava, Distrito Federal through the Programa Universitario de Medio Ambiente, UNAM (2009–2010). As well as the support of the Dirección General de Asuntos del Personal Académico through the Programa de Apoyo a Proyectos de Investigación e Innovación Tecnológica, PAPIIT IN219809. The funders had no role in study design, data collection and analysis, decision to publish, or preparation of the manuscript.

Competing Interests: The authors have declared that no competing interests exist.

* Email: mazari@unam.mx

Introduction

In megacities with populations exceeding ten million inhabitants, water is a scarce resource that is in high demand. The exploration and application of new, more sustainable water use and reuse approaches are urgently needed. Water managers must change the perspectives and policies that assume that water can only be used once by considering the benefits obtained from the natural environment surrounding the urban areas.

In emerging economies, cities grow before sufficient hydraulic infrastructure has been developed to handle the water supply and wastewater disposal requirements [1]. This lack of infrastructure significantly affects the surface and groundwater systems, making waterways vulnerable to pollution and leading to deleterious effects on water sources and human health. There is a need for more adequate approaches regarding processes that occur at the basin level, in addition to directly considering hydrological ecosystem services, such as water provision and water quality.

During the evolution of what is now the Mexico City Metropolitan Area (MCMA), lakes and rivers have been transformed into drains or converted into sewers in the vicinity of modern avenues, and freeways transit over piped waterways [2]. The hydraulic system of the Basin of Mexico, where the MCMA and its 22 million inhabitants are located, has been irreversibly altered. Of the 45 rivers in this region, most are piped to avoid flooding and unsanitary conditions.

In the Basin of Mexico, the current water demand is 77 m^3/s for first use, of which 71% is groundwater extracted through wells, 2% is derived from springs and surface water, 21% is from the Cutzamala surface water system, and 6% is from the Lerma groundwater system [3,4]. The subject of this research, the Magdalena-Eslava river system, represents the most important surface water source for human use in Mexico City.

Rivers flow from their origins in high areas to low-lying areas, exposing these water bodies to the integrated effects of human activities along their course and throughout the basin landscape. Anthropogenic impacts have impaired the ability of many river ecosystems to provide the goods and services upon which society depends. Watercourses may be degraded and rendered unable to support healthy aquatic communities such that natural processes are hampered and social value is lost.

Balancing the interactions between natural and constructed systems in urban areas is crucial for the future supply of water for large human settlements. Interest in river rehabilitation and restoration exists at the global level. River restoration is defined as an improvement in the integrity of a river through human intervention. The aim of river restoration is either to recover the hydrologic and ecological processes or to achieve rehabilitation, which means recovering a system that has not returned to its original state or is as healthy as it would have been if it had been fully restored [5].

The Magdalena-Eslava river system is one of the last non-piped rivers remaining in the MCMA. Given its contribution to the water supply in its upper reaches and the potential to recover some of the environmental services provided by the system, there is significant interest in the possibility of improving the water quality in the lower parts of its course. The aim of this study was to evaluate the Magdalena-Eslava river system in terms of the amount and quality of water, based on physicochemical and microbiological attributes that it offers the community. Accordingly, the study was performed based on historical and current data. As one of the last urban rivers, this study provides an opportunity to pose alternatives for more efficient water use.

Study area description

The study area, which is located in Mexico City, comprises the Magdalena and Eslava rivers. The river system covers two climatic regions. The lower area, which reaches to 3,050 masl, presents a temperate sub-humid climate with a summer rainy season and average temperatures that oscillate between 12 and 18°C. The upper basin has a semi-cold climate and a rainy season during the summer in areas higher than 3,050 masl, with an annual mean temperature between 5 and 12°C [6]. The rainy season ranges from May to October, with a monthly precipitation in excess of 250 mm in July [7]. The annual precipitation in most of the area is between 1,200 and 1,500 mm [8].

The Magdalena River originates in *Sierra de Las Cruces* from springs and runoff from the *La Palma*, *El Gavilán* and *El Muñeco* mountains in a location called *Cieneguillas*. The Magdalena River Basin is actually a sub-basin of the Basin of Mexico (hereafter called the Magdalena River Basin) and belongs to what is legally known as the *Zona Protectora Forestal Cañada de Contreras*, which was declared in 1932. The protected zone includes a fringe of 12 km from the river's origin, with a 500 m buffer on each side of the river, as declared in 1947 [9,10].

The Magdalena river basin extends from 2,570 to 3,870 masl [11], covering an area of 29.80 km^2. The Eslava River, which is the basin's largest tributary, flows through an area of 24.05 km^2, with an altitude descending to 2,510 masl at the urban limit.

The Magdalena River is 28.2 km long and flows 14.8 km through a natural forested area known as the Conservation Soil, with an additional 13.4 km crossing through the urban area. In the urban section, 8.8 km of river flows as an open canal, and 4.6 km is piped [12]. The Eslava River is 13.7 km long, with 10.96 km in the Conservation Soil and 2.74 km in urban areas [12]. The basins of the Magdalena and Eslava rivers are shown in Figure 1.

The Magdalena Basin, known as *Cañada de Contreras* in *Sierra de las Cruces*, is the only drainage of significant magnitude [13] within the Basin of Mexico and is primarily classified as temperate sub-humid. Despite its location in a peri-urban area, the Magdalena Basin serves as an important refuge for biodiversity and is home to 1,175 species, of which 212 are considered useful and 39 are listed in the at-risk category [14].

Three vegetation communities are present in the area, according to Ávila-Akerberg (2009) [8] and Nava (2003) [15]. The dominant vegetation types are *Abies religiosa* forest (46%) from 2,750 to 3,500 masl, *Pinus hartwegii* forest (29%) from 3,420 to 3,800 masl, *Quercus* (8.3%) and mixed forest (1.3%) from 2,600 to 3,370 masl, grassland (7.2%), and cloud forest (0.2%). Each type of forest produces different rates of runoff, as calculated by Jujnovsky et al. (2010) [16]. The *Abies* area produces 10,944,800 m^3 of water per year, the *Pinus* forest produces 6,878,000 m^3 of water per year, and the mixed oak forest generates 3,217,500 m^3 of water per year.

The natural vegetation cover has been altered as a result of human activities, resulting in an increase in erosion and the reduction of water and soil humidity retention. Therefore, water flows rapidly during flood events in the lower parts of the basin [13]. The forested area southwest of Mexico City represents the most important surface water source and groundwater recharge area. This area supplies water and regulates both flood events and the microclimate in the MCMA. Urban expansion represents a threat to the Conservation Soil area, which currently provides ecosystem services, such as fresh water.

At the request of the City, the *Gobierno del Distrito Federal* through its Environmental Ministry (*Secretaría del Medio Ambiente*), the two main universities located in Mexico City, the National Autonomous University of Mexico (UNAM) and the Metropolitan Autonomous University (UAM), generated a Master Plan to Rescue the Magdalena and Eslava Rivers [12]. This plan was developed by an interdisciplinary team that included engineers, architects, ecologists, biologists, architects, geographers, and sociologists, who proposed to rehabilitate the river as an urban area that could be used as a community recreation area while recovering the water quality and the watercourse running through the city [12]. A second project developed a system of indicators to measure the improvements resulting from the actions taken as a result of the Master Plan [17], which was not completely implemented.

The topic of river ecosystem recovery and rehabilitation has grown in relevance, receiving increasing attention in the last two decades. From an urban perspective, rivers are crucial for water supply, flood and drought regulation, storm-water management, water quality improvement, and erosion control; additionally, rivers serve as urban wildlife refuges. Overall, rivers can serve a range of purposes from supplying water to offering locations for social and recreational activities; hence, rivers often account for a significant fraction of the few recreational green areas in large cities [18,19].

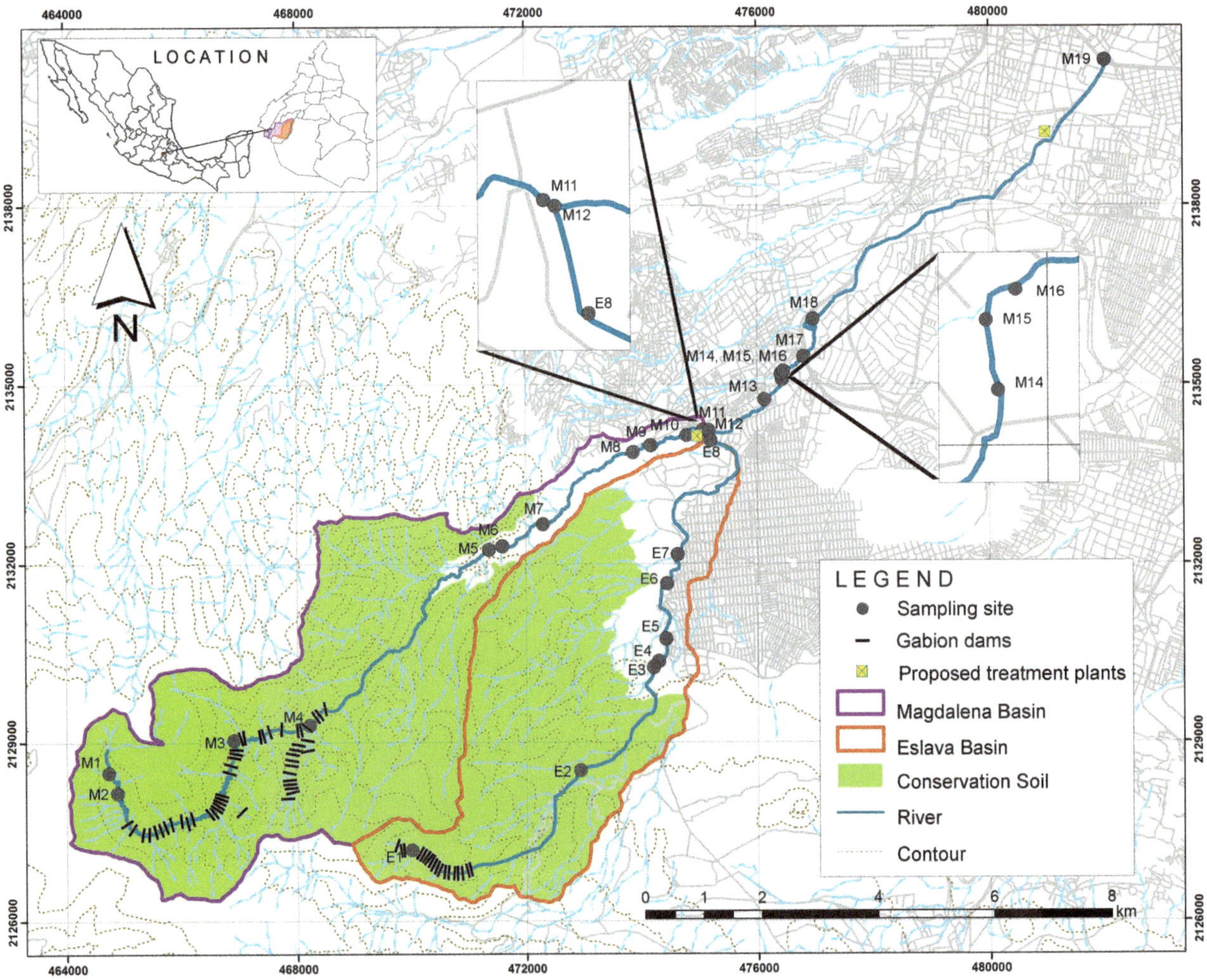

Figure 1. Study area, including the Magdalena and Eslava rivers within the Basin of Mexico and the Mexico City Metropolitan Area.

Methods

To determine the water flow, historical data were collected and analyzed (1973–2010); this approached allowed us to discern important variations due to seasonal or management effect in the river system. Based on current water quality measurements (2008–2012), we describe the actual condition of the river based on physicochemical and microbiological variables. The methods described in this section were applied.

Water flow

A flow analysis was conducted based on historical data recorded at the Santa Teresa hydrometric station (26440) to understand the water quantity and variability of flow at the only existing station of the National Water Commission. This hydrometric station is located after the junction of the Magdalena and Eslava rivers. As mentioned above, the Eslava is the main tributary of the Magdalena River, which also receives inflows from additional small creeks and natural run-off, combined with the discharge of wastewater from the surrounding urban area.

Daily information was obtained from the official flow surveys recorded at the Santa Teresa hydrometric station (26440) and was published in the National Data Bank of Surface Water [20] of the National Water Commission (*Comisión Nacional del Agua*). The data were analyzed for consistency and to detect and correct deviations from the historic series based on the double mass curve method [21,22]. The series of daily data spanning 38 years were then plotted, and two periods for the flow regime were identified.

We calculated the average monthly flow for each regime, i.e., the natural and the regulated areas, so we could compare the regulating effect of water infrastructure on the river. Finally, we plotted the average, minimum and maximum annual flows to observe the river's evolution over time.

Experimental design

After a general survey of the area, sampling sites for water quality evaluation were selected at places with little human or animal influence in the natural area and at others with inputs of urban runoff and wastewater in the urban area. Based on an initial diagnosis performed in 2008 and 2009 [12,17,23], the sampling locations were evaluated for redundancy in terms of the information provided. We computed a correlation matrix for the 19 sampling locations. After 2009, the sampling protocols included ten sampling sites. If two locations had a Pearson's correlation above 0.6 and were neighbors of each other, one was eliminated. This procedure allowed us to reduce the number of sampling locations from 19 to eight stations. We then added two sampling locations a few meters upriver of the intake of the water treatment plants that are located along the river. These ten stations were sampled from 2010 to 2012. The sampling covered the dry and rainy seasons, as shown in Table 1. The selected sampling sites, identified as representative of the Magdalena and Eslava rivers, are depicted in Figure 1. Fieldwork permission was given by the *Secretaría del Medio Ambiente* (SMA) and by the *Comisión de Recursos Naturales* (CORENA), the local authority that belongs to the *Gobierno del Distrito Federal* (GDF).

The behaviors of the two rivers were plotted independently, where the Magdalena River (main course) is presented with a five-year data set (2008–2012), while its tributary, the Eslava River, is shown with a two-year data set (rainy season 2010, 2011–2012) in Table 1.

Physicochemical analyses

As part of a long-term water quality monitoring program in the Magdalena Eslava river system, the attributes of temperature (°C), pH, electric conductivity (μS/cm), turbidity (NTU), total dissolved solids (TDS, mg/L), dissolved oxygen (DO, mg/L), DO saturation percentage, total phosphorus (total P, mg/L), orthophosphate (P-PO_4, mg/L), total nitrogen (total N, mg/L), ammonia (N-NH_3, mg/L), and nitrates (N-NO_3, mg/L) were measured.

The physicochemical variables of temperature, pH, electric conductivity, turbidity, total dissolved solids and dissolved oxygen were measured *in situ* over the course of five years. All measurements were made using a Multiparameter Water Quality Sonde YSI 6600-M (Yellow Springs, OH, USA). Five-hundred-milliliter water samples were taken and analyzed in triplicate for nutrients using a spectrophotometer HACH Model DR2400 (Loveland, CO, USA) following the HACH manual [24]. Total P was measured following the method of PhosVer3 by acid digestion USEPA (0.06 to 3.50 mg/L P) and the molybdovanadate method (1 to 100 mg/L PO_4^{3-}). Orthophosphate was measured using the amino acid method (0.23 to 30.00 mg/L PO_4^{3-}), and ascorbic acid was measured using PhosVer3 USEPA (0.02 to 2.50 mg/L PO_4^{3-}). Total N was measured using the persulfate digestion method (10 to 50 mg/L N) and (0.5 to 25.0 mg/L N). Ammonia was measured following the silicate method (0.02 to 2.50 mg/L NH_3-N), and nitrates were measured using the cadmium reduction method USEPA (0.01 to 0.50 mg/L NO_3-N) [24].

Microbiological analyses

Samples for microbiological analyses were taken at the same time and location as those collected for the physicochemical analyses.

Bacteria counts. One-liter samples for bacteriological analyses were collected from selected sites using polypropylene sterile flasks. Samples taken in triplicate were transported and stored at 4°C according to the American Public Health Association standard procedures [25,26]. Microbiological samples were processed within 24 h of collection. All samples were analyzed in triplicate, and the results were reported in colony-forming units (CFUs) [27,28].

Two bacteriological variables, including fecal coliform (FC, UFC/100 mL) and fecal enterococci (FE, UFC/100 mL), were measured constantly during the period from 2008–2012. Total coliforms (TC, UFC/100 mL) were determined only for the period 2008–2009. Positive samples were processed to identify the pathogenic and opportunistic bacteria.

Water samples were analyzed following the standard membrane filtration procedures to enumerate three bacterial types, namely, FC, FE, and TC. Membrane filters (0.45 μm cellulose acetate, Millipore MF type HA) were placed on M-FC agar for FC, KF Streptococcus agar for streptococci/enterococci, and m-Endo agar MF for total coliforms. Incubation was performed with a WTB Binder brand incubator at 35°C for 24 h for enterococci and total coliforms and at 44.5°C for 24 h for fecal coliforms, according to APHA [25,26]. The results are reported as colony forming units (CFUs).

Bacteria identification. Gram-stain and biochemical tests were performed to identify bacteria using a semi-automatic DADE MicroScan, AutoSCAN-4 (DADE International. West Sacramento, CA) [26]. Positive samples from water filtration isolates were selected based on morphology. Five colonies of each specific morphology were selected and identified by the MicroScan Auto SCAN-4 DADE (West Sacramento, CA).

Additional microbiological variables were analyzed during the period 2008–2009 and during the rainy season of 2010. We evaluated these variables because of their importance and relation to gastrointestinal diseases as a result of their prevalence in water in Mexico. These microorganisms included coliphage, enterovirus (EV), adenovirus F human serotypes 40 and 41 (AdV F), *Cryptosporidium parvum* oocysts and *Giardia lamblia* cysts.

Table 1. The sampling design for the Magdalena-Eslava river system.

Variable	Dry season	Rainy season
Magdalena River		
Physicochemical	2008, 2009, 2010, 2012	2008, 2009, 2010, 2011, 2012
Indicator bacteria	2008, 2009, 2010, 2012	2008, 2009, 2010, 2011, 2012
Pathogens	2008, 2009	2008, 2009, 2010
Coliphages	2008, 2009	2008, 2009
Virus	2008, 2009	2008, 2009
Protozoa	2008, 2009	2008, 2009
Eslava River		
Physicochemical	2011, 2012	2010, 2011, 2012
Indicator bacteria	2011, 2012	2010, 2011, 2012
Pathogens		2010

Coliphage. Coliphages were quantified in duplicate following the double-layer agar method based on the International Standard Water quality-Detection and enumeration of bacteriophages (ISO 10705-1, 1995) [29]. The plaques were quantified after the incubation period, and the results were reported in plaque forming units per milliliter (PFU/mL) [30].

The method detection limit was determined following the double-layer agar technique. Serial dilutions were performed using an MS2 phage with a known titer of 10^8 PFU/mL and *E. coli* K12– Hfr 3000 (ATCC 23631) as the host bacteria. The detection limit for this method was determined to be between 10 and 1 PFU [31].

The coliphage quantification with the *E. coli* Hs (pFamp) as host bacteria was performed in duplicate using the adaptation USEPA "Method 1601: Male-specific (F+) and Somatic Coliphage in Water by Two-step Enrichment Procedure)" [29,30].

Virus determination. For the detection of viruses and protozoa, 10-L water samples collected from the natural sites were concentrated by ultrafiltration following the method of Polaczyk et al. (2008) [32].

Enterovirus (EV) quantification was performed in duplicate on the water samples using RNA extraction techniques based on the RNA viral QIAamp (QIAGEN) mini kit. To quantify the genome number of EV, the basic protocol reported by Monpoeho et al. (2000) [33] was followed.

To obtain the AdV F genome number, the protocol by Xagoraraki et al. (2007) [34] was followed.

Protozoa quantification. *Cryptosporidium parvum* and *Giardia lamblia* were detected and quantified by immunofluorescence microscopy.

One milliliter of each sample (direct for wastewater or concentrated for clean samples) was analyzed in single samples using an indirect immunofluorescence technique in liquid phase according to Method 1623 [35], as modified by Rangel-Martínez [36] and Tapia-Palacios [37]. To count the (oo)cysts in the samples, an Axiostar Plus (Carl Zeiss, Gottingen, Germany) fluorescence microscope was used. Total scans of each slide were performed at 20X and 40X and, in special cases, at 100X for confirmation.

Statistical Analyses

The data gathered during the study were analyzed in a spatial and temporal context. Because the sampling locations were located along river courses, we assumed that the spatial coordinates were one-dimensional. In this coordinate system, we identified each site by a site code and by the distance from the first sampling location in the uppermost part of each river. We fitted generalized linear models (GLMs) with the adequate link function for each attribute [38]. For example, for each variable whose histogram clearly skewed to the left, we fitted GLMs with Gamma error and a logarithmic link function. These analyses were performed to explore the association between the physicochemical and biological variables, using site, season and year as covariates. Such analysis allowed us to detect significant effects of the covariates in each of the attributes considered in the analysis. We also created interaction plots to graphically depict the spatial behavior of the values of the variables along the downhill path of the Magdalena and Eslava rivers and at their confluence. Separating the courses in this way allowed us to compare the Magdalena and Eslava rivers before their merging point and to detect whether differences were present in the water quality variables relative to downriver points.

Results and Discussion

Water flow

Despite their relatively small size, the Magdalena River and its main tributary, the Eslava River, provide water to part of the southern area of Mexico City and are the main surface water source for the city, representing more than half of the surface water used. Several studies have been conducted to estimate the amount of water provided by the Magdalena River. In the 1950s, when water sources for the city were investigated, the Magdalena River registers showed a mean flow of 0.33 m^3/s or 10,400,000 m^3/yr [39]. The mean water volume for the Magdalena River was reported in 1975 as 6,970,000 m^3, equivalent to 0.22 m^3/s, and that for the Eslava River was 3,578,000 m^3, equivalent to 0.11 m^3/s [40]. These volumes total 0.33 m^3/s, with a total volume for potential use of 10,548,000 m^3/yr.

A first view of the hydrologic ecosystem services in terms of water quantity and quality with runoff estimates per environmental unit gives an annual production for 2002–2003 of 21,538,250 m^3, equivalent to 0.67 m^3/s, which is similar to the 0.58 m^3/s reported at the hydrometric station in 1999. From 1999–2001 [41], the average flow was 0.70 m^3/s, of which 81% corresponded to baseflow and 19% to surface runoff, providing water for approximately 118,237 inhabitants. In a complementary study, Jujnovsky et al. (2012) [42] assessed the water supply from the Magdalena River Watershed, determining the water balance based on the SWAT model, and obtained a water provision of 18,400,000 m^3/yr. The baseflow contribution accounted for 85%, while the surface runoff was approximately 15%, providing drinking water for approximately 78,500 inhabitants and being capable of supplying 153,203 potential beneficiaries.

Natural and regulated flow regimes. Our study shows the mean annual flow in the hydrometric station to be 0.78 m^3/s, with contributions of 0.45 m^3/s (57%) from the Magdalena River and 0.34 m^3/s (43%) from the Eslava River. The daily regimen is shown in Figure 2, based on historical data from 1973 to 2010, clearly depicting two periods. The first period is from 1973 to 1989, which represents the natural flow regime (NFR), and the second period is from 1990 to 2010, which represents a regulated flow regimen (RFR). This graph shows a decreasing flow and less seasonal variability in the maximum and minimum under the RFR.

The natural flow regime achieved rates of 0.90 m^3/s per year, surpassing the 2 m^3/s during 16 of the 17 years identified as NFR. Starting in 1990, the maximum flow rarely reached 2 m^3/s. The minimum regulated flow of the second period is three times higher than the natural minimum flow, varying from 0.13 to 0.37 m^3/s, that is, the minimum flow increased by 176%. In contrast, the maxima were reduced by 69% (Figure 2). Regulation structures, such as gabion dams, increase the minimum flow due to retention and water storage. At present, more than 60 regulation structures have been registered [12], as shown in Figure 1. These structures were constructed for different objectives, such as decreasing the flow velocity, creating micro wetlands and diminishing sediment volumes, among others [12]. These hydraulic structures are not recommended because they function as frontiers that break the biological continuity of lotic systems [17].

The hydrometric information is depicted in the curves of average monthly flow (Figure 3). The NFR includes the dry season (January to May), with a mean measurement of 0.29 m^3/s and a lower flow in April of 0.26 m^3/s. The rainy season shows an increase in flow from June to August and a decline in December (0.66, 1.72 and 0.46 m^3/s, respectively). At the end of the 1980s,

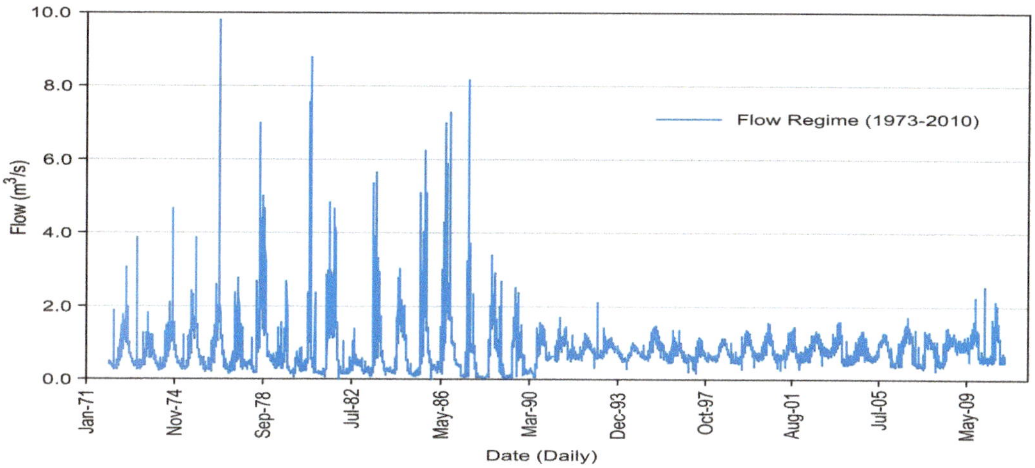

Figure 2. Daily volumetric flow regime (1973–2010) measured at the hydrometric station Santa Teresa (26440).

the basins of the Magdalena and Eslava rivers experienced flow variations, as observed in Figure 3. In the dry season, the RFR curve increased 128% compared with the natural regime (NFR). The RFR value for April was 0.60 m^3/s, compared with 0.26 m^3/s for the NFR in the same month. In the rainy season, the RFR curve diminished with respect to the NFR by 30% from July to October.

Figure 3 shows the changes experienced by the river before and after the hydraulic regulation in the periods mentioned previously, which is defined for four average monthly flow regimes. The inertial behavior of the minimum natural flow regime (mNFR), shows that the dry period (January-May) minimum flow was 0.06 m^3/s, except in March-April, when the minimum flow was 0.04 m^3/s. Starting in the rainy period, the minimum flow responded to the natural increase from June-August (a maximum flow of 0.57 m^3/s) until declining over the period of September to December from 0.51 to 0.09 m^3/s.

The minimum regulated flow regime (mRFR) differs from the natural condition (mNFR curve), with a greater effect observed during the dry season. In March and April, the minimum natural flows increased to 490% (from 0.04 to 0.24 m^3/s) and then increased 202% in January, February and May. From July to October, the mRFR curve increased 64% from its original condition (1973 to 1989); in November and December, the flows increased from 0.20 and 0.09 to 0.57 and 0.47 m^3/s, respectively. This increase can be attributed to the hydraulic structures and to external factors, such as wastewater discharge from irregular settlements in the lower drainage basins and surface runoff during the rainy season. The hydrological histogram (Figure 3) shows the discrepancy in hydrological conditions: under low-flow conditions, the mRFR curve corresponds to the NFR conditions. Therefore, the minimum regulated behavior mimics a natural condition.

Figure 4 depicts the annual flow over a 38-year period. In the first period (1973–1989), the maximum flow reached 10 m^3/s, with a mean annual maximum of 4.94 m^3/s. In the 1990s, the development of hydraulic infrastructure in the river caused a laminar flow effect. During the second period, from 1990 to 2010, the maximum flow decreased 69% (relative to 2.53 m^3/s as the maximum), with a mean annual maximum flow of 1.52 m^3/s.

The behavior of the mean annual flows (Figure 4) also reflects the effect of regulation. Although the mean flow in both periods is very similar, namely, 0.77 and 0.79 m^3/s, its volume has not changed significantly. Over time, extraction from the river has been balanced by wastewater inputs into the river, which has led

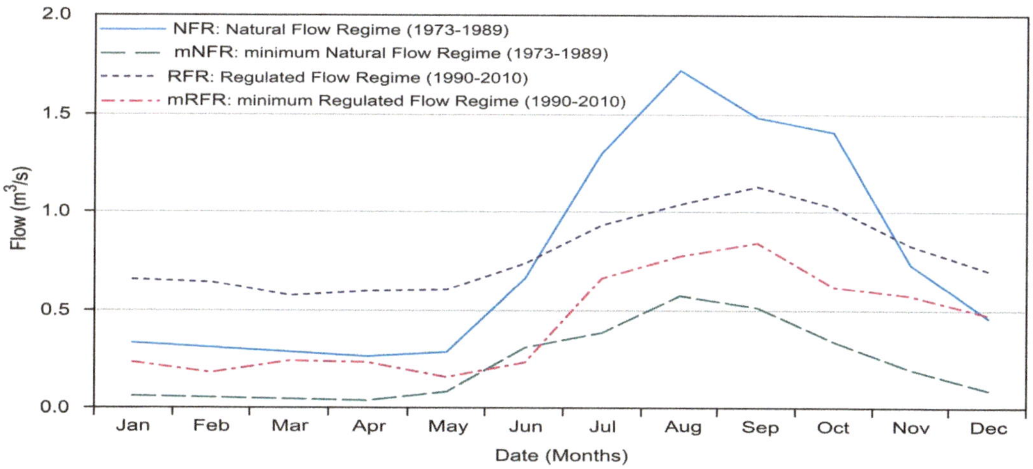

Figure 3. Interannual variation of the NFRs, RFRs, mNFRs and mRFRs measured at hydrometric station 26440.

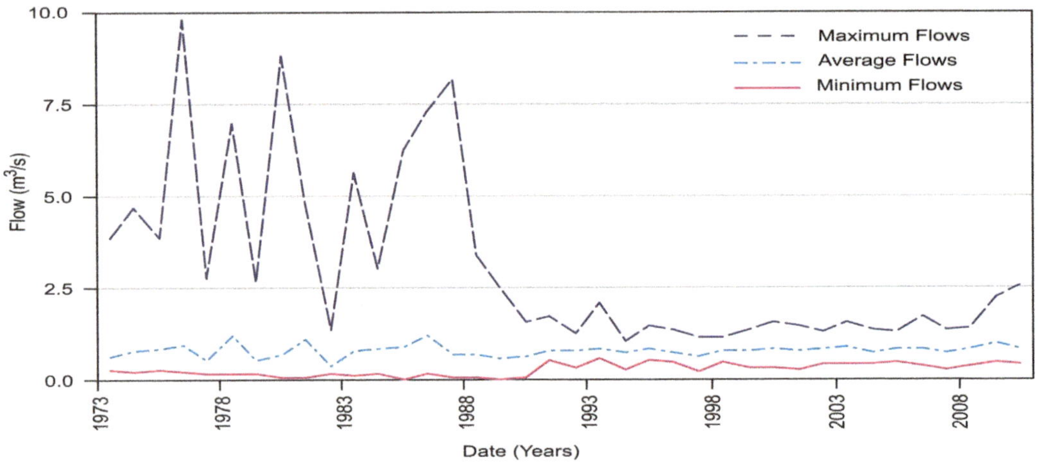

Figure 4. Annual volumetric flow regime at the outlet of the Magdalena and Eslava river basins. Average, maximum and minimum flows.

to an increase of 0.02 m^3/s. The exchange of clean water for untreated wastewater has significantly degraded the environmental condition of the river.

There are only two direct discharge points into the river over the course of the distance from its origin to the treatment plant; in contrast, after the treatment plant, 58 inputs are known [12]. Irregular settlements are problematic because they often lack a sewer system, thereby allowing wastewater to flow directly into the rock or soil and hence toward the lower discharge area where the river is located.

In Figure 5, a historical annual comparative analysis illustrates the natural flow from 1973 to 1989 in two flow-duration curves; during this time, the annual mean flow at the lower point of the Magdalena river was at least 0.54 m^3/s more than 90% of the time, which represents the probable flow that the river used to supply along the year. High flows with flows of approximately 1.14 m^3/s are present during a very short period (10% of the times). In the 1990s, the development of hydraulic infrastructure in the river caused a laminar flow effect; therefore, in the curve for the regulated period from 1990–2010, the mean flow at the river is at least 0.71 m^3/s for 90% of the time, which is an increase of 31% in comparison with the previous natural minimum; consequently, a 25% decrease in high flows is observed. It is also possible to observe changes to the slope of the natural flow curve (reference) with respect to the regulated flow curve, specifically, changing from 0.73 to 0.24%.

The behavior of the natural curve of collected frequencies (1973 to 1989) depicted in Figure 5 is representative of mountain rivers; therefore, based on the flow duration curve observed for the regulated period (1990 to 2010), a new behavior is observed. That is, the river is behaving if it were a floodplain river, where changes are less apparent. Generally, this response is due to changes to the flow of the Magdalena River as a result of anthropogenic activities, with an associated increase in the concentration of several water quality variables.

Figure 6 shows the distribution of water flux for the rainy and dry seasons for the years 2008–2010, starting with the dry season from January-April 2008, followed by the rainy season of 2008, the dry season from November 2008–April 2009, and so on. The figure demonstrates that the mean value is relatively constant for dry and rainy seasons, while the variability changed during those same years; hence, no pattern can be observed for the changes in variability. This is related to the high variability of the amount of rain falling in the upper parts of the basin, particularly during the rainy season of 2010. The low variability for the dry season from November–December 2010 is due in part to the small amount of data for this time lapse.

Water quality

Physicochemical and microbiological analyses. The spatial and temporal analyses for the physicochemical variables and nutrients are presented in Figures 7 and 8. These variables are those most representative of river water quality, such as electrical conductivity, dissolved oxygen, ammonia, total N, orthophosphates and total P, as well as fecal coliforms and fecal enterococci. These variables are considered representative because they have the highest absolute value in the factor loadings of the first three principal components. The factor loadings for these attributes account for more than 80% of the total variability of the Magdalena River system [43].

Although the number of samples at each site was not the same, it only affects the variance of the mean values of the measured water properties. This effect has been considered and corrected in all of our analyses by dividing the mean value estimates by $\sigma/\sqrt{n}$.

The GLM models fitted to the water quality variables [38] indicate statistically significant differences for all of the variables, except dissolved oxygen, in the upper and lower sections of both rivers. The point of change was the location at which human activities and settlements began on either or both river banks. Water quality variables sampled upstream of these urban areas fell within acceptable limits for use as a water supply. The values gradually fell to lower quality as the sampling sites became increasingly affected by the wastewater generated in the irregularly inhabited areas. The spatial trends along the river gradient for the eight most-relevant water quality attributes are shown as box plots in Figure 7 for the Magdalena River and in Figure 8 for the Eslava River.

The plot shows a change point at site M9, which is located at the border between the natural and urban areas: here, increasing concentrations of all values can be observed, except that of dissolved oxygen, which decreases. The highest concentrations were detected at site M15, where part of the water is piped to the city's drainage system. These values then decrease slightly at sites M17 and M18, where other effluents with residual water from the southern area of the city join the waterway.

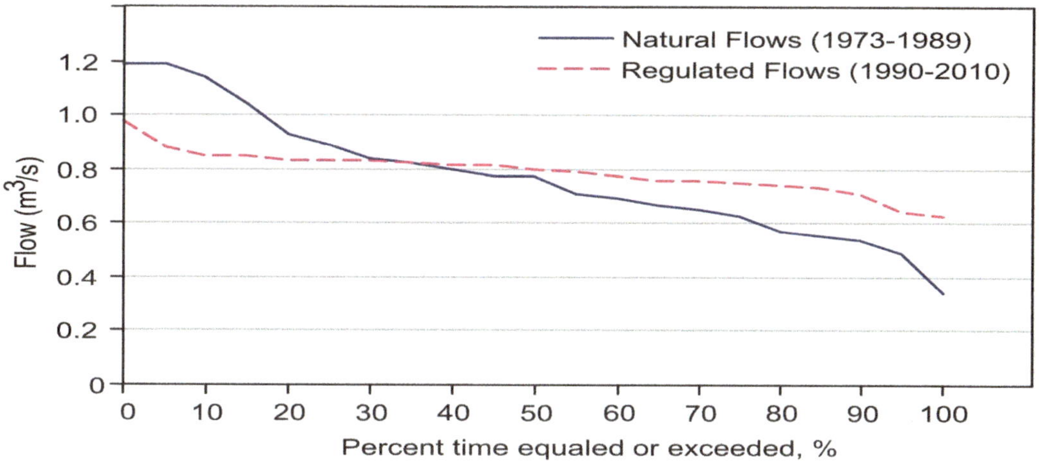

Figure 5. Duration curve of annual natural and regulated flows, Magdalena river basin, 1973–2010.

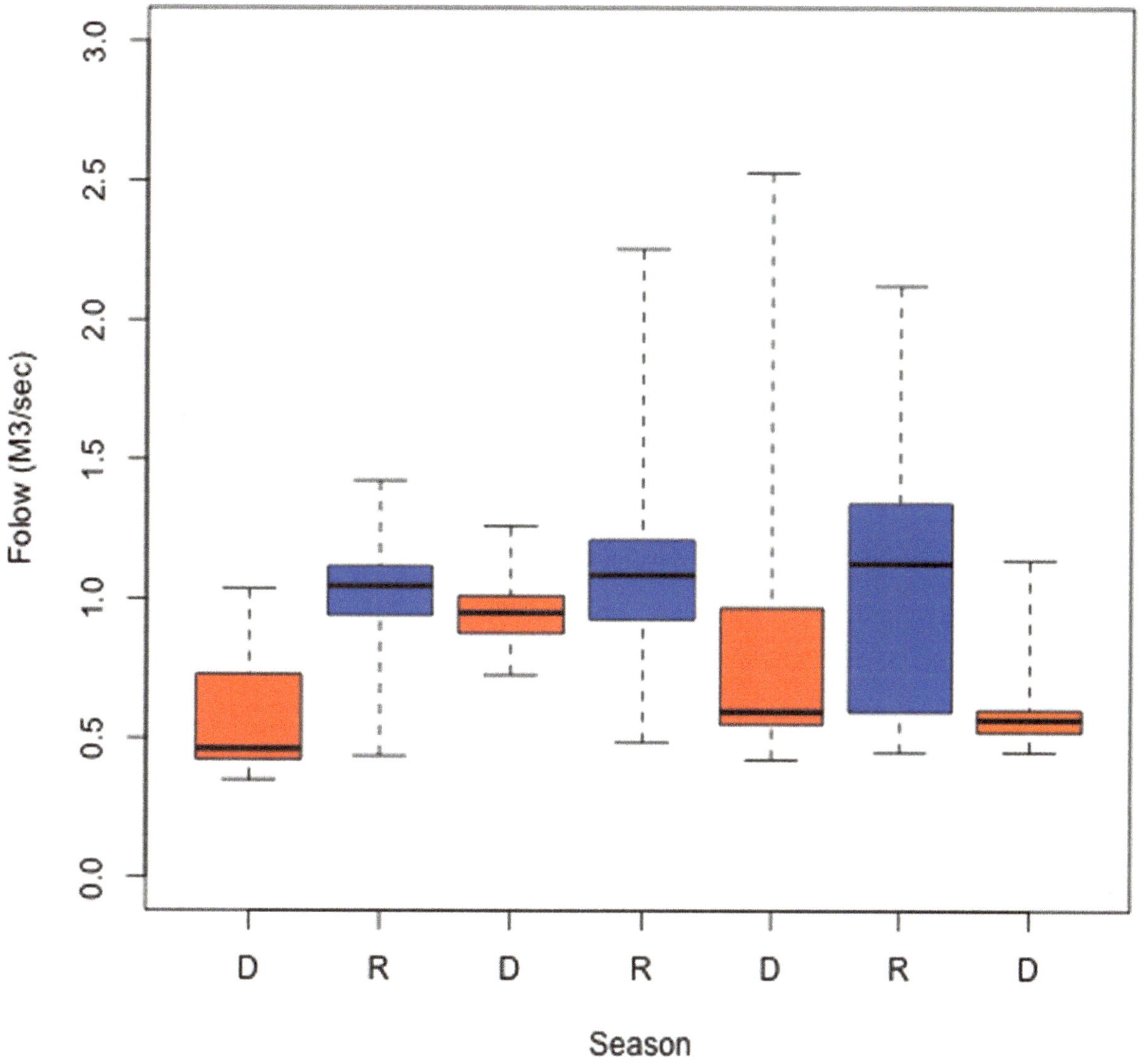

Figure 6. Water flux distribution from 2008–2010 grouped by season.

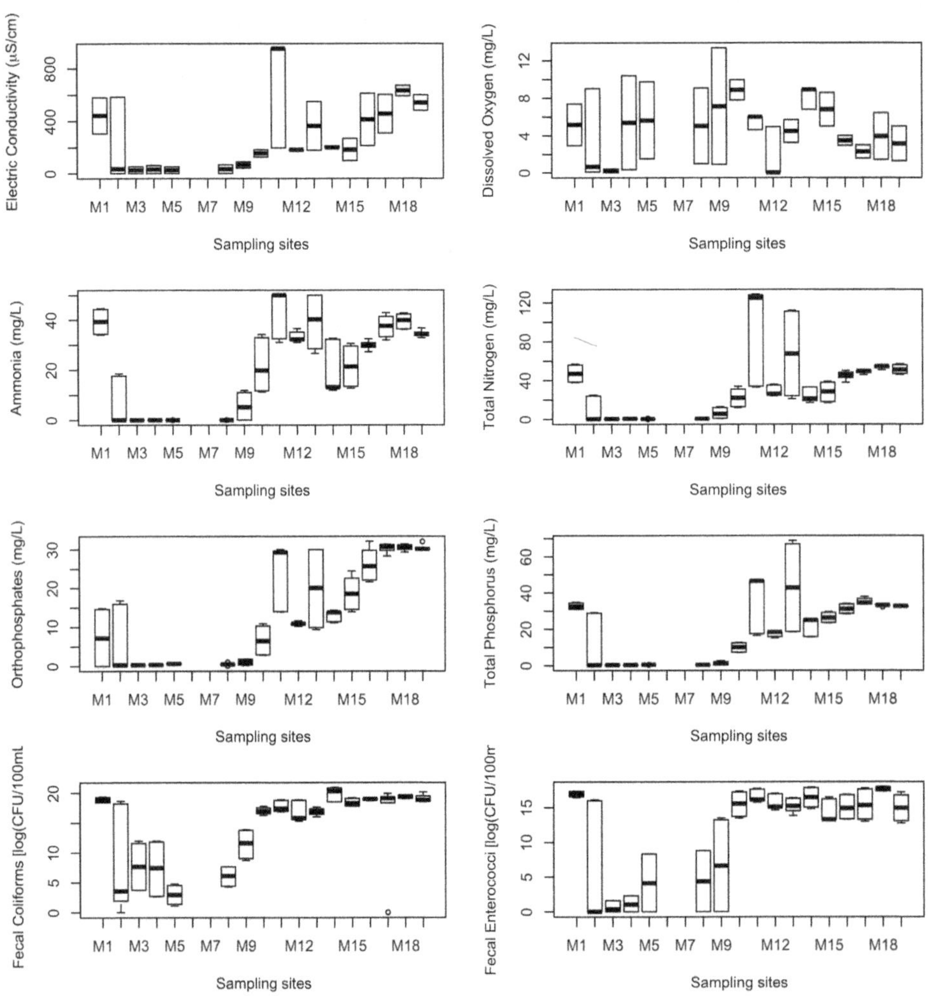

Figure 7. Spatial trend for the eight main water quality variables in the Magdalena River, based on 19 sampling stations.

In the Eslava River, all variables except dissolved oxygen are significantly different between the natural and urban areas, a change that occurs somewhere between E5 and E6. Further, all variables exhibit stable values until the confluence point. The Eslava River has lower water quality than the Magdalena River does at the confluence point, particularly in terms of electrical conductivity, the value of which is six times higher in the Eslava River; additionally, the value of total N is three times higher, while that of total P is twice the concentration of that in the Eslava River. The fecal enterococci values were three times as large on a log scale, which means that this measurement is 1,000 times greater. Please note this difference in the values of the axes on the graphs.

According to Mexican standards, several attributes surpass the NOM-001-SEMARNAT-1996 [44] limits for water discharged to Mexican water bodies. The total N exceeds these levels from M15 to M18 and is double the limit at E8 in the Eslava River. At E7, the values are slightly above the limit.

Because both rivers are used as surface water sources for human use and consumption, the LFDMA [45] limit is 1,000 FC per 100 mL. Taking this into account, the fecal coliforms surpass the limit and require treatment before distribution through the southern parts of the city. The standards are met at the Potabilization Plant La Magdalena, but there are many water intakes near settled populations in an irregular land ownership situation, which means that these populations are exposed to high levels of bacteria; this is an important risk factor that should be taken into further account.

Bacteria identification. The 164 bacterial isolates and identifications obtained from 516 water samples taken from the Magdalena River in the period 2008 to 2010 are presented in Table 2. These belong to nine different families and 18 genera [26,46]. The most abundant genera were *Enterococcus*, *Escherichia coli*, *Enterobacter* and *Klebsiella*. From the Eslava River, 27 bacterial isolates and identifications were obtained from 120 water samples taken during 2012. These belong to four different families and seven genera, as reported in Table 3.

Therefore, a variety of bacteria belonging to a total of 20 different genera were identified in the river system crossing the southern area of Mexico City. The most prevalent bacteria were from the Enterococcaceae family, *Enterococcus*, and from the Enterobacteriaceae family, *Escherichia*, *Enterobacter* and *Klebsiella*. Although these results serve as an initial approach to the identification of bacteria, it is important to note that all of these bacteria are common to the human gut and can be opportunistic pathogens that may cause illness to immune-compromised people who access this water when untreated.

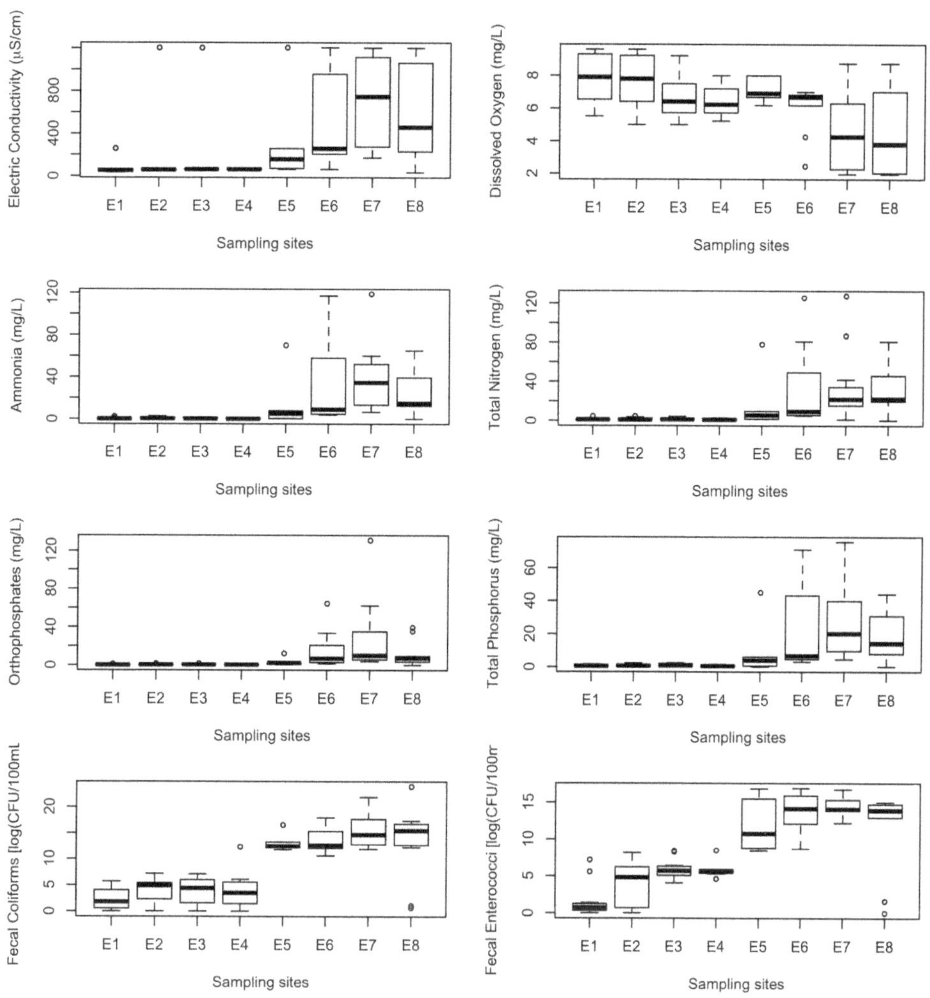

Figure 8. Spatial trend for the eight main water quality variables in the Eslava River, based on eight sampling stations.

Coliphage, virus and parasite analyses. The presence of coliphage was measured by bacterial hosts in samples from the urban zone; this finding is consistent with the inputs of raw and untreated wastewater to the Magdalena River (Table 4). As host bacteria, *E. coli* K12-Hfr registered the higher sensitivity, although larger counts were obtained for *E. coli* Hs (pFamp). It is important to note that there are two infection mechanisms of interest, either by adherence to the cell wall or infection via pilli, as in the coliphage F-specific [47,48,49,50]. Therefore, the host is more susceptible to infection and shows higher counts.

In the natural area sites, no coliphage presence was detected, which suggests that perturbation does not have a detectable impact using this method. Generally, the presence of a phage indicates fecal contamination by a virus; however, this level of analysis is not conclusive because it cannot establish the contaminants' origin, namely, whether human or animal. In urban areas, coliphages are present from the transition zone between the natural and urban zones, where the highest density was detected at site M19; hence, the number of phages increases with human perturbation and influence.

Both adenovirus (AdV F) and enterovirus (EV) exclusively affect humans; their presence indicates human fecal contamination [51,52]. Once these viral particles have been excreted into the environment, their number cannot increase because the presence of the cell host is required for their replication [53]. Therefore, the detection of AdV and EV reflects the fact that in the human population, the virus is circulating and replicating. This constitutes a threat that must be considered regarding the river because these viruses can cause infections to susceptible individuals, yielding viral gastroenteritis.

AdV F replicates quickly in the gastrointestinal track, so the amount of excreted viral particles is on average 1×10^{10}/g in feces [53]. Such infections mainly occur in children under five years of age because adults who were infected as young children often acquire immunity [52]. Nevertheless, if adults are exposed to viral particles, they can be asymptomatically reinfected, with a lower amount of viral particles excreted in their feces. This can favor the dispersion of AdV F because these individuals will not be cautious in their behaviors because they are asymptomatic.

People can present infections and acute gastroenteritis as a result of AdV F [54]. It has been estimated that this virus is the cause of 9 to 12% of acute gastrointestinal cases in children under five years of age in Mexico; however, based on the lack of follow up regarding this type of pathogen and the lack of updated epidemiological information (few studies have been conducted since 1985), this may be an underestimation [55]. Due to the relevance of this virus as a gastroenteritis-causing agent, it is considered the second-most important virus after rotavirus

Table 2. The bacteria identified in the Magdalena River water samples in 2008, 2009 and 2010.

Isolated microorganisms	Sites M1	M2	M3	M4	M5	M6	M7	M8	M9	M10	M11	M12	M13	M14	M15	M16	M17	M18	M19	Total
n	24	24	6	24	42	42	18	18	42	42	6	6	42	24	42	24	42	24	24	
Moraxellaceae																				
Acinetobacter spp.																		1		1
Aeromonadaceae																				
Aeromonas sp.				1										1		1	1	1		5
Burkholderiaceae																				
Burkholderia sp.								1												1
Alcaligenaceae																				
Bordetella sp.								2												2
Enterobacteriaceae																				
Citrobacter spp.													1				1			2
Enterobacter spp.	1	1			1			3	1		2	1	3	2		1		1	2	19
Escherichia sp.	2	1			2			1	2	1	2	3	2	1	1	3	3	3		27
Hafnia sp.	1																			1
Klebsiella spp.								1	1		1	2	1		1	2	2	2		13
Proteus sp.					1															1
Providencia sp.			1	1																2
Salmonella sp.								1												1
Serratia spp.									2											2
Yersinia sp.						1		1												2
Leuconostocaceae																				
Leuconostoc sp.					1															1
Pseudomonadaceae																				
Pseudomonas sp.			2	3	1				1		1				1					9
Vibrionaceae																				
Vibrio sp.									1			2	1	2	1	1				8
Enterococcaceae																				
Enterococcus spp.	4	2		5	5	1	1	6	5	2	6	3	2	6	2	6	6	5		67
Total	8	4	3	10	11	2	1	16	13	3	12	11	10	12	6	14	13	13	2	164

Table 3. The bacteria identified in the Eslava River water samples in 2010.

	Sites								
Isolated microorganisms	**E1**	**E2**	**E3**	**E4**	**E5**	**E6**	**E7**	**E8**	**Total**
n	**15**	**15**	**15**	**15**	**15**	**15**	**15**	**15**	
Burkholderiaceae									
Ralstonia spp.				1					1
Bacillaceae									
Bacillus spp.	1								1
Enterobacteriaceae									
Escherichia spp.		1	1	1	1	1	1	2	8
Enterobacter spp.				1					1
Klebsiella spp.						1	1	2	4
Yersinia spp.								1	1
Enterococcaceae									
Enterococcus spp.	1	1	1	1	1	1	1	4	11
Total	2	2	2	4	2	3	3	9	27

Table 4. The geometric mean, standard deviation in log scale, maximum and minimum of Coliphage, Enterovirus, Adenovirus, *Cryptosporidium parvum* oocysts and *Giardia lamblia* cysts in the Magdalena River water from 2008 to 2009.

		Sites							
		M1	**M5**	**M8**	**M11**	**M12**	**M17**	**M18**	**M19**
Coliphage with host	Geometric mean	<1	<1		39	52	57	22	114
K12	Standard deviation	<1	<1		17	18	58	51	14
PFU/mL	Min-Max	<1	<1		29–65	27–65	9–140	7–110	97–127
	n	4	4		4	4	6	6	6
Coliphage with host	Geometric mean	<1	<1	<1	4	7	3	3	33
Hs	Standard deviation	<1	<1	<1	10	7	10	3	26
PFU/mL	Min-Max	<1	<1	<1	<1–18	<1–17	<1–20	<1–7	18–80
	n	4	4	2	4	4	6	6	6
Enterovirus Gen/mL	Geometric mean	<1	<1	<1	<1	<1	<1	<1	<1
	Standard deviation	<1	<1	<1	<1	<1	<1	<1	<1
	Min-Max	<1	<1	<1	<1	<1	<1	<1	<1
	n	3	5	6	4	4	6	4	4
Adenovirus Gen/mL	Geometric mean	10	<1	<1	12974	7299	98294	7305	97084
	Standard deviation	577	<1	<1	46297	28951	253772	10157	35432
	Min-Max	<1–1000	<1	<1	1130–930464	982–63008	24930–557800	1618–25186	71486–144840
	n	3	5	4		4	6	4	4
Oocysts	Geometric mean	<1	32	<1	6784	21	575	4244	<1
Cryptosporidium	Standard deviation	<1	707	<1	14849	5196	9504	2121	<1
parvum/L*	Min-Max	<1	0–1000	<1	2000–230002	<1–9000	<1–19000	3000–6000	<1
	n	2	2	1		3	3	2	2
Cysts *Giardia*	Geometric Mean	1000	71	<1	18	18	30	89	<1
lamblia/L*	Standard deviation	<1	3536	<1	2828	3464	16166	5657	<1
	Min-Max	1000–1000	<1–5000	<1	<1–4000	<1–6000	<1–28000	<1–8000	<1
	n	2	2	1	2	2	3	2	2

Cryptosporidium parvum and Giardia lamblia detection limit 1 (oo)cyst/mL [37]; Adenovirus (AdV) detection limit 10 genome copies [34]; Enterovirus (EV) detection limit 10 genome copies [74]; FRNA specific bacteriophage detection limit 1 PFU/mL [30], somatic bacteriophage detection limit 1 PFU/mL [30,34].

worldwide [56]; therefore, it is important to monitor the prevalence of this virus both in the human population and in residual water as an indicator of what is infecting a given population.

According to our results (Table 4), AdV F is present at both the natural and urban sites, indicating human fecal contamination in both locations. The presence of AdV F at the river origin (M1) could represent an isolated contamination event because it was only detected at a low density in the samples from this site. The presence of human fecal contamination coincides with the irregular settlements that discharge their wastewater directly to the river. Because the system once flowed continuously, the viral contamination issue is likely a recent phenomenon. Wastewater discharge is constant; unsurprisingly, the higher amounts of AdV F viral particles were detected in the urban zone due to wastewater discharge without proper treatment [54,57].

It is important to note that the number of AdV F particles can represent an important mechanism of exposure to the population, primarily for children under five, who may be more likely to come in direct contact with the water in urban areas; accordingly, this exposure can lead to moderate to severe gastroenteritis [53]. For the exposure dose, a point estimate was reported by Crabtree *et al.* [58], who considered AdV 4 as a surrogate because no AdV 40/41 dose-response has been published. Based on this finding, a 2 infectious viral particle can be considered the probability for infection based on an average dose of 1.66 [59].

Considering that the detection limit of the protocol applied for AdV 40/41 quantification was ten genome copies, as determined by serial dilutions from the positive control, it would not be possible to identify a hazard related to AdV 40/41. Nevertheless, the fact that the genome detection rate is commonly larger than the number of viral particles [60] clearly demonstrates that the detection capacity of 10 genome copies is adequate for the identification of adenovirus as a hazard.

Cryptosporidium parvum and *Giardia lamblia* are the most common intestinal protozoa in the environment, commonly infecting both humans and domestic and wild animals [61]. These protozoa are among the main disease-causing agents of diarrhea in the world [62]. Both species form resistance structures that allow them to survive outside the host; they can thus be transmitted through water [63]. Additionally, both species have a low infective dose (ID_{50}), specifically, 132 *Cryptosporidium* spp. oocysts [64] and 25–100 *Giardia lamblia* cysts [65]. Thus, a potential host consuming a low number of these microorganisms can readily develop an infection.

Based on these facts, the detection of *C. parvum* and *G. lamblia* in 50% and 44% of the Magdalena River samples, respectively (Table 4), means that the population and animals that use the water for diverse activities are exposed to parasites, which eventually could lead to an outbreak or epidemic. Children between 0 and 14 years of age and immunodeficient persons are susceptible.

For the Cuitzmala river, a rural area in Jalisco, Mexico, Tapia-Palacios (2012) [37] reported a range of 0–87 oocysts/L *C. parvum* and 0–2,655 cysts/L of *G. lamblia*, which in contrast with the results of this analysis for the urban area, though both species were detected in the two areas. The range in the natural zone was lower for both species (0–1,000 oocysts of *C. parvum* and 1,000–5,000 cysts/L of G. lamblia), which is in contrast with the urban areas that exhibited higher ranges (0–23,000 oocysts/L of *C. parvum, and* 0–28,000 cysts/L of *G. lamblia*). This can be explained by the input of wastewater from the urban area [12]. Untreated water could be increasing the number of oocysts present in the zone, leading to a potential population exposure risk because the population densities range from 1–2 log units more than the infective dose.

Because these rivers represent the only surface water source from within the Basin for Mexico City, particular concern should be held regarding the presence of certain microorganisms, such as the protozoa *Giardia* and *Cryptosporidium* and adenovirus, which are relatively resistant to the chlorination process in comparison to other microorganisms. Additionally, the presence of opportunistic and/or potentially pathogenic bacteria in a surface water source that is chlorinated and distributed to the population suggests that there should be additional funds to consider preventive measures, consistent with the practices carried out in other countries concerning water intended for human use.

Physical removal is critical to the control of *Cryptosporidium* because the organism is highly resistant to standard disinfection practices [66,67]. As mentioned by Betancourt and Rose (2004) [68], the effectiveness of conventional disinfection by chlorination and alternative disinfection procedures, such as chlorine dioxide, ozonation and ultraviolet irradiation (UV), toward inactivating *Cryptosporidium* has been the focus of current research. Recent USEPA criteria regarding disinfection processes for water containing this pathogen suggest disinfection with ultraviolet light or ozone and/or filtration [69]. The goals for systems that treat surface water must be disinfection and filtration to remove or inactivate 99.9% of *Giardia lamblia*, 99% of *Cryptosporidium* oocysts and 99.99% of viruses [69]. In Mexico and other developing countries, modern treatment processes should be applied, given the presence of these microorganisms in the surface waters.

One of the first steps when dealing with rivers is understanding their flow and any variations that occur throughout the year and across different years. In Mexico, monitoring practices were initiated but have since been abandoned for many rivers throughout the country. A sustainable approach to water management, especially in light of climate change, would be to monitor and report environmental and social conditions to provide a guide toward a more sustainable water management strategy.

The regimes of the Magdalena and Eslava rivers belong to homonymous basins that have been modified, and both exhibit two distinct historical periods. The periods are defined by the construction of excess hydraulic infrastructure works, such as gabion dams, 50 on the Magdalena and 18 on the Eslava (Figure 1), and marginal drains. The management of the flows in the Magdalena and Eslava rivers has caused significant fluctuations in flow magnitude (69%), has increased the frequency of maximum ordinary flows and has caused a significant increase (176%) in the minimum flow magnitude. This NFR alteration will certainly have consequences for the river communities in the natural area and for system rehabilitation.

These regime changes may contribute to the observed water quality degradation and contamination of the systems because the loss of velocity along a slope and other discontinuities in the river affect the inherent auto-depuration processes. As recorded in our fieldwork, the observed higher concentrations of the physicochemical variables indicate the presence of higher levels of contaminants in the lower section of both basins, especially as the urban zone is entered.

Wastewater discharges also contribute to the degradation of water quality. Hence, even if discharge does not contribute significantly to flow volume, its effect on water quality is magnified due to the decrease in flow velocity. Further, wastewater discharge increases the concentrations of various water quality attributes.

In some rivers, contaminants are diluted through their tributaries. However, in this case, the Eslava River contributes its own contaminant load, which for some variables is orders of magnitude higher than the Magdalena's water, thereby hampering the auto-depuration process.

It is necessary to treat the residual wastewater dumped into the river and recover the fluvial processes and function of the rivers as ecosystems. This approach would provide several advantages. For example, there would be more water produced in the higher basin, which could be disinfected and distributed to the population. Additionally, a decrease in the amount of water that flows to the drainage system and that must be pumped out of the Basin of Mexico would be realized, as well as additional amounts of water once the wastewater has been treated and reused for irrigation near the river.

In this study, we demonstrate a practical and representative approach to achieving a detailed water quality program that could allow for the detection of change points. Additionally, we identify the problems at some locations and various methods for improving the respective environmental conditions. In this particular case, it is also of important that locations for the required wastewater treatment plants be suggested, as well as the type of plant and the installed capacity.

The water to be treated is considered domestic wastewater based on its Biochemical Oxygen Demand (DBO_5) and Chemical Oxygen Demand (COD), according to Metcalf and Eddy (1996) [70].

It is important to note the need for treatment plants to improve water quality. The wastewater discharge along the river has high concentrations of nitrogen and phosphorus forms, total dissolved solids, high conductivity, and a BOD_5 of approximately 130 mg/L to 240 mg/L (results not shown). These factors, combined with the high microorganism counts, argue in favor of the installation of several wastewater treatment plants (WWTPs) [71]. A general concept for viable plants and processing methods was presented by Orta de Velasquez (2008) [23], although treatability tests would be required, along with a specific design for each plant with an associated project feasibility study.

The wastewater treatment plants considered for this river, in all cases, included tertiary treatment. This treatment is implemented by biological secondary treatment, and the advanced treatment is implemented by physico-chemical treatment. The first treatment of activated sludge helps remove nutrients (N and P) and both soluble and colloidal biodegradable organic matter. Tertiary filtration is used to treat wastewater intended for reuse to remove any suspended matter that was not removed in the previous operations. Microbial standards can be met using both filtration (physical removal) and disinfection (inactivation). The removal and inactivation capabilities vary depending on the filtration technology or disinfectant applied. The first wastewater treatment plant constructed in the area of the Magdalena River was finished at the end of 2013, not clear if it is in operation at the moment. With the goal of removing both nutrients and pathogens, advanced treatment, which includes activated sludge, tertiary filtration and ultraviolet disinfection, was implemented. In this case, an ultraviolet light disinfection process was applied, which can effectively inactivate most viruses (111 mW*s/cm^2 for 6-log inactivation), *Giardia* (121 mW*s/cm^2 for 2-log inactivation) and *Cryptosporidium* (1,900 mW*s/cm^2 for 2-log inactivation) [72]. With this plant being operational, the river must have a basic flow level along its path to preserve some system functions.

At least two WWTP are proposed. The first could be placed near sites M11 and E8, which could treat low-quality water flowing from both rivers. A second WWTP could be placed before site M19, where the treated water could be used for irrigation in one of the few large green areas of the southern part of the city, known as *Viveros de Coyoacán*. The potential locations for these two plants are presented in Figure 1.

From the perspective of the ecosystem services provided by this lotic system, the amount of water provided by the Magdalena-Eslava river system accounts for the ecological flow in the natural area, after which the water goes through a potabilization process and is distributed for human use and consumption in the southern area of the city. The amount of water that is contaminated when clean water from the natural areas of both rivers combines with untreated wastewater is not accounted for, but it is necessary to consider the financial resources that are being wasted on these relatively small rivers.

The current average flow of the Magdalena river is 0.45 m^3/s, based on measured flow, and the cost per m^3 with the water subsidy is $8.17 (equivalent to 0.628 USD at an exchange rate of US $1 = $12.99 Mexican peso) Mexican pesos; without subsidy, the cost is $25.97 (equivalent to US $1.99) (SACM, Official Receipt. *Derechos por el suministro de agua. Magdalena Contreras. Manzana Baja*. September, 2013) [73]. Therefore, the amount of water lost accounts for 14,191,200 m^3 per year, which means a loss of $115,942,104 Mexican pesos (equivalent to US $8,925,489) at the subsidized cost, or $368,545,464 Mexican pesos (equivalent to US $28,371,475). In this example, we have only considered the Magdalena River, which has better water quality and more abundant flow. An equivalent of this at Mexico City prices for bottled water at $5.66 per liter ($5,660 per m^3) is an estimated $79,470,720,000 Mexican pesos (equivalent to US $6,117,838,337).

As part of any megacity water management strategy, there is a need to consider the amount of water lost and the financial capital that those losses represent. There is an urgent need to rehabilitate and recover the flow of the river and its water quality, which is an approach that could be replicated in more than 40 other rivers around Mexico City. The ecosystem service of clean water is being lost in the megacity, where water must be pumped at a very high cost from either groundwater systems or distant basins. These rivers serve as one example of similar situations that have arisen for other rivers in the Basin of Mexico, where water could be managed in a more sustainable way in the vicinity of this megacity. Doing so would yield enormous environmental and economic benefits in the long term.

From the perspective of providing water, water quality represents a fundamental issue. Better use of the existing water sources can be achieved in the MCMA. The rehabilitation of the Magdalena-Eslava river system could provide an example for other local and remote locations to follow.

Health issues are at stake, so action must be taken to ensure water quality in this particular system. The contaminated water flowing through the urban area should be improved to avoid future deleterious effects to the population.

Considering the treatment of wastewater that flows through the urban area, this water could also be used for irrigation in the green areas in the southern area of Mexico City.

Conclusions

The flow regimes of the Magdalena and Eslava rivers have been modified by the increase of hydraulic infrastructure constructed along the river, resulting in decreased maximum flows and increased minimum flows. These changes in flow regimes are directly related to the quality of the river water, due to the loss of velocity in the flow and the loss of continuity of flow.

The water quality of this system deteriorates as the river flows from a natural forested area to a semi-urban area, passing irregular settlements without basic water and drainage services before flowing then into an urbanized area.

The water quality is worst in the Eslava River at the confluence point, M12, after which either a dilution effect or an artifact of the complex drainage system arises, leading to an apparent improvement in the wastewater quality downstream. Organic matter, such as that found in the discharge, affects the oxygen balance and is usually accompanied by microbiological contamination.

The domestic and agricultural wastes contribute phosphates and ammonia nitrogen. Agriculture also contributes to the dissolved solid levels measured in the river.

The Magdalena Eslava river system could be used more efficiently. The river receives domestic discharge, and a portion of the flow passes through the *Delegación Magdalena Contreras*, where no industry is present.

The construction of at least two wastewater treatment plants is proposed, with the goal of restoring the river's water quality and maintaining the flow in the system. The suggested placement of these treatment plants is based on the low flow that occurs during the dry season and the sewer system conditions. By considering the landscape design to be a linear park, the treated wastewater flow was defined to rely on water flow and high water quality for recreational use. Additionally, the potential to reuse this water for green area irrigation and car washing exists. Treating the wastewater in this area of Mexico City would be relatively simple in comparison to treating the wastewater in other industrial areas of this metropolitan area.

Human activities, which have increased significantly in recent decades, have impacted the environment. The scale of urbanization, industrial operation and agricultural production has reached a point at which these activities impact the quality of the hydraulic resources. This is a global phenomenon to which Mexico City is not an exception.

Wastewater in the cities of the developing world transports a combination of liquid and solid residues from regular and irregular residential areas, public institutions, and industrial and commercial establishments. These residues eventually reach pluvial, surface and groundwater. Due to the diversity of compounds and microorganisms in wastewater, such as those observed in this study, it is not possible to provide details for all compounds but only to give an idea of the water technology required for improving water quality, with a view toward reuse and sustainability. Undoubtedly, there exist other compounds that were not measured as part of this evaluation, such as endocrine disruptors or organic compounds; these compounds must be evaluated, especially in the urban section of the lotic system, to develop a complete picture.

The amounts of several types of microorganisms in the urban zone, each with a different resistance to water and wastewater treatment, must be considered from the public health perspective because these microorganisms represent an important exposure factor for the local population.

Mexico City has lost the majority of its rivers due to mismanagement. The case of the Magdalena and Eslava rivers represents an opportunity to rehabilitate a major river system and could serve as a model for other waterways.

Acknowledgments

We thank Lucía Almeida Leñero for her enthusiastic collaboration throughout the years in the study area and the support received for this work. We thank Erick Hjort Colunga, Rosa Isabel Amieva, Ana Cecilia Espinosa, and María del Jazmín Aguilar for their invaluable assistance in the field and with laboratory work. We thank Alma de los Ríos for assistance in creating the figures.

Author Contributions

Conceived and designed the experiments: MMH MTOL CDA FAV. Performed the experiments: GPO MAT RSO MAS IYN. Analyzed the data: CDA MMH GPO TLV MTOL FAV. Contributed reagents/materials/analysis tools: MMH MTOL YLV CDA FAV. Wrote the paper: MMH GPO CDA FAV MATOL YLV MAT RSO. Performed the field work: GPO MAT MMH. Lab work: GPO MAT RSO YLV MAS.

References

1. Foster S, Lawrence A, Morris B (1998) Groundwater in urban development. Assessing management needs and formulating policy strategies. The World Bank. Washington, D.C.
2. Legorreta J (2009) Ríos, lagos y manantiales del Valle de México. Universidad Autónoma Metropolitana-Gobierno del Distrito Federal. Artes Impresas Eón, S.A. de C.V. México, D.F. 365 p.
3. Jiménez B (2009) Risks and integrated management of the urban water cycle in megacities of the developing world: Mexico City. In: Water and Urban Development Paradigms. Towards an integration of engineering, design and management approaches. Feyen J, Shannon K, Neville M (Eds.). CRC Press. Taylor & Francis Group. London. pp. 387–396.
4. Domínguez-Mora R (2011) Diagnóstico resumido de los problemas de abastecimiento de agua y de inundaciones en el Valle de México. Revista Digital Universitaria 12(2): 1067–6079.
5. Bradshaw AD (2002) Introduction and Philosophy. In: Perrow MR, Davy AJ. Handbook of Ecological Restoration. Principles of Restoration. Cambridge University Press. Cambridge, U.K.
6. García E (1988) Modificaciones al sistema de clasificación climática de Köppen. Instituto de Geografía, UNAM, México, D.F. 33 p.
7. Álvarez-Román KE (2000) Geografía de la Educación Ambiental: algunas propuestas de trabajo en el Bosque de Los Dinamos, Área de Conservación Ecológica de la Delegación Magdalena Contreras. Tesis Licenciatura en Geografía. Facultad de Filosofía y Letras, UNAM. México. 127 p.
8. Ávila-Akerberg V (2009) Forest quality in the southwest of Mexico City. Assessment towards ecological restoration of ecosystem services. Ph.D. Dissertation. Institute für Landespflege. Universität Freiburg. Germany. 138 p.
9. Secretaría de Medio Ambiente y Recursos Naturales, Instituto Nacional de Ecología (SEMARNAT-INE) (2003a) Acuerdo que declara Zona Protectora Forestal los bosques de La Cañada de Contreras, D.F. In: Áreas naturales protegidas de México con decretos federales. México, D.F. 302 p.
10. Secretaría de Medio Ambiente y Recursos Naturales, Instituto Nacional de Ecología (SEMARNAT-INE) (2003b) Declaración de Zona Protectora Forestal del Río Magdalena In: Áreas Naturales Protegidas de México con Decretos Federales. México, D.F. 317 p.
11. Ávila-Akerberg V (2004) Autenticidad de los bosques en la cuenca alta del río Magdalena. Diagnóstico hacia la restauración ecológica. Tesis Maestría. Facultad de Ciencias, UNAM. México. 112 p.
12. Programa Universitario de Estudios sobre la Ciudad-Gobierno del Distrito Federal (PUEC-GDF) (2008) Plan Maestro de Manejo Integral y Aprovechamiento Sustentable de la Cuenca del río Magdalena, Distrito Federal. Diagnóstico Integral. 90 p.: http://www.sma.df.gob.mx/sma/links/dowload/biblioteca/diagnostico_integral.pdf Accessed 2013 Aug 20.
13. Rzedowski GC de, Rzedowski J (2001) Flora fanerogámica del Valle de México. 2ª ed. Instituto de Ecología, A.C. y Comisión Nacional para el Conocimiento y Uso de la Biodiversidad. Pátzcuaro, Michoacán. 1406 p.
14. Cantoral-Uriza EA, Almeida-Leñero L, Cifuentes-Blanco L, Paniagua-León L, Luis-Martínez MA, et al. (2009) La diversidad de una cuenca en la Ciudad de México. Ciencias 94: 28–33.
15. Nava M (2003) Los bosques de la cuenca alta del río Magdalena, D.F., México. Un estudio de vegetación y fitodiversidad. Tesis Licenciatura Biología. Facultad de Ciencias. UNAM. México. 64 p.
16. Jujnovsky J, Almeida-Leñero L, Bojorge-García M, Monges YL, Cantoral-Uriza E,et al. (2010) Hydrologic ecosystem services: water quality and quantity in the Magdalena River, Mexico City. Hidrobiológica 20(2): 113–126.
17. Programa Universitario de Medio Ambiente-Gobierno Distrito Federal (PUMA-GDF) (2009) Sistema de Indicadores para el rescate de los ríos Magdalena y Eslava. México, D.F. 134 p. Available: http://www.sma.df.gob.mx/sma/links/download/archivos/sistema_indicadores.pdf. Accessed 2013 Aug 20.
18. Riley AL (1998) Restoring Streams in Cities. A Guide for Planners, Policymakers, and Citizens. Island Press. Washington, D.C.
19. González Reynoso AE, Hernández Muñoz L, Perló Cohen M, Zamora Saenz I (2010) Rescate de ríos urbanos. Propuestas conceptuales y metodológicas para la restauración de ríos. Universidad Nacional Autónoma de México. Programa de Estudios sobre la Ciudad. Ed. Color, S.A. de C.V. México, D.F. 109 p.

20. Banco Nacional de Datos de Aguas Superficiales (BANDAS) (2006) *National Data Bank Surface Waters*. Comisión Nacional del Agua. Available: <ftp://ftp.conagua.gob.mx/Bandas/Bases_Datos_Bandas> Accessed 2013 March 18.
21. Searcy JK, Hardison CH (1960) Double-mass curves. U.S. Geological Survey Water-Supply Paper 1541-B. USGS, Washington, D.C.
22. Yuefeng Y, Tijiu C, Xiaohua W, Mingfang Z, Cunyong J (2011) Effect of forest recovery on summer streamflow in small forested watersheds, Northeastern China. Hydrol Process 26 (8): 1208–1214. DOI: 10.1002/hyp.8204.
23. Orta de Velásquez MT, Yáñez Noguez I, Mazari Hiriart M, Ríos Soriano B, Velázquez Pedroza K, et al. (2008) Restauración de un río urbano. Aspectos ambientales e hidrológicos. XXXI Congreso Interamericano de Ingeniería Sanitaria y Ambiental. Santiago de Chile, Chile. 12–15.
24. HACH (2002) Water analysis Handbook. 4th ed. HACH Co. CA. 1260 p.
25. American Public Health Association (APHA), American Water Works Association, Water Environmental Federation (2005) Standard Methods for Examination of Water and Wastewater. 21a ed. United Book Press. Washington. D.C.
26. Murray PR, Baron EJ, Pfaller MA, Tenover FC, Yolken RH (Eds.) (1995) Manual of Clinical Microbiology. American Society for Microbiology. Washington D. C. 1482 p.
27. Gerba CP (2000) Indicator microorganisms. In: Maier RM, Pepper IM, Gerba CP (Eds.) Environmental Microbiology. Academic Press. San Diego, California. pp. 491–503.
28. Toranzos GA, McFeters GA, Borrego JJ, Savill M (2007) Detection of microorganisms in environmental freshwaters and drinking water. In: Manual of Environmental Microbiology. Edited by Hurst CJ, Crawfortd RL, Garland JL, Lipson DA, Mills AL, et al. Washington D. C. ASM Press. pp. 249–264.
29. ISO 10705-1 (1995), Water quality -Detection and enumeration of bacteriophages-part 1: Enumeration of F-specific RNA bacteriophages. Geneva, Switzerland: International Organization for Standardization.
30. Solano-Ortiz R (2011) Avances metodológicos para la detección del virus de la hepatitis A y bacteriófagos en agua. Tesis Licenciatura, Biología. Facultad de Ciencias, UNAM. México. 68 p.
31. Fonseca-Salazar MA (2010) Bacteriófagos como indicadores de la calidad del agua en Ciudad Universitaria. Tesis Licenciatura, Biología. Facultad de Ciencias, UNAM. México, D.F. 62 p.
32. Polaczyk LA, Narayanan J, Cromeans LT, Hahn D, Roberts MJ, et al. (2008) Ultrafiltration-based techniques for rapid and simultaneous concentration of multiple microbe classes from 100-L tap water samples. J Microbiol Methods 73: 92–99.
33. Monpoeho S, Dehee A, Mignotte B, Schwartzbrod L, Marechal V, et al. (2000) Quantification of enterovirus RNA in sludge samples using single tube real-time RT-PCR. BioTechniques 29: 88–93.
34. Xagoraraki I, Kuo DH, Wong K, Wong M, Rose JB (2007) Occurrence of human adenoviruses at two recreational beaches of the Great Lakes. Appl Environ Microbiol 73: 7874–7881.
35. US Environmental Protection Agency (USEPA) (2005) Method 1623: *Cryptosporidium* and *Giardia* in Water by Filtration/IMS/FA, Cincinnati, OH, December 2005. 68 p.
36. Rangel-Martínez C (2010) Identificación de *Blastocytis hominis* en muestras de agua obtenidas de Ciudad Universitaria en la Ciudad de México. Tesis Licenciatura, Biología. Universidad Simón Bolívar. México D.F. 62 p.
37. Tapia Palacios MA (2012) Detección de *Cryptosporidium parvum* y *Giardia lamblia* en agua del río Cuitzmala, Jalisco. Tesis Licenciatura, Biología. Facultad de Ciencias, UNAM. México. 64 p.
38. McCullagh P, Nedler JA (1989) Generalized Linear Models. Monographs on Statistics and Probability, Vol. 37, Chapman and Hall, New York.
39. Hiriart F, Marsal RJ, Cruickshank G, Key F (1952) Contribución de la Comisión Federal de Electricidad a la Solución del Problema de Abastecimiento de Agua a la Ciudad de México. México. 132 p.
40. Departamento del Distrito Federal (DDF) (1975) Memoria de las Obras del Sistema de Drenaje Profundo del Distrito Federal. Tomo I. DDF-Talleres Gráficos de la Nación. México.
41. González-Martínez MT (2008) Modelación hidrológica como base para el pago por servicios ambientales en la microcuenca del río Magdalena, Distrito Federal. Tesis Maestría. Universidad Autónoma de Querétaro. México. 137 p.
42. Jujnovsky J, González-Martínez TM, Cantoral-Uriza EA, Almeida-Leñero L (2012) Assessment of Water Supply as an Ecosystem Service in a Rural-Urban Watershed in Southwestern Mexico City. Environ Manage 49 (3): 690–702. DOI 10.1007/s00267-011-9804-3.
43. Krzanowski WJ (1988) Principles of multivariate analysis: A user's perspective. Oxford University. Clarendon Press. New York. 563 p.
44. Diario Oficial de la Federación (DOF) (1997) NOM-001-SEMARNAT-1996. Que establece los límites máximos permisibles de contaminantes en las descargas de aguas residuales en aguas y bienes nacionales. 6 enero 1997.
45. Ley Federal de Derechos en Materia de Agua (LFDMA) (2012) Ley Federal en Materia de Agua. México.
46. Bergey's (2012) Bergey's Manual of Systematic Bacteriology (2001–2012), 2nd ed. Springer Verlag. New York.
47. Long CS, El-Khoury SS, Oudejans GS, Sobsey DM, Vinjé J (2005) Assessment of sources and diversity of male-specific coliphages for soured tracking. Environ Eng Sci 22(3): 367–377.
48. Schaper M, Jofre J (2000) Comparison of methods for detecting genotypes of F-specific RNA bacteriophages and fingerprinting the origin of feacal pollution in water samples. J Virol Methods 89: 1–10.
49. Havelaar HA, Hogeboom MW (1984) A method for the enumeration of male-specific bacteriophages in sewage. J Appl Bacteriol 56: 439–447.
50. Mooijman AK, Bahar M, Muniesa M, Havelaar HA (2002) Optimisation of ISO 10705-1 on enumeration of F-specific bacteriophages. Journal of Virological Methods 103: 129–136.
51. Muir P, Van Loon AM (1997) Enterovirus infections of the central nervous system. Intervirology 40: 153–66.
52. Wadell G (1984) Molecular Epidemiology of Human Adenoviruses. Curr Top Microbiol Immunol 110: 191–220.
53. Maier RM, Pepper IL, Gerba CP (2009) Environmental Microbiology. China. Academic Press. 598 p.
54. Hierholzer JC (1992) Adenoviruses in the immunocompromised host. Clin Microbiol Rev 5: 262–274.
55. Luna MM, Leboreiro JI, Zapata IB, Rendón-Macías ME (2013) Los adenovirus como causa de gastroenteritis aguda en los niños. Revista Mexicana de Pediatría 80: 98–104.
56. Enriquez CE, Hurst CJ, Gerba CP (1995) Survival of the enteric adenoviruses 40 and 41 in tap, sea, and waste water. Water Res 29: 2548–2553.
57. Ciruela P, Izquierdo C, Oviedo M, Domínguez A (2004) Rotavirus y adenovirus 40–41 como causantes de gastroenteritis infantil. Vacunas 5: 16–23.
58. Crabtree PA, Gerba CP, Rose JB, Hass CN (1997) Waterborne adenovirus a risk assessment. Water Science and Technology 35 (11–12): 1–6.
59. Kundu A, McBride G, Wuertz S (2013) Adenovirus-associated health risk for recreational activities in a multi-use coastal watershed based on site-specific quantitative microbial risk assessment. Water Res 47: 6309–6325.
60. El-Senousy WM, Osman GA, Melegy AA (2014) Survival of Adenovirus, Rotavirus, Hepatitis A Virus, Pathogenic Bacteria and Bacterial Indicators in Ground Water. World Applied Sciences Journal 29 (3): 337–348. DOI: 10.5829/idosi.wasj.2014.29.03.13849
61. Xiao L, Fayer R (2008) Molecular characterization of species and genotypes of Cryptosporidium and Giardia and assessment of zoonotic transmission. Int J Parasitol 38: 1239–1255.
62. Reynolds KA, Mena KD, Gerba CP (2008) Risk of waterborne illness via drinking water in the United States. Rev Environ Contam Toxicol 192: 117–158.
63. Plutzer J, Karanis P, Domokos K, Törökné A, Márialigeti K (2008) Detection and characterization of Giardia and Cryptosporidium in Hungarian raw, surface and sewage water samples by IFT, PCR and sequence analysis of the SSUrRNA and GDH genes. Int J Hyg Environ Health 211: 524–533.
64. Okhuysen PC, Chappell CL, Crabb JH, Sterling CR, Dupont HL (1999) Virulence of three distinct Cryptosporidium parvum isolates for healthy adults. J Infect Dis, 180: 1275–81.
65. Rendtorff RC (1979). The experimental transmission of Giardia lamblia among volunteer subjects. In:W. Jakubowski and J.C. Hoff (eds) Waterborne Transmission of Giardiasis. Report No. EPA-600/9-79-001, pp. 64-8. US Environmental Protection Agency. Environmental Research Centre, Cincinnati.
66. US Environmental Protection Agency (USEPA) (2001a) *Cryptosporidium*: Drinking Water Health Advisory. Office of Science and Technology EPA-822-R-01-009.Environmental Protection Office of Water: 31 p.
67. US Environmental Protection Agency (USEPA) (2001b) Drinking Water Criteria Document Addendum: *Cryptosporidium*. United States Environmental Protection Agency, Office of Water, Washington, DC.
68. Betancourt WQ, Rose JB (2004) Drinking water treatment processes for removal of *Cryptosporidium* and *Giardia*. Vet Parasitol 126 (1–2): 219–234.
69. US Environmental Protection Agency (USEPA) (2013) Basic Information about Pathogens and Indicators in Drinking Water. December 13, 2013. http://water.epa.gov/drink/contaminants/basicinformation/pathogens.cfm Accessed 2014 Jun 21.
70. Metcalf & Eddy Inc (1996) Ingeniería de Aguas Residuales Tratamiento, Vertido y Reutilización 1: 53–55, 145–46. México.
71. Velázquez PK (2009) Inventario de descargas y propuestas de tratamiento para el saneamiento del Río Magdalena. Tesis Ingeniería Química. FES Zaragoza. Universidad Nacional Autónoma de México.
72. US Environmental Protection Agency (USEPA 1999) *Alternative Disinfectants and Oxidants. Guidance Manual*. Ch. 8. Ultraviolet Radiation. Environmental Protection Office of Water EPA 815-R-99-014. USA.
73. Sistema de Aguas de la Ciudad de México (SACM) (2013) Official Receipt. Derechos por el suministro de agua. Magdalena Contreras. Manzana Baja. September, 2013.
74. Silva-Magaña MA (2013) Uso de indicadores virales para la evaluación de la calidad de agua potable y de reúso en Ciudad Universitaria, México. Tesis Licenciatura. Biología. Facultad de Ciencias, Universidad Nacional Autónoma de México. México: 70 pp.

7

The Sequential Application of Macroalgal Biosorbents for the Bioremediation of a Complex Industrial Effluent

Joel T. Kidgell, Rocky de Nys, Nicholas A. Paul, David A. Roberts*

MACRO – the Centre for Macroalgal Resources and Biotechnology, and School of Marine and Tropical Biology, James Cook University, Townsville, Queensland, Australia

Abstract

Fe-treated biochar and raw biochar produced from macroalgae are effective biosorbents of metalloids and metals, respectively. However, the treatment of complex effluents that contain both metalloid and metal contaminants presents a challenging scenario. We test a multiple-biosorbent approach to bioremediation using Fe-biochar and biochar to remediate both metalloids and metals from the effluent from a coal-fired power station. First, a model was derived from published data for this effluent to predict the biosorption of 21 elements by Fe-biochar and biochar. The modelled outputs were then used to design biosorption experiments using Fe-biochar and biochar, both simultaneously and in sequence, to treat effluent containing multiple contaminants in excess of water quality criteria. The waste water was produced during ash disposal at an Australian coal-fired power station. The application of Fe-biochar and biochar, either simultaneously or sequentially, resulted in a more comprehensive remediation of metalloids and metals compared to either biosorbent used individually. The most effective treatment was the sequential use of Fe-biochar to remove metalloids from the waste water, followed by biochar to remove metals. Al, Cd, Cr, Cu, Mn, Ni, Pb, Zn were reduced to the lowest concentration following the sequential application of the two biosorbents, and their final concentrations were predicted by the model. Overall, 17 of the 21 elements measured were remediated to, or below, the concentrations that were predicted by the model. Both metalloids and metals can be remediated from complex effluent using biosorbents with different characteristics but derived from a single feedstock. Furthermore, the extent of remediation can be predicted for similar effluents using additive models.

Editor: Paul Jaak Janssen, Belgian Nuclear Research Centre SCK•CEN, Belgium

Funding: This research is part of the MBD Energy Research and Development program for Biological Carbon Capture and Storage with the co-operation of Stanwell Energy Corporation. This project is supported by the Advanced Manufacturing Cooperative Research Centre (AMCRC), funded through the Australian Government's Cooperative Research Centre Scheme, and the Australian Renewable Energy Agency (ARENA). Stanwell Energy Corporation provided effluents for the study. The other funders had no role in study design, data collection and analysis, decision to publish, or preparation of the manuscript.

Competing Interests: The authors have the following interests. This research is part of the MBD Energy Research and Development program for Biological Carbon Capture and Storage with the co-operation of Stanwell Energy Corporation. Stanwell Energy Corporation provided effluents for the study. The data contained within this study form a component of International Patent 2736PC2-MLE/MAR "Biosorbent and methods of use," James Cook University. There are no further patents, products in development or marketed products to declare.

* Email: david.roberts1@jcu.edu.au

Introduction

Coal-fired energy generation produces metal-contaminated waste water with high concentrations of regulated elements such as Al, As, Cd, Mo, Se, V and Zn. The effluent is created when water is used to wash residual ash from combustion chambers and flue stacks [1]. As these effluents are typically too contaminated to be discharged, and current treatment options are restricted by cost [2], the effluent from coal-fired power stations is often stored in large Ash Dams (ADs). However, high bioavailability and rapid bioaccumulation of toxic elements from Ash Dam Water (ADW) has resulted in cases of ecotoxicity to vertebrates in ADW catchments [3,4] and, therefore, sustainable water treatment technologies are required. Biosorption, the use of dried biological material to passively bind contaminants from waste water, is an option for the treatment of industrial effluents [5,6]. We have recently demonstrated that dried freshwater macroalgae are an effective feedstock for the production of biosorbents, and native species from industrial facilities can be intensively cultivated to provide biomass for biosorption [6,7]. This circumvents one of the critical constraints to algal-based biosorption, which is sourcing a sustainable feedstock for production of biosorbents that does not compete with established markets for cultivated macroalgae [8].

Biosorption exploits charge-based interactions between dissolved elements and biosorbents. Negatively charged functional groups on the surface of dried macroalgae makes it an effective biosorbent with which to remove positively charged metal cations from solution and various macroalgal biosorbents have proven effective against a range of dissolved metals [9–11]. However, existing biosorption research is largely limited to synthetic effluents with only one or a few elements targeted for remediation [5,7,12,13]. In reality, industrial effluents contain a myriad of coexisting elements and the complexities of treating these effluents are not replicated in experiments with synthetic solutions [14]. A continued focus on the kinetics of metal uptake from synthetic effluents by biosorbents, at the expense of empirical data on the performance of biosorbents in real-world effluents, has arguably limited the application of algal-based biosorption at scale [6,15].

The treatment of complex effluent is difficult because the dissolved ions in the effluent have a variety of properties, and hence, affinities for biosorbents [6,16,17]. Macroalgae are in general effective against a range of dissolved cations and the affinity of dried macroalgae for metals (e.g. Ni^{2+}) can be enhanced by converting biomass to biochar through slow pyrolysis [14,18]. However, both dried biomass and biochar have low affinities for metalloids (e.g. SeO_4^{2-}) that co-occur with metals in industrial waste water as oxyanions [19–21]. An additional conversion procedure, the treatment of biochar with an iron (Fe^{3+}) solution, can enhance the affinity of biochar for metalloids [22]. The resulting Fe-biochar binds a range of metalloids including Se, As and Mo at a rate that is independent of pH but also releases a suite of metals (cations), that are natural components of the biomass, back into solution in the process [14,22]. While the release of metals during deployment of Fe-biochar for the purposes of metalloid biosorption may appear counter-productive to the aims of treating waste water, sequential application of Fe-biochar to remove metalloids and then biochar to remove metals can, in principle, treat a complex effluent far more comprehensively than if any one biosorbent was applied individually. The application of multiple biosorbents to holistically treat complex waste water through biosorption has never been empirically tested.

Here we test the applicability of a multiple biosorbent treatment technique, using Fe-biochar and biochar produced from freshwater macroalgae, to treat a complex effluent containing both metalloids and metals. We first develop a predictive model for the removal of metalloids and metals by Fe-biochar and biochar, respectively, from ADW collected from Tarong power station (Queensland, Australia). We then apply the model to predict the quantity of each biosorbent that is required to maximise the removal of metalloids and metals from ADW, accommodating for any leaching of contaminants from the biomass during the use of Fe-biochar. Finally, we produce Fe-biochar and biochar from cultivated green macroalgae *Oedogonium*, that was originally isolated from the AD of Tarong power station, and quantitatively compare the efficacy of Fe-biochar and biochar deployed simultaneously, and in sequence, to remediate metalloids and metals from the ADW.

Materials and Methods

Industrial effluent

This study focused on ADW from Tarong coal-fired power station in south-east Queensland, Australia (26.76°S, 151.92°E). Tarong is one of Queensland's largest coal-fired power stations with a generation capacity of 1400 MW and a 46,000 ML AD to store waste water from ash disposal processes on site. The ADW was sourced directly from the AD and transported to James Cook University (JCU), Townsville in clean plastic 1000 L Intermediate Bulk Containers (IBCs) and stored at ambient temperature in 12,000 L storage tanks until use. The ADW was collected and shipped with the permission and assistance of Stanwell Energy Corporation.

Algal biosorbents production and preparation

Oedogonium sp. (Genbank: KF606974 [23]) (hereafter *Oedogonium*), was used as the feedstock for the biosorbents. *Oedogonium* is a filamentous, freshwater macroalga that is native to Tarong AD. The biomass for this study was cultivated in f/2 media in 2500 L tanks during the austral summer months (January – March) in the aquaculture facility on the Townsville campus of JCU (19.33°S, 146.76°E). Prior to experiments, 2 kg of algae was harvested from the tanks and oven dried at 60°C for 48 hours (h). The biomass was then converted into biochar by slow pyrolysis under conditions previously described [24]. Briefly, *Oedogonium* was suspended in a muffle furnace and purged with N_2 gas at 4.0 L min^{-1} while being heated to 450°C for 1 h. A sample of the biochar was converted to Fe-biochar by soaking it in a 5% Fe^{3+} solution (diluted Sigma Aldrich 45% w/v $FeCl_3$ stock solution) at a density of 25 g L^{-1} for 24 h on a shaker plate (100 rpm) at 20°C. The Fe-biochar was filtered from the $FeCl_3$ solution and rinsed three times with deionized (DI) water at a rate of 20 ml g^{-1} and then dried at 60°C for 48 h.

Derivation of predictive sorption model

A model was developed using data from a previous study with Tarong ADW [14] (Table S1) to predict the change in concentration of 21 elements after the deployment of each biosorbent. Most metal and metalloid sorption occurs within the first hour of exposure and the most effective sorption occurs when the pH of the ADW is un-manipulated (pH ~7.1) [14]. The biosorption data collected under these conditions was used to construct the predictive model. Following 1 h of exposure of Fe-biochar or biochar to ADW the q-value (the mass of an ion [μg] adsorbed from, or released into, solution per unit of biosorbent [g]), was calculated according to Volesky (2007) [7]. The q-value was determined for each combination of biosorbent (Fe-biochar and biochar) and element (Al, As, Ba, B, etc.) to derive a model that predicted the additive effect of multiple biosorbents, applied sequentially or simultaneously, on the remediation of dissolved elements within a single effluent:

$$M_i = (C \times V) - \sum [q_x \times S_{x,i}]$$

Where, M_i is the mass of an element in solution (μg) following treatment number *i*; C is the initial concentration of the element in solution (μg L^{-1}); V is the volume of effluent solution (L); q_x is the q-value of the element for biosorbent, *x* (μg g^{-1}); $S_{x,i}$ is the mass (g) of biosorbent, *x*, added for treatment, *i*.

Biosorption experiments

The aim of the biosorption experiment was to maximise the removal of oxyanionic elements (As, Mo and Se) with Fe-biochar, followed by the removal of cationic contaminants with secondary, or simultaneous, deployment of biochar. As described above, this sequential treatment strategy is logical because the targeted removal of metalloids by Fe-biochar also contributes some metals into solution. This requires a second phase of biosorption with biochar to remove existing and leached metals in ADW.

In total there were 6 biosorption treatments: (1) Fe-biochar ("FeBC"); (2) biochar ("BC"); (3) sequential Fe-biochar ("FeBC → FeBC"); (4) sequential biochar ("BC → BC"); (5) sequential Fe-biochar and biochar ("FeBC → BC"); and (6) simultaneous Fe-biochar and biochar ("FeBC + BC") (Figure 1). As the aim of the biosorption experiment was to achieve remediation of metalloids and metals, the primary treatments of interest are the sequential and simultaneous application of Fe-biochar and biochar (treatments 5 and 6 respectively, Figure 1). Treatments 1–4 were included as controls to assess the effects of Fe-biochar and biochar in isolation for comparison with the sequential and simultaneous treatments. Each treatment strategy was repeated 3 times to evaluate the model predictions.

The predictive model was used to select the loading densities of each biosorbent for the two main treatments and controls. Metalloids were targeted for removal by using a density of Fe-

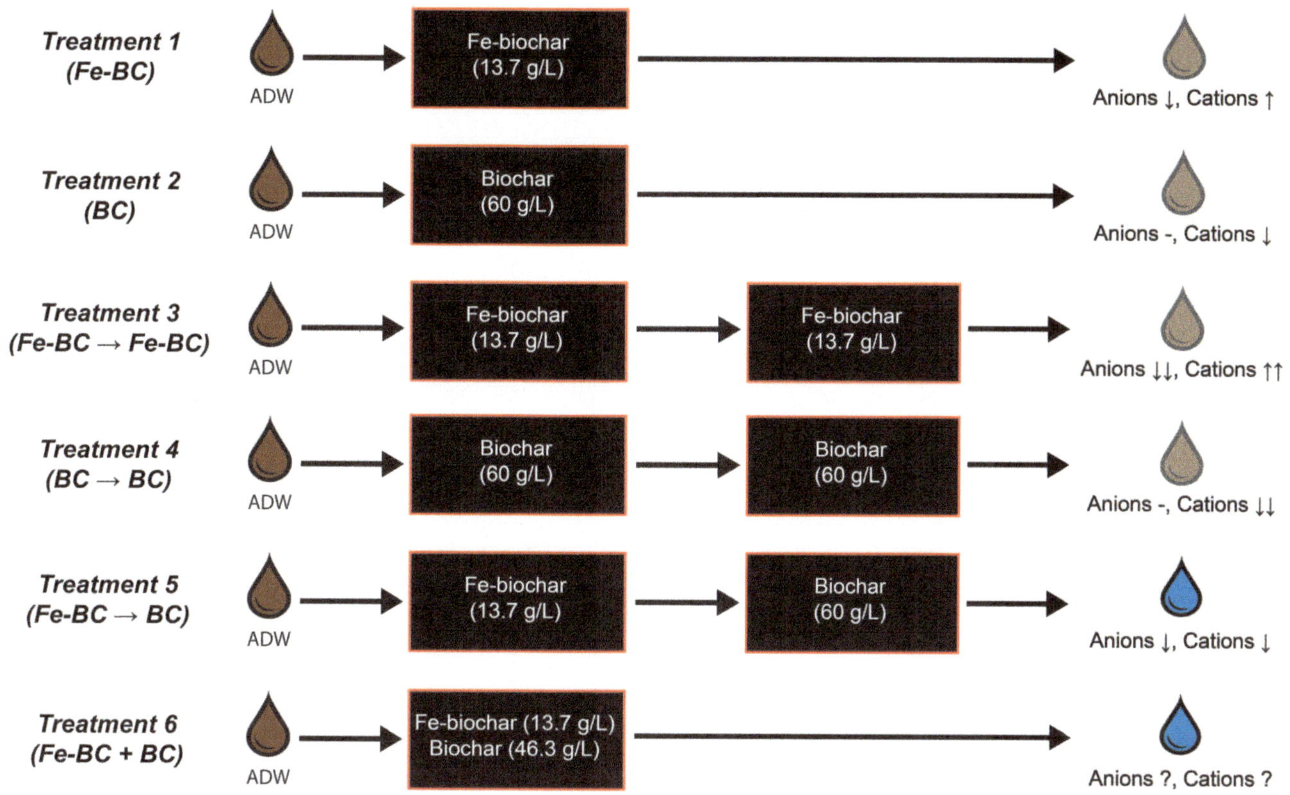

Figure 1. Experimental treatments and predicted changes in anion and cation concentrations in ADW. Arrows signify the direction and magnitude of predicted change (up = increase, down = decrease, dash = no change, ? = unknown response). Six biosorption treatments were tested: (1) Fe-biochar ("FeBC"); (2) biochar ("BC"); (3) sequential Fe-biochar ("FeBC → FeBC"); (4) sequential biochar "BC → BC"; (5) sequential Fe-biochar and biochar ("FeBC → BC" and (6) simultaneous Fe-biochar and biochar ("Fe-BC + BC").

biochar that was predicted by the model to result in removal of Mo, the most abundant metalloid, to below ANZECC water quality criteria. The result was that Fe-biochar would be used at a density of 13.7 g L^{-1} (see further justification in results section "*Development of treatment scenarios using the predictive model*"). A limitation of the experiment was that the biochar density predicted by the model to remove all metals (110 g L^{-1}), is not feasible as the maximum possible loading density of biochar or Fe-biochar was 60 g L^{-1}. Therefore, all biochar treatments were loaded at a density of 60 g L^{-1} and all Fe-biochar treatments were loaded at a density of 13.7 g L^{-1}. The simultaneous "Fe-BC + BC" treatment included 13.7 g L^{-1} Fe-biochar and 46.3 g L^{-1} biochar to give the maximum 60 g L^{-1} of biosorbent (Figure 1).

Each replicate consisted of a 250 ml plastic beaker with 50 ml of ADW and the appropriate mass of biosorbent. The pH of the ADW was unaltered (7.06±0.01) for all treatments. Fe-biochar or biochar (depending on treatment) was added to the ADW and placed in a shaker cabinet (Eppendorf Innova 44R) at 100 rpm in 20°C for 1 h. After 1 h, the biosorbent was separated from ADW by centrifugation (7000 rcf, 5 min), followed by two stages of filtration (75 µm nylon filter paper, then 0.45 µm syringe filter). The filtered solution was transferred to another 250 ml plastic beaker for the next treatment (biochar or Fe-biochar depending on the scenario). The process of the second sequential treatment was identical to the first, except that the mass of biosorbent was adjusted to account for any loss of solution during the separation process to ensure the same density was applied. After the second treatment the solution was processed as described and the water samples retained for analysis.

It is important to note that testing multiple treatments in the biosorption experiments provides an opportunity to validate the performance of the model when initial elemental profiles differ from those under which the model was produced. For example, after the first stage of the "BC → BC" treatment, the effluent is predicted to have low metal, but unchanged metalloid, concentration. If the model accurately predicts the performance of the second BC treatment this indicates the model is robust to variations in initial effluent characteristics. Such a finding would support the use of the model across diverse effluents with a wide range of elemental profiles.

Elemental analysis

The concentrations of metals and light metal ions (Al, Ba, Ca, Cd, Co, Cr, Cu, Fe, K, Mn, Mg, Na, Ni, Pb, Sr, V, and Zn) and metalloids (As, B, Mo, Se) were measured using a Bruker 820-MS Inductively Coupled Plasma Mass Spectrometer (ICP-MS; Al, Ba, Cd, Co, Cr, Cu, Fe, Mn, Ni, Pb, Sr, V, Zn) or a Varian Liberty series II Inductively Coupled Plasma Optical Emissions Spectrometer (ICP-OES; Ca, K, Mg, Na). An external calibration strategy was used for both instruments, where a standard solution of 0.45 µm filtered ADW was used as the vector to calculate the concentration of elements. Collisional Reaction Interface (CRI) was used for As (CRI gas: H_2) and V (CRI gas: He), while ^{82}Se isotope was used for Se quantification, to eliminate polyatomic interferences for these elements. A 1% HCl solution was spiked with 1 ppb As, Se and V and measured three times for quality control; recovery between 98.5 and 110% indicated no significant interferences. All analyses were conducted at the Advanced Analytical Centre at JCU, Townsville.

Data analysis

Multivariate patterns in biosorption were visualized using non-metric Multi-Dimensional Scaling (nMDS) from a Bray-Curtis similarity matrix following forth-root transformation. A Permutational Multivariate Analysis of Variance (PERMANOVA) was conducted in Primer 6.1.14 between the elements and the factor of treatment (Figure 1). Univariate analysis took the form of one-way Analysis of Variance (ANOVA), with the final concentration for each sequential treatment analysed. Data were examined for normality and homogeneity of variance using normal-probability plots of raw residuals and predicted-residual scatter plots [25]. When necessary the data were log-transformed. The PCA and ANOVA tests were conducted in Statistica (Ver. 10, Statsoft Inc.). The log-predicted final concentrations of each element from the biosorption model were plotted against the log-observed values on a scatterplot for validation of the model. A trend line of $y = x$ was plotted to indicate where the points would be expected to lie if the observed and perfect values were in perfect agreement. Any point with a residual value $> \pm 1$ was highlighted as a deviation from the predicted concentration.

Results

Characteristics of ADW

Twelve (Al, As, B, Cd, Cr, Cu, Pb, Mn, Mo, Ni, Se, and Zn) of the 21 elements measured in the ADW have trigger levels established by the Australian and New Zealand Environmental Conservation Council (ANZECC, Table 1) [26]. Nine of these elements (Al, As, B, Cd, Cu, Mo, Ni, Se and Zn) were present in untreated ADW at concentrations in excess of the ANZECC trigger values for the protection of aquatic life at the 95% level (Table 1). These elements are therefore the focus of the following results and discussion.

Development of treatment scenarios using the predictive model

The model was used to predict the amount of Fe-biochar and biochar required to achieve comprehensive remediation of metalloids and metals to below ANZECC trigger values. Allowing for a 15% deviation in removal capacity, the model predicted 13.7 g L^{-1} of Fe-biochar would be sufficient to reduce the concentrations of As, Mo and Se in ADW to below the trigger values (Figure 2a, c and e). As the Fe-biochar leaches some metals into solution [14,22], it was applied first in the sequential treatments so that the subsequent treatment of biochar could adsorb both the existing and leached metals from ADW (Figure 2b, d, f). The density of biochar required during the second treatment for removal of the metals was therefore calculated assuming 13.7 g L^{-1} Fe-biochar had been used in the first stage of treatment. A density of 110 g L^{-1} biochar was predicted for the removal of the leached and existing metals. Given physical limitations the maximum stocking density that could be achieved was 60 g L^{-1}. Consequently, all biochar treatments were applied at this density (Figure 1). The simultaneous treatments were tailored to result in the maximum removal of both metalloids and metals within the physical constraints of the biochar system. Therefore, Fe-biochar was applied at a density of 13.7 g L^{-1} and biochar was added at 46.3 g L^{-1}, giving the maximum stocking density of 60 g L^{-1} in treatment 6 ("FeBC + BC", Figure 1). The simultaneous treatment ("FeBC +BC", Figure 1) was included despite the need for different stocking densities to determine if a mixed biosorbent consisting of Fe-biochar and biochar has potential as a single-step biosorption treatment for complex effluents. As biosorption and metal leaching occurs relatively rapidly (15 min), it is possible there are rapid interactions between leached and bound elements in the simultaneous exposure that are not evident in the sequential treatments.

Evaluation of treatment scenarios with biosorption experiments

Treatment of ADW with Fe-biochar and biochar resulted in a significantly different final composition of dissolved elements than treatment with only Fe-biochar or only biochar (PERMANOVA: "treatment" pseudo-$F_{(3,8)} = 79.1$, $p = 0.001$). ADW treated with Fe-biochar (treatment 1 "FeBC" and treatment 3 "FeBC → FeBC", Figure 1) had lower concentrations of metalloids (As, Mo and Se) than untreated ADW (Figure 3). ADW treated with biochar (treatment 2 "BC" and treatment 4 "BC → BC", Figure 1) had lower alkali, alkaline and transition metal concentrations (Na, Mg, Sr, Ca, Cr, Pb, Fe, Al, Ni, Cu, Zn and Co) than untreated ADW (Figure 3). When ADW was treated with both Fe-biochar and biochar (treatment 5 "FeBC → BC" and treatment 6 "FeBC + BC", Figure 1), it had a unique elemental composition compared to the other treatments, achieving significantly lower concentrations of both metalloids and metals than untreated ADW (Figure 3), with the best outcome being from the sequential treatment (treatment 5 "FeBC → BC") which attained the lowest concentrations of metalloids and metals of all the treatment scenarios (Table 2).

a) Sorption by Fe-biochar alone ("FeBC" and "FeBC → FeBC"). The metalloids (As, Mo and Se) were removed from ADW by Fe-biochar (Figure 4a–c; Table 2). Two applications of Fe-biochar in sequence (treatment 2 "FeBC → FeBC", Figure 1) resulted in greater removal of metalloids than any other treatment, but also resulted in the highest concentrations of dissolved metals being released (Figure 4, Figure S1). For example, Mo decreased from 3913 to 43 µg L^{-1} in ADW (Figure 4b, Table 2), while Zn increased from 36 to 2167 µg L^{-1} (Figure 4e, Table 2). The first deployment of Fe-biochar was more efficient at removing the metalloids than the second sequential deployment, which yielded lower removal rates of As, Mo and Se per unit Fe-biochar used (Figure 4a–c). Therefore, ADW exposed to "FeBC" and "FeBC → FeBC" treatments was characterised by lower metalloid and higher metal concentrations than untreated ADW (Table 2).

b) Sorption by biochar alone ("BC" and "BC → BC"). Metals (Al, Cd, Cu, Mn, Ni, Pb and Zn) were removed from ADW by biochar (Figure 4d–f, Table 2). The final concentrations of Al, Cu, Mn, Ni, Pb, and Zn were lowest following the sequential biochar treatment ("BC → BC") (Figure 4d–e, Table 2). Al, Ni and Zn were all decreased to below their trigger levels in ADW during the sequential treatment "BC → BC" (Figure 4d–e, Table 2). As was also reduced in concentration in the "BC → BC" treatment, however, the final concentration was higher than the final concentration in the "FeBC → FeBC" treatment (Figure 4a, Table 2). The change in the concentration of K was the opposite to the other metals, with substantial leaching occurring, resulting in a total increase in the concentration in ADW of over 1,800,000 µg L^{-1} in the "BC → BC" treatment (Figure S1, Table 2).

c) Sequential and simultaneous treatments ("FeBC → BC" and "FeBC + BC"). The sequential treatment of ADW using Fe-biochar to remove metalloids followed by biochar to target the existing and leached metals (treatment 5 "FeBC → BC") resulted in the most comprehensive treatment of the ADW

Table 1. Concentration of elements in Ash Dam Water and associated ANZECC trigger values.

Element	ANZECC Trigger (μg L^{-1})	Initial Concentration [μg L^{-1} (±SE)]	
Aluminium	**55**	**89**	**(2.1)**
Arsenic	**13**	**34**	**(0.2)**
Boron	**370**	**3823**	**(18)**
Cadmium	**0.2**	**2.1**	**(0.01)**
Chromium	1	0.1	(0.02)
Copper	**1.4**	**2.1**	**(0.03)**
Lead	3.4	0.03	(<0.01)
Manganese	1900	0.9	(0.2)
Molybdenum	**34**	**3913**	**(7.2)**
Nickel	**11**	**38**	**(0.3)**
Selenium	**11**	**97**	**(0.3)**
Zinc	**8**	**36**	**(2.0)**
Barium	-	100	(0.3)
Calcium	-	335,333	(12,012)
Cobalt	-	0.3	(<0.01)
Iron	-	5.0	(<0.01)
Magnesium	-	99,800	(3348)
Potassium	-	44,867	(2178)
Sodium	-	446,333	(11,169)
Strontium	-	4080	(21)
Vanadium	-	982	(8.1)

Bold values exceed the ANZECC trigger value for protection of aquatic life at the 95% level.

(Table 2). As described above, metalloids were removed by the initial application of Fe-biochar while the metals, Al, Cr, Cu, Ni, Pb and Zn all leached off the biosorbent into solution (Figure 4). However, during the subsequent application of biochar, most of these metals were adsorbed to below initial concentrations (e.g. Figure 4d–e), with the exception of Cu and Ni (Table 2). Of the eight ANZECC metals (Al, Cd, Cr, Cu, Mn, Ni, Pb, Zn), four (Al, Cd, Pb, Zn) were reduced in ADW to the lowest concentrations measured across all treatments, and an additional three (Cu, Ni, Mn) were reduced to the second lowest concentrations in ADW in the "FeBC → BC" treatment (Table 2, Table S2).

When Fe-biochar and biochar were used simultaneously (treatment 6 "FeBC + BC") there were higher final concentrations of As, Mo and Se in ADW compared to "FeBC → FeBC" and "FeBC → BC" treatments (Figure 4a–c). B was reduced to its lowest concentration in the "FeBC + BC" treatment, however, this only constituted a 4% drop from the initial concentration (Table 2, Table S2). Similarly, Cr was reduced to the lowest concentration following a treatment with "FeBC + BC" of 1.2±0.04 μg L^{-1}, however, this concentration is still an order of magnitude higher than the initial concentration of Cr in ADW, 0.1±0.02 μg L^{-1} (Table 2, Table S2). Both Cu and Ni had equivalent concentrations between the treatments of "FeBC → BC" and "FeBC + BC" (Table 2, Table S2). An interesting result was the response of V, which reduced in concentration following sequential and simultaneous treatments, but increased slightly in concentration with individual treatments of biochar and Fe-biochar (Figure S1).

Evaluation of the model

The final concentrations of elements following the "FeBC" and FeBC → FeBC" treatments were close to those predicted by the model (r = 0.97 and 0.96 respectively, Figure 5a–b). The treatment of ADW with Fe-biochar delivered greater than expected reductions in Mo, but was less effective than predicted for As (in both single and sequential treatments) and Se (in sequential treatments) (Figure 5a and b). The model was slightly less accurate for predicting the concentrations following "BC" and "BC → BC" treatments, as demonstrated by the weaker correlations between predicted and observed concentrations (r = 0.88 and 0.69 respectively, Figure 5c, d). Despite the weaker correlation, the concentration of 14 of the 21 elements was accurately predicted for a single application of biochar (Figure 5c). The treatment of ADW in the "BC" treatment resulted in higher than predicted concentrations of B and V and lower than expected concentrations of Mn (Figure 5c). The deviations from model prediction were compounded in the "BC → BC" treatment as 11 elements (As, B, Ba, Cr, Fe, Mn, Na, Ni, Se, V & Zn) out of the 21 had residual values > ±1. Na had the greatest deviation from expected concentration (Figure 5d; Table 2). B, V and Mn had similar deviations from predicted concentrations in the "BC → BC" treatment (Figure 5d).

In the "FeBC → BC" treatment 17 of the 21 elements were remediated close to, or below, predicted concentrations (Figure 6). Twelve elements out of 21 (Al, As, B, Co, Cr, Cu, Fe, Mo, Pb, Se, V & Zn) had residuals > ±1 (Figure 6). However, of these most were removed from ADW more effectively than was predicted by the model. As, B, Se and V had higher concentrations than predicted after the "FeBC → BC" treatment (Figure 6). Despite the high rate of leaching of Fe from the Fe-biochar treatment, the

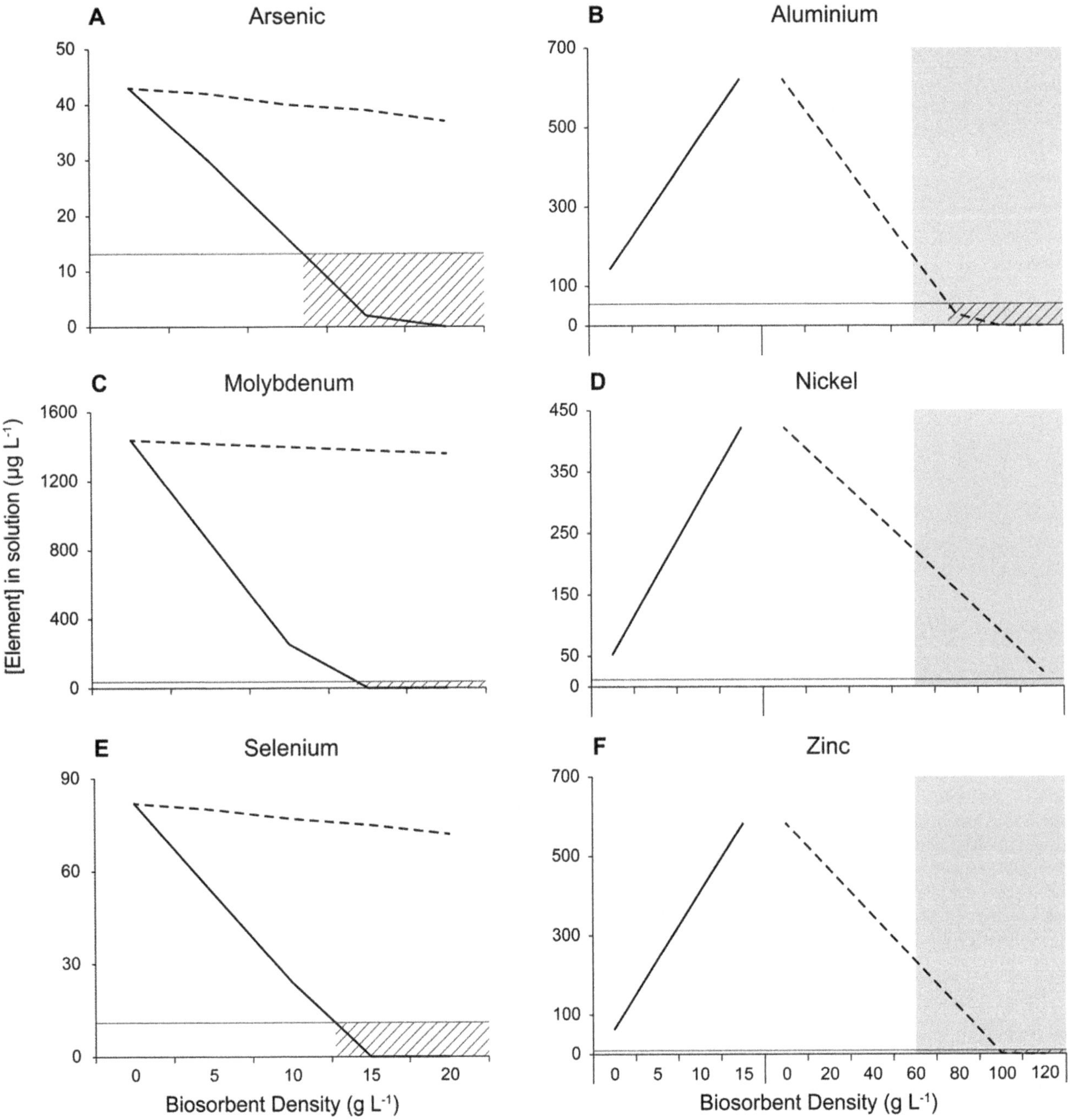

Figure 2. Modelled changes in concentrations of (a) As, (b) Al, (c) Mo, (d) Ni, (e) Se and (f) Zn in ADW treated with Fe-biochar and biochar (solid and dashed lines respectively). ANZECC trigger levels and stocking densities which result in reduction below the ANZECC concentration indicated by horizontal grey line and dashed area, respectively. The initial concentration of Al, Ni and Zn for the biochar treatment in plots (b), (d) and (f) are as predicted following application of 13.7 g L^{-1} Fe-biochar. The grey shaded area indicates biosorbent densities above the physical limit for experimentation (60 g L^{-1}).

subsequent biochar treatment removed all of this Fe, resulting in a concentration (5 µg L^{-1}) much lower than was predicted (39,451 µg L^{-1}) (Figure 6; Table 2).

The predictive model is included as supporting information to this publication, allowing users to insert metal concentrations and desired effluents to see predicted treatment using the sequential approach (Table S3; instructions and assumptions available in the table caption). The elements are treated independently in the model with the assumption that the q-value for each element is independent of initial concentration. The model also assumes the effects of Fe-biochar and biochar use are additive. Therefore, the model output will show linear increases in the requisite Fe-biochar stocking densities as the initial concentration of each target metalloid in solution increases. The amount of biochar required in the second step of the sequential treatment will also increase linearly with the stocking density of Fe-biochar used. Instructions for use are supplied along with Table S3.

Table 2. Change in element concentrations in ADW after exposure to Fe-biochar and biochar biosorbents (see Figure 1 for explanation of treatments).

Element	Initial concentration	Final concentration					
		FeBC	FeBC → FeBC	BC	BC → BC	FeBC → BC	FeBC +BC
Al	89±2.1	505±3.0	872±13	0.4±0.2	0.3±0.01	0.3±0.01	69±9.8
As	34±0.1	14±0.2	7.4±0.2	25±0.2	16±0.2	12±0.3	28±0.3
B	3823±18	4520±21.0	4450±4.7	4337±14	4207±22	4133±19	3650±29
Ba	100±0.3	193±5.2	623±29	94±0.4	55±1	295±13	1700±25
Ca	335,333±12,012	363,667±5683	412,333±11,703	261,667±3569	150,667±2419	423,000±2625	47,433±10,607
Cd	2.1±0.01	1.6±0.01	1.7±0.01	1.1±0.01	1.1±0.01	0.2±0.01	0.8±0.01
Co	0.3±0.01	6.8±0.2	15±0.5	0.2±0.01	0.3±0.01	1.9±0.1	5.2±0.2
Cr	0.1±0.02	13±0.8	49±9.4	2.3±0.01	3.8±0.1	3.6±0.1	1.2±0.04
Cu	2.1±0.03	20±0.3	46±1	2.1±0.02	2.3±0.1	2.9±0.1	3.3±0.1
Fe	5.0±0.01	32,833±1674	84,433±1554	8.2±2.6	5.0±0.01	5.0±0.01	281±115
K	44,867±2178	56,933±504	71,400±1576	1,000,000±9798	1,850,000±24,944	1,096,667±5443	981,000±36,003
Mg	99,800±3348	117,333±272	170,333±981	91,100±1594	74,433±542	133,000±816	14,266±981
Mn	0.9±0.2	1570±33	3880±118	0.9±0.1	1.1±0.04	2887±181	4827±60
Mo	3913±7.2	242±5.7	43±2	4597±28	4617±26	298±8.6	2300±37.4
Na	446,333±11,169	459,667±8060	507,667±3031	321,000±5312	224,000±2625	400,333±3209	446,667±18,764
Ni	38±0.2	924±35	1777±78	11.6±0.4	7.3±0.3	89±5.8	96±6.6
Pb	0.03±0.01	2.7±0.1	6.7±0.1	< LOD	< LOD	0.03±0.01	0.03±0.01
Se	97±0.3	37.3±0.2	24±0.9	112±0.7	112±0.5	40±0.4	50.4±0.8
Sr	4080±21	5413±22	5810±36	3690±45	2227±17	6343±47	5807±28
V	982±8.1	1390±12	1323±7	1333±2.7	1313±12	542±6.4	316±2.6
Zn	36±2.0	438±17	2167±28	2.5±0.1	5.7±0.2	2.5±0.01	53±3.6

All data at mean concentrations ($\mu g\ L^{-1}$ ± S.E.).

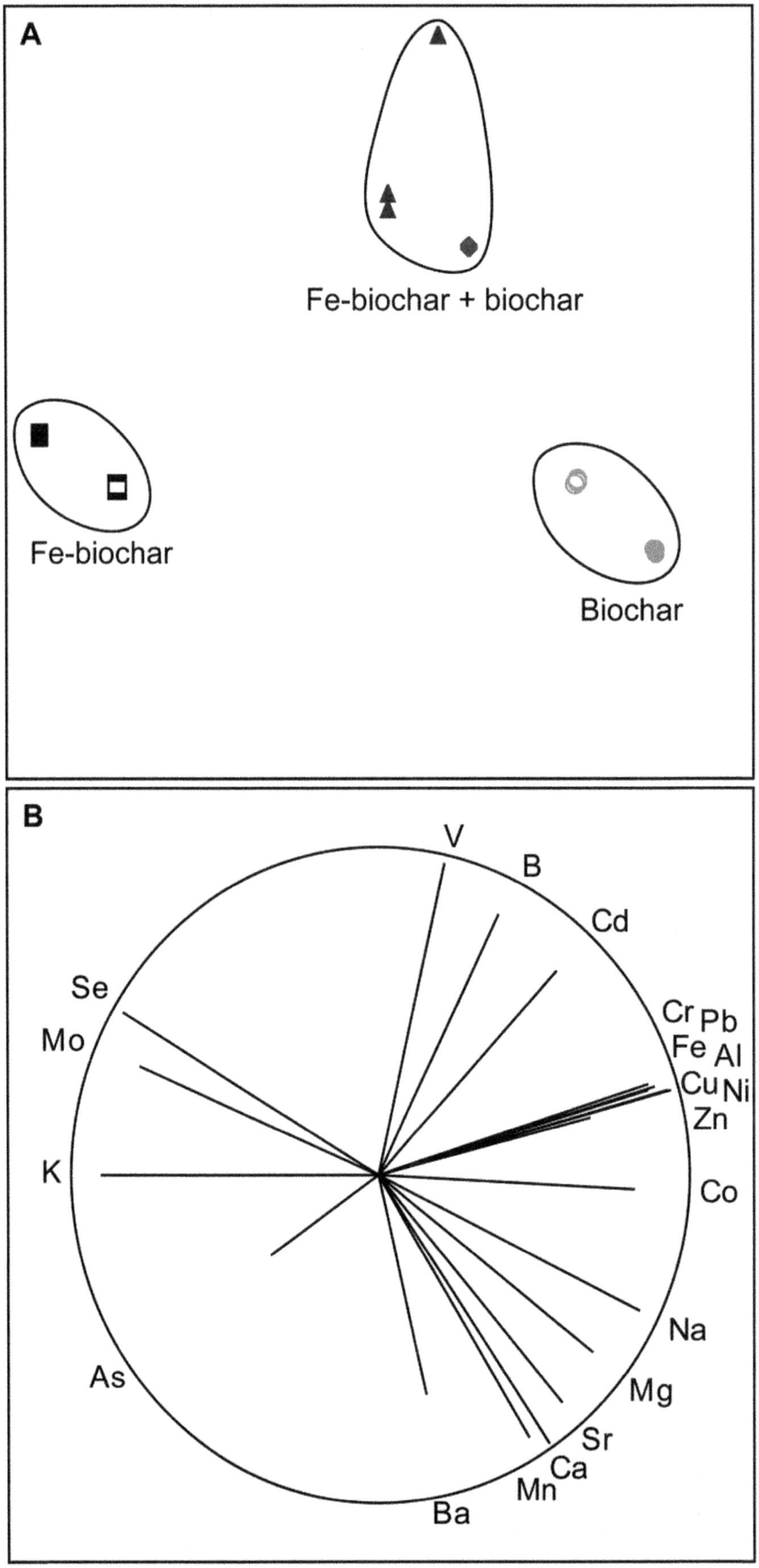
A
Fe-biochar + biochar
Fe-biochar
Biochar
B
V
B
Cd
Se
Mo
Cr
Pb
Fe
Al
Cu
Ni
Zn
K
Co
Na
As
Mg
Sr
Ca
Mn
Ba

Figure 3. Non-metric Multi-Dimensional Scaling (nMDS) demonstrating the sorption of elements by Fe-biochar, biochar, and a combination of the two biosorbents in single and sequential treatments. (A) nMDS (Stress <0.01) with cluster analysis superimposed. Open and closed squares and circles represent single and sequential applications of Fe-biochar and biochar, respectively. Triangles represent sequential application of Fe-biochar followed by biochar and diamonds represent simultaneous application of Fe-biochar and biochar. (B) nMDS with vectors superimposed, the direction and length of indicate the strength of correlation with the treatment.

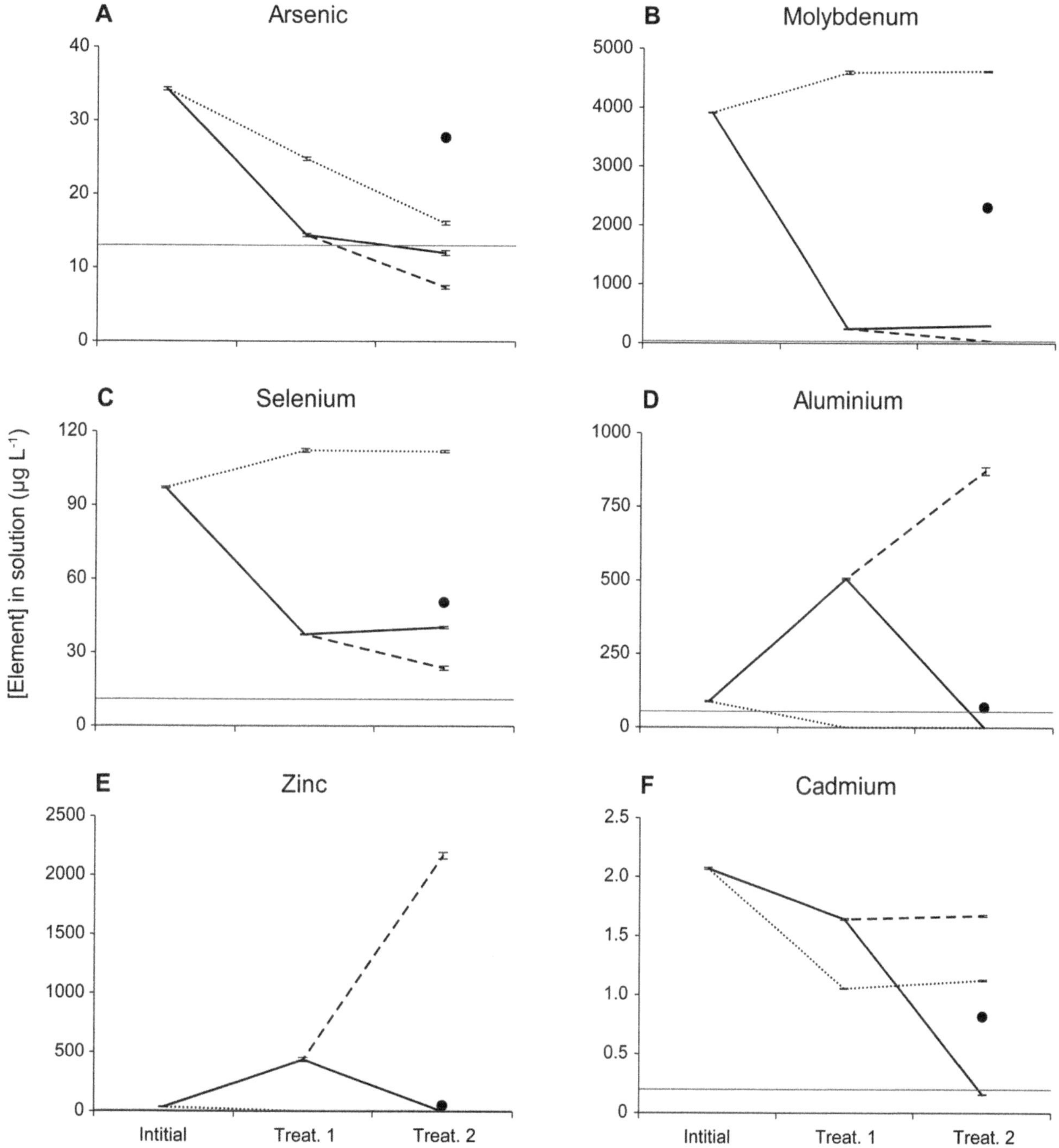

Figure 4. Change in solution concentration of (a) As, (b) Mo, (c) Se, (d) Al, (e) Zn, and (f) Cd in ADW after sequential exposure to Fe-biochar and biochar. The "BC", "FeBC" and "FeBC → BC" treatments are represented as dotted, dashed and solid lines, respectively. The "FeBC + BC" treatment is represented by the circle in each panel. The "FeBC + BC" result is placed under treatment 2 to compare with the final concentrations of the other treatments. ANZECC trigger level is represented by a horizontal grey line.

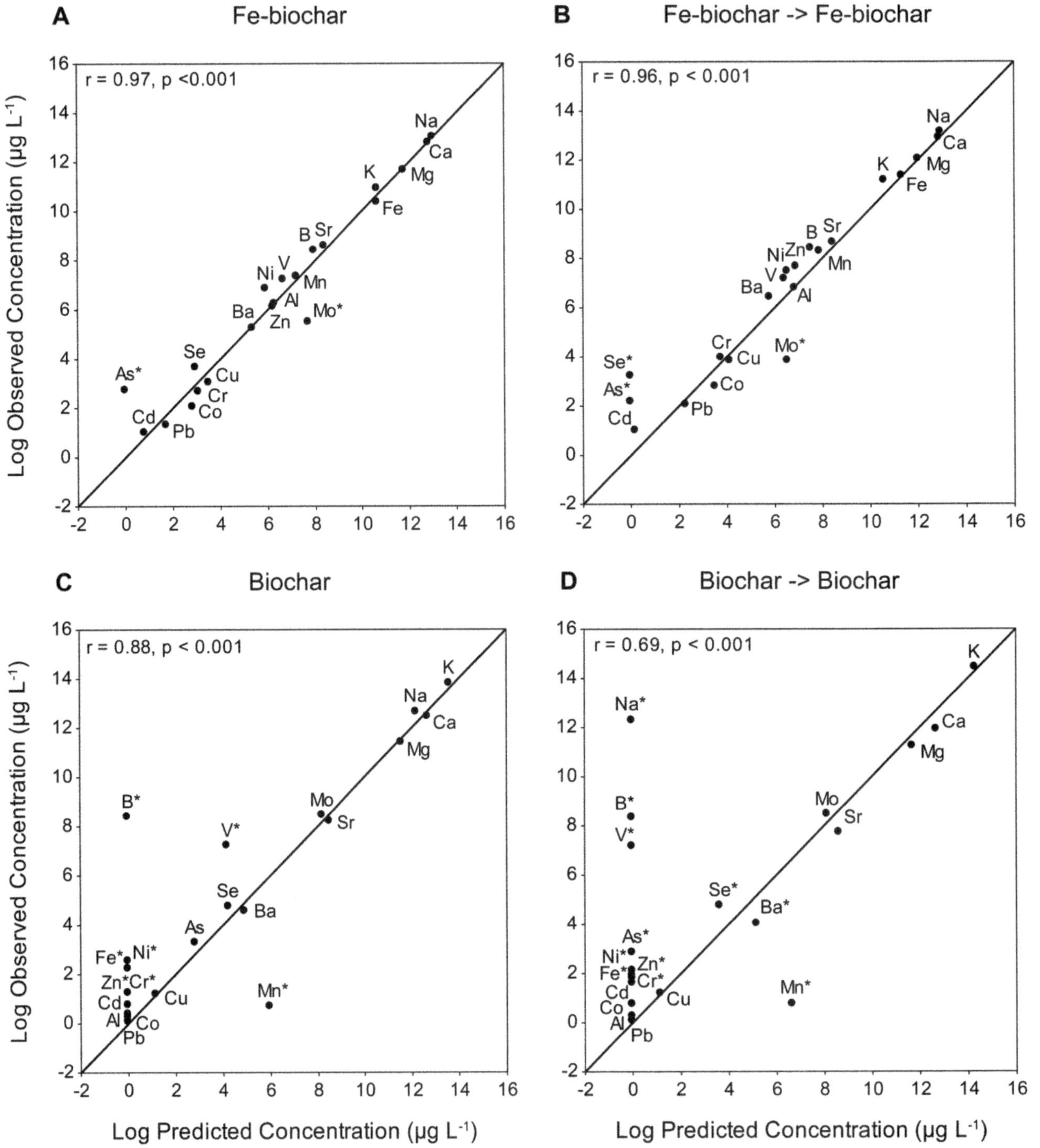

Figure 5. Scatterplots of predicted vs. observed concentrations of elements in ADW (μg L^{-1}) following exposure to (a) "FeBC", (b) "FeBC → FeBC", (c) "BC", and (d) "BC → BC" treatments. The data have been log transformed. The plotted line shows the expected relationship if all observed values are identical to predicted values. Points falling above the line were overestimated by the model (higher than expected final concentrations). Points falling below the line were underestimated by the model (lower than expected final concentrations). Asterisks indicate elements with a residual of greater than ±1.

Discussion

We have demonstrated that the sequential application of multiple biosorbents provides a comprehensive treatment of a complex effluent. The combination of targeted remediation of metalloids by Fe-biochar and metals by biochar resulted in a greater number of elements being treated compared to the application of either biosorbent independently. Furthermore, the reduction of metalloids followed by metals, by the sequential application of Fe-biochar then biochar, demonstrates that the effects of these biosorbents are additive. Our model predicted the concentration of most elements following treatments with Fe-

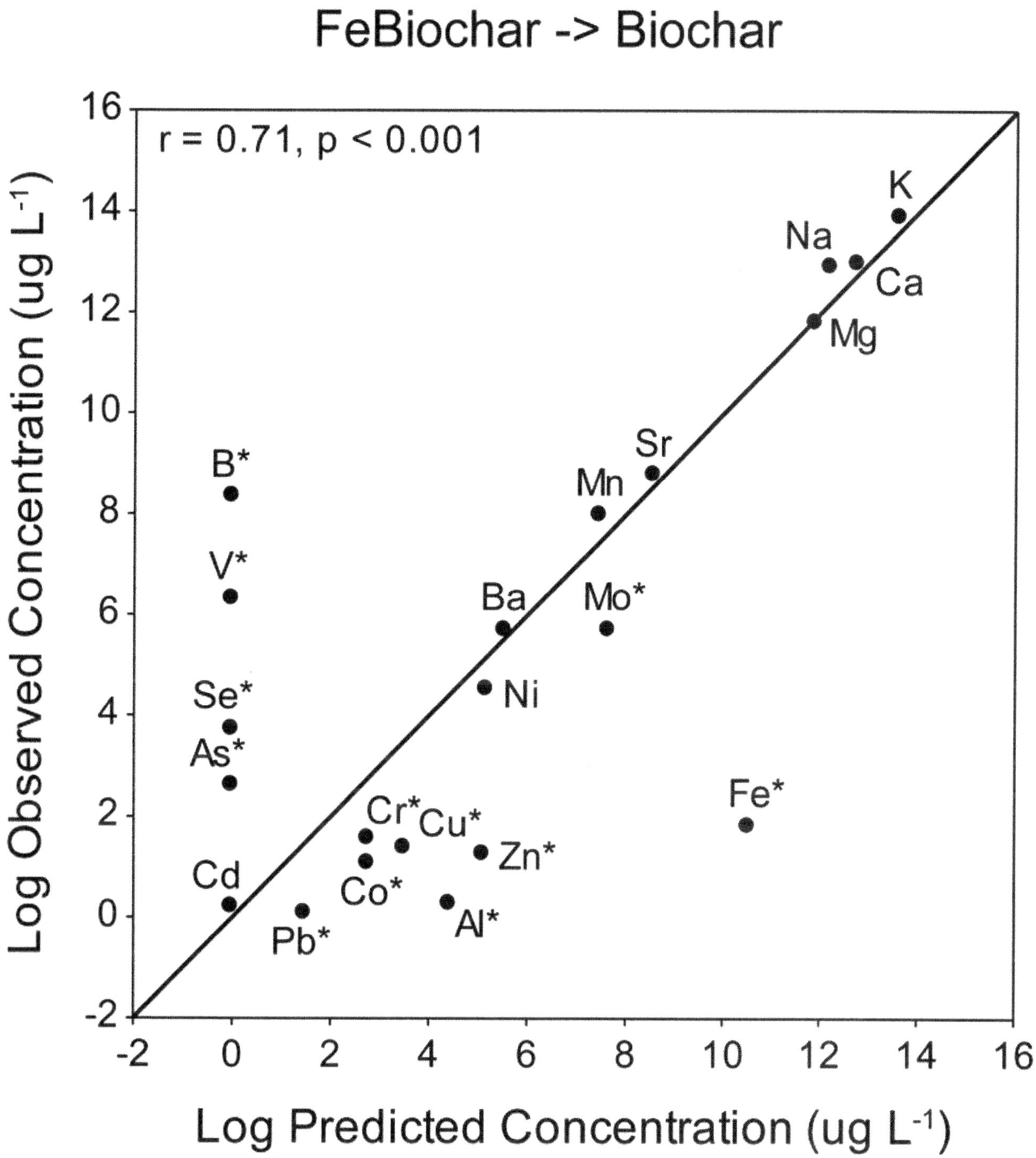

Figure 6. Scatterplot of predicted vs. observed concentrations of elements in ADW ($\mu g\ L^{-1}$) following the "FeBC → BC" treatment. Data have been log transformed. The plotted line shows the expected relationship if all observed values are identical to predicted values. Points falling above the line were overestimated by the model (higher than expected final concentrations). Points falling below the line were underestimated by the model (lower than expected final concentrations). Asterisks indicate elements with a residual of greater than ±1.

biochar, although the model was slightly less effective at predicting the concentrations of elements when biochar was involved in the sequence. Regardless, the selectivity of the biosorbents was consistent and accurately modelled across the six water types (i.e. ADW treated by Fe-biochar, biochar, and combinations thereof) in which the initial element composition varied greatly. This provides the basis to predict the efficacy of these treatments (Fe-biochar and biochar in any order of application) in remediating the effluents from other industries.

Treatments that combined Fe-biochar and biochar ("FeBC → BC" and "FeBC + BC") achieved the most comprehensive remediation of the greatest number of elements. However, sequential treatments with Fe-biochar ("FeBC → FeBC") and biochar ("BC → BC") were more effective at removing their target constituents (metalloids and metals, respectively). In the same way that the removal of a given element will be greater in isolation than when that element is present in a complex mixture, removing multiple elements from an effluent occurs at the expense of the sorption capacity of the biosorbent for each individual element [27]. Although treatment of a wastewater with a sequential application of macroalgal biosorbents has not been done previously, it has proven successful with other biosorbents. For example, in an analogous experiment, a sequential approach was taken to the treatment of wastewater using anaerobic fermented biological sludge, first using raw sludge to remove metals, then using sludge that had been pre-treated with a cation detergent (either Al^{3+} or Fe^{3+}) to remove the metalloids [28]. We found no benefit in terms of efficacy for simultaneous application of the biosorbents, and some negatives regarding the loading ratios per unit volume, meaning water should be treated in two-step process of Fe-biochar followed by biochar.

Predicting the concentrations of elements following treatments of Fe-biochar and biochar with the model was very successful, with all of the sequential and single treatment scenarios resulting in at least a moderately strong correlation between predicted and observed values. Modelling has been used extensively in biosorption studies, however often only to predict and quantify the mechanism by which the element is being absorbed onto the biosorbent using Langmuir and Freundlich isotherms [12,16,29,30]. Our empirical model performed particularly well when considering that the starting concentrations of most elements were significantly different to those in the effluent that was used to derive the model, and that the water quality after each treatment was unique, spanning 6 types of water treated with different combinations of biosorbents, resulting in a broad range of initial concentrations (Table 2). In this context, the model is as a robust tool for predicting the treatment of a broad range of waste water profiles by biosorbents derived from *Oedogonium*, and as a tool to develop working treatment strategies for a variety of waste water sources with known elemental compositions.

The model presented here intrinsically has some assumptions and limitations. The model is based on q-values from single deployment of the biosorbents in untreated ADW. The key assumption of the model is that there is a linear relationship between the extent of biosorption and biosorbent density that is independent of the initial concentration of each element in solution [16]. The model was slightly more effective at predicting final concentrations of elements after a single deployment of Fe-biochar or biochar than after two deployments. This suggests that the q-value is not entirely independent of the initial element concentrations because the biosorbents tend to remove slightly less of each element than predicted when initial concentrations are lower than the conditions under which the model was derived [31,32]. However, the magnitude of difference between predicted and observed final concentrations are small despite the large variation in initial conditions across the six treatments. This reiterates the importance of assessing the efficacy of biosorbents under realistic initial conditions, rather than deriving models of biosorbent kinetics in single-element system with inordinately high initial metal concentrations. Regardless, our approach to determining q-values of biosorbents for multiple elements in a real-world effluent provides a robust metric for use in biosorption modelling and with additional testing under different scenarios, further accuracy and confidence limits may be applied.

Conclusions

There is substantial literature on macroalgal biosorption but the majority of existing data has assessed elements in isolation, while an industrial effluent typically contains a multitude of co-existing, interacting elements. Given the large quantities of biomass that will inevitably be required to treat industrial effluents, it will also be necessary to select and cultivate biomass locally for the express purposes of bioremediation. We have demonstrated that different types of biosorbents (Fe-biochar and biochar) can be produced from a single feedstock that is native to an industrial facility (the freshwater macroalga *Oedogonium*, see also [18,33]) and used to sequentially treat metalloids and metals from a complex effluent. Through the use of a predictive model we demonstrate that sequential biosorption is largely predictable and consistent across a wide range of initial conditions. Targeting metalloids that do not have a natural affinity for biochar requires Fe-treatment, which inherently results in a suite of additional metals being released into solution. This metal leaching can in turn be addressed through the use of biochar as a final treatment. Therefore, while biosorption has been widely cited as a sustainable and cost-effective means of treating waste waters, our data clearly show that the reality is far more complex than is typically acknowledged, but remains achievable.

Supporting Information

Figure S1 Change in solution concentration of (a) potassium, (b) manganese, and (c) vanadium following sequential exposure to Fe-biochar and biochar. Sequential exposure of the ADW to biochar, Fe-biochar, and Fe-biochar followed by biochar and represented as dotted, dashed and solid lines, respectively. Exposure of ADW to both Fe-biochar and biochar simultaneously is represented as a round dot. Simultaneous exposure only had one application yet is placed under treatment 2 to compare with the final concentrations of the other treatments. ANZECC trigger level represented by a horizontal grey line. Error bars show standard error.

Table S1 Change in dissolved elemental concentration ($\mu g\ L^{-1}$) in ADW following treatment with biochar and Fe-biochar for 1 h at a solution pH of 7.1, and the derived q-values ($\mu g\ g^{-1}$). Data were collected during experiments for Kidgell et al (in press).

Table S2 Summary table for ANOVA tests run on each of the 21 elements investigated. One-way analysis of variance tests were run on final elemental concentration with the factor of Treatment. Type III sum of squares was used. All tests met the assumption of homogeneity of variance, normality of residuals and independence. Transformation of the data were required for some elements, the transformation applied is listed next to the title. Factors in bold indicate significance under alpha of 0.05.

Table S3 Predictive biosorption model for treatment of industrial waste water through sequential Fe-biochar and Biochar applications. ***Instructions to users.*** 1. Cells that can be edited by the user are shaded orange. 2. Input initial dissolved concentrations of elements and the volume of effluent requiring treatment in the orange shaded cells at Step 1. The model will calculate the recommended Fe-biochar stocking density which is displayed at Step 2. 3. Select the maximum Fe-biochar stocking density for elements of interest from Step 2 and enter it in the orange shaded cell at Step 3. The model will calculate the stocking density of biochar required to remove leached and existing metals in the effluent after Fe-biochar deployment which is played at Step 4. 4. Select the maximum biochar stocking density for elements of interest from Step 4 and enter it in the orange shaded cell at Step 5. The model will calculate the predicted final concentrations of metalloids and metals after the sequential application of Fe-biochar and biochar in the effluent at Step 6.

Acknowledgments

This research is part of the MBD Energy Research and Development program for Biological Carbon Capture and Storage with the co-operation of Stanwell Energy Corporation. This project is supported by the Advanced Manufacturing Cooperative Research Centre (AMCRC), funded through the Australian Government's Cooperative Research Centre Scheme, and the Australian Renewable Energy Agency (ARENA). We thank Stanwell Energy Corporation for providing effluents for the experimental studies.

Author Contributions

Conceived and designed the experiments: JK RdN NP DR. Performed the experiments: JK. Analyzed the data: JK DR. Contributed reagents/materials/analysis tools: RdN. Contributed to the writing of the manuscript: JK RdN NP DR.

References

1. Volesky B (1999) Biosorption for the next century. In: Amils R, Ballester A, editors. Biohydrometallurgy and the environment: Toward the mining of the 21st century. Amsterdam: Elsevier. pp. 161–170.
2. Frankenberger WT Jr, Amrhein C, Fan TWM, Flaschi D, Glater J, et al. (2004) Advanced treatment technologies in the remediation of seleniferous drainage waters and sediments. Irrig Drain Syst 18: 19–42.
3. Hamilton SJ (2004) Review of selenium toxicity in the aquatic food chain. Sci Total Environ 326: 1–31.
4. Sappington KG (2002) Development of aquatic life criteria for selenium: a regulatory perspective on critical issues and research needs. Aquat Toxicol 57: 101–113.
5. Davis TA, Volesky B, Mucci A (2003) A review of the biochemistry of heavy metal biosorption by brown algae. Water Res 37: 4311–4330.
6. Gadd GM (2009) Biosorption: critical review of scientific rationale, environmental importance and significance for pollution treatment. J Chem Technol Biotechnol 84: 13–28.
7. Volesky B (2007) Biosorption and me. Water Res 41: 4017–4029.
8. Saunders RJ, Paul NA, Hu Y, de Nys R (2012) Sustainable sources of biomass for bioremediation of heavy metals in waste water derived from coal-fired power generation. PLOS One 7: e36470.
9. Esmaeili A, Ghasemi S, Sohrabipour J (2010) Biosorption of copper from wastewater by activated carbon preparation from alga *Sargassum* sp. Nat Prod Res 24: 341–348.
10. Pavasant P, Apiratikul R, Sungkhum V, Suthiparinyanont P, Wattanachira S, et al. (2006) Biosorption of Cu^{2+}, Cd^{2+}, Pb^{2+}, and Zn^{2+} using dried marine green macroalga *Caulerpa lentillifera*. Bioresour Technol 97: 2321–2329.
11. Sheng PX, Ting YP, Chen JP, Hong L (2004) Sorption of lead, copper, cadmium, zinc, and nickel by marine algal biomass: characterization of biosorptive capacity and investigation of mechanisms. J Colloid Interface Sci 275: 131–141.
12. Moghaddam MR, Fatemi S, Keshtkar A (2013) Adsorption of lead (Pb^{2+}) and uranium (UO_2^{2+}) cations by brown algae; experimental and thermodynamic modeling. Chem Eng J 231: 294–303.
13. Sulaymon AH, Mohammed AA, Al-Musawi TJ (2013) Competitive biosorption of lead, cadmium, copper, and arsenic ions using algae. Environ Sci Pollut Res Int 20: 3011–3023.
14. Kidgell JT, de Nys R, Hu Y, Paul NA, Roberts DA (2014) Bioremediation of a complex industrial effluent by biosorbents derived from freshwater macroalgae. PLOS One. In press.
15. Fomina M, Gadd GM (2014) Biosorption: current perspectives on concept, definition and application. Bioresour Technol. Accepted 03.01.2014.
16. Mehta SK, Gaur JP (2005) Use of algae for removing heavy metal ions from wastewater: Progress and prospects. Crit Rev Biotechnol 25: 113–152.
17. Naja G, Volesky B (2011) The mechanism of metal cation and anion biosorption. In: Kotrba P, Mackova M, Macek T, editors. Chemistry and ecology. Amsterdam: Springer Netherlands. pp. 19–58.
18. Gupta VK, Rastogi A (2008) Equilibrium and kinetic modelling of cadmium(II) biosorption by nonliving algal biomass *Oedogonium* sp. from aqueous phase. J Hazard Mater 153: 759–766.
19. Murphy V, Hughes H, McLoughlin P (2008) Comparative study of chromium biosorption by red, green and brown seaweed biomass. Chemosphere 70: 1128–1134.
20. Kratochvil D, Pimentel P, Volesky B (1998) Removal of trivalent and hexavalent chromium by seaweed biosorbent. Environ Sci Technol 32: 2693–2698.
21. Deng L, Zhang Y, Qin J, Wang X, Zhu X (2009) Biosorption of Cr(VI) from aqueous solutions by nonliving green algae *Cladophora albida*. Miner Eng 22: 372–377.
22. Roberts DA, Paul NA, Dworjanyn SA, Hu Y, Bird MI, et al. (2014) *Gracilaria* waste biomass (sampah rumput laut) as a bioresource for selenium biosorption. J Appl Phycol. Accepted 09.05.2014.
23. Lawton RJ, de Nys R, Skinner S, Paul NA (2014) Isolation and selection of *Oedogonium* species and strains for biomass applications. PLOS One 9: e90223.
24. Bird MI, Wurster CM, de Paula Silva PH, Bass AM, de Nys R (2011) Algal biochar—production and properties. Bioresour Technol 102: 1886–1891.
25. Quinn GP, Keough MJ (2002) Experimental design and data analysis for biologists. Cambridge: Cambridge University Press. 556 p.
26. ANZECC (2000) An introduction to the Australian and New Zealand guidelines for fresh and marine water quality. Canberra: Australian and New Zealand Environment and Conservation Council. 314 p.
27. Figueira MM, Volesky B, Ciminelli VS (1997) Assessment of interference in biosorption of a heavy metal. Biotechnol Bioeng 54: 344–350.
28. Geuens L (1987) Waterwater purification process. Patent US 06/901,649.
29. Ali I, Gupta VK (2006) Advances in water treatment by adsorption technology. Nat Protoc 1: 2661–2667.
30. Yang JB, Volesky B (1999) Removal and concentration of uranium by seaweed biosorbent. In: Amils R, Ballester A, editors. Biohydrometallurgy and the environment. Toward the mining of the 21st Century. Amsterdam:Elsevier. pp. 483–492.
31. Plazinski W (2013) Binding of heavy metals by algal biosorbents. Theoretical models of kinetics, equilibria and thermodynamics. Adv Colloid Interface Sci 197: 58–67.
32. Plazinski W (2013) Equilibrium and kinetic modeling of metal ion biosorption: on the ways of model generalization for the case of multicomponent systems. Adsorption 19: 659–666.
33. Bakatula E, Cukrowska E, Weiersbye I, Mihaly-Cozmuta L, Peter A, et al. (2014) Biosorption of trace elements from aqueous systems in gold mining sites by the filamentous green algae (*Oedogonium* sp.). J Geochemical Explor. In press.

8

Changes in Microbial Communities, Including both Uncultured and Culturable Bacteria, with Mid-Ocean Ballast-Water Exchange during a Voyage from Japan to Australia

Akiko Tomaru[1*¤], Masanobu Kawachi[2], Mikihide Demura[2], Yasuwo Fukuyo[1]

1 Asian Natural Environmental Science Center, The University of Tokyo, Tokyo, Japan, **2** Environmental Biology Division, National Institute for Environmental Studies, Ibaraki, Japan

Abstract

We assessed changes in the microbial communities in ballast water during a trans-Pacific voyage from Japan to Australia that included a mid-ocean ballast-water exchange. Uncultured (i.e., total) and culturable bacteria were counted and were characterized by using denaturing gradient gel electrophoresis (DGGE). There was a clear decrease over time in numbers of uncultured microorganisms, except for heterotrophic nanoflagellates, whereas the abundance of culturable bacteria initially decreased after the ballast-water exchange but then increased. The increase, however, was only up to 5.34% of the total number of uncultured bacteria. Cluster analysis showed that the DGGE profiles of uncultured bacteria clearly changed after the exchange. In contrast, there was no clear change in the DGGE profiles of culturable bacteria after the exchange. Multidimensional scaling analysis showed changes in microbial communities over the course of the voyage. Although indicator microbes as defined by the International Convention for the Control and Management of Ships' Ballast Water and Sediments were occasionally detected, no coliform bacteria were detected after the exchange.

Editor: Arga Chandrashekar Anil, CSIR- National institute of oceanography, India

Funding: Support was provided by the Global Environment Research Funds (http://www.env.go.jp/policy/kenkyu/). The funders had no role in study design, data collection and analysis, decision to publish, or preparation of the manuscript.

Competing Interests: The authors have declared that no competing interests exist.

* E-mail: akotoma@gmail.com

¤ Current address: Division of Microbiology, National Institute of Health Sciences, Setagaya-ku, Tokyo, Japan

Introduction

The history of non-indigenous aquatic organisms being dispersed by cargo vessels goes back more than 100 years [1]. One of the more famous reports by Ostenfeld [2] is about a bloom of the diatom *Biddulphia* (now *Odontella*) in the North Sea in 1903–1907. Ostenfeld also cited the hypotheses of Fabre-Domergue [3] and Cotton [4] that exotic species of algae might be introduced by currents, directly by vessels, or with mixed cultivation of young oysters or lobsters.

The cargo ship *John Bowes* was the first to have built-in water ballast tanks installed, in 1852 [5]. Even in the 1800s, several investigators attempted to determine whether planktonic organisms would survive in ballast tanks during voyages (reviewed in [6]). Beginning in 1973, as a result of the International Conference on Marine Pollution, the International Maritime Consultative Organization began investigating the organisms transported by ballast water [7]. At that conference, ballast-water problems were considered and Resolution 18 was adopted. This resolution was based on a draft resolution calling for research into the effects of ballast-water discharge containing epidemic-causing disease bacteria; the draft had appeared in response to an outbreak of gastroenteritis in children in about 1972 (Y. Sasamura, personal communication). In the early 1970s, however, outbreaks of gastroenteritis were caused not by bacteria but by norovirus and rotavirus pathogens [8,9]. Nevertheless, this was the impetus for surveys that focused on the relationship between bacteria and ballast water.

Although ballast water has been sampled and studied since vessels were first built with ballast-water tanks, probably the first samples known to have been collected from a ship's ballast tanks at the end of a voyage were those reported by Medcof (1975) [10]. A subsequent review by Rosenthal (1976) [11], as quoted by the International Maritime Organization (IMO) (1998) [12], pointed out that ballast water was only one of the candidate risk factors for the accidental introduction of exotic organisms; aquaculture operations were just as likely to spread exotic species.

The control of exotic marine species became a global movement in 1982 with the United Nations Convention on the Law of the Sea [13]. The IMO began a concrete and substantial effort to address the ballast-water problem presented by Canada in a paper at the 26th session of the Marine Environment Protection Committee (MEPC26) in September 1988 concerning the presence and implications of foreign organisms in ship ballast water discharged into the Great Lakes [12]. As if to reconfirm this mission, in Rio de Janeiro in 1992 the United Nations Conference on Environment and Development recognized the issue of the introduction of non-indigenous species via ballast water in ships (Agenda 21, Chapter17) [14]. At MEPC33 in October 1992, the committee looked at the extent to which the 1991 Guidelines for

Preventing the Introduction of Unwanted Organisms and Pathogens from Ships' Ballast Waters and Sediment Discharges were being implemented [12].

It is believed that in the last 20–30 years many invasive species have been introduced via ballast water [6,15–18]. Thus, the movement of ballast water and ballast tank sediments is currently regarded as one of the most important mechanisms for the transfer of aquatic non-indigenous species [18,19]. Although the introduction of exotic organisms has a major economic impact on the aquatic environment, aquaculture, and other industries in some countries, it is not clear which invasive species are transported by which vectors. Nevertheless, because the IMO suspected and recognized ballast water in ships as a potential vector for invading species, in February 2004 it adopted the International Convention for the Control and Management of Ships' Ballast Water and Sediments (BWM) [20]. This convention includes technical standards and requirements for the control and management of ships' ballast water that are currently awaiting ratification. The IMO standard will be phased in over time, depending on the age and size of vessels. It is based on an allowable discharge of viable organisms within specified size categories [20].

During the time when BWM was under negotiation, ballast-water exchange was being used to reduce the risk of introducing alien invaders. Ballast-water exchange is the replacement of ballast water taken up in coastal areas with water from the open ocean. Oceanic organisms generally do not survive when released into the coastal or fresh waters of a destination port [21]. This method will continue to be used during the phase-in period until all vessels meet the IMO standard.

Ship-borne microbial transport could result in the introduction of toxic dinoflagellates, which was one of the concerns of the MEPC27 [12,22], although it is not clear whether these species have been transported in the past. An epidemic caused by *Vibrio cholerae* O1 in Latin America in 1991 [23] accelerated concerns over the problem of ballast water and pathogenic microorganisms; cholera was then placed on the "Ten Most Unwanted" list of diseases in the Global Ballast Water Management Program (GloBallast) [24]. Even though the *V. cholerae* O1 responsible for the 1991 outbreak was thought to have come from ballast water, this could not be proven [25]. Officials of the Pan American Health Organization believed that the bacteria first arrived with a freighter, which apparently released contaminated bilge water into the harbor at Lima, Peru [26]. Bilge water is defined in "ANNEX I: Regulations for the prevention of pollution by oil" of the International Convention for the Prevention of Pollution from Ships, 1973 (MARPOL); "sewage" is defined in "ANNEX IV: Regulations for the prevention of pollution by sewage from ships" [27]. Although there is no chance today of bilge water having pathogens, there was no retroactive application of Annex IV at the time of its adoption. It is therefore possible that bilge water was mixed with sewage at the time of discharge into the harbor at Lima, Peru. Ultimately, there no clear cause was identified for the *Vibrio cholerae* O1 outbreak in Latin America.

With this as background, the Annex to BWM, "Regulation D-2, Ballast Water Performance Standard," regulates not only plankton but also microbes. The indicator microbes are toxicogenic *Vibrio cholerae* (O1 and O139), *Escherichia coli* and intestinal *Enterococci*. These bacteria are basically detected by live counting methods using plate counts [20]. Because bacterial control was an addendum to BWM, the IMO delayed its adoption of BWM [28,29].

In assessing microbes in ballast water, most studies have focused solely on pathogens (e.g., [19,30–34]). Tomaru et al. (2010) [35] studied the effect of mid-ocean ballast-water exchange on bacterial abundance and community structure in ballast water, but their research was limited to uncultured bacteria. In this study, we investigated the qualitative and quantitative changes in members of the microbial loop community resulting from a mid-ocean ballast-water exchange during a trans-Pacific voyage from Japan to Australia. We observed both uncultured and culturable bacteria, including those regulated by BWM.

Materials and Methods

Ethics statement

No specific permits were required for the field studies. The ship's owners, the company operating the ship, the customers transporting cargo on the ship, and Fukushima Prefecture's local fishermen's union reviewed and approved our experimental protocol before we initiated sampling. No permit was necessary for sampling on the high seas because of "freedom of the seas," a principle in the United Nations Convention on the Law of the Sea (1982). The field studies did not involve endangered or protected species. In addition, our terms of agreement with the ship's owners, the company operating the vessel, and the customers transporting cargo on the vessel prohibited us from publishing any details or identifiable cargo information, such as the names of ports or the name of the vessel. All of the information about this voyage that we are permitted to present is in Table 1.

Sample collection

This study was conducted aboard a coal-carrier (58,098 t gross) on a voyage from Japan to Australia (Table 1). Ballast water was taken on at a port in Japan and then discharged before the reloading of ballast water in mid-ocean (17°42′5″N–17°23′8″N) on day 5 of the voyage. The water in each ballast tank was exchanged. Ballast-tank water samples ("BW" series) were collected with a vacuum pump; samples of ballast-water from the surface (BWS) and the bottom (BWB) were collected from a manhole and an on-deck sounding pipe by using hoses or pipes, respectively (Figure S1). Seawater samples ("SW" series) were taken with a bucket over the side of the vessel. The ballast-water sampling schedule is shown in Table 1. Day 5 samples were collected after the ballast-water exchange. The main tank sampled was no. 7 TST P (capacity, 760 m^3) (Figure S2). Samples were numbered in sequence.

For bacterial analyses, about 15 L of seawater or ballast water was collected into disposable sterilized plastic bags (a single sample). For plankton analysis (data not included in this study), about 300 L of water was filtered through 40-μm and 100-μm-mesh-size plankton nets, or about 20 L was collected directly into buckets. Although ballast water samples were shared by several researchers, there were only two or three researchers aboard, because merchant vessels generally have no space for research personnel. Our ballast-water sampling systems were washed with clean fresh water after each ballast-water sampling.

Temperature and salinity

Water samples for temperature and salinity measurements were collected into buckets from the vacuum pump. Temperature and salinity were immediately measured with a portable conductivity/pH meter (WM-22EP; DKK-TOA Co., Osaka, Japan).

Direct counts of microbes

For direct counts of microbes or uncultured bacteria, seawater subsamples were fixed with glutaraldehyde (final concentration, 1% by volume) within 1 h of sampling. For counting virus-like particles (VLPs), 0.1–1 mL of fixed sample was filtered onto a

Table 1. Sampling schedule and sample identifiers.

Date	Day of voyage	Latitude	Longitude	SW series	BW series	
				(Surface only)	Surface	Bottom
18 June	0	37°53′N	141°01′E	SW01		
20 June	1	37°53′N	141°01′E	SW02	BW01S	BW01B
21 June	2	34°54′N	142°40E		BW02S	BW02B
22 June	3	29°18′N	142°40E		BW03S	BW03B
23 June	4	23°42′N	145°56E		BW04S	BW04B
24 June*	5	17°44′N	147°53E	SW03	BW05S	BW05B
25 June	6	12°49′N	149°27E		BW06S	BW06B
26 June	7	07°15′N	150°60E		BW07S	BW07B
27 June	8	01°09′N	152°31E		BW08S	BW08B
28 June	9	04°28′S	153°21E		BW09S	BW09B
29 June	10	10°24′S	154°25E		BW10S	BW10B

Seawater (SW) series and ballast-water (BW) series. Asterisk (*) indicates mid-ocean ballast-water exchange on day 5.

0.02-μm-pore-size Anodisc filter (Whatman International Ltd., Maidstone, UK) and stained with SYBR Gold (Invitrogen Corporation, Carlsbad, CA, USA) [36]. For total bacterial counts (TBCs), 2 mL of fixed sample was stained with 4′,6-diamidino-2-phenylindole dihydrochloride n-hydrate (DAPI; Wako Pure Chemical Industries, Ltd., Osaka, Japan) and passed through a 0.2-μm-pore-size black Nuclepore filter (Whatman International Ltd., Kent, UK) [37]. Heterotrophic nanoflagellates (HNFs) and autotrophic nanoflagellates (ANFs) were counted by using a double-staining epifluorescence technique. For this technique, 5–100 mL of fixed sample was stained with DAPI and fluorescein-4-isothiocyanate (FITC; Dojindo Laboratories, Kumamoto, Japan) and then filtered through a 0.8-μm-pore-size black Nuclepore filter (Whatman International Ltd., Kent, UK) [38,39]. All filters were placed on glass slides and stored at −20°C until the microbes were counted. VLPs and bacterial cells on filters were counted at 1000× magnification under an epifluorescence microscope (BX60; Olympus, Tokyo, Japan). HNFs and ANFs on filters were counted at 400× magnification under an epifluorescence microscope (BX60; Olympus). Counts were to at least 400 cells or over 20 fields of view.

Counts of live microbes

Here, we define culturable bacteria are those bacteria detectable by plate count methods, because BWM specifies that plate count methods be used for detection and control of indicator microbes [20]. For counts of culturable bacteria, we followed the procedure proposed by the Marine Environment Protection Committee of Japan [40,41], briefly summarized here. Plate counts of heterotrophic bacteria (HPCs) were performed by pipetting 100 μL of seawater sample or seawater sample diluted with sterilized phosphate buffer solution [42] containing NaCl (final concentration 0.3%) onto a Petri dish containing solid media, spreading the liquid sample by using sterilized spreaders, and allowing the plate to dry before inverting and incubating at room temperature (around 25°C). Colonies were counted 5 days later. Two media were used: Marine Agar 2216 (BD Difco, Detroit, MI, USA) for plate counts on seawater-based medium (HPC_{sw}), and R2A Agar (BD Difco) for plate counts on freshwater-based medium (HPC_{fw}). Because freshwater bacteria are able to survive in seawater, our aim was to investigate the fate of the bacteria culturable on freshwater-based medium after the ballast-water exchange.

Our counting protocol was based on the methods for counting coliforms (including *Escherichia coli*), enterococci, and *V. cholerae* O1 and O139, defined as indicator bacteria by BWM. However, we could not count enterococci onboard during the voyage because the solid medium used for their culture (m-Enterococcus Agar; Difco) does not keep for long periods of time. To measure the abundance of other indicator bacteria as defined by BWM, we did plate counts of coliforms, *E. coli*, and *V. cholerae* O1 and O139.

For counts of BWM indicator bacteria, water samples (1–100 mL) were passed through sterilized 0.2-μm-pore-size membrane filters (Millipore Corp., Bedford, MA USA). Each filter was placed on a Petri dish containing solid medium. In some cases, samples were not filtered but instead 100 μL of sample was spread on a Petri dish containing solid medium. Colonies of coliforms and *E. coli* were counted on XM-G Agar (Nissui Pharmaceutical Co. Ltd, Tokyo, Japan) after 24 h of incubation at 35±0.5°C. Possible *Vibrio cholerae* O1 and O139 colonies were counted on TCBS Agar (Nissui Pharmaceutical Co. Ltd) after 24 h of incubation at 35±0.5°C. These strains form small, yellow colonies on this medium as a result of sucrose fermentation; they are collectively referred to here as "the cholera group (CG)". Five replicates at each dilution were spread on each type of agar plate. Bacterial colonies were counted onboard during the voyage.

Marine *Vibrio* species were identified in our laboratory onshore at the University of Tokyo. The yellow colonies were isolated and inoculated into 1% peptone broth (pH 8.6; BD Difco) at 35±0.5°C for 8 h to inhibit the growth of marine bacteria until the samples could be analyzed. Potential colonies of *V. cholerae* O1 and O139 were isolated and suspended in a saline solution. The cells in suspension were identified by reaction with O1 and O139 antisera by using the kits *Vibrio cholerae* AD "SEIKEN" and Cholerae immunoserum "SEIKEN", respectively (Denka Seiken Co., Ltd., Tokyo, Japan).

Denaturing gradient gel electrophoresis

Denaturing gradient gel electrophoresis (DGGE) was conducted by using the methods of Tomaru et al. (2010) [35]. The samples of uncultured bacteria used for DGGE were the same as those in the

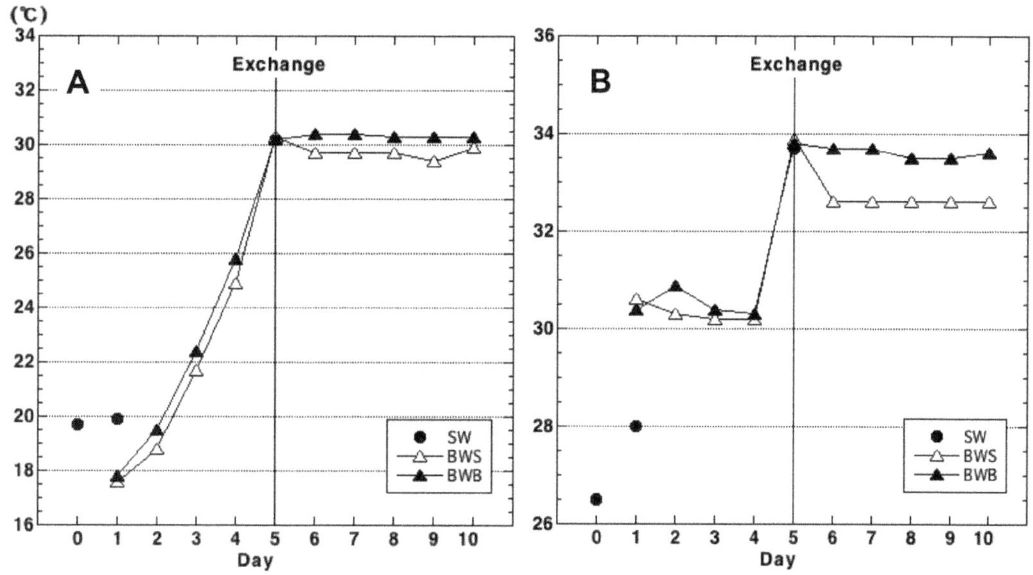

Figure 1. Temperature (A) and salinity (B) of surface seawater and ship's ballast water during a cruise from Japan to Australia in June 2005, roughly along longitudes 140°E to 155°E. Surface seawater (SW), closed circles; surface ballast water (BWS), open triangles; bottom ballast water (BWB), closed triangles. Mid-ocean ballast water exchange occurred on day 5 (Exchange).

TBC sample series. Water samples of 0.5–5 L were passed through 47-mm-diameter, 10.0-μm-pore-size Nuclepore filters (Whatman International Ltd) within 2 h of sample collection. The filtrate was then passed through 47-mm-diameter, 0.2-μm-pore-size Nuclepore filters. The filters were frozen at −20°C until DNA extraction.

The culturable bacteria used for DGGE were colonies from Marine Agar 2216, the same as the HPC_{sw} series (see "Counts of live microbes" in the Results section). After the determination of HPC_{sw}, these colonies were frozen at −20°C until DNA extraction.

For DNA extraction, the 0.2-μm filters or frozen colonies were sonicated in 0.2% sodium dodecyl sulfate – TE buffer (10 mM Tris [pH 8.0], 1 mM EDTA) for 5 min. The filter was then removed and DNA was extracted from the solution by using a Fast DNA Kit (Qbiogene, Inc., Carlsbad, CA, USA).

Bacterial 16S rDNA was amplified by PCR using primers GC341-F [43] and 907-R [44], which are intended to be specific for bacteria. The cycling conditions for PCR analysis were as follows: an initial denaturing step at 94°C for 1 min; 5 cycles of 94°C for 1 min, 63°C for 1 min, and 72°C for 1 min; 5 cycles of 94°C for 1 min, 58°C for 1 min, and 72°C for 1 min; 20 cycles of 94°C for 1 min, 55°C for 1 min, and 72°C for 2 min; and a final extension at 72°C for 7 min.

PCR products were purified with a QIAquick PCR Purification Kit (Qiagen, Valencia, CA, USA) and then loaded onto 6% polyacrylamide gels (acrylamide:N,N′-methylenebisacrylamide, 37:1) with a denaturant gradient of 35% to 65% (where 100% is defined as 7 M urea and 40% [vol/vol] formamide). Electrophoresis was performed with the D Gene System (Bio-Rad, Hercules, CA, USA) using 1× tris-acetate-EDTA (TAE) running buffer (Bio-Rad) at 60°C for 13 h at 70 V. Gels were stained for 30 min in SYBR Green nucleic acid gel stain (1:10,000 dilution; Molecular Probes, Eugene, OR, USA) and photographed with UV transillumination. DGGE banding patterns were converted to a binary matrix, and a Euclidean distance matrix was calculated from the binary data.

Statistical analysis

The variations in bacterial communities over time leading up to, resulting from, and following the mid-ocean ballast-water exchange were assessed by multidimensional scaling (MDS) for composition changes and by unweighted pair-wise group arithmetic averages (UPGMA) for grouping. All calculations were performed with the R 2.3.1 statistical software package (http://www.r-project.org/). DGGE banding-pattern analyses were performed according to the methods of van Hannen et al. (1999) [45].

The effects of the mid-ocean water exchange on the bacterial community were analyzed by Wilcoxon test using the R 2.3.1 statistical software package (http://www.r-project.org/). Before the analysis, all data were square-root transformed [46].

Results

Water temperature and salinity profiles

The temperature and salinity of both surface and bottom ballast water showed similar trends over the course of the voyage (Fig. 1). Ballast-water temperature increased until the ballast-water exchange near the Equator and then remained constant at around 30°C. Salinity of both surface and bottom ballast water stayed at about 30.5 before the exchange and suddenly increased just after the exchange (day 5). The salinity remained at about 33.6 in the bottom water from day 6 onward, whereas it suddenly dropped in the surface water on day 6 and then remained constant at about 32.5 for the rest of the voyage.

The temperature of ambient seawater on day 1 was higher than those of day-1 ballast-water samples BW01S and BW01B (Fig. 1A), because sample SW02 (day 1) was collected from the surface of the ocean in the daytime. The salinity of SW02 was lower than those of BW01S and BW01B (Fig. 1B), possibly reflecting the effect of a river at the coast.

Direct counts of microbes

We used direct methods for TBCs and to count VLPs, HNFs, and ANFs. The results of direct counts showed similar trends for

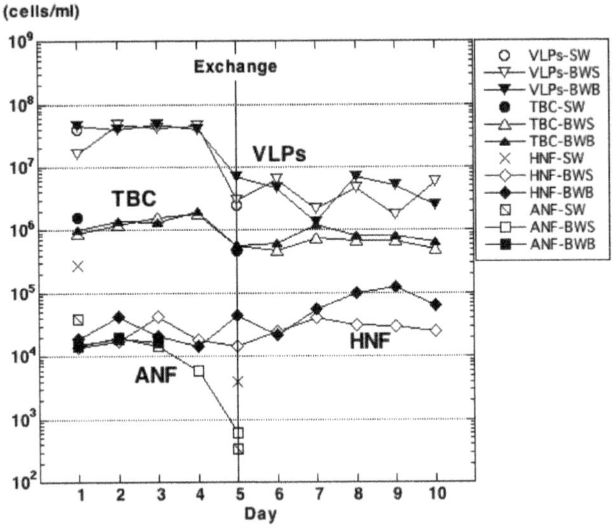

Figure 2. Microbial abundance as determined by direct counts, including virus-like particles (VLPs), total bacterial counts (TBCs), heterotrophic nanoflagellates (HNFs), and autotrophic nanoflagellates (ANFs). Microbes were counted in samples of seawater (SW) and in ballast-water surface (BWS) and bottom (BWB) samples. The mid-ocean ballast-water exchange occurred on day 5 (Exchange). Day 5 water samples were collected after the water exchange. No ANFs were detected in bottom ballast water after day 4 or in surface ballast water after day 6.

both surface and bottom ballast water throughout the voyage (Fig. 2).

The trends of VLP and TBC abundances were similar: constant before the ballast-water exchange and suddenly decreasing just after the exchange (samples BW05S and BW05B). The lower abundances of these groups remained constant after the exchange. Before the mid-ocean exchange of ballast water, TBCs were steady at approximately 1×10^6 cells mL^{-1}. After the exchange on day 5 of the voyage, TBCs rapidly decreased by one order of magnitude to approximately 1×10^5 cells mL^{-1} (samples SW03, BW05S, and BW05B) and were fairly constant from day 8 onward. The difference between ballast-water TBCs before and after the exchange was significant (Wilcoxon test: surface, $W=24$, $P<0.01$; bottom, $W=23$, $P<0.05$). Before the mid-ocean water exchange, densities of VLPs were steady at approximately 1×10^7 cells mL^{-1}. After the exchange, VLP densities also suddenly decreased one order of magnitude to approximately 1×10^6 cells mL^{-1} (samples SW03, BW05S, and BW05B) and were almost constant from day 8 onward. The decrease in VLPs after the exchange was significant (Wilcoxon test: surface, $W=24$, $P<0.01$; bottom, $W=24$, $P=0.01$).

There was a significant increase in HNF densities after the ballast-water exchange (Wilcoxon test: surface, $W=1$, $P<0.05$; bottom, $W=1$, $P<0.05$). Even though the densities remained around the same order of magnitude, the abundances in both surface and bottom samples gradually increased after the exchange.

Densities of ANFs in surface ballast-water samples gradually decreased leading up to the ballast-water exchange and then further decreased immediately after the exchange (BW05S). ANFs were not detected in day 6 (BW06S) and later surface samples. In bottom samples, ANFs could be detected only in the first three samples (BW01B, BW02B, and BW03B) before the exchange.

Counts of live microbes

We used plate-counting methods to monitor numbers of live bacteria. Plate counts were obtained for heterotrophic bacteria that were culturable on seawater-based medium (HPC_{sw}) and freshwater-based medium (HPC_{fw}). BWM indicator microbes were grouped as the cholera group or coliforms. On this voyage, we did not detect *V. cholerae* O1 or O139 or *E. coli* in any samples. The exchange of ballast water temporally affected HPC_{sw} and HPC_{fw}: the abundance of both groups dropped just after the exchange in both surface and bottom ballast-water samples, and then the numbers increased (Fig. 3). There were no significant differences between densities before and after the exchange (Wilcoxon test. HPC_{sw}: surface, $W=12$, $P=1.00$; bottom, $W=13$, $P=0.914$; HPC_{fw}: surface, $W=6$, $P=0.257$; bottom, $W=15$, $P=0.610$). We calculated the ratios of plate (live) counts to direct (total) counts (HPC_{sw}:TBC or HPC_{fw}:TBC; Table 2). The ratio was higher in the HPC_{sw} series (0.24% to 9.84%) than in the HPC_{fw} series (0.01% to 0.20%).

BWM indicator microbes were rarely found before the ballast-water exchange. Coliforms were detected in only one sample (BW04S), and the cholera group was detected sporadically in several surface and bottom samples (BW01S, BW04S, BW02B, BW03B, BW04B) (Fig. 3). After the ballast-water exchange only the cholera group was detected, and in only one sample (BW08B).

DGGE profiles

DGGE band patterns of uncultured bacteria in ballast-water samples are shown in Figure 4-A and B. The samples of uncultured bacteria used for DGGE had the same community composition as the samples for TBCs. The UPGMA results clearly show two clusters for both the surface (Fig. 5-A) and the bottom (Fig. 5-B) water samples: one from before the water exchange and the other after the water exchange.

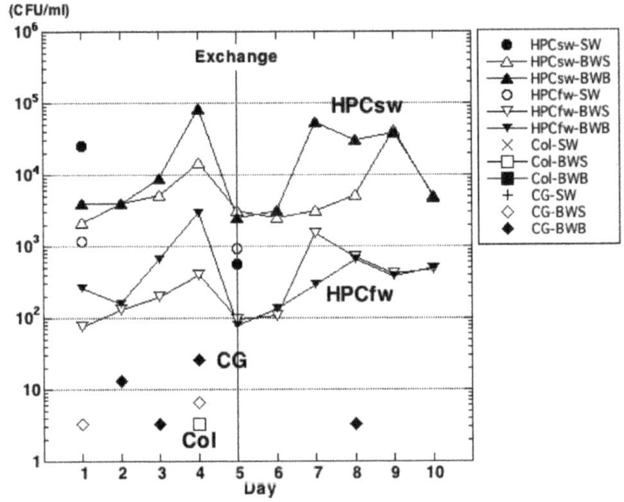

Figure 3. Densities of culturable bacteria in ballast water and natural seawater samples. Bacterial heterotrophic plate counts are classified as from seawater-based medium (HPC_{sw}), freshwater-based medium (HPC_{fw}), coliforms (Col), or members of the cholera group (CG). Microbes were counted in samples of ambient seawater (SW) and in ballast-water surface (BWS) and bottom (BWB) samples. Mid-ocean ballast-water exchange occurred on day 5 (Exchange). Missing symbols indicate days when no members of the coliform or cholera groups were detected. CFU, colony-forming unit.

Table 2. Numbers of heterotrophic bacteria as determined by the heterotrophic plate count (HPC) method, expressed as percentages of the direct total bacterial count (TBC).

Day of voyage	HPCsw/TBC (%)			HPCfw/TBC (%)		
	SW series	BW series		SW series	BW series	
		Surface	Bottom		Surface	Bottom
1	1.57	0.24	0.40	0.07	0.01	0.03
2		0.33	0.28		0.01	0.01
3		0.33	0.66		0.01	0.05
4		0.81	4.16		0.02	0.15
5*	1.19	0.57	0.45	0.02	0.02	0.01
6		0.52	0.50		0.02	0.02
7		0.43	4.45		0.20	0.02
8		0.78	3.87		0.11	0.08
9		6.13	4.89		0.06	0.05
10		9.84	7.65		0.10	0.08

SW, seawater samples; BW, ballast-water samples. Asterisk (*) indicates mid-ocean ballast-water exchange on day 5. Heterotrophic bacterial counts are from both freshwater-based medium (HPC_{fw}) and seawater-based medium (HPC_{sw}).

MDS analysis of DGGE band patterns of uncultured bacteria in surface-water samples (Fig. 6-A) revealed three groups: seawater from the port of departure (SW01 and SW02), samples taken before the exchange (BW01S to BW04S), and samples taken after the exchange (BW05S to BW10S), including mid-ocean water (SW03). The same analysis of the bottom water samples (Fig. 6-B) showed two groups: the samples taken before the water exchange (BW01B to BW04B) were grouped close to the port seawater samples (SW01 and SW02), and the samples taken after the exchange (BW05B to BW10B) were grouped with the mid-ocean water sample (SW03). The bacterial communities of both surface and bottom water changed slightly over time within each of these groups.

The DGGE band patterns of culturable bacteria (Fig. 7-A and B) were clearly different from those of uncultured bacteria. These bacteria were from the colonies of culturable bacteria after the determination of HPC_{sw}. The results of UPGMA suggest, but do not clearly show, two separate clusters (i.e. before and after the

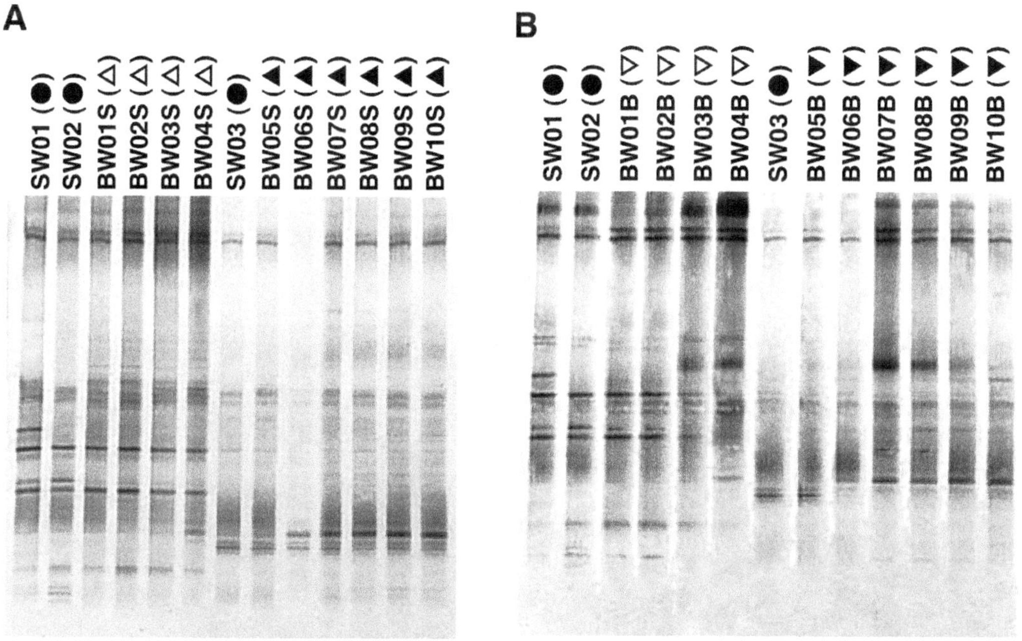

Figure 4. DGGE profiles of uncultured bacteria: DGGE band patterns. Refer to Table 1 for sample details. Surface (A) and bottom (B) samples, with surface seawater samples (SW01, SW02, SW03) included for comparison. Surface samples (A): seawater series (SW), closed circles; ballast-water (BW) series before mid-ocean exchange, open triangles; BW series after mid-ocean exchange, closed triangles. Bottom samples (B): seawater series (SW; surface only), closed circles; BW series before mid-ocean exchange, open inverted triangles; BW series after mid-ocean exchange, closed inverted triangles.

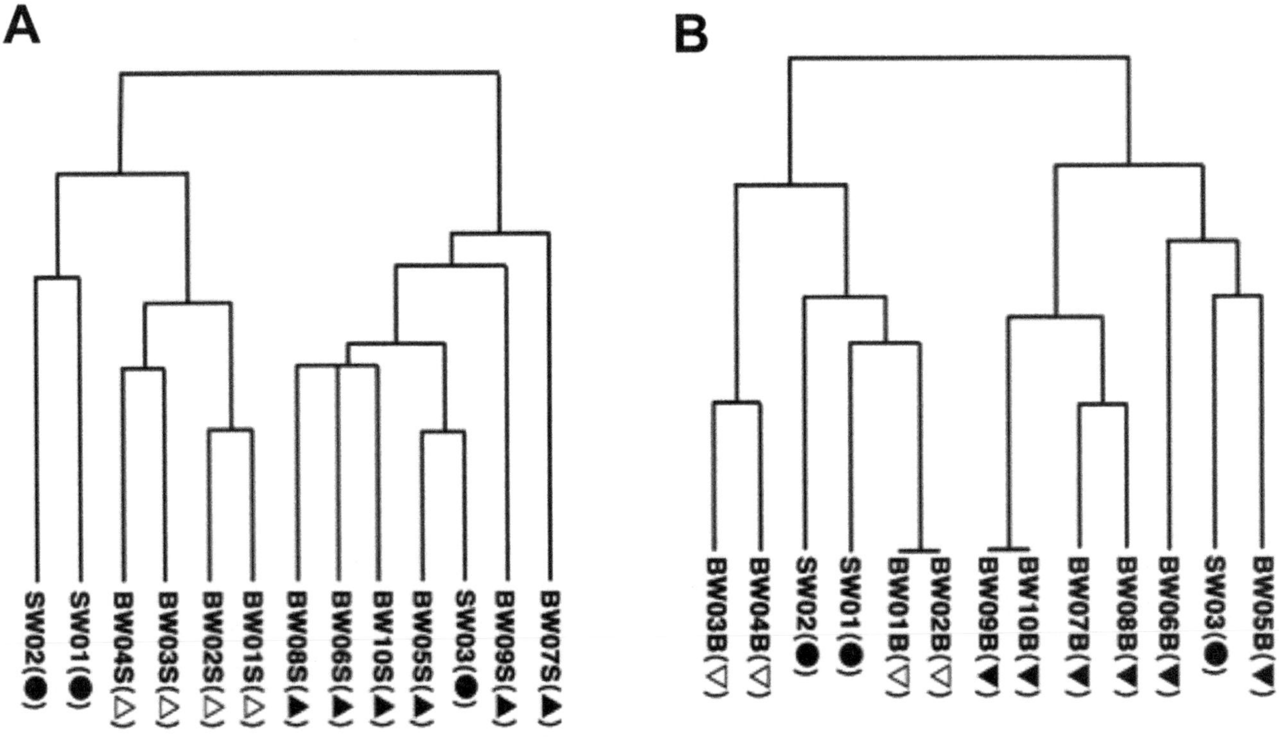

Figure 5. DGGE profiles of uncultured bacteria: unweighted pair-wise group arithmetic averages (UPGMA). Refer to Table 1 for sample details. Symbols as defined in Figure 4.

ballast-water exchange) for both the surface (Fig. 8-A) and the bottom (Fig. 8-B) samples. At the highest level, surface samples were separated into the home-port seawater samples (SW02 and SW03) along with BW01S, and the rest of the samples. At the next level they were roughly separated into samples from before and after the exchange, and these branches were nested. The grouping of the bottom samples was more complicated than that of the surface samples, and there was no clear trend relative to ballast-water exchange.

The results of MDS analysis of culturable bacteria from both surface (Fig. 9-A) and bottom (Fig. 9-B) samples did not show any clear effect of the ballast-water exchange. However, the shifting sample positions in this analysis reflect changes over time in each bacterial community.

Discussion

During the voyage, coastal ballast water was taken on before the start of the voyage and exchanged offshore well into the trip. The TBCs were about 1×10^6 cells mL^{-1} before the ballast-water exchange and decreased to about 1×10^5 cells mL^{-1} after the exchange (Fig. 2). There are extensive published data based on the direct counting method (e.g. [47]), and direct counts of hetero-

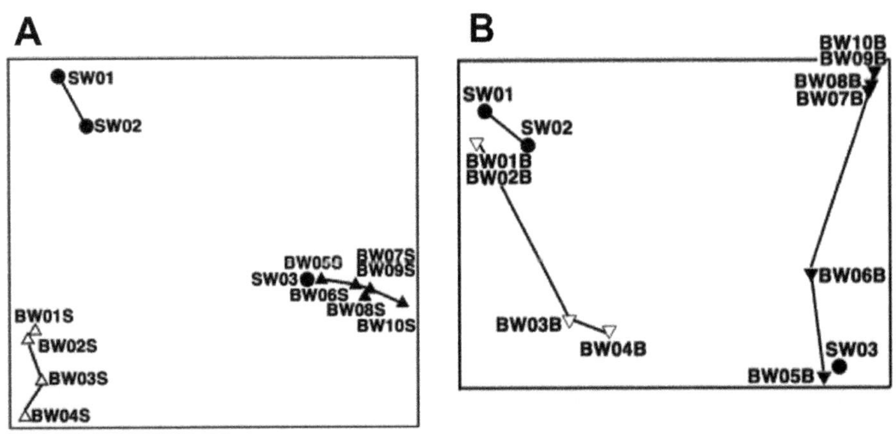

Figure 6. DGGE profiles of uncultured bacteria: multidimensional scaling analysis (MDS). Refer to Table 1 for sample details. Symbols as defined in Figure 4.

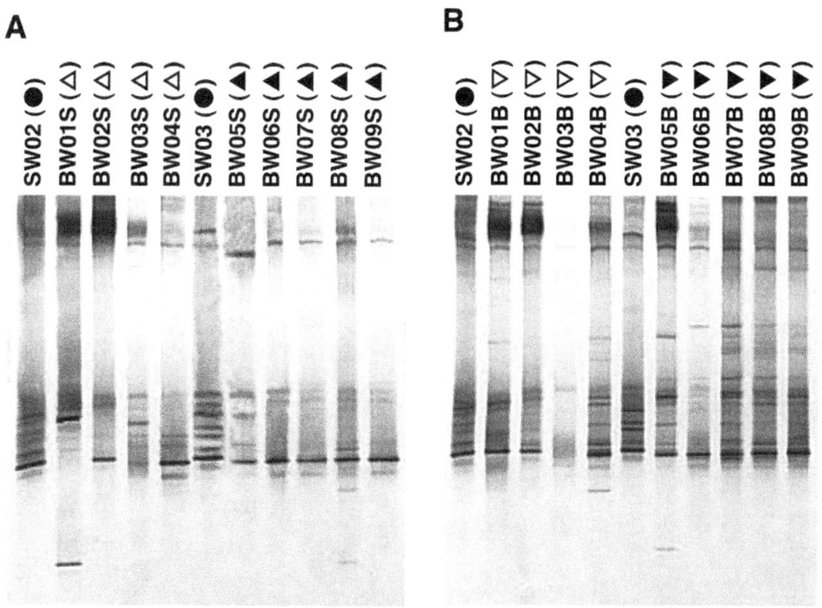

Figure 7. DGGE profiles of culturable bacteria: DGGE band patterns. Refer to Table 1 for sample details. Symbols as defined in Figure 4.

trophic bacteria are generally around 5×10^6 cells mL^{-1} in estuaries, 1 to 5×10^6 cells mL^{-1} in coastal waters, 5×10^4 to 1×10^6 cells mL^{-1} in offshore waters, and 1×10^4 cells mL^{-1} in deep ocean waters [48,49]. Our estimates of TBCs in ballast water and seawater are consistent with these values. The patterns of change in TBCs were similar in surface and bottom samples, and there were significant decreases in both types of sample after the ballast-water exchange. These trends are similar to those we observed in a series of samples collected during a winter voyage over the same track [35].

Drake et al. (2002) [50] reported no significant decrease in bacterial abundance after a ballast-water exchange in the North Atlantic Ocean. However, the abundances both before and after the exchange in their study were lower than in our present study (around 1×10^5 cells mL^{-1}), and thus no statistically significant differences were detected. If the concentration of organic matter

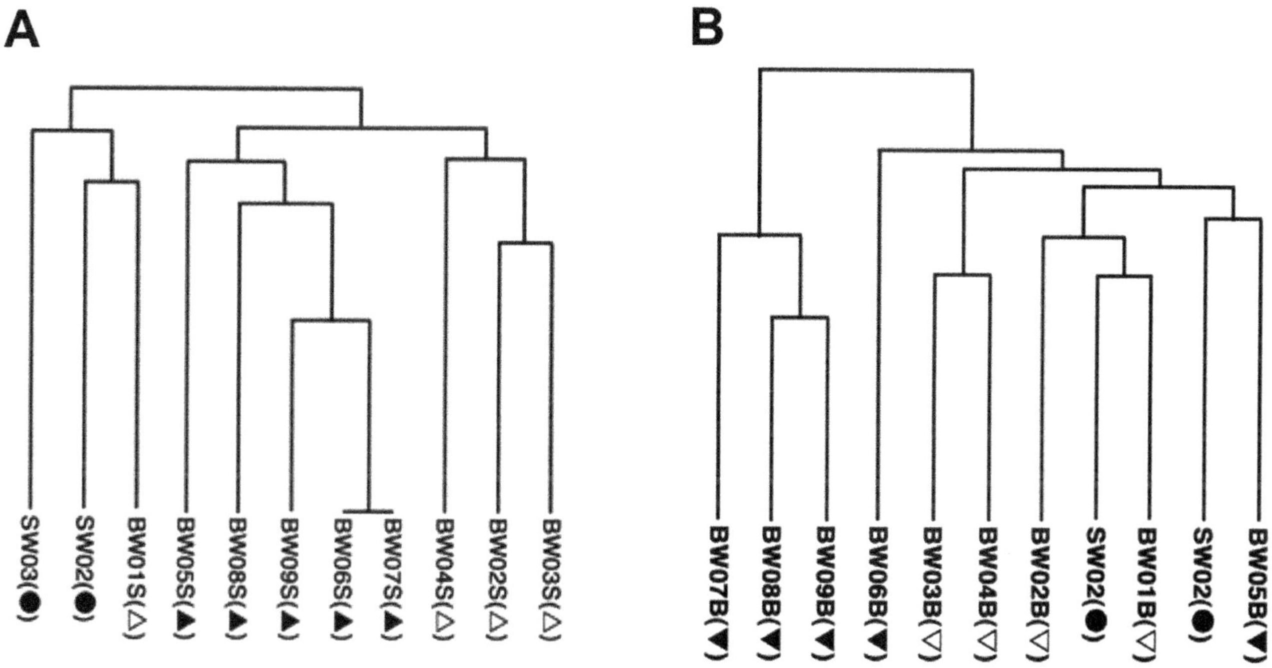

Figure 8. DGGE profiles of culturable bacteria: unweighted pair-wise group arithmetic averages (UPGMA). Refer to Table 1 for sample details. Symbols as defined in Figure 4.

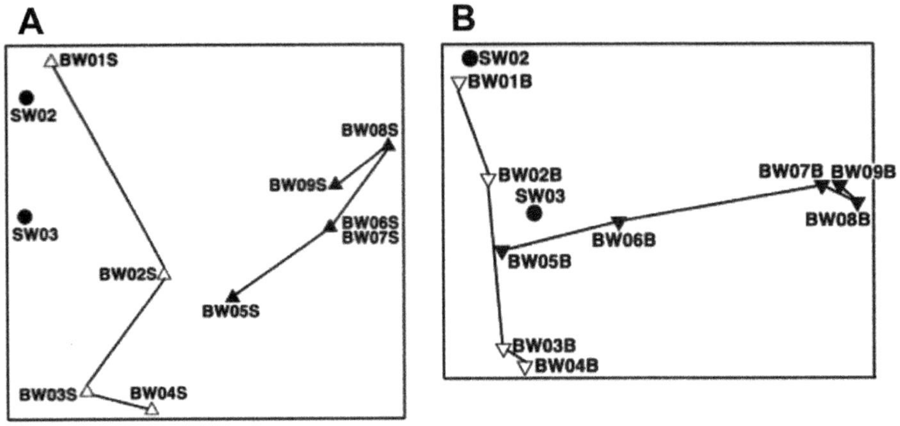

Figure 9. DGGE profiles of culturable bacteria: multidimensional scaling analysis (MDS). Refer to Table 1 for sample details. Symbols as defined in Figure 4.

was low in ballast-tank seawater in their study, the microbes were kept in a state of starvation. Bacteria in this state are known to decrease in size [51], and these smaller bacteria may not have been detected by the flow cytometer used in their study. Other surveys of ballast water had results similar to ours, with densities around 1×10^5 to 1×10^6 cells mL^{-1}, which are reasonable numbers for marine environments [48,49] and ballast water (e.g. [19,35,52]).

Generally, the concentration of VLPs in seawater ranges between 1×10^3 and 1×10^9 mL^{-1} [53]. Fuhrman (1999) [54] reported that the abundance of VLPs is typically 1×10^{10} L^{-1} in surface waters, or about 5 to 25 times the bacterial abundance. As expected, we observed similar dynamics for VLPs and TBCs, as did Drake et al. (2002) [50].

The changes that we observed in ANFs during this voyage are reasonable (Fig. 2). Because this voyage began after the ballast tanks were filled with coastal water, the abundance of ANFs in the tanks decreased over time in the absence of light, until the exchange of ballast water. In short, autotrophic species started dying earlier. Although mixotrophic species survived in the dark of the ballast-water tanks, they lost photosynthetic pigments and the abundance of ANFs decreased dramatically after day 3.

ANFs were next detected just after the water exchange; however, this time they disappeared more quickly than at the start of the voyage. One reason could be the low concentration of chlorophyll in mid-ocean water [53]; lower initial numbers after the water exchange would have resulted in ANFs reaching undetectable levels sooner by our counting method. Previous studies of ballast water report both increases and decreases in the abundance of phytoplankton (e.g. [55–58]). We interpret this to mean that the survival of ballast-water populations depends on the source of the water and the duration of the voyage.

Our results for HNFs show a significant difference between samples taken before and after the water exchange. However, the abundance in surface ballast water increased by a factor of 1.21 and in the bottom by 2.86 (Fig. 2). Some protists can survive for several weeks in ballast tanks (e.g. [59,60]), and some, including algae, can transform to resting stages in ballast tanks (reviewed in [61]). Many species of protists form cysts (e.g. [62]), and this life-cycle stage generally plays an important role in either protection or dispersal of the species.

It is also possible that some protists, including HNFs, can survive and grow in ballast water because they are heterotrophic and detritus feeders. The ballast-water incubation hypothesis suggests that, in the darkness of ballast tanks, photosynthesis ceases, phytoplankton die, and zooplankton in turn starve and die; subsequently, this detritus in the system fuels bacterial production (e.g., [50]). We suggest that this detritus also fuels heterotrophic protists.

Some researchers contend that free-living microorganisms have no biogeography and are global in their distributions (e.g., [63,64]); they therefore cannot be considered to "invade" environments. This idea has been explored with respect to ship-borne microorganisms [65]. We will not reiterate that discussion here, but in short, it contends that resolution of the microbial-ubiquity hypothesis is highly relevant to considerations of ballast management. Although the role of protists in ecosystems may not differ substantially among species, we must accept the possibility that protists in cyst form "hitchhike" in sea-going vessels and settle and grow in other areas as introduced species.

Our plate count abundance values for both HPC_{sw} and HPC_{fw} were higher than in previous reports (Table 2; reviewed in [66]). The increases in HPC_{sw} and HPC_{fw} before the ballast-water exchange closely paralleled the increasing water temperature. Although we did not measure dissolved or particulate organic matter concentrations, the concentration of dissolved organic matter is generally lower in open-ocean water than in coastal waters [67]; therefore, we speculate that water temperature controlled the increases in bacterial abundance. Numbers of both groups dropped in both surface and bottom ballast water just after the ballast-water exchange; both groups then increased, although not at the same rate (Fig. 3). Although the HPC_{sw} increased by about one order of magnitude after the ballast tanks were first filled (Fig. 3), this increase accounted for up to only 5.34% of TBC by day 9 of the voyage (Table 3). In short, the organisms detected by HPC_{sw} contributed a very small proportion of the TBCs. A study by Mimura et al. (2005) [33] using plate-counting methods could not confirm the effectiveness of ballast-water exchange in reducing the number of culturable bacteria, except on one of six voyages. Although their water samples were kept at 4°C until arrival in Japan, and the number and composition of bacteria in the samples at the time of analysis might have differed from those just after sample collection [68], our results were similar to theirs.

One of our goals was to detect indicator microbes by focusing on three pathogen groups targeted by the IMO for control. We could not detect enterococci, because the culture medium for this

Table 3. Daily increase in numbers of heterotrophic culturable bacteria in ballast water as a percentage of direct total bacterial counts (TBC).

Day of voyage	Daily increase (%) relative to total bacterial counts in ballast water			
	HPCsw		HPCfw	
	Surface	Bottom	Surface	Bottom
2	0.15	0.00	0.00	−0.01
3	0.08	0.37	0.00	0.04
4	0.53	3.71	0.01	0.11
5	−2.12	−14.29	−0.05	−0.51
6	−0.14	0.11	0.00	0.01
7	0.09	4.19	0.19	0.01
8	0.31	−2.91	−0.12	0.05
9	5.34	1.00	−0.05	−0.04
10	1.57	1.68	0.02	0.02

Heterotrophic bacteria were counted by using the plate-count (HPC) method. Asterisk (*) indicates the day of the mid-ocean ballast-water exchange (day 5). Heterotrophic bacteria were counted on both freshwater-based medium (HPC_{fw}) and seawater-based medium (HPC_{sw}).

group of bacteria could not be preserved for inoculation during our voyage. Attention was focused instead on the other two BWM control pathogens: the cholera group and the coliforms—the group of enteric bacteria (including *E. coli*) that inhabit the intestinal tracts of humans and other warm-blooded animals [69]. The presence of these bacteria in water samples indicates fecal pollution. Taking this into consideration, our failure to detect coliforms in samples taken after the ballast-water exchange (Fig. 3) is reasonable.

Members of the genus *Vibrio* are highly abundant in aquatic environments, including in estuaries, marine coastal waters, and sediments, and in aquaculture settings worldwide [70]. We sporadically detected members of the cholera group in ballast-water samples (Fig. 3). This is reasonable, because the cholera group, including members of the genus *Vibrio*, contains sucrose-fermenting Gram-negative bacteria that are not likely to be difficult to culture from seawater. We could not, however, detect *V. cholerae* O1 or O139 at our laboratory after this voyage in any of the cholera-group samples cultured on TCBS agar. Other studies have obtained similar results [32,33,52], although Ruiz et al. (2000b) [34] reported detecting *V. cholerae* in plankton samples from all ships examined. We believe that the detection of *V. cholerae*, including serotypes O1 and O139, depends on the ballast water history.

The DGGE profiles of uncultured bacteria showed clear differences between samples taken before and after the ballast-water exchange (Figs. 4, 5 and 6), confirming that the ballast water had been replaced. DGGE profiles of culturable bacteria showed no clear difference between samples taken before and after the ballast-water exchange (Figs. 7, 8 and 9). The culture method used for detection seems to selectively detect cosmopolitan species and reveal artificial microbial communities. We suggest that the culture detection method is of limited value for assessing the effects of ballast-water exchange on the structure of heterotrophic bacterial communities.

Although the IMO adopted BWM in 2004, this convention has not yet come into effect. One of the reasons why it is still awaiting ratification is the absence of a proven ballast water management system [71], because just before the adoption of BWM it was decided to include bacterial controls [72,73]. The indicator microbes identified by BWM are mainly pathogenic bacteria. Because individual countries control ship-borne pathogenic bacteria by using quarantine regulations, the pathogenic bacteria control provisions of BWM should be revised. It seems that both previous reports (e.g., [19,30–34]) and our results show a very low probability of ship-borne pathogen distribution. Because it is difficult to revise the convention before its entry into force, additional ratifications by the interested administrations would be desired to amend the clause for bacteria criteria appropriately.

Supporting Information

Figure S1 Ballast water sampling. Sampling at no. 7 TST P (capacity, 760 m3) ballast water tank (A); sampling surface ballast water (B); and bottom ballast water (C). In (B), 300 L of surface ballast water is being filtered through a plankton net for plankton analysis (data not included in this study). Water samples for bacteria were not filtered; instead 15 L of unfiltered ballast water was collected into a sterilized plastic bag. Ballast water samples were shared by several researchers, although only two or three scientists were aboard because merchant vessels have little extra room for survey personnel. Our sampling should not be against their voyage, so we could not have the control ballast tank (no ballast water exchange tank).

Figure S2 Schematic of ballast water tank sampling. Top side tank (TST), Water ballast tank (WBT), Port (P).

Acknowledgments

We thank the anonymous owners of the ship and the anonymous customers transporting cargo for their permission to conduct this study. We also thank the officers and crews of the NYK Line for their contributions to the study. We are grateful to Mr. Bin Okamura of the Japan Ship Technology Research Association for arranging for us to search the documents pertaining to meetings of the International Convention for the Prevention of Pollution from Ships, 1973 (MARPOL73), at the IMO library in London, and to Mr. Yoshio Sasamura of the Japan Ship Technology Research Association and to Mr. Shin-ichi Hanayama of Ocean Policy Research Foundation for conversations about the situation at the MARPOL73 meeting. We are also grateful to Dr. Yoshihito Omura, National Museum of Nature and Science, and Ms. Takako Tanoi, National Institute for Environmental Studies, for guidance and help with the DGGE method.

Author Contributions

Conceived and designed the experiments: AT MK MD YF. Performed the experiments: AT MK MD. Analyzed the data: AT. Contributed reagents/materials/analysis tools: MK YF. Wrote the paper: AT MK MD YF.

References

1. Elton CS (1958) Chapter 5. Change in the sea. In: The ecology invasion by animals and plants: Methuen and Co. Ltd., London, John Wiley and Sons Inc., New York. pp. 94–108.
2. Ostenfeld CH (1908) On the immigration of *Biddulphia sinensis* Grev. and its occurrence in the North Sea during 1903–1907 and on its use for the study of the direction and rate of flow of the currents. Medd Komm Havunde. Ser Plankton 1(6): 1–44.
3. Fabre-Domergue M (1906) Une invasion d'algues méridionales sur les huîres de la rivière de Vannes. Compt Rendus Acad Sc Paris 142: 1223–1225.
4. Cotton AD (1908) The appearance of Colpomenia sinuosa in Britain. Roy Bot Kew Bull 2: 73–77.
5. Ueno K (1980) 4.2 (5) Tetuhune no koûzoû p.302–311. In Hune no sekaisi Joû kan (in Japanese), Kazi Co., Ltd. Tokyo.
6. Carlton JT (1985) Transoceanic and interoceanic dispersal of coastal marine organisms: the biology of ballast water. Oceanogr Mar Biol Ann Rev 23: 313–371.
7. International Maritime Consultative Organization (IMCO) (1973a) International conference on marine pollution. 1973. Committee III, Text of Draft Resolutions 3, 4, 5, 6, 6A, 6B and 7 as agreed by the Committee. MP/CONF/DR/1; 23 October 1973, p.10.
8. Bishop RF, Davidson GP (1973) Virus particles in epithelial cells of duodenal mucosa from children with acute non-bacterial gastroenteritis. Lancet 302: 1281–1283.
9. Kapikian AZ, Wyatt RG, Dolin R, Thornhill TS, Kalica AR, et al. (1972) Visualization by immune electron microscopy of a 27-nm particle associated with acute infectious nonbacterial gastroenteritis. J Virol 10: 1075–1081.
10. Medcof JC (1975) Living marine animals in a ship's ballast water. Proc Natl Shellfish Ass 65: 11–12.
11. Rosenthal H (1976) Implications of transplantations to aquaculture and ecosystems, FAO Technical Conference on Aquaculture. Rome, FAO, FIR:AQ/Conf. 76/E.67, 1–19.
12. International Maritime Organization (IMO) (1998) Alien invaders–putting a stop to the ballast water hitch-hikers. Focus on IMO. October 1998, 161–177.
13. United Nations (UN) (1982) Article 196. Use of technologies or introduction of alien or new species. In United Nations Convention on the Law of the Sea (UNCLOS). Available: http://www.un.org/depts/los/convention_agreements/texts/unclos/unclos_e.pdf. Accessed 2014 Jan 20.
14. United Nations Conference on Environment and Development (UNCED) (1992) Chapter 17 Protection of the oceans, all kinds of seas, including enclosed and semi-enclosed seas, and coastal areas and the protection, rational use and development of their living resources. In Agenda 21. Available: http://www.un.org/depts/los/consultative_process/documents/A21-Ch17.htm. Accessed 2013 Aug 23.
15. Carlton JT, Geller JB (1993) Ecological roulette: the global transport of non-indigenous marine organisms. Science 261: 78–82.
16. Cohen AN (1998) Ships' ballast water and the introduction of exotic organisms into the San Francisco estuary: current problem status and management options. Report for California Bay Delta Authority, Sacramento, CA, USA.
17. Cohen AN, Carlton JT (1998) Accelerating invasion rate in a highly invaded estuary. Science 279: 555–558.
18. Ruiz GM, Carlton JT, Grosholz ED, Heines AH (1997) Global invasions of marine and estuarine habitats by non-indigenous species: mechanisms, extent and consequences. Am Zool 37: 621–632.
19. Ruiz GM, Fofonoff PW, Carlton JT, Wonham MJ, Hines AH (2000a) Invasion of coastal marine communities in North America: apparent patterns, processes, and biases. Ann Rev Ecol Syst 31: 481–531.
20. IMO (2005) The International Convention for the Control and Management of Ships Ballast Water and Sediments. IMO PUBLICATION, London, p134.
21. National Research Council (NRC) (1996) Stemming the tide: Controlling introductions of nonindigenous species by ship's ballast water. National Academy Press, Washington, DC, USA.
22. Hallegraeff GM, Bolch CJ, Bryan J, Koerbin B (1990) Microalgal spores in ship's ballast water: A danger to aquaculture. In Granéli E, Sundström B, Edler L, editors. Toxic marine phytoplankton: proceeding of the Fourth International Conference on Toxic Marine Phytoplankton held June 26–30, 1989, in Lund, Sweden: Elsevier, New York pp. 475–480.
23. Centers for Disease Control and Prevention (CDC) (1991) Cholera—Peru, 1991. MMWR 40: 108–110.
24. GloBallast (2002) Global Ballast Water Management Programme. Available: http://globallast.imo.org. Accessed 2013 Aug 23.
25. Coleman W (2003) 4. Transmission and epidemiology of cholera. In: Alcamo IE, editor, Cholera, 2nd edition (deadly diseases and epidemics): Chelsea house publishers, New York, USA. pp. 36–45.
26. Anderson C (1991) Cholera epidemic traced to risk miscalculation. Nature 354: 255.
27. IMCO (1973b) International convention for the prevention of pollution from ships, 1973 (MARPOR73).
28. IMO (1999) Report of the marine environment protection committee on its forty-third session. MEPC 43/21.
29. IMO (2001) Report of the marine environment protection committee on its forty-sixth session. MEPC 46/23.
30. Aridgides LJ, Doblin MA, Berke T, Dobbs FC, Matson DO, et al. (2004) Multiple PCR allows simultaneous detection of pathogens in ships ballast water. Mar Pollut Bull 48: 1096–1101.
31. Lee SKY, Wang HZ, Law SHW, Wu RSS, Kong RYC (2002) Analysis of the 16S–23S rDNA intergenic spacers (IGSs) of marine Vibrios for species-specific signature DNA sequences. Mar Pollut Bull 44: 412–420.
32. McCarthy SA, Khambaty FM (1994) International dissemination of epidemic Vibrio cholerae by cargo ship ballast and other nonpotable waters. Appl Environ Microbiol 60: 2597–2601.
33. Mimura H, Katakura R, Ishida H (2005) Changes of microbial populations in a ship's ballast water and sediments on a voyage from Japan to Qatar. Mar Pollut Bull 50: 751–757.
34. Ruiz GM, Rawlings TK, Dobbs FC, Drake LA, Mullady T, et al. (2000b) Global spread of microorganisms by ships. Nature 408: 49–50.
35. Tomaru A, Kawachi M, Demura M, Fukuyo Y (2010) Denaturing gradient gel electrophoresis shows that bacterial communities change with mid-ocean ballast water exchange. Mar Pollut Bull 60: 299–302.
36. Noble RT, Fuhrman JA (1998) Use of SYBR Green I for rapid epifluorescence counts of marine viruses and bacteria. Aquat Microb Ecol 14: 113–118.
37. Porter KG, Feig YS (1980) The use of DAPI for identifying and counting aquatic microflora. Limnol Oceanogr 25: 943–948.
38. Sherr B, Sherr E (1983a) Double-staining epifluorescence technique to assess frequency of dividing cells and bacteriovory in natural populations of heterotrophic microprotozoa. Appl Environ Microbiol 46: 1388–1393.
39. Sherr B, Sherr E (1983b) Enumeration of heterotrophic microprotozoa by epifluorescence microscopy. Est Coast Shelf Sci 16: 1–7.
40. Marine Environment Protection Committee (MEPC) (2004) Harmful aquatic organisms in ballast water. Quantitative analytical methods of bacteria used for test-bed tests. Proposal for development of the guidelines for approval of ballast water management systems. Submitted by Japan. MEPC 52/2/7. 9 July 2004.
41. The Ship Equipment Inspection Society of Japan Nippon hakuyohin kentei kyokai, HK (2007) 4.3. Saikin pp. 32–40. In Barasuto sui kainnri shisutemu no shoukin no sai no seibutsu bunseki houhou Kaitei ban. (in Japanese). Available: http://www.hakuyohin.or.jp/seibuthubunseki2009.pdf. Accessed 2013 Aug 23.
42. Japanese Industrial Standards Committee (JIS) (2002) Testing methods for detection and enumeration of heterotrophic bacteria in industrial water and wastewater. JIS K 0350-30-10. Japanese Standards Association, Tokyo.
43. Muyzer GE, de Waal C, Uitterlinden AG (1993) Profiling of complex microbial populations by denaturing gradient gel electrophoresis analysis of polymerase chain reaction-amplified genes coding for 16S rRNA. Appl Environ Microbiol 59: 695–700.
44. Wilmotte A, der Auwera GV, Wachter RD (1993) Structure of the 16 S ribosomal RNA of the thermophilic cyanobacterium chlorogloeopsis HTF ('mastigocladus laminosus HTF') strain PCC7518, and phylogenetic analysis. FEBS Lett 317: 96–100.
45. van Hannen EJ, Mooij W, van Agterveld MP, Gons HJ, Laanbroek HJ (1999) Detritus-dependent development of the microbial community in an experimental system: qualitative analysis by denaturing gradient gel electrophoresis. Appl Environ Microbiol 65: 2478–2484.
46. Zar J (1999) Data transformation. In Biostatistical analysis, 4th ed., New Jersey: Prentice-Hall. pp. 273–281.
47. Pomeroy LR, Wiebw WJ (1988) Energetics of microbial food webs. Hydrobiologia 159: 7–18.
48. Azam F, Fenche T, Gray JG, Meyer-Reil LA, Thingstad T (1983) The ecological role of water-column microbes in the sea. Mar Ecol Prog Ser 10: 257–263.
49. Es FB, Meyer-Reil LA (1982) Biomass and metabolic activity of heterotrophic bacteria. Adv Microb Ecol 6:111–170.
50. Drake LA, Ruiz GM, Galil BS, Mullady TL, Friedmann DO, et al. (2002) Microbial ecology of ballast water during a transoceanic voyage and the effects of open-ocean exchange. Mar Ecol Prog Ser 233: 13–20.
51. Byrd JJ (2000) Chapter 2. Morphological changes leading to the nonculturable state. In: Colwell RR, Grimes DJ editors. Nonculturable microorganisms in the environment, ASM Press, Washington DC. pp. 7–18.
52. Burkholder JM, Hallegraeff GM, Melia G, Cohen A, Bowers HA, et al. (2007) Phytoplankton and bacterial assemblages in ballast water of U.S. military ships as a function of port of origin, voyage time, and ocean exchange practices. Harmful Algae 6: 486–518.

53. Lalli CM, Parsons TR (1997) Chapter 5. Energy flow and mineral cycling. In: Biological oceanography: an introduction: Elsevier Butterworth-Heinemann, California, pp. 112–146.
54. Fuhrman JA (1999) Marine viruses and their biogeochemical and ecological effects. Nature 399: 541–548.
55. Dickman M, Zhang F (1999) Mid-ocean exchange of container vessel ballast water. 2: Effects of vessel type in the transport of diatoms and dinoflagellates from Manzanillo, Mexico, to Hong Kong, China. Mar Ecol Prog Ser 176: 252–262.
56. McCollin T, Shanks AM, Dunn J (2007) The efficiency of regional ballast water exchange: changes in phytoplankton abundance and diversity. Harmful Algae 6: 531–546.
57. Rigby G, Hallegraeff G (1994) The transfer and control of harmful marine organisms in shipping ballast water: behaviour of marine plankton and ballast water exchange trials on the MV "Iron Whyalla". J Marine Environ Eng 1: 91–110.
58. Zhang F, Dickman M (1999) Mid-ocean exchange of container vessel ballast water. 1: Seasonal factors affecting the transport of harmful diatoms and dinoflagellates. Mar Ecol Prog Ser 176: 243–251.
59. Galil BS, Hülsmann N (1997) Protist transport via ballast water – biological classification of ballast tanks by food web interactions. Eur J Protistol 33: 244–253.
60. Hüsmann N, Galil BS (2001) The effect of fresh water flushing on marine heterotrophic protists – implications for ballast water management. Mar Pollut Bull 42: 1082–1086.
61. Hallegraeff GM (1998) Transport of toxic dinoflagellates via ships' ballast water: bioeconomic risk assessment and efficacy of possible ballast water management strategies. Mar Ecol Prog Ser 168: 297–309.
62. Lee JJ, Leedale GF, Bradbury P, editors (2000) An illustrated guide to the protozoa, 2nd ed., vol. 1 and 2. Society of Protozoologists, Kansas, U.S.A.
63. Finlay BJ (2002) Global dispersal of free-living microbial eukaryote species. Science 296: 1061–1063.
64. Finlay BJ, Fenchel T (2004) Cosmopolitan metapopulations of free-living microbial eukaryotes. Protist 155: 237–244.
65. Dobbs FC, Rogerson A (2005) Ridding ships' ballast water of microorganisms. Environ Sci Technol 39: 259A–264A.
66. Amann RI, Ludwig W, Schleifer K-H (1995) Phylogenetic identification and in situ detection of individual microbial cells without cultivation. Microbiol rev 59: 143–169.
67. Hansell DA, Carlson AC, editors (2002) Biogeochemistry of marine dissolved organic matter: Academic Press, Tokyo.
68. Heinemann S, Dobbs FC (2006) Correspondence: Microbiological 'bottle effects' are not to be ignored. Mar Pollut Bull 52: 1310 (a comment on Mimura et al., 2005).
69. Madigan MT, Martinko JM, Parker J (2006) Brock biology of microorganisms. 11th edition. Prentice-Hall International, Inc, San Francisco, USA.
70. Thompson FL, Iida T, Swings J (2004) Biodiversity of vibrios. Microbiol Mol Biol Rev 68: 403–431.
71. IMO (2010) Report of the Review Group on Ballast Water Treatment Technologies (BWRG). MEPC 61/WP.8.
72. IMO (2003a) Proposed amendments to the regulations of the draft international convention for the control and management of ships ballast water and sediments. MEPC49/2/2.
73. IMO (2003b) Report of the second intersessional meeting of the MEPC ballast water working group. MEPC49/2/3.

9

Effects of Land Use, Topography and Socio-Economic Factors on River Water Quality in a Mountainous Watershed with Intensive Agricultural Production in East China

Jiabo Chen[1], Jun Lu[1,2]*

1 Department of Natural Resources, College of Environment and Natural Resources, Zhejiang University, Hangzhou, Zhejiang Province, China, 2 China Ministry of Education Key Lab of Environment Remediation and Ecological Health, Zhejiang University, Hangzhou, Zhejiang Province, China

Abstract

Understanding the primary effects of anthropogenic activities and natural factors on river water quality is important in the study and efficient management of water resources. In this study, analysis of Variance (ANOVA), Principal component analysis (PCA), Pearson correlations, Multiple regression analysis (MRA) and Redundancy analysis (RDA) were applied as an integrated approach in a GIS environment to explore the temporal and spatial variations in river water quality and to estimate the influence of watershed land use, topography and socio-economic factors on river water quality based on 3 years of water quality monitoring data for the Cao-E River system. The statistical analysis revealed that TN, pH and temperature were generally higher in the rainy season, whereas BOD_5, DO and turbidity were higher in the dry season. Spatial variations in river water quality were related to numerous anthropogenic and natural factors. Urban land use was found to be the most important explanatory variable for BOD_5, COD_{Mn}, TN, DN, NH_4^+-N, NO_3^--N, DO, pH and TP. The animal husbandry output per capita was an important predictor of TP and turbidity, and the gross domestic product per capita largely determined spatial variations in EC. The remaining unexplained variance was related to other factors, such as topography. Our results suggested that pollution control of animal waste discharge in rural settlements, agricultural runoff in cropland, industrial production pollution and domestic pollution in urban and industrial areas were important within the Cao-E River basin. Moreover, the percentage of the total overall river water quality variance explained by an individual variable and/or all environmental variables (according to RDA) can assist in quantitatively identifying the primary factors that control pollution at the watershed scale.

Editor: Miguel Lopez-Ferber, Ecole des Mines d'Alès, France

Funding: Chinese National Key Technology R&D Program (Grant No. 2012BAC17B01). The funders had no role in study design, data collection and analysis, decision to publish, or preparation of the manuscript.

Competing Interests: The authors have declared that no competing interests exist.

* Email: jlu@zju.edu.cn

Introduction

The deterioration of river water quality has become a primary environmental concern due to unsustainable anthropogenic activities; the demand for freshwater has been rapidly increasing in many developing countries, especially in China [1–3]. River water quality is controlled by complex anthropogenic activities and natural factors at both the river and watershed scales [1], [7–9]. Understanding the temporal and spatial variations in river water quality and estimating the primary regional factors that affect water quality can assist researchers in establishing priorities for sustainable water management [2], [10–12]. The relationships between water quality parameters and land use/cover, population density and point source discharge have been frequently studied [1], [7], [9], [13–15]. Simultaneously, topography and animal waste discharge are considered important factors that affect watershed river water quality [7], [13–16]. Particularly, several countries, e.g., China, contain large mountainous areas [17] and extensive animal production that lack strict management techniques [16], [18,19].

Recently, Pearson correlations have been widely employed to determine the relationships between environmental variables and river water quality [1], [7]. This method is simple and provides quantitative information; however, Pearson correlations lack visualization. Multiple regression analysis (MRA) is a useful tool that is commonly used to determine that watershed characteristics that best explain the spatial variability of an individual river water quality variable [7], [20]. This method lacks general information regarding pollution types. Moreover, principal component analysis (PCA) has become a widely accepted method in river water quality assessment and source apportionment studies in the last decade [2]. This method is commonly used to obtain the types of specific pollution sources without explanatory variables; however, PCA can only provide preliminary information regarding pollution types. Redundancy analysis (RDA), a multivariate statistical analysis method, has been proved useful for qualitative analysis of the interactions between river water quality and watershed characteristics in highly complex systems [13], [21,22]. RDA can reveal the influences of environmental factors on overall water quality, not only on single water quality variables [20]. Several commonly used statistical methods for pollution source identification are complementary (i.e., they have their own benefits and limits). Summary of commonly used statistical methods on pollution source identification in recent years is listed in **Table S1**. However, comprehensive applications of different statistical methods to evaluate the effects of environmental variables on water quality have not been fully explored in river studies in China. Furthermore, few studies have used RDA analysis to quantitatively evaluate the effects of watershed characteristics on the overall river water quality.

The Cao-E River is located upstream of the Qiantang Estuary, which is a major riverine system in Zhejiang Province, China, and is the primary source of industrial, agricultural and domestic water supplies [10]. Furthermore, the blue algae bloom that occurred in July 2004 occurred in the Qiantang Estuary [23,24].

In this study, ANOVA, Pearson correlation analysis, MRA, PCA and RDA were applied to investigate the effects of sub-watershed land use, topography and socio–economic factors (including animal waste discharge) on river water quality at the watershed scale for the Cao-E River system in a GIS environment. The objectives of this study were threefold: (1) to examine the temporal and spatial variations in water contamination within the study area, (2) to investigate the relationships between the river water quality parameters and land use, topography and socio-economic factors, and (3) to identify the primary pollution sources to estimate the possible sources that affect the water quality parameters in the Cao-E River basin. The results can be helpful for water conservation in the Cao-E River basin, and they provide a valuable tool for water quality agencies to develop assessment strategies for effective water quality management and rapid solutions for water pollution problems at the watershed scale.

Materials and Methods

1 Ethics statement

No specific permits were required for the described field studies; the sampling did not cause any disturbance to the environment or to the protected species at the sampling sites.

2 Study area

The Cao-E River (29°08′-30°15′ N, 120°30′-121°15′ E) is located in Zhejiang Province, East China. The Cao-E River is one of the main tributaries of the Qiantang Estuary, which flows into the East China Sea. The river system contains a main stream and several major tributaries (i.e., the CT River, the XC River, the CL River, the HZ River, the XS River and the YT River; **Figure 1**). The main stream is ~197 km long with an average slope of 3.0‰. The Cao-E River basin has a typical subtropical monsoon climate with an average annual rainfall of 1500 mm and an average ambient temperature of 16.2°C [25]. The watershed is within a mountainous region. Specifically, mountains and hills cover 2/3 of the area; the remaining area is covered by plains with intensive agricultural production [26].

3 Data Sources

Water quality data. The stream water samples (the number of samples used for statistical analysis is 36) were collected from July 2003 to June 2006 at 20 sampling sites throughout the watershed (Figure 1). Specific sampling stations that were largely influenced by point source pollution were excluded from further analysis. These specific sampling sites should meet the following conditions: (1) there was an obvious outlet of point pollution discharge near the sampling site, and (2) The standard deviation of the water quality parameter (BOD_5) at the specific sampling sites were triple more than the standard deviation for total sampling sites. The samples were collected once a month between 9:00 am and 16:00 pm. Moreover, the water samples were collected at approximately 0.3 m below the water surface in the central stream, placed into plastic bottles (2.5 L), transported to the laboratory and stored at 0~4°C for subsequent chemical analysis. The chemical measurements were performed in the laboratory within 24 h after collecting the water samples. Moreover, dissolved oxygen (DO), water temperature (T), pH and electrical conductivity (EC) were measured when the water samples were collected with a hand-held multi-parameter instrument (Multi 340i SETs, The Merck Co. Ltd., Germany). The turbidity was measured with a hand-held Turbiquant 1000IR (Multi 340i, The Merck Co. Ltd., Germany). The following chemical and biological water quality parameters were measured: chemical oxygen demand (COD_{Mn}), 5-day biochemical oxygen demand (BOD_5), total nitrogen (TN), total phosphorus (TP) and dissolved phosphorus (DP), which were measured using standard methods [27]. The ammonium nitrogen (NH_4^+-N) content was measured using an Astoria analyzer system (AAS, Brown Rupee CO. Ltd., Germany) after filtration through a 0.45-μm filter (Hailing Medicine, Zhejiang Province, China). The nitrate nitrogen (NO_3^--N) content was measured from the absorbance of the sample at 220 nm (A220) and 275 nm (A275) using a visible and ultraviolet spectrophotometer (Shanghai Spectrum Instruments, Shanghai, China) after filtration through a 0.45-μm filter; the results were determined by the following relationship: A220 - 2×A275 [27].

Land use/cover, topography and socio-economic data and GIS analysis. Latitude and longitude were measured using a hand-held GPS. The catchment boundaries were delineated on a 30-m spatial resolution digital elevation model (DEM); the stream network was represented using the Soil and Water Assessment Tool (SWAT) model. ArcGIS 9.3 Desktop GIS software was used to calculate the watershed characteristics for each sampling site. All datasets were converted to a common digital format (WGS 1984) and a common coordinate system (Albers Equal Area). Hydrologic units were not solely defined by the watershed above an individual point [28]; a hydrologic unit on the Cao-E River may cover 2/3 or more of the entire watershed above that sampling station (most hydrologic units on tributaries encompass the entire watershed above a sampling station).

The socioeconomic data, including land use change, the gross domestic product, animal husbandry output and human population in the watershed, were obtained from the Shengzhou (SZ), Xinchang (XC), Shangyu (SY) and Shaoxing (SX) County Statistical Yearbooks for the period 2003–2006 and were adjusted with some typical investigation in several towns within the watershed. The temporal variations in land use were considered to be minimal. The annual changes in forest areas and water areas were less than 0.05%, the annual decrease rate in cropland areas was less than 1% and the annual increase rate in urban areas was less than 2.5%., respectively, for the entire watershed and throughout the study period. Thus, Landsat Thematic Mapper imagery from 2006 was used to map the land cover in the study area [29]. The land cover was categorized into the following 4 classes: forest, water, cropland and urban land. The annual increases in human population density, gross domestic product per capita and animal husbandry output per capita were less than 1.5%, 10% and 2%, respectively, for the entire watershed and throughout the study period. Therefore, the averages of the socio-economic data for the entire study period were selected in this study. The slope map was derived from the DEM using a GIS spatial analyst tool (ESRI, 2006). The land use, slope, DEM and

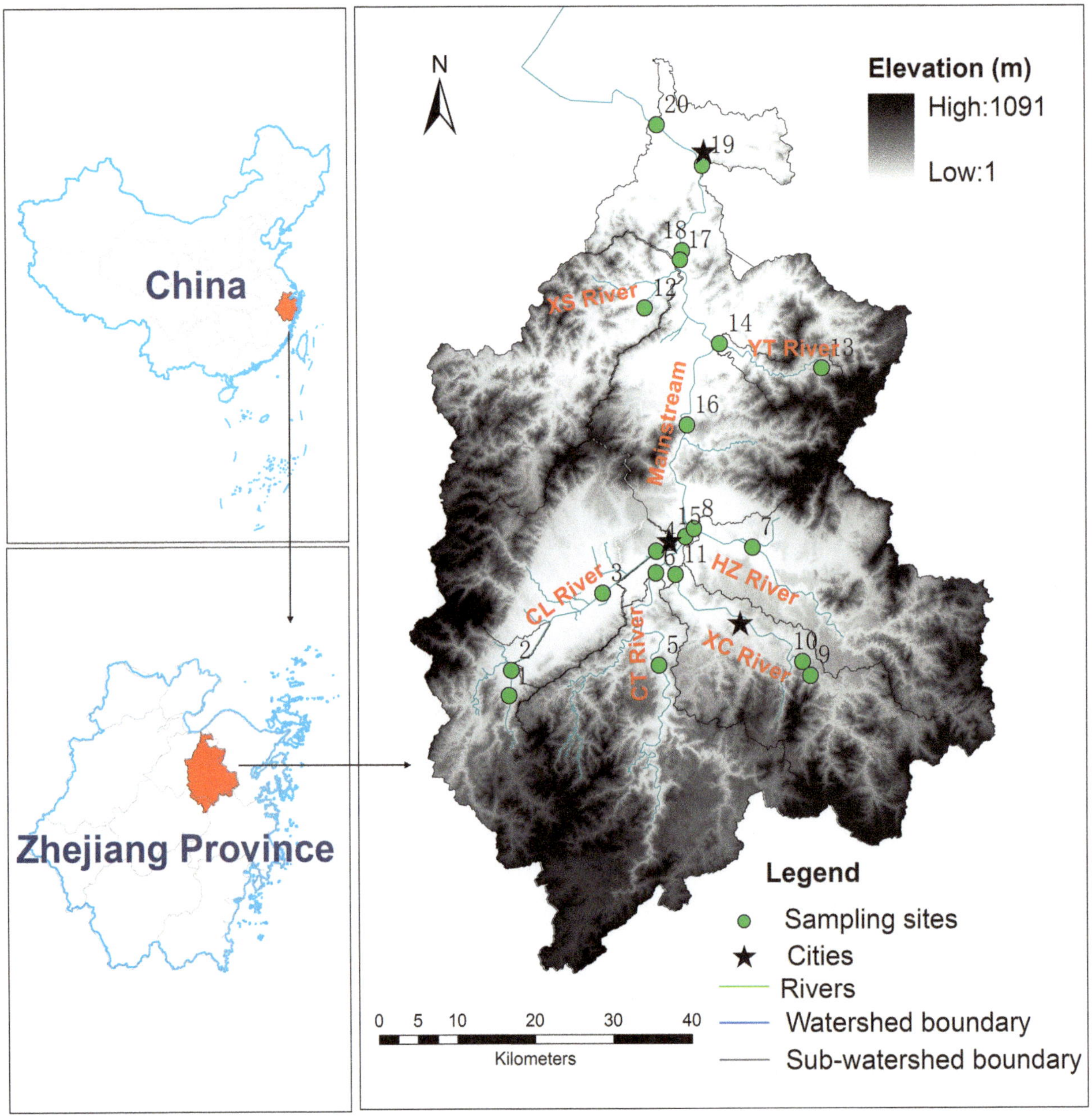

Figure 1. The Geographical Location Of The Sampling Sites In The Cao-E River System. (Sampling Stations 1-Cl1, 2- Cl2, 3-Cl3, 4-Cl4, 5- Ct 1,6-Ct2,7- Hz1,8- Hz2,9- Xc1,10- Xc2,11- Xc3,12- Xs1,13- Yt1,14- Yt2,15- M1,16- M2,17- M3,18- M4,19- M5, 20- M6).

administrative map (with attribute data, including the gross domestic product, animal husbandry output and human population for every town) were used to calculate the respective land use area (forest, water, urban land and cropland), mean elevation, mean slope, area, human population density, gross domestic product per capita and animal husbandry output per capita within each sub-watershed using a GIS spatial analyst tool. The abbreviations and statistics of the sub-watershed characteristics in the Cao-E River basin are summarized in **Table 1**.

4 Data analysis

Analysis of variance (ANOVA) was used to compare river water quality variations between different seasons (i.e., rainy and dry) via the Student-Newman-Keuls multiple-range test ($P<0.05$). Relationships among the considered variables were tested using Pearson's coefficient with statistical significance set at $P<0.05$. Major gradients and principal patterns in the water quality data between wet and dry seasons were detected using principal component analysis (PCA). PCA transforms a dataset consisting of p variables (analytical constituents) that are interrelated or correlated to various degrees into a new dataset containing p new orthogonal and uncorrelated variables; these variables are called the principal components (PCs) [30], [35]. The PCs are linear functions of the original variables in which the sum of their variances is equal to that of the original variables [30]. The PCs following a descending ordered according to their variances with PC1 corresponding to the variable with the largest variance. Algebraically, for p original variables (i.e., $x_1, x_2,\ldots, x_p$). Additional

Table 1. Abbreviations And Basic Statistical Information Of The Sub-Watershed Characteristics In The Cao-E River Basin.

	Parameters	Abbreviation	Range	Mean ± Sd
Land Use	Forest Percentage (%)	Frt	25.72–74.50	50.31±12.36
	Water Percentage (%)	Wt	0.13–3.39	1.87±0.87
	Urban Percentage (%)	Ub	0.45–14.44	4.09±3.52
	Cropland Percentage (%)	Crp	21.58–57.73	43.85±10.18
Topography	Mean Elevation (M)	Elv	125.12–556.54	306.17±116.17
	Mean Catchment Slope (°)	Slp_{mean}	10.20–17.72	13.74±2.29
	Catchment Area (Km^2)	Ar	38.39–2286.21	764.53±674.58
Socio-Economic Factors	Human Population Density (Persons/Km^2)	Hp_d	123.01–576.68	316.11±119.12
	Gross Domestic Product Per Capita (10^5 Chinese Yuan)	Gdp_{pc}	135.51–2151.16	582.38±427.88
	Animal Husbandry Output Per Capita (10^5 Chinese Yuan)	Aho_{pc}	8.07–39.40	20.56±8.61

PCs up to p can be formulated in a similar manner. The variances of the PCs are the eigenvalues; the coefficients or weights, are the eigenvectors extracted from the covariance or correlation matrix. In the example PCA, the correlation matrix was used in all cases. The goal of PCA is that the first k PCs (where $k << p$) retain most of the information in all of the p original variables, which effectively reduces the practical dimensionality of the dataset. More specifically, if the correlations are high among many of the original variables, the first few PCs contain (or explain) a large percentage of the total variance and may be used to describe multivariate patterns or water quality variations across the watershed nearly as well as using the complete set of p original variables. These patterns are often related to specific sources of contamination [30].

The stepwise regression analysis was performed to determine the environmental or socio-economic factors (TP, $P<0.10$ and the others, $P<0.05$) that best explained the spatial variability of an individual river water quality variable [20]. Variables that were strongly intercorrelated with others (variance inflation factor >10) in the initial analysis, were removed and a further analysis was carried out with the remaining environmental variables for MRA and RDA [20], [31,32]. Variance inflation factor (VIF) is a common way for detecting multicollinearity, a general rule is that the Variance Inflation Factor should not exceed 10 [49]. RDA was performed using Canoco 4.5 software to evaluate the relationships between river water quality and the environmental variables [33]. RDA was chosen because previous inspection of the data revealed a linear response rather than a unimodal response in the primary river water quality variables [34]. A Monte Carlo permutation test was used to verify the significance of the models [29]. Two series of matrices (i.e., for river water quality and environmental variables) from the RDA results were visualized as ordination diagrams using the CanoDraw software package for Windows. The river water quality parameters and the environmental/socio-economic variables were represented with arrows. Correlations between the watershed characteristics and/or the river water quality parameters were obtained by projecting the watershed characteristics onto each river water quality parameter in which higher values indicated higher correlations [34]. Application summary of 5 statistical methods in this study was listed in **Table S2**. The statistical analyses, except RDA, were completed using the SAS 9.1 software for Windows.

Results

1 Physico-chemical water quality in the Cao-E River basin

ANOVA was used to compare river water quality variations between the different seasons (i.e., rainy and dry). Temporal and spatial variations in the physico-chemical parameters are shown in **Table 2** and **Figure 2**. More specifically, BOD_5, TN, DO, pH, T and turbidity exhibited significant temporal variations (**Table 2**, $p<0.10$). Moreover, TN, pH and T were generally higher in the rainy season (April to September), whereas higher values for BOD_5, DO and turbidity occurred in the dry season (October to March). Most of the water quality parameters, except pH and T, exhibited considerable spatial variations. BOD_5, COD_{Mn} and NH_4^+-N were higher in urban areas compared with surrounding rural areas, whereas DO exhibited a reverse response to the extent of urbanization. TN, DN, NO_3^--N, TP and DP were higher in the lower part of the basin where more land has been developed and were lower in the upper-eastern part of the Cao-E River basin. Turbidity and EC were higher in the lower part of the basin (**Figure 2**) where the reach is often affected by tides and/or point source pollution [36,37].

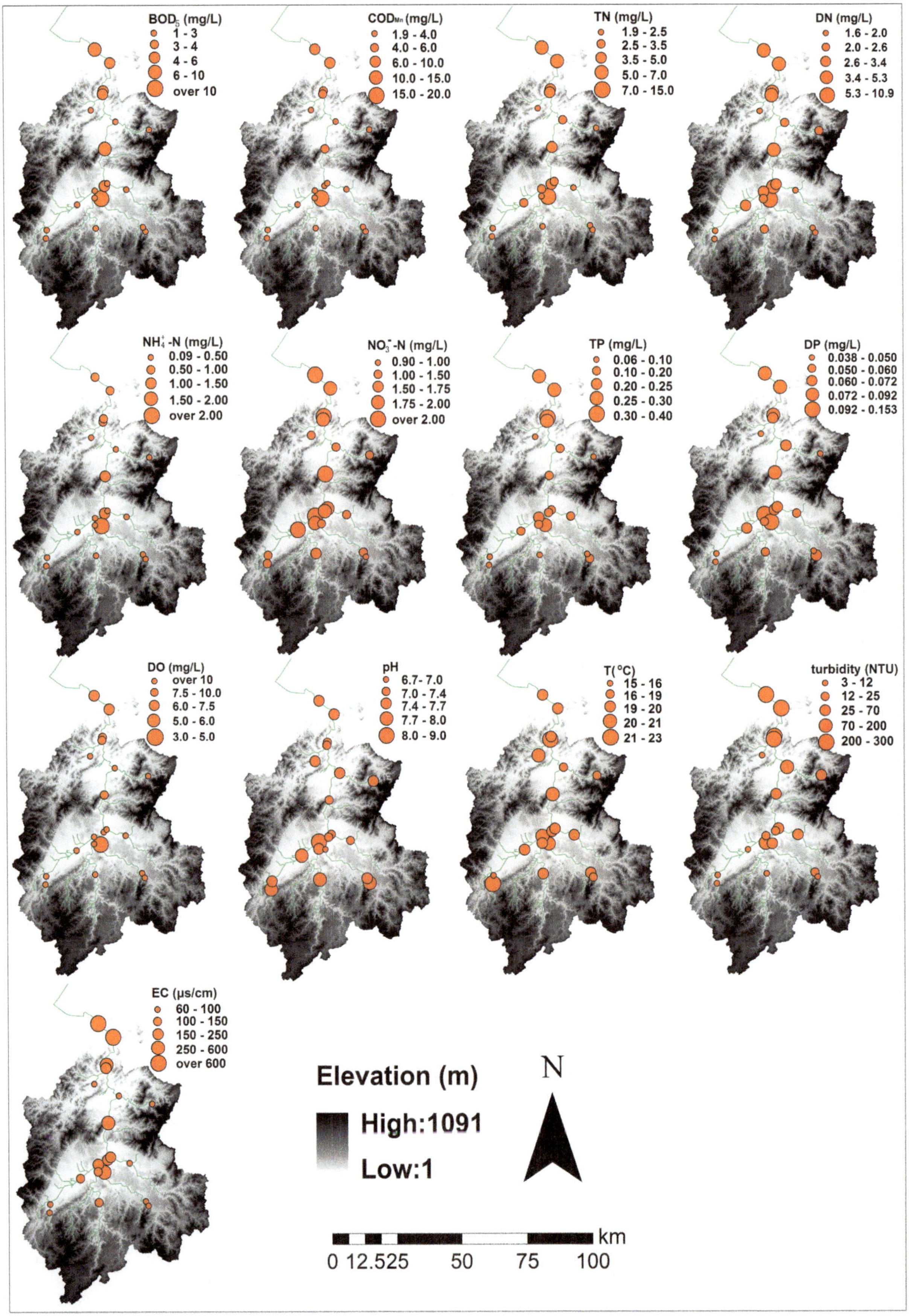

Figure 2. Spatial Variations In Bod5, Codmn, Tn, Dn, Nh_4^+-N, No_3^--N, Tp, Dp, Do, Ph, T, Turbidity And Ec In The Study Area.

Table 2. Seasonal Averages, Coefficients Of Variation [Cv (%)] And River Water Quality Parameter Ranges.

Water Quality Parameters	Rainy Season		Dry Season		
	Average [Cv (%)]	Range	Average [Cv (%)]	Range	*P-Value*
Bod_5 (Mg/L)	5.912 [183]	0.05–173.5	7.850 [181]	0–255.00	0.052
Cod_{mn} (Mg/L)	3.845 [87]	0.39–43.87	4.0361 [99]	0.49–47.75	0.557
Tn (Mg/L)	4.793 [103]	0.29–45.85	4.208 [73]	0.56–31.09	0.049
Dn (Mg/L)	3.549 [90]	0.06–22.53	3.295 [80]	0.03–30.07	0.219
Nh_4^+-N (Mg/L)	0.605 [195]	0–15.87	0.759 [2099]	0–19.58	0.156
No_3^--N (Mg/L)	1.677 [41]	0.01–4.30	1.606 [69]	0.08–15.08	0.283
Tp (Mg/L)	0.150 [108]	0–1.88	0.168 [113]	0–3.86	0.317
Dp (Mg/L)	0.061 [91]	0–0.49	0.076 [253]	0–3.49	0.163
Do (Mg/L)	7.164 [33]	0–19.55	8.045 [42]	0–17.86	<0.001
Ph	7.538 [7]	5.33–9.34	7.449 [8]	5.45–10.12	0.055
T (°C)	24.967 [24]	8.40–37.20	14.262 [64]	4.70–105.00	<0.001
Turbidity (Ntu)	70.011 [105]	0.83–586.10	92.939 [113]	0.91–1111	0.002
Ec (μs/Cm)	306.38 [347]	14–14420	348.56 [282]	14–11230	0.557

The Probabilities Associated With The Student-Newman-Keuls Multiple-Range Test Are Also Provided.

Major gradients and principal patterns in the water quality data between the wet and dry seasons were detected using PCA. The PCA based on only the dry season water quality data indicated that 4 significant factors (i.e., PCs with an eigenvalue >1) explained 87.8% of the total variance (**Table 3**). The first factor accounted for 50.2% of the total variance and had a strong positive correlation with BOD_5, COD_{Mn}, TN, DN, NH_4^+-N, and T. The PCA of the wet season water quality data showed that the eigenvalues for the 2 most significant factors were 8.442 and 1.817, accounting for 78.9% of the total variance. Factor 1 accounted for 64.9% of the total variance and was positively correlated with BOD_5, COD_{Mn}, TN, DN, TP and DP (**Table 3**).

2 Relationships between watershed characteristics and river water quality

Relationships between the river water quality variables and their corresponding sub-watershed land use, topography and socio-economic factors were explored using a Pearson correlation test, MRA and RDA. All environmental variables, including land use, topography and socio-economic factors, were utilized to

Table 3. River Water Quality Variables (13) On Rotated Pcs For Datasets During The Dry And Wet Seasons.

Water Quality Parameters	Dry Season				Wet Season	
	Pc1	Pc2	Pc3	Pc4	Pc1	Pc2
Bod_5 (Mg/L)	0.354	−0.110	0.008	0.076	0.312	−0.041
Cod_{mn} (Mg/L)	0.350	0.200	−0.187	0.128	0.310	0.263
Tn (Mg/L)	0.380	0.001	0,077	0.083	0.333	−0.054
Dn (Mg/L)	0.362	−0.044	0.147	0.127	0.311	−0.197
Nh_4^+-N (Mg/L)	0.345	−0.215	0.019	0.084	0.267	−0.346
No_3^--N (Mg/L)	0.275	−0.020	0.425	−0.079	0.283	0.174
Tp (Mg/L)	−0.221	0.409	0.077	−0.476	0.314	0.200
Dp (Mg/L)	0.097	0.329	0.609	−0.090	0.300	−0.018
Do (Mg/L)	−0.270	0.361	0.091	−0.204	0.272	0.349
Ph	−0.146	0.328	0.268	0.614	−0.184	0.546
T (°C)	0.418	−0.420	0.119	−0.212	0.010	0.246
Turbidity (Ntu)	0.222	0.340	−0.416	−0.312	0.283	0.246
Ec (μs/Cm)	0.241	0.308	−0.335	0.382	0.247	0.395
Eigenvalues	6.522	2.344	1.518	1.028	8.442	1.817
Percentage Of Variance	0.502	0.180	0.117	0.079	0.649	0.140
Cumulative Percentage Of Variance	0.502	0.682	0.799	0.878	0.649	0.789

Pcs Refer To Principal Components. The Significant Factors (I.E., Pcs With An Eigenvalue >1) Are List In The Table.

Table 4. The Pearson Correlation Coefficients For The River Water Quality Parameters Compared With The Corresponding Sub-Watershed Characteristics.

	Frt	Wt	Ub	Crp	Elv	Slp_{mean}	Ar	Hp_d	Gdp_{pc}	Aho_{pc}
Bod_5	−0.52*	0.11	0.76**	0.35	−0.31	−0.40	−0.08	0.59**	0.32	0.09
Cod_{mn}	−0.61**	0.31	0.90**	0.40	−0.58	−0.64**	0.10	0.78**	0.60**	0.42
Tn	−0.72**	0.27	0.96**	0.51*	−0.61**	−0.74**	0.37	0.87**	0.60**	0.48*
Dn	−0.70**	0.17	0.91**	0.51*	−0.53*	−0.68**	0.34	0.81**	0.48*	0.40
Nh_4^+-N	−0.60**	0.15	0.84**	0.42	−0.39	−0.51*	0.14	0.69**	0.38	0.19
No_3^--N	−0.54*	0.13	0.41	0.50*	−0.52*	−0.76**	0.70**	0.60**	0.45*	0.70**
Tp	−0.63**	0.37	0.77**	0.46*	−0.82**	−0.76**	0.46**	0.77**	0.66**	0.80**
Dp	−0.60**	0.08	0.61**	0.50*	−0.48*	−0.63**	0.30	0.64**	0.45*	0.48*
Do	0.48*	−0.43	−0.75**	−0.29	0.45*	0.59**	−0.63**	−0.71**	−0.44	−0.36
Ph	0.53*	0.01	−0.66**	−0.42	0.37	0.33	−0.44	−0.53*	−0.16	−0.15
T	−0.43	0.18	0.44	0.36	−0.44	−0.49*	0.46	0.53*	0.23	0.33
Turbidity	−0.28	0.40	0.42	0.15	−0.71**	−0.51*	0.36	0.45*	0.58**	0.81**
Ec	−0.21	0.61**	0.52*	0.02	−0.64**	−0.60**	0.20	0.57**	0.88**	0.78**

Abbreviations Are Provided In Table 1. The Figures In The Table Are Correlation Coefficients.
*Indicates Significance At The 0.05 Probability Level.
**Indicates Significance At The 0.01 Probability Level.

evaluate the response of river water quality to environmental gradients. The correlation analysis indicated that CRP was positively correlated with TN, DN, NO_3^--N, TP and DP ($r>0.45$, $p<0.05$; r refers to the correlation coefficient). Moreover, UB was positively correlated with BOD_5, COD_{Mn}, TN, DN, NH_4^+-N, TP, DP and EC ($r>0.50$, $p<0.05$) and negatively correlated with DO and pH ($r<-0.65$, $p<0.01$). FRT was positively correlated with DO and pH ($r>0.47$, $p<0.05$) and exhibited a negative correlation with BOD, COD, TN, DN, NH_4^+-N, NO_3^--N TP, and DP ($r<-0.50$, $p<0.05$). Furthermore, SLP_{mean} was positively correlated with DO ($r=0.59$, $p<0.01$) and negatively correlated with COD_{Mn}, TN, DN, NH_4^+-N, NO_3^--N, TP, DP, T, turbidity and EC ($r<-0.48$, $p<0.05$). The elevation was positively correlated with DO ($r=0.45$, $p<0.05$) and negatively correlated with COD_{Mn}, TN, DN, NO_3^--N, TP, DP, turbidity and EC ($r<-0.47$, $p<0.05$). Additionally, GDP_{pc} exhibited a positive correlation with COD, TN, DN, NO_3^--N, TP, DP, turbidity, and EC ($r>0.44$, $p<0.05$). AHO_{pc} was positively correlated with TN, NO_3^--N, TP, DP, turbidity and EC ($r>0.47$, $p<0.05$). Lastly, HP_d exhibited a negative correlation with DO and pH ($r<-0.52$, $p<0.05$) and a positive correlation to all other river water quality parameters ($r>0.44$, $p<0.05$) (**Table 4**).

MRA demonstrated that no individual environmental factor was able to describe the overall water quality; however, most of the physico-chemical water parameters could be sufficiently predicted using 1 to 3 environmental factors (**Table 5**). Specifically, BOD_5 could be predicted using UB, AR and GDP_{pc}; COD_{Mn} with UB and AR; TN, DN, pH and DO with UB; NH_4^+-N with UB and HP_d; DP with HP_d; turbidity with AHO_{pc}; EC with GDP_{pc}; TP with UB and AHO_{pc}; and NO_3^--N with UB and SLP_{mean}.

The RDA of the overall water quality as the dependent variable suggested that FRT, UB, CRP, ELV, SLP_{mean}, AR, HP_d, GDP_{pc} and AHO_{pc} explained 15.7–50.8% of the river water quality spatial variation, and a combination of the topography, land use and socio-economic factors explained 86.1% of the variation in overall water quality (**Table 6**). The ordination diagram of overall water quality and topography, land use and socio-economic factors indicated that the first RDA axis displayed a pollution gradient (e.g., COD_{Mn} and TN increased along the axis), which was positively correlated with UB and HP_d and negatively correlated with FRT and SLP_{mean} (**Figure 3**), accounting for 52.4% of the total variance in water quality (**Table 7**). The second axis was related to AHO_{pc} and only explained 16.5% of the total variance (**Table 7**).

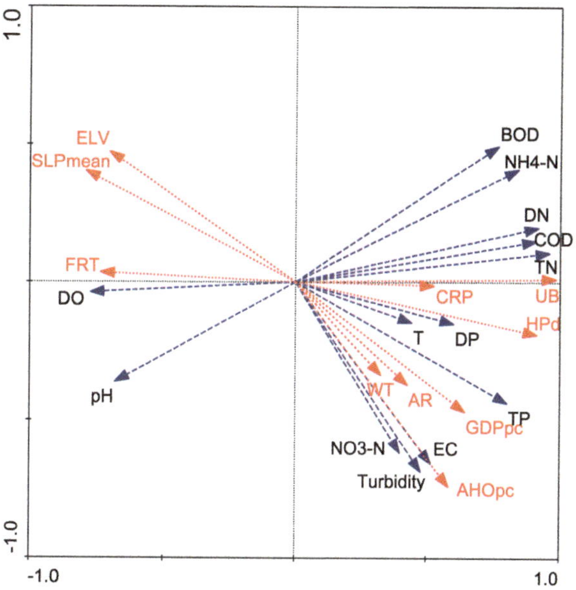

Figure 3. Biplots Of The River Water Quality Parameters And Watershed Characteristics In The Cao-E River Basin According To The Redundancy Analysis.

Table 5. Stepwise Multiple Regression Models For The River Water Quality Parameters And Watershed Characteristics In The Cao-E River Basin In Eastern China.

Parameters	Independent Variables	Regression Equation	R^2	Adjusted R^2	Significance
Bod_5	Ub, Ar, Gdp_{pc}	$1.053+566.924ub-0.012ar-0.015gdp_{pc}$	0.856	0.830	**
Cod_{mn}	Ub, Ar	1.451+88.727ub-0.002ar	0.903	0.892	**
Tn	Ub	1.668+69.168ub	0.912	0.907	**
Dn	Ub	1.231+53.629ub	0.821	0.811	**
Nh_4^+-N	Ub, Hp_d	$1.407+61.854ub-1.030hp_d$	0.805	0.782	**
No_3^--N	Slp_{mean}, Ub	$5.161012-0.233slp_{mean}-7.675ub$	0.720	0.670	**
Tp	Ub, Aho_{pc}	$0.012+1.093ub+0.050aho_{pc}$	0.794	0.770	*
Dp	Hp_d	$0.022+0.015hp_d$	0.406	0.373	**
Do	Ub	9.802-50.534ub	0.562	0.538	**
Ph	Ub	7.724012-5.814ub	0.439	0.408	**
Turbidity	Aho_{pc}	$-115.264+96.209aho_{pc}$	0.662	0.643	**
Ec	Gdp_{pc}	$-261.761+1.009gdp_{pc}$	0.778	0.766	**

Abbreviations Are Provided In Table 1. The Parameters Without Regression Models Are Not Listed.
*Indicates Significance At The 0.05 Probability Level.
**Indicates Significance At The 0.01 Probability Level.

Table 6. The Rda Results For The Percentage Of The Overall River Water Quality Variance Explained By An Individual Variable And All The Explanatory Variables.

Explanatory Variables	Variance Explained%	*P* Value
Land Use		
Frt	29.6	0.002
Wt	8.8	0.16
Ub	50.8	0.002
Crp	16.2	0.016
Topography		
Elv	29.8	0.004
Slp_{mean}	36.3	0.002
Ar	15.7	0.032
Socio-Economic Factors		
Hp_d	44.6	0.002
Gdp_{pc}	26.2	0.032
Aho_{pc}	26.9	0.002
Land Use + Topography + Social-Economic Factors	86.1	0.002

Abbreviations Are Provided In Table 1.

Discussion

1 Seasonal effects

Seasonal variations in the flow caused by the subtropical monsoon climate can partially explain the river water quality temporal variations [7], [38]. The relatively high TN concentrations in the rainy season were attributed to a flushing effect [36]. BOD_5, which corresponds to point source pollution, were lower in the rainy season due to the dilution by precipitation [2]. Relatively high T in the rainy season resulted from the specific climatic conditions during this period (T is higher in the rainy season than in the dry season; see **Table 2**). The lower values of DO and higher values of pH in the rainy season may related to some environmental factors variations, such as flow rate, water temperature, aquatic plant growth and so on [50], [39–41]. The reasons are complex and the further research is needed in the future. The turbidity was higher in the dry season, which may be primarily because the reduced flow during this period caused the river water to be easily influenced by tides. Moreover, Odokuma and Okpokwasili [40] reported that the other parameters such as phosphate showed significantly higher values in the rainy season than in the dry season in the New Calabar River, Nigeria. These different results compared with our study were mainly due to the regional climate difference. However, the absence of a significant difference in the other physico-chemical parameters between the dry and wet seasons indicated mixed and irregular influences, e.g., point sources and diffusion may have played an important role [1], [42].

2 Land cover effects

In the study area, land-use types could play important roles in affecting major river water quality parameter distributions, as reflected by their considerable variability (**Figure 2**). This result is corroborated by the strong positive correlations between urban land and BOD_5, COD_{Mn}, TN, DN, NH_4^+-N and TP (**Table 5**) and the strong negative correlations with DO and pH. Moreover, most of the nutrient variables exhibited lower concentrations in the forest-dominated region, whereas the cropland-dominated region had high nutrient concentrations. The region downstream of the urban area exhibited higher concentrations of BOD_5, COD, NH_4^+ N, TP and DP and lower pH and DO (**Figure 2**), this finding agrees with many related studies [21], [43,44]. Studies have also shown that certain river water quality parameters (e.g., TN and TP) are primarily determined by agricultural land use at

Table 7. The Rda Results Showing The Eigenvalues And Percentage Of The Overall River Water Quality Parameter Variance Explained By The First Two Rda Axes And All The Rda Axes.

	Axis 1	Axis 2	All Axes
Eigenvalues	0.524	0.165	0.861
Percentage Of Water Quality Variance Explained	52.4	16.5	86.1
Dominant Environmental Factors	Ub(0.967)	Aho_{pc}(−0.699)	
	Hp_d (0.897)		
	Slp_{mean}(−0.780)		
	Frt (−0.725)		

Abbreviations Are Provided In Table 1. The Number In Parentheses Indicates The Canonical Coefficients Of The Environmental Variables With The First Two Rda Axes.

the watershed scale in many parts of the world [1], [20]. However, this study found that agricultural land use in watersheds plays a minor role in explaining the spatial variations in river water quality (explaining 16.2% of the total overall water quality variance) in the Cao-E River basin. The different result is primarily due to the enhanced effects of urban runoff and incompletely treated industrial waste and domestic pollution on river water quality compared to agricultural runoff. In addition, vegetated buffers adjacent to the croplands can substantially mitigate agricultural runoff of nutrients and other contaminants via deposition, absorption, and denitrification [20].

Urban areas are primarily located along river networks in the Cao-E River basin. Therefore, urban effects on the river water quality were expected. Urban land use comprised a much smaller percentage of the Cao-E River basin than the cropland in this study. However, as observed from the regression model (BOD_5, COD_{Mn}, TN, DN, NH_4^+-N, NO_3^--N, TP, DO and pH) and the RDA analysis (explaining 50.8% of the total overall water quality variance) results, urban areas play a pivotal role in influencing water quality (**Tables 5 and 6**). This observation is most likely due to two factors: (1) the deficient capacity of urban sewage treatment plants leads to domestic pollution and industrial wastewater inputs into the Cao-E River system, and (2) the high percentage of impervious surfaces and over fertilization of grassy areas can increase discharge rates, sedimentation and pollutant runoff to streams [44].

3 Topography effects

Topographical factors played important roles in explaining spatial variations in river water quality within the Cao-E River basin (SLP_{mean} was 36.3% and ELV was 29.8%). Topography largely regulated the river water quality parameters. Most related studies have shown that higher SLP_{mean} and/or ELV lead to higher erosion rates, which subsequently increase the rate at which particulate matter enters a water body [13], [21], [45]. However, our results indicated an inverse relationship. Here, SLP_{mean} and ELV were both negatively correlated with TN, DN, NO_3^--N, TP and DP; SLP_{mean} exhibited a highly significant negative correlation with NO_3^--N values (**Figure 3 and Table 5**). The slope effects on water chemistry varied. Watershed physical characteristics, such as soil properties (soil texture and soil drainage), morphological variables (drainage density and elongation) and particularly surficial debris, largely affected water chemistry in river waters [37], [46]. In addition, these negative correlations existed primarily because ELV and SLP_{mean} were strongly correlated with land cover (**Table 8**). High elevations and/or steep catchments are primarily occupied by forest land cover; however, cropland is primarily located in regions in which the topography is relatively flat and has a low ELV. Therefore, the forested mountain area, which has higher SLP_{mean} and/or ELV, may export fewer nutrients than flat land (e.g., cropland). Similar results were published regarding water quality spatial variations in the Han River basin in South Korea [7] and non-point source effects on stream nutrient concentrations in the Seattle region of the USA [47].

4 The effects of socio-economic factors

The socio-economic factors commonly have different extent effects on river water quality, especially in developing countries such as China. Our study shows that socioeconomic factors have significant effects on river water quality during the study period (**Table 5**) and that HP_d, AHO_{pc}, and GDP_{pc} explain 44.6%, 26.9% and 26.2% of the total overall river water quality variance, respectively (**Table 6**). Furthermore, HP_d is a fundamental

Table 8. The Pearson Correlation Matrix For The Watershed Characteristics Of The Cao-E River Basin In Eastern China.

	Frt	Wt	Ub	Crp	Elv	Slp_{mean}	Ar	Hp_d	Gdp_{pc}	Aho_{pc}
Frt	1.00									
Wt	0.09	1.00								
Ub	−0.79**	0.31	1.00							
Crp	−0.96**	−0.31	0.59**	1.00						
Elv	0.57**	−0.48*	−0.71**	−0.41	1.00					
Slp_{mean}	0.83**	−0.33	−0.82**	−0.70**	0.74**	1.00				
Ar	−0.44	0.17	0.42	0.37	−0.44	−0.56*	1.00			
Hpd	−0.79**	0.37	0.95**	0.60**	−0.79**	−0.91**	0.55*	1.00		
Gdp_{pc}	−0.38	0.49*	0.69**	0.18	−0.65**	−0.67**	0.30	0.76**	1.00	
Aho_{pc}	−0.47*	0.20	0.55*	0.36	−0.69**	−0.72**	0.47*	0.64**	0.72**	1.00

*Indicates That The Correlation Was Significant At The 0.05 Probability Level.
**Indicates That The Correlation Was Significant At The 0.01 Probability Level.

parameter for predicting DP and AHO_{pc} for TP and turbidity (**Table 5**). This result can be attributed to high population density, the massive animal production and the lack of strict management regulations for animal waste in the study area. The population density exceeds 380 people km^{-2}, the annual animal output value exceeds 2.0 billion Yuan and the annual hog production exceeds 1.0 million heads for the entire basin [48]. Most human and animal waste with incomplete treatment is discharged into the surrounding water bodies. Moreover, EC was positively correlated with GDPpc, which was most likely related to point source pollution [37].

In addition, some researches [4–6] point out that climate condition is also an important factor of influencing river water quality. Thus, we will further study it in our future work.

Conclusions

The primary results of this study can be summarized as follows:

1) The in-stream water quality in the Cao-E River basin streams suggested that TN, pH and T were generally higher in the rainy season, whereas BOD_5, DO and turbidity were higher in the dry season.
2) Spatial variations in river water quality are typically associated with several anthropogenic and natural factors. Urban land cover was determined to be the most important explanatory variable for BOD_5, COD_{Mn}, TN, DN, NH_4^+-N, NO_3^--N, TP, DO and pH. Moreover, Animal husbandry output per capita was an important predictor for TP and turbidity, and Gross domestic product per capita largely determined the spatial variations in EC. The remaining unexplained variance resulted from other factors, such as topography.
3) 3) Pollution control for animal waste discharge in rural settlements was important in the study area. Moreover, agricultural runoff, industrial pollution and domestic pollution in urban and industrial areas were also important factors within the Cao-E River basin.
4) The percentage of total overall river water quality variance explained by individual variables and/or all environmental variables in the study area (as determined using RDA) can assist in quantitatively identifying primary pollution control factors at the watershed scale.

This study improved our understanding of the anthropogenic activities and natural factors that affect river water quality and can assist in the design of efficient strategies for controlling river water pollution at the watershed scale. Moderate-resolution remote sensing data were adopted in this study; future investigations will require high-resolution DEM maps and additional land use classes to better evaluate the effects of specific environmental variables on overall river water quality.

Supporting Information

Table S1 Summary of commonly used statistical methods on pollution source identification in recent years.

Table S2 Application summary of 5 statistical methods in this study.

Acknowledgments

We would like to acknowledge the members of the Zhejiang University soil physical laboratory for assistance in the field and laboratory.

Author Contributions

Conceived and designed the experiments: JL JC. Performed the experiments: JC. Analyzed the data: JC JL. Wrote the paper: JL JC.

References

1. Li S, Gu S, Liu W, Han H, Zhang Q (2008) Water Quality In Relation To Land Use And Land Cover In The Upper Han River Basin, China. Catena 75: 216–222.
2. Zhang Y, Guo F, Meng W, Wang X (2009) Water Quality Assessment And Source Identification Of Daliao River Basin Using Multivariate Statistical Methods. Environ Monit Assess 152: 105–121.
3. Zhou N, Westrich B, Jiang S, Wang Y (2011) A Coupling Simulation Based On A Hydrodynamics And Water Quality Model Of The Pearl River Delta, China. J Hydrol 396: 267–276.
4. Astaraie-Imani M, Kapelan Z, Fu G, Butler D (2012) Assessing The Combined Effects Of Urbanisation And Climate Change On The River Water Quality In An Integrated Urban Wastewater System In The Uk. J Environ Manage 112: 1–9.
5. Costa-Cabral M, Coats R, Reuter J, Riverson J, Sahoo G, et al. (2013) Climate Variability And Change In Mountain Environments: Some Implications For Water Resources And Water Quality In The Sierra Nevada (Usa). Clim Change 116: 1–14.
6. Aubert Ah, Gascuel-Odoux C, Merot P (2013) Annual Hysteresis Of Water Quality: A Method To Analyse The Effect Of Intra-And Inter-Annual Climatic Conditions. J Hydrol 478: 29–39.
7. Chang H (2008) Spatial Analysis Of Water Quality Trends In The Han River Basin, South Korea. Water Res 42: 3285–3301.
8. Hoorman J, Hone T, Sudman Jr T, Dirksen T, Iles J, et al. (2008) Agricultural Impacts On Lake And Stream Water Quality In Grand Lake St. Marys, Western Ohio. Water Air Soil Pollut Int 193: 309–322.
9. Ahearn Ds, Sheibley Rw, Dahlgren Ra, Anderson M, Johnson J, et al. (2005) Land Use And Land Cover Influence On Water Quality In The Last Free-Flowing River Draining The Western Sierra Nevada, California. J Hydrol 313: 234–247.
10. Huang F, Wang X, Lou L, Zhou Z, Wu J (2010) Spatial Variation And Source Apportionment Of Water Pollution In Qiantang River (China) Using Statistical Techniques. Water Res 44: 1562–1572.
11. Chang N (2005) Sustainable Water Resources Management Under Uncertainty. Stoch Environ Res Risk Assess 19: 97–98.
12. Kolovos A, Christakos G, Serre Ml, Miller Ct (2002) Computational Bayesian Maximum Entropy Solution Of A Stochastic Advection - Reaction Equation In The Light Of Site - Specific Information. Water Resour Res 38: 51–54.
13. Sliva L, Dudley Williams D (2001) Buffer Zone Versus Whole Catchment Approaches To Studying Land Use Impact On River Water Quality. Water Res 35: 3462–3472.
14. Heathcote Aj, Filstrup Ct, Downing Ja (2013) Watershed Sediment Losses To Lakes Accelerating Despite Agricultural Soil Conservation Efforts. Plos One 8: E53554.
15. Keatley Be, Bennett Em, Macdonald Gk, Taranu Ze, Gregory-Eaves I (2011) Land-Use Legacies Are Important Determinants Of Lake Eutrophication In The Anthropocene. Plos One 6: E15913.
16. Wei R, Ge F, Huang S, Chen M, Wang R (2011) Occurrence Of Veterinary Antibiotics In Animal Wastewater And Surface Water Around Farms In Jiangsu Province, China. Chemosphere 82: 1408–1414.
17. Xianghao Z (2008) Study Of Protection And Construction Of Mountain Ecological Security Barrier In China. J Mount Sci (In Chinese) 1: 4–5.
18. Jiang X, Sommer Sg, Christensen Kv (2011) A Review Of The Biogas Industry In China. Energy Policy 39: 6073–6081.
19. Tong L, Li P, Wang Y, Zhu K (2009) Analysis Of Veterinary Antibiotic Residues In Swine Wastewater And Environmental Water Samples Using Optimized Spe-Lc/Ms/Ms. Chemosphere 74: 1090–1097.
20. Liu W, Li S, Bu H, Zhang Q, Liu G (2012) Eutrophication In The Yunnan Plateau Lakes: The Influence Of Lake Morphology, Watershed Land Use, And Socioeconomic Factors. Environ Sci Pollut Res Int 19: 858–870.
21. Ye L, Cai Q, Liu R, Cao M (2009) The Influence Of Topography And Land Use On Water Quality Of Xiangxi River In Three Gorges Reservoir Region. Environ Geol 58: 937–942.
22. Tian C, Lu X, Pei H, Hu W, Xie J (2013) Seasonal Dynamics Of Phytoplankton And Its Relationship With The Environmental Factors In Dongping Lake, China. Environ Sci Pollut Res Int 185: 2627–2645.
23. Yang X, Wu X, Hao H, He Z (2008) Mechanisms And Assessment Of Water Eutrophication. J Zhejiang Univ Sci B 9: 197–209.

24. Chen D, Lu J, Shen Y, Gong D, Deng O (2011) Spatio-Temporal Variations Of Nitrogen In An Agricultural Watershed In Eastern China: Catchment Export, Stream Attenuation And Discharge. Environ Pollut 159: 2989–2995.
25. Jin S, Lu J, Chen D, Shen Y, Shi Y (2009) Relationship Between Catchment Characteristics And Nitrogen Forms In Cao-E River Basin, Eastern China. J Environ Sci (China) 21: 429–433.
26. Shen Y, L U J, Chen D, Shi Y (2011) Response Of Stream Pollution Characteristics To Catchment Land Cover In Cao-E River Basin, China. Pedosphere 21: 115–123.
27. Sepbc (2002) Water And Wastewater Analysis Method. Prepared And Published By: State Environment Protection Bureau Of China (Sepbc). Beijing: China Environmental Science Press. 784 P.
28. Dodds Wk, Oakes Rm (2004) A Technique For Establishing Reference Nutrient Concentrations Across Watersheds Affected By Humans. Limnol Oceanogr Methods 2: 333–341.
29. Chen J, Lu J (2013) Establishment Of Reference Conditions For Nutrients In An Intensive Agricultural Watershed, Eastern China. Environ Sci Pollut Res Int 4: 2496–2505.
30. Olsen Rl, Chappell Rw, Loftis Jc (2012) Water Quality Sample Collection, Data Treatment And Results Presentation For Principal Components Analysis–Literature Review And Illinois River Watershed Case Study. Water Res 46: 3110–3122.
31. Myers Rh (1990) Classical And Modern Regression With Applications: Duxbury Press Belmont, Ca. 350p.
32. Miserendino Mial, Casaux R, Archangelsky M, Di Prinzio Cy, Brand C, et al. (2011) Assessing Land-Use Effects On Water Quality, In-Stream Habitat, Riparian Ecosystems And Biodiversity In Patagonian Northwest Streams. Sci Total Environ 409: 612–624.
33. Braak Ct, Šmilauer P (2002) Canoco Reference Manual And Canodraw For Windows User's Guide: Software For Canonical Community Ordination (Version 4.5). Section On Permutation Methods. Microcomputer Power, Ithaca, New York. 300p.
34. Lepš J, Šmilauer P (2003) Multivariate Analysis Of Ecological Data Using Canoco. New York: Cambridge University Press. 283 P.
35. Johnson Ra, Wichern Dw (2002) Applied Multivariate Statistical Analysis: Prentice Hall Upper Saddle River, Nj. 755p.
36. Chen D, Lu J, Yuan S, Jin S, Shen Y (2006) Spatial And Temporal Variations Of Water Quality In Cao-E River Of Eastern China. J Environ Sci (China) 18: 680–688.
37. Li S, Xia X, Tan X, Zhang Q (2013) Effects Of Catchment And Riparian Landscape Setting On Water Chemistry And Seasonal Evolution Of Water Quality In The Upper Han River Basin, China. Plos One 8: E53163.
38. Shah Vg, Dunstan Rh, Geary Pm, Coombes P, Roberts Tk, et al. (2007) Comparisons Of Water Quality Parameters From Diverse Catchments During Dry Periods And Following Rain Events. Water Res 41: 3655–3666.
39. Karaer F, Küçükballi A (2006) Monitoring Of Water Quality And Assessment Of Organic Pollution Load In The Nilüfer Stream, Turkey. Environ Monit Assess 114: 391–417.
40. Odokuma Lo, Okpokwasili Gc (1997) Seasonal Influences Of The Organic Pollution Monitoring Of The New Calabar River, Nigeria. Environ Monit Assess 45: 43–56.
41. Kannel Pr, Lee S, Lee Y (2008) Assessment Of Spatial–Temporal Patterns Of Surface And Ground Water Qualities And Factors Influencing Management Strategy Of Groundwater System In An Urban River Corridor Of Nepal. J Environ Manage 86: 595–604.
42. Chen Js, Gao Xm, He Dw, Xia Xh (2000) Nitrogen Contamination In The Yangtze River System, China. J Hazard Mater 73: 107–113.
43. Pratt B, Chang H (2012) Effects Of Land Cover, Topography, And Built Structure On Seasonal Water Quality At Multiple Spatial Scales. J Hazard Mater 209: 48–58.
44. Mehaffey Mh, Nash Ms, Wade Tg, Ebert Dw, Jones Kb, et al. (2005) Linking Land Cover And Water Quality In New York City's Water Supply Watersheds. Environ Monit Assess 107: 29–44.
45. Xiao B, Wang Q, Wu J, Huang C, Yu D (2010) Protective Function Of Narrow Grass Hedges On Soil And Water Loss On Sloping Croplands In Northern China. Agric Ecosyst Environ 139: 653–664.
46. Lindell L, Aa Str O M M, O Berg T (2010) Land-Use Change Versus Natural Controls On Stream Water Chemistry In The Subandean Amazon, Peru. Appl Geochem 25: 485–495.
47. Brett Mt, Arhonditsis Gb, Mueller Se, Hartley Dm, Frodge Jd, et al. (2005) Non-Point-Source Impacts On Stream Nutrient Concentrations Along A Forest To Urban Gradient. Environ Manage 35: 330–342.
48. Shaoxing City Statistical Bureau (2005) Shaoxing Statistical Yearbook. Shaoxing: Shaoxing City Press. 300p.
49. Belsey Da, Kuh E, Welsch Re (1980) Regression Diagnostics: Identifying Influential Data And Sources Of Collinearity: John Wiley. 320p.
50. Zhao J, Fu G, Lei K, Li Y (2011) Multivariate Analysis Of Surface Water Quality In The Three Gorges Area Of China And Implications For Water Management. J Environ Sci (China) 23: 1460–1471.

10

Evaluation of the Ecotoxicity of Sediments from Yangtze River Estuary and Contribution of Priority PAHs to Ah Receptor-Mediated Activities

Li Liu[1], Ling Chen[1], Ying Shao[1,2], Lili Zhang[1], Tilman Floehr[2], Hongxia Xiao[2], Yan Yan[2], Kathrin Eichbaum[2], Henner Hollert[1,2,3,4], Lingling Wu[1]*

1 Key Laboratory of Yangtze Water environment, Ministry of Education, Tongji University, Shanghai, China, **2** Department of Ecosystem Analysis, Institute for Environmental Research (Biology V), Aachen Biology and Biotechnology, RWTH Aachen University, Aachen, Germany, **3** College of Resources and Environmental Science, Chongqing University, Chongqing, China, **4** School of Environment, Nanjing University, Nanjing, China

Abstract

In this study, in vitro bioassays were performed to assess the ecotoxicological potential of sediments from Yangtze River estuary. The cytotoxicity and aryl hydrocarbon receptor (AhR)-mediated toxicity of sediment extracts with rainbow trout (*Oncorhynchus mykiss*) liver cells were determined by neutral red retention and 7-ethoxyresorufin-*O*-deethylase assays. The cytotoxicity and AhR-mediated activity of sediments from the Yangtze River estuary ranged from low level to moderate level compared with the ecotoxicity of sediments from other river systems. However, Yangtze River releases approximately 14 times greater water discharge compared with Rhine, a major river in Europe. Thus, the absolute pollution mass transfer of Yangtze River may be detrimental to the environmental quality of estuary and East China Sea. Effect-directed analysis was applied to identify substances causing high dioxin-like activities. To identify unknown substances contributing to dioxin-like potencies of whole extracts, we fractionated crude extracts by open column chromatography. Non-polar paraffinic components (F1), weakly and moderately polar components (F2), and highly polar substances (F3) were separated from each crude extract of sediments. F2 showed the highest dioxin-like activities. Based on the results of mass balance calculation of chemical toxic equivalent concentrations (TEQs), our conclusion is that priority polycyclic aromatic hydrocarbons indicated a low portion of bio-TEQs ranging from 1% to 10% of crude extracts. Further studies should be conducted to identify unknown pollutants.

Editor: Aditya B. Pant, Indian Institute of Toxicology Reserach, India

Funding: This work was supported by a grant from National Natural Science Foundation of China (No. 41101499), the Fundamental Research Funds for the Central Universities of China (No. 0400219213) and Chinese 111 Program. The work was also supported by the research cluster "Pollutants/Water/Sediment-Impacts of Transformation and Transportation Processes on the Yangtze Water Quality" sponsored by the Federal Ministry of Education and Research, Germany (BMBF) and supported by a cooperation project with Chinese colleagues sponsored by the BMBF DLR. The funders had no role in study design, data collection and analysis, decision to publish, or preparation of the manuscript.

Competing Interests: The authors have declared that no competing interests exist.

* Email: wulingling@tongji.edu.cn

Introduction

Marine ecosystem contamination has been greatly affected by human activities. As a coastal transitional system, an estuary serves as a recipient of environmental contaminants derived from land and rivers as well as the atmosphere. Sediments in estuaries are the main sinks of numerous potential chemical and biological pollutants. These sediments may also be considered as a secondary source of water pollution or be absorbed by benthic organisms via bioaccumulation. Thus, sediments as contaminants in rivers and estuaries may pose a potential threat to aquatic organisms and human health.

Toxicity biotests are necessary to assess sediment quality, because they can determine the effect of chemicals in sediments on organisms directly [1]. Some studies have shown many species, such as bacteria, microalgae, yeast, and fish, are widely used to assess the acute toxicity of sediments [2,3]. To obtain comprehensive insights into potential ecotoxicological effects, the specific effects such as mutagenic, genotoxic, and dioxin-like responses should also be assessed. In vitro cell-based bioassays are of great benefit to characterize mechanism-specific activities of environmental contaminants because they are sensitive and require a smaller amount of sample than whole animal experiments [3]. Fish cell lines have been successfully used as a biological alternative of animal tests to assess the toxic effects of sediment extracts [4,5]. For example, rainbow trout (*Oncorhynchus mykiss*) liver cell line (RTL-W1) exhibits high sensitivity to the cytotoxicity of pure substances [5,6]; moreover, this cell line can highly express cytochrome P4501A (CYP1A)-based EROD activity upon exposure to dioxin-like compounds [1]. As such, RTL-W1 is commonly used to determine the acute cytotoxicity and CYP1A-based EROD activity of sediment extracts [2,7].

Effect-directed analysis (EDA) is a powerful tool used to identify toxic substances in complex environmental samples [8]. This method is based on a combination of biotests, fractionation procedures, and chemical analysis. EDA can be performed to evaluate the toxic potencies of substances in sediments. This method was applied to analyze toxic chemicals in complex environmental samples. Tetrabromobisphenol A diallyl ether was identified as an emerging neurotoxicant in environmental samples by bioassay-directed fractionation and high-performance liquid chromatography-atmospheric pressure chemical ionization-tandem mass spectrometry [9]. Applying EDA to evaluate sediment extracts in Bitterfeld, Germany, where dinitropyrenes and 3-nitrobenzanthrone are quantitatively identified as the main mutagens [10]. Other studies have also successfully applied EDA to identify EROD-inducing compounds in sediments [11,12].

Polycyclic aromatic hydrocarbons (PAHs) have been extensively investigated because these substances are widely distributed in sediments and may function as mutagens and carcinogens [13]. As dioxin-like compounds, PAHs can induce the biotransformation of CYP1A enzyme in cells by binding its ligand to aryl hydrocarbon receptor (AhR) [14]. PAH concentrations have increased as a result of the continuous and rapid development of China's economy; these compounds have been detected in the sediments of coastal embayment, continental shelf in China, and Yangtze River estuary [15,16].

The Yangtze River estuary, one of the largest estuaries in the world, is located in the east of China and adjacent to the East China Sea. With agricultural and industrial developments in this region, numerous pollutants have been discharged into this estuary [17]. The increasing fraction of wastewater in the Yangtze River increases the levels of nutrients, heavy metals, and dissolved organic carbon; in addition to wastewater, industrial organic chemicals at concentrations of 500 kg to 3,500 kg per day [18]. Approximately 206 hazardous organic chemicals have been detected in water, and 106 chemicals have been found in sediments; among these chemicals, 17 are listed as priority controlled pollutants in America [19]. Studies have also indicated that various organic contaminants, such as PAHs [16], polychlorinated dibenzo-*p*-dioxins (PCDDs)/polychlorinated dibenzofurans (PCDFs) [20], and polychlorinated biphenyls (PCBs) [21], are present in sediments from the Yangtze River estuary. In addition, newly emerging contaminants, such as polybrominated diphenyl ethers and perfluorinated compounds, have been extensively investigated because of their presence in the Yangtze River [22,23]. Despite the polluted state of this river, studies have been rarely conducted regarding the ecotoxicological potential of Yangtze River sediments [17,24].

In this study, RTL-W1 cells were used to determine the acute cytotoxicity and CYP1A-based EROD activity of the sediment extracts from the Yangtze River estuary. EDA was applied to identify the potential hazardous substances causing AhR-mediated activities. PAHs were chemically analyzed to obtain comprehensive information on these sediments and estimate the extent to which these sediment extract chemicals elicit EROD-inducing effects. The main objectives of this study are listed as follows: (1) to assess the ecotoxicological potential of sediments from the Yangtze River estuary by evaluating the cytotoxic and AhR-mediated effects on RTL-W1 cells; (2) to identify the sediment fractions causing such AhR-mediated effects; and (3) to determine the contribution of priority PAHs to EROD induction of sediment samples.

Materials and Methods

2.1 Sampling

No specific permissions were required for the completion of this study as the field measurements did not involve endangered or protected species nor were conducted in a specified protected area. The surface layer of sediment samples (0 cm to 5 cm) were collected from nine locations in the Yangtze River estuary in March 2012. For each sampling locations, four samples was collected and then mixed to form one sample. The sampling locations are shown in Figure 1. Samples were collected along the salinity gradient and location information is presented in Table S1. Samples Y1 to Y3 were fresh water dominated sediments. Sites Y4 and Y5 were located in the turbidity maximum zone and the samples were brackish water dominated sediments. Sites Y6 to Y9 were located in the river plume zone and the samples were marine sediments. Sediments were collected from each location by using a stainless steel grab sampler. Samples were shock-frozen at −20°C immediately until further processing. Afterward, these samples were freeze-dried at −50°C and sieved using a 100-mesh sieve. The samples were then stored in combusted glass with Teflon lined lids at −20°C in the dark until extraction.

2.2 Sediment extraction procedure

The samples were prepared according to previously described methods [25]. The freeze-dried sediment samples (20 g) were separately extracted with acetone (Merck, Darmstadt, Germany, HPLC) for 48 h by using standard reflux (Soxhlet) extractors with six cycles per hour. Elemental sulfur was removed by treating the extracts with activated copper. The extracts were then reduced in volume to obtain a final volume of 2 mL by rotary evaporation. The solvent was divided into two aliquots and concentrated close to dryness with gentle nitrogen stream. One aliquot was replaced with 1 mL of dimethylsulfoxide (DMSO; Sigma-Aldrich Chemie GmbH, Steinheim, Germany) and used for in vitro biotests, resulting in a final concentration of 10 g of sediment dry weight per mL of DMSO (10 g/mL). The other aliquot was dissolved in *n*-hexane (Merck, HPLC) and used for multilayer fractionation. The extracts were then stored at −20°C until further analysis.

2.3 Multilayer fractionation

To identify unknown substances that contribute to the dioxin-like potencies of whole extracts, we selected the sediments used for fractionation on the basis of maximum AhR agonist activities in crude extracts. Fractionation was performed according to previously described methods with slight modification [26]. Fractionation was also performed using a silica gel/aluminum oxide column. The column was prepared according to the following procedures. Silica gel (60 mesh to 200 mesh, Merck, Darmstadt, Germany) and aluminum oxide (50 mesh to 200 mesh, Merck, Darmstadt, Germany) were activated at 180°C for 24 h. Sodium sulfate was baked at 450°C for 5 h before use to eliminate possible production residues, such as phthalates. Glass columns (inner diameter = 10 mm and height = 50 cm) were filled from the bottom with 10 g of silica gel, 5 g of aluminum oxide, and 2 g of sodium sulfate. The gels were then washed twice with *n*-hexane. Afterward, the extracts were placed in the columns and eluted with different solvents at increasing polarities: F1 containing non-polar paraffinic components were eluted with 50 mL of *n*-hexane; F2 characterized by weakly and moderately polar components were eluted with 70 mL of *n*-hexane/dichloromethane (Merck, HPLC; 7:3, v/v); and F3 containing highly polar components were eluted with 50 mL of acetone/methanol (Merck, HPLC; 1:1, v/v). Using a rotary evaporator, we decreased the volumes of the eluates to

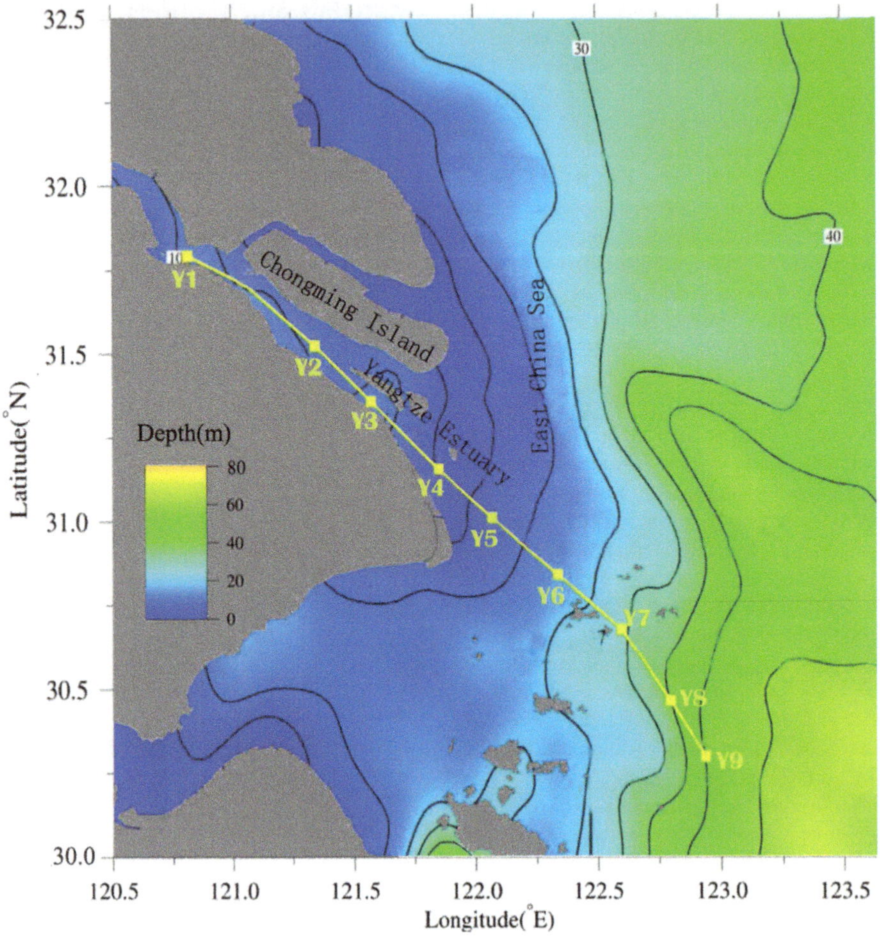

Figure 1. Map of the sampling locations in the Yangtze River estuary. Nine samples Y1–Y9 were collected along the estuary in March 2012.

2 mL and evaporated close to dryness with gentle nitrogen stream. Afterward, the eluates were replaced with 1 mL of DMSO. These fractions were then stored at –20°C in the dark until analysis.

2.4 Cell culture

Cytotoxicity was assessed using the fibroblast-like permanent cell line RTL-W1 isolated from the liver of a female *O. mykiss* [6]. The cells were kindly provided by Drs. Niels C. Bols and Lucy Lee (University of Waterloo, Canada) and cultured according to the method described by Klee et al. [27].

2.5 Neutral red retention (NR) assay

The acute cytotoxicity of sediment extracts on RTL-W1 cells were assessed by NR assay according to Borenfreund and Puerner [28] with slight modifications by Klee et al. [27]. Sediment extracts were serially diluted with L15 medium to obtain a concentration range of 1.57 mg to 100 mg of dry sediment per mL medium with six internal replicates. 3,5-Dichlorophenol (Sigma, Seelze, Germany) was used as a positive control sample at a concentration of 80 mg/L in each test plate. All of the experiments were performed in independent triplicates. The cytotoxic potentials of individual extracts that induced 50% mortality after 48 h (48 h NR_{50}) were calculated accordingly. The unit of NR_{50} was designated as mg dry weight sediment equivalent (SEQ) per mL test medium.

2.6 EROD induction assay

The dioxin-like activity of sediment extracts was determined by EROD induction assay according to previously described methods [29]. The highest test concentration of the EROD assay to determine the enzyme activity and avoid the cytotoxic effects based on NR_{80} obtained from a preliminary NR assay with RTL-W1 cells. The cells were seeded into 96-well microtiter plates (TPP, Trasadingen, Switzerland) and exposed to sediment extracts in eight dilution steps with six internal replicates. 2,3,7,8-Tetrachlorodibenzo-*p*-dioxin (TCDD, Promochem, Wesel, Germany) was serially diluted to obtain a final concentration ranging from 3.13 pM to 100 pM in two separate rows of each plate as a series of positive control sample. After the samples were incubated for 72 h, induction was terminated by removing the growth medium and freezing at –70°C to lyse the cells. The deethylation of exogenous 7-ethoxyresorufin was initiated by adding 7-ethoxyresorufin to each well; the resulting reaction mixture was then incubated in the dark at room temperature for 10 min before NADPH (Sigma) in PBS (Sigma) was added. The plates were further incubated for 10 min and the reaction was terminated by adding fluorescamine (Sigma) dissolved in acetonitrile (Merck, HPLC). EROD activity was determined fluorometrically after another 15 min at excitation and emission wavelengths of 544 and 590 nm, respectively, by using an Infinite M200 plate reader (Tecan, Crailsheim, Germany). The amount of protein was fluorometrically determined using the fluorescamine method at excitation and emission wavelengths of 360 and 465 nm, respectively [30]. The concentration-response curves of EROD induction in the RTL-W1 bioassay were designed by non-linear regression (Prism 4.0, GraphPad, San Diego, USA) using the classic sigmoid curve or Boltzmann curve as model equations.

AhR agonist activities were determined using a fixed-level approach [31]. Extract EC_{25TCDD} of each sediment extract was obtained and normalized to that of the positive control 2,3,7,8-TCDD as biological toxic equivalent concentrations (bio-TEQs) [29]. EC_{25TCDD} was used to compare the samples and calculate TCDD-equivalent concentrations in the samples. Bio-TEQ concentrations were calculated as follows:

$$[bio-TEQ](pg/g(d.w.) = \frac{TCDDEC_{25}(pg/mL)}{extractEC_{25TCDD}(g(d.w.)/mL)} \quad (1)$$

where TCDD EC_{25} (pg/mL) is the concentration of the TCDD positive control sample causing 25% of EROD induction and extract EC_{25TCDD} (g/mL) is the concentration of the sediment extract equivalent causing 25% of EROD induction. EC_{25TCDD} was considered as a more appropriate measure than EC_{50} in this study because EC_{50} has not been well defined by dose-response curve in several cases [31].

2.7 Chemical analysis

A total of 16 priority PAHs in the crude sediment extracts described by Environmental Protection Agency (EPA) were quantified by Agilent 6890B gas chromatograph coupled to a mass selective Agilent 5977A MSD detector, which was operated in the selective ion monitoring mode. The sum parameter was calculated on the basis of 16 priority PAHs as follows: naphthalene; acenaphthylene; acenaphthene; fluorene; phenanthrene; anthracene; fluoranthene; pyrene; benzo[a]anthracene; chrysene; benzo[b,j]fluoranthene; benzo[k]fluoranthene; benzo[a]pyrene; dibenzo[a,h]anthracene; benzo[g,h,i]perylene; and indeno[1,2,3-cd]pyrene. 2-Fluorobiphenyl and terphenyl-d14 were used as surrogate standards and added before Soxhlet extraction for each sample. The detector was equipped with 30 m×0.25 μm film HP-5MS fused silica capillary column (Agilent Technologies), and 1 μL of each fraction was injected in a splitless mode. The initial temperature was set at 45°C for 2 min and then increased to 265, 285, and 320°C at a rate of 20, 6, and 10°C/min. Afterward, temperature was held constant at 320°C for 4 min. Instrumental analysis was conducted after internal standards (naphthalene-d8, acenaphthene-d10, phenanthrened10, chrysene-d12, and perylene-d12) were added.

During the analytical procedure, a procedural blank, a spiked blank, and a duplicate sample were processed. Targets were not detected in procedural blanks. The surrogate recoveries of 2-fluorobiphenyl and terphenyl-d14 were 88% (±5%) and 102% (±4%), respectively. The recoveries of targets ranged from 88% to 103% in spiked blank samples. The reported concentrations were not corrected by surrogate standard recoveries.

2.8 Chem-TEQ calculation

To determine the contribution of EPA-PAHs to the overall AhR-mediated activity, we determined chem-TEQs based on PAH potencies relative to 2,3,7,8-TCDD [32]. Chem-TEQs were calculated by multiplying the concentration of each AhR-active chemical by specific relative potency (REP) of RTL-W1 cells (Eq. 2) [29,32]. Chem-TEQ concentrations at picogram PAH per gram of SEQ were calculated as picogram per gram (pg/g).

$$Chem-TEQ[pgPAH/gSEQ] = \sum_{i=1}^{n} concentrationPAH_i(\frac{pg}{g}dwSEQ) \times REP_i \quad (2)$$

2.9 Data analysis

Data were expressed as mean ± SD. Statistical analyses were performed using SPSS 17.0 (SPSS Inc., Chicago, IL, USA). Maximum concentrations were used to assess the worst pollution scenario at each section compared with other sediment systems.

Results and Discussion

3.1 Cytotoxicity of crude extracts and multilayer fractions

The results of the NR assay on the crude extracts are shown in Figure 2. The sediment extracts from the study sites Y1, Y2, and Y3 showed very low potency and failed to elicit cytotoxic effects at a concentration of 100 mg/mL. By contrast, extracts from the six other sites revealed evident cytotoxicity to RTL-W1 cells with NR_{50} ranging from 4.1 mg/mL to 43.4 mg/mL (Table S2). The extracts from sites Y5, Y8, and Y9 revealed relatively greater cytotoxicity than those from other sites. The lowest cytotoxic potency was obtained from extracts collected from sites Y4, Y6, and Y7. The cytotoxicity of the sediment extracts from downstream regions generally showed higher activities than those from upstream regions.

In the multilayer fractionations, only fraction F1 from sites Y5 and Y9 elicited cytotoxic effects with NR_{50} of 63.2 and 60.4 mg/mL, respectively. No cytotoxic potential was detected in F2 and F3.

According to a sediment classification system of sediments found in Germany [33], the threshold values of NR biotest system are as follows: $NR_{50}>80$ mg/mL, non-toxic; $80\geq NR_{50}\geq 31$ mg/mL, moderately toxic; and $NR_{50}<31$ mg/mL, strongly toxic. On the basis of this threshold values, we classified the crude sediment extracts from sites Y4 and Y7 as moderately toxic and those from sites Y5, Y6, Y8, and Y9 were strongly toxic. The cytotoxicity effects elicited by the sediment extracts in this study were compared with those observed in sediment extracts from other Asian, European, and South American locations (Table 1). The cytotoxic potential of the sediment extracts from the Yangtze River estuary was comparable to that of the suspended particulate matter (SPM) and sediment extracts from the upper Danube River and the surface sediments and sediment core extracts from the upper Rhine River [5] and the Mecklenburg Bight (Western Baltic Sea) [2]. Furthermore, the cytotoxic potential of these extracts from the Yangtze River estuary was higher than that of the sediment extracts from Tietê River [34].

The cytotoxic effect of the sediment extracts may be attributed to the organic contaminants previously polluting the Yangtze River estuary with the rapid development of industrialization. In 2006, more than 10,000 chemical enterprises, or equivalent to approximately half of the total number of enterprises in China, were situated by the river [35]. Many petroleum and chemical plants and docks were also located along the river. In this study, high PAH concentrations, pharmaceuticals, and other organic compounds have been frequently detected in the study region [36,37,38]. Organic compounds, such as PAHs with two or three rings (naphthalene, acenaphthylene, acenaphthene, fluorene and phenanthrene), can elicit cytotoxic effects on rainbow trout (gill) cells [39]. Pharmaceuticals, such as substituted phenols, are also

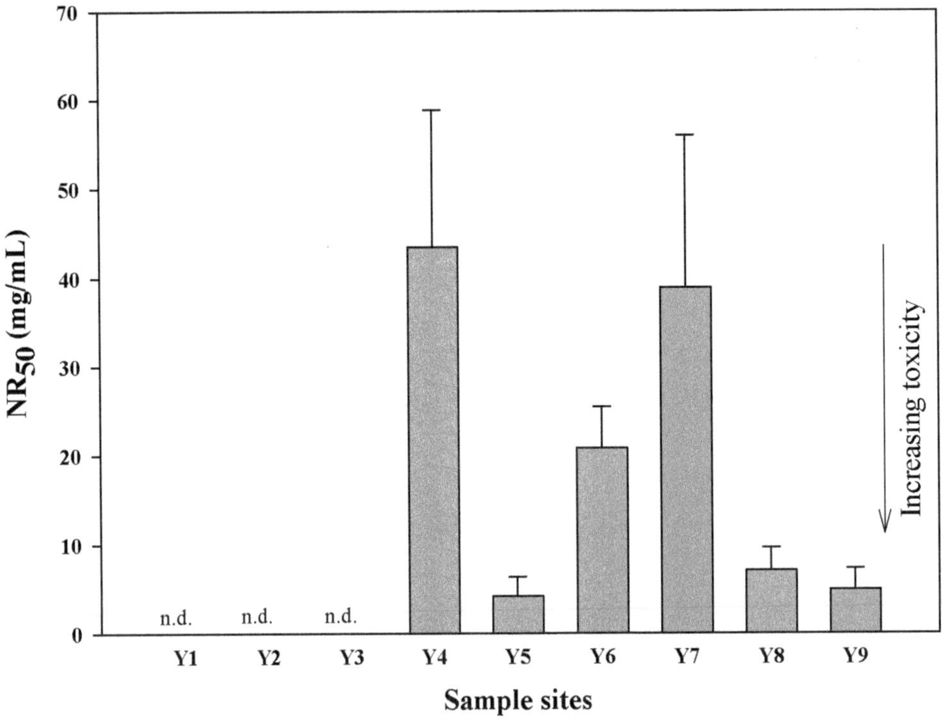

Figure 2. Cytotoxicity of sediment extracts from the Yangtze River estuary. n.d. = not detectable, indicating that no toxic effect of sediment extracts at a concentration of 100 mg/mL with RTL-W1 cells. NR_{50} values for sediment extracts are given in mg sediment equivalent per mL medium (mg/mL). All the values are expressed as means ± SD.

potential inducers of cytotoxicity [40]. In this study, toxic compounds may have accumulated in the Yangtze River estuary and may be accounted for the cytotoxic potential of sediment extracts.

3.2 AhR-mediated activities in crude sediment extracts

Figure 3 shows the results of the EROD assay of the AhR agonists in the crude sediment extracts. The dioxin-like potentials of the sediment extracts among samples varied. In particular, bio-TEQ of crude extracts ranged from 38.9 pg/g dw to 323.5 pg/g dw (Table S2). A relatively high EROD induction was observed in sediment extracts from sites Y2, Y5, and Y9. By contrast, bio-TEQs of the sediments from Y3 showed relatively low EROD induction.

We obtained low to moderate values (Table 2) and corresponding AhR-mediated activity of the investigated sediments compared with bio-TEQs from other river systems. The bio-TEQs of the sediments from different locations were compared on the basis of the reported results regardless of the difference between the extraction methods. EROD induction potential of the sediment extracts from the Yangtze River estuary was comparable to that of the sediment extracts from River Elbe estuary [41] and Saginaw Bay [42]. The sediment extracts from the Yangtze River estuary showed higher EROD induction potential than those from Masan Bay [43], Tai Lake [44], and Yellow Sea [45]. The EROD induction potential of the sediment extracts from the Yangtze River estuary was also lower than that of the sediments from Tietê River [34], Morava River [14], Dagu Rivers, and Haihe [46]. Keiter et al. [33] developed a classification system for AhR-mediated toxicity of German river systems. According to this evaluation scheme, the sediment samples of the Yangtze River estuary revealed only minor AhR-mediated effects. The concentrations of organic pollutants may be lower in Yangtze River than in Rhine River; however, a comparably higher mass transport of water and particulate matter in the Yangtze River system contributes to relatively larger amounts of organic pollutants in the receiving water body of Yangtze River [17]. The mean water discharge (30,200 m^3/s) (http://china.org.cn/english/eng-

Table 1. Cytotoxicity of sediment extracts of the Yangtze River estuary and that of other sediment extracts from previous studies.

Sampling Locations	NR_{50} (mg/ml)	References
Mecklenburg Bight (Western Baltic Sea)	≥13.5	[3]
Upper Rhine (Germany)	≤50	[6]
Upper Danube River (Germany)	<40	[6]
Tietê River (Sao Paulo, Brazil)	29~225	[35]
Yangtze River estuary (China)	4.1~43.4	In this study

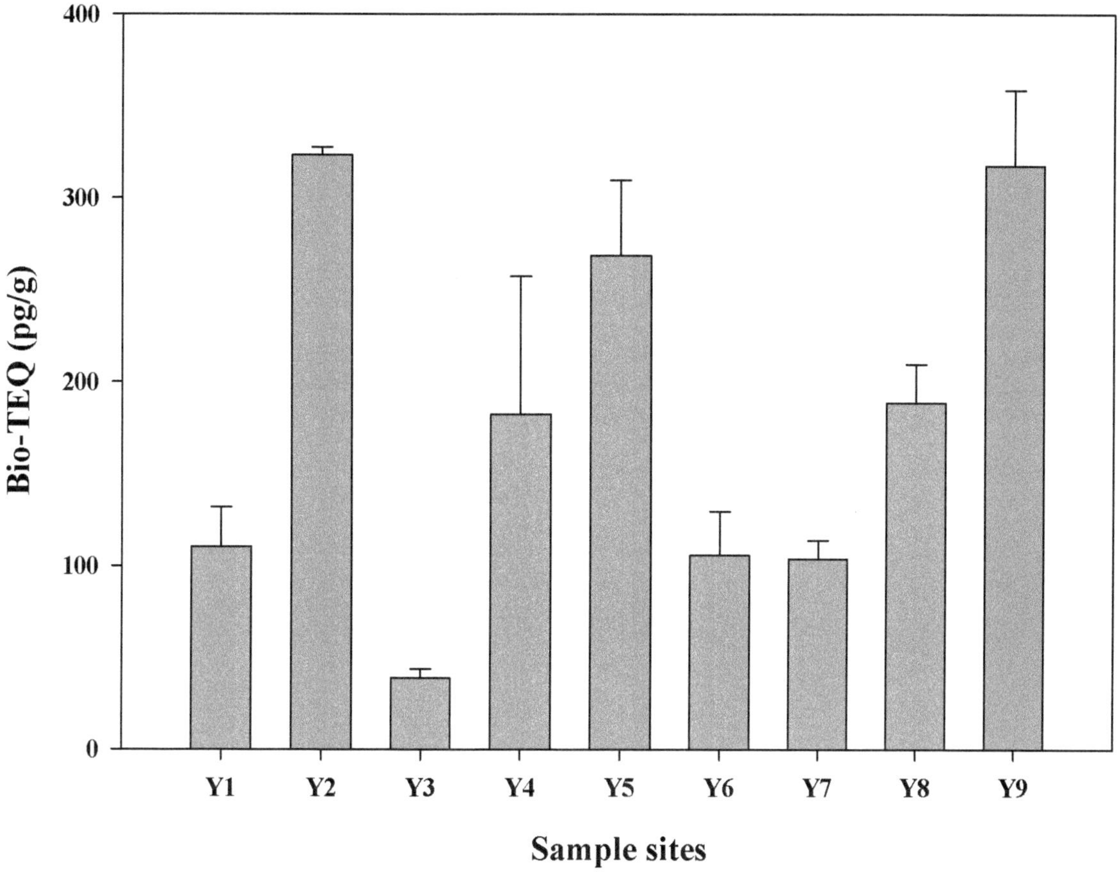

Figure 3. AhR-mediated activity of sediment extracts from the Yangtze River estuary. The values as determined by the EROD assay with RTL-W1 cells are expressed as biological toxicity equivalents (Bio-TEQ; pg/g dw). Data are given as means of 3 replicates ± SD.

shuzi2003/gq/dili5.htm) Yangtze River is approximately 14 times greater than that of Rhine River (2,200 m^3/s) [47]. As such, the total mass transfer of AhR compounds has been extensively observed in the China Sea because Yangtze River releases a higher discharge than European waters.

NR assay results showed that the Yangtze River estuary is polluted by a large variety of chemicals. A high concentration of organic pollutants, such as PAHs [48], PCBs, and PCDDs/PCDFs [49], which are potential inducers of the cytochrome P450 system, were detected in the sediments collected from Yangtze River estuary. These sediment-bound contaminants may affect the AhR activity of the samples. Discharges from the industries along the Yangtze River estuary should be dealt with, and measures should

Table 2. AhR-mediated activity of sediment extracts of the Yangtze River estuary and those of other rivers described in previous studies.

Sampling Locations	Bio-TEQ (pg/g)	References
Masan Bay (Korea)	17~275	[44]
River Elbe Estuary (Germany)	15.5~322	[42]
Tietê River (Brazil)	n.d.~24170	[35]
Saginaw Bay (USA)	11~348	[43]
Morava River (Czech Republic)	1~17000	[15]
Tai Lake (China)	n.d.~114.5	[45]
Dagu Rivers (China)	1200~13900	[47]
Haihe (China)	330~930	[47]
Yellow Sea (China)	3.4~28	[46]
Yangtze River estuary (China)	38.9~323.5	In this study

Note: n.d. = not detectable or below the detection limit.

be established to reduce the input of toxic effluents into the estuary [18].

3.3 AhR-mediated activities in multilayer fractions

Based on EROD induction assay results of the crude extracts, the sediment samples from sites Y2, Y4, Y5, Y7, Y8, and Y9, which had the highest bio-TEQ values, were selected for further identification of AhR-mediated activities using multilayer fractionation. As shown in Figure 4, AhR-mediated activities caused by multilayer fractions (F1 to F3) were compared with crude sediment extract inductions.

Fractions F1, which contained non-polar paraffinic components, showed no or very low AhR-mediated activities. Primary fractions F2, which contained weakly and moderately polar components, showed the highest AhR-mediated activities. In addition, fractions F3, which contained more polar components, also showed significantly increased activities. The results were consistent with those of other studies performed previously. Some non-priority PAHs and more polar compounds, such as heterocyclic compounds containing nitrogen, sulfur, or oxygen heteroatoms (NSO-Het), also show dioxin-like activity [50]. The highest AhR agonistic effect of the SPM collected during a flood event from the rivers Neckar and Rhine was found in fractions containing high molecular weight PAHs with more than 16 aromatic C-atoms [51]. Engwall et al. [52] indicated that the polyaromatic fractions of the bottom sediment and suspended particulate matter showed relatively higher dioxin-like potencies than the other fractions. Therefore, weakly and moderately polar components and more polar components are major inducers of AhR-mediated activities and require further study.

However, AhR-mediated activities showed site-specific differences among locations (Figure 4). Sample sites Y4, Y7, and Y8 (especially site Y4) showed significantly greater bio-TEQs for the sum of primary fractions F1 to F3 than the crude extracts, and the percentages of the exceeded parts were 62%, 40%, and 32%, respectively (Table S3). Sample site Y5 exhibited comparable bio-TEQs of crude extracts to the added primary fractions.

Interactions, such as synergistic or antagonistic effects, exist in different substances and may lead to changes in the effects of certain substances. Low EROD activities do not necessarily indicate low induction potency but may have been caused by high concentrations of inducers and non-inducers that inhibit EROD activity [53]. In crude extracts, the low bio-TEQs might be explained by EROD antagonistic and inhibiting effects because sediments contain a broad range of unspecified compounds [2]. EROD inductions with more than 100% induction in fractions compared with crude extracts may be caused by the retention of humic substances during the fractionation procedure [54]. The separation of aliphatic and polar compounds from nonpolar aromatics resulted in an activity of a fraction that exceeds that of the crude sediment extract in the Neckar River basin [11]. Thus, in this study, some antagonistically acting substances were

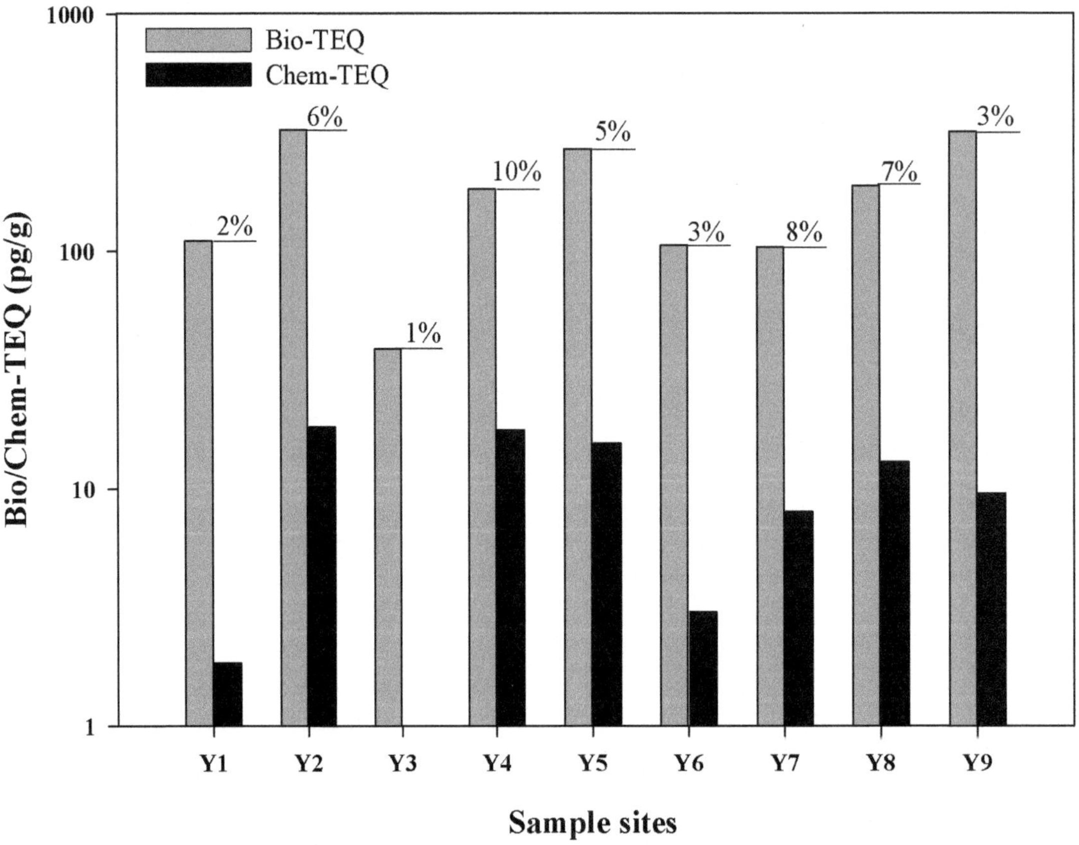

Figure 4. AhR-mediated activities of crude and multilayer fractions of sediment extracts from the Yangtze River estuary. The results are based on EROD inductions with RTL-W1 cells and presented as bio-TEQs. F1 contains non-polar paraffinic components. F2 contains weakly and moderately polar components. F3 contains highly polar components. The numbers in percentage were calculated as (F1+F2+F3–CE)/CE×100%. The difference between the combined F1 to F3 and the induced crude extracts induction is given. Bio-TEQs are given as means of $n=3$ independent experiments. CE, crude extracts.

probably separated in the process of elution, and thus, agonists could completely display activity.

In contrast, sites Y2 and Y9 showed decreased induction when comparing bio-TEQs of F1 to F3 with the crude extract. This phenomenon can be explained as follows: fractionation of mixtures of bioactive compounds into multiple fractions may have reduced the individual bioactivity to below detection limits, thereby efficiently rendering some of the active compounds non-detectable by the bioassay [55]. Thomas et al. [56] applied the EDA method to analyze the estrogen receptor agonist of water extracts. After fractionation, the sum of activity of the individual fractions was less than 20% of the activity of the crude extract. The crude extracts are a complex mixture of unknown substances, and thus, the full recovery of all effect compounds within the samples was difficult. The AhR-mediated activities of the sediment extract fractions are no longer observed in the present study, indicating that the reduced complexity of the mixture also reduces toxicity.

3.4 Concentrations of PAHs

Chemical analysis was applied to determine the concentrations of the 16 EPA-PAHs listed in Table 3. Most of the 16 EPA-PAHs were detected in all the sediment samples from the Yangtze River estuary. The highest concentrations of total PAHs were found in site Y2, and the lowest concentrations were found in site Y1 and Y3. The concentrations of total PAHs ranged from 21.5 ng/g dw to 190.5 ng/g dw sediment. Some studies reported the levels of PAHs in the sediment from different sections of the Yangtze River. Concentrations of total PAHs at the Chongqing section was 257 ng/g dw to 723 ng/g dw, and at Jialing River, a tributary of the upper Yangtze River, was 132 ng/g dw to 349 ng/g dw [57]. Concentration of PAHs in the Wuhan section was high, ranging from 303 ng/g to 3,995 ng/g in the main stream and 4,121 ng/g to 4,262 ng/g in the tributaries [58]. Compared with other sections of the Yangtze River, the concentration of PAHs in sediment from the Yangtze River estuary was relatively low [17]. The dilution from the East China Sea may be one of the reasons for such low concentration [59]. In addition, Wang et al. [60] showed that the PAH composition in the Yangtze River estuary, which was dominated by four-ring to six-ring PAHs, was mainly caused by petroleum combustion, vehicle emission, and biomass combustion (mainly coal) in the nearshore area, whereas PAHs composition, which was dominated by two-ring to three-ring PAHs, in the farther shore zone originated from petroleum combustion of shipping processes and shoreside discharges.

3.5 Correlation between bio-TEQs and chemical analyses

According to Bols et al. [32], chem-TEQs were calculated by multiplying the specific REP factors for RTL-W1 cells with corresponding compound concentrations. The total biological response in the EROD assay (Bio-TEQs) and chem-TEQs of PAHs are shown in Figure 5. For sites Y1, Y3, Y6, and Y9, less than 5% of the induction could be explained by known priority PAHs, which were expressed as chem-TEQs. For sediment extracts from sites Y4 and Y7, 10% and 8% of the induction could be attributed to the presence of analyzed PAHs, respectively.

Several studies have shown that halogenated aromatic hydrocarbons and PAHs are chemicals that are usually related to dioxin-like activity, and can act as AhR agonists [11,61]. Generally, chemical analysis focused on substances that were considered to be a priority or are more relevant, and other non-priority substances were ignored. However, bio-TEQs are not explained by the chem-TEQs calculated from the concentrations of the analyzed priority compounds. In this study, less than 10% of the bio-TEQs could be attributed to the EPA-PAHs, whereas non-priority substances were shown as high inducers.

Concentrations of pollutants at different matrixes were based on reported results, and differences between the analytical methods have been neglected. According to Keiter et al. [7], 55% to 88% of AhR-mediated responses observed in upper Danube River sediments were not due to the measured priority PAHs, PCBs, and PCDDs/PCDFs. The highest bio-TEQ of sediments core layers (2.26×10^5 pg/g) was sampled at the historically contaminated dumping site in the Baltic Sea. This value was much higher than that obtained in this study, but it was lower than expected from the PAH concentration in the samples from the dumping site [2]. Special interactions of some contaminants or of contaminants with sediment components might be responsible for such low value. In contrast, the majority of bio-TEQ (approximately 58%) measured from the extracts of sediments from the Elbe River by EROD assays could be due to the priority PAHs [41]. In a number of studies involving sediments collected from the Elbe, Tietê, and Danube Rivers, the chemical measurements of priority PAHs or other persistent organic pollutants, such as PCBs and PCDDs/PCDFs, could only account for a small portion of the AhR-mediated potency [7,11,34]. The results from these above mentioned studies were consistent with those obtained in the present study. Kaisarevic et al. [12] showed that only minor portions of biologically derived TCDD-TEQs from waste water canal sediments could be due to the monitored PAHs with known relative potencies. In a previous study, PCBs and PCDDs/PCDFs were responsible for a minor portion of the total AhR-mediated activities of SPM in two floods [29]. The non-priority pollutants mainly mediated the high induction rates. Therefore, the main induction in the biotest systems was caused by non-priority pollutants [11,62]. Some studies showed that the non-priority substances might have caused AhR-mediated activity. Hinger et al. [50] showed that non-priority PAHs and NSO-Het are very potent AhR agonists. Heterocyclic polyaromatic compounds, including dinaphthofurans, 2-(2-naphthalenyl) benzothiophene, methylated chrysene, and benz[a]anthracene, were identified and confirmed as major cytochrome P4501A (CYP1A)-inducing compounds in a contaminated sediment of Bitterfeld (Germany) [63]. Furthermore, Brack et al. [11] demonstrated that PAHs, especially PAHs with a molecular weight between 228 and 252 g/mol, could explain the major dioxin-like potencies of sediment extracts from the Neckar river basin.

The potential contribution of non-priority pollutants to environmental hazards was indicated in this study. To better assess the environmental samples, a broader range of substances should be considered, and studies should not only focus on prioritized pollutants. Identification of unknown pollutants that cause the main AhR-mediated activity is recommended. EDA is a suitable tool for identifying the compounds.

Conclusion

The present study assessed the ecotoxicological hazard potential of the sediments from the Yangtze River estuary using NR retention and EROD induction assays with RTL-W1 cells. The results showed that cytotoxicity and AhR-mediated activity of sediment from Yangtze River estuary were at a minor to medium level when compared with those from other river systems. Concentrations of organic pollutants may be lower in the Yangtze River than in the Rhine River because of the comparably higher mass transport of water and particulate matter in the former, but such mass transport still results in comparably larger amounts of organic pollutants that end up in the Yangtze River's receiving

Table 3. Concentrations of the 16 US EPA-polycyclic aromatic hydrocarbons (PAHs; ng/g dw) in sediment samples from the Yangtze River estuary.

Sampling site	Y1	Y2	Y3	Y4	Y5	Y6	Y7	Y8	Y9
Naphthalene	6.0	47.0	11.0	22.0	7.0	4.0	26.0	7.0	8.0
Acenaphthylene	n.d.	4.0	n.d.	1.0	2.0	1.0	2.0	2.0	2.0
Acenaphthene	n.d.	2.0	n.d.	2.0	2.0	n.d.	n.d.	1.0	2.0
Fluorene	1.0	5.0	1.0	3.0	6.0	2.0	3.0	4.0	4.0
Phenanthrene	2.0	15.0	1.0	14.0	12.0	5.0	6.0	11.0	9.0
Anthracene	2.0	4.0	1.0	3.0	4.0	1.0	2.0	3.0	2.0
Fluoranthene	2.0	17.0	1.0	20.0	13.0	4.0	6.0	10.0	7.0
Pyrene	1.0	8.5	0.5	10.0	7.0	2.5	3.0	5.5	4.5
Benzo[a]anthracene	1.0	13.0	1.0	14.0	10.0	3.0	5.0	8.0	6.0
Chrysene	2.0	14.0	1.0	13.0	11.0	3.0	5.0	9.0	6.0
Benzo[b]fluoranthene	2.0	19.0	1.0	16.0	18.0	6.0	9.0	14.0	12.0
Benzo[k]fluoranthene	1.0	5.0	n.d.	6.0	5.0	1.0	3.0	4.0	3.0
Benzo[a]pyrene	n.d.	14.0	1.0	13.0	10.0	n.d.	4.0	9.0	6.0
Indeno[1,2,3-cd]pyrene	1.0	10.0	1.0	9.0	8.0	2.0	4.0	7.0	5.0
Dibenz[a,h]anthracene	n.d.	3.0	n.d.	2.0	2.0	n.d.	1.0	2.0	1.0
Benzo[g,h,i]perylene	1.0	10.0	1.0	8.0	8.0	2.0	4.0	7.0	5.0
Sum of EPA-PAHs	**22.0**	**190.5**	**21.5**	**156.0**	**125.0**	**36.5**	**83.0**	**103.5**	**82.5**

Note: n.d. = not detectable or below the detection limit.

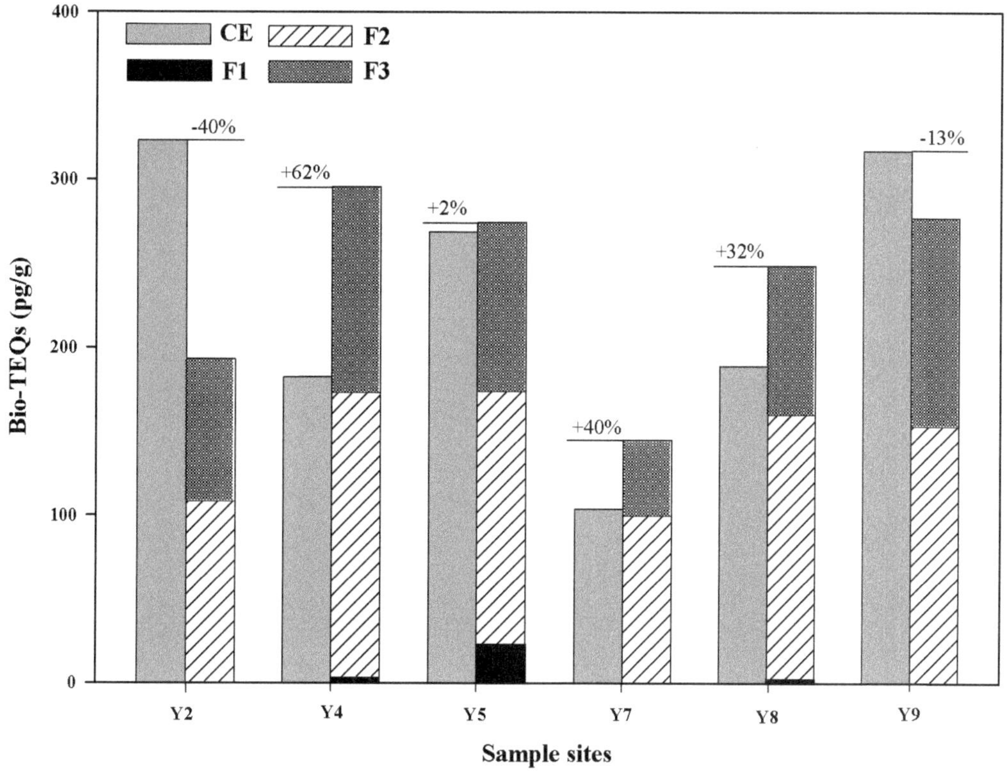

Figure 5. The contribution of Chem-TEQs of PAHs to Bio-TEQs. The total biological response of the crude extracts in the EROD assay with RTL-W1 cells are expressed as bio-TEQs. The chem-TEQs of the 16 measured EPA PAHs were calculated by multiplying compounds concentrations and relative equivalency potencies and are given in pg/g dw. Numbers in percent indicated the calculated contribution of these PAHs to the EROD induction in each crude extract.

water body. At the same contamination levels in both rivers, the 14-fold amount of toxic substances would still enter the East China Sea. Therefore, the effect of the Yangtze River on the East China Sea needs to be considered.

Results of the fractionation showed that weakly and moderately polar components and polar components showed the highest AhR-mediated activities and can maximum exceed about 60% of the crude extract. A combined analysis of chemical measurements of PAHs and the results from bioassays revealed that priority EPA-PAHs contributed only a minor portion of the determined AhR-mediated activities. Furthermore, identification of unknown pollutants causing the high biological AhR-mediated potency should be the focus of future research. EDA is a suitable tool for identifying the unknown pollutants and can be used in further studies to better protect the estuary and to serve as reference for environmental monitoring in this region. Moreover, for the protection of the Yangtze River estuary, some non-priority pollutants, which cause high ecotoxicity, should also be monitored.

Supporting Information

Table S1 Locations for the nine samples in this study.

Table S2 Cytotoxicity and dioxin-like activities of the crude sediment extracts in the RTL-W1 cells. NR_{50} values for sediment extracts are given in mg sediment equivalent per mL medium (mg/mL). Dioxin-like activity expressed as biological toxicity equivalents (Bio-TEQ) in pg/g dw.

Table S3 The dioxin-like activity of the multilayer fractions in the RTL-W1 cells and expressed as biological toxicity equivalents (Bio-TEQ) in pg/g dw.

Acknowledgments

The authors would like to express their thanks to Drs. Niels C. Bols and Lucy Lee (University of Waterloo, Canada) for providing RTL-W1 cells.

Author Contributions

Conceived and designed the experiments: HH LW. Performed the experiments: LL YS LZ YY. Analyzed the data: LL LC TF HX KE HH LW. Contributed reagents/materials/analysis tools: HH LW. Contributed to the writing of the manuscript: LL HH LW.

References

1. Hallare A, Seiler T-B, Hollert H (2011) The versatile, changing, and advancing roles of fish in sediment toxicity assessment–a review. Journal of Soils and Sediments 11: 141–173.
2. Wölz J, Borck D, Witt G, Hollert H (2009) Ecotoxicological characterization of sediment cores from the western Baltic Sea (Mecklenburg Bight) using GC–MS and in vitro biotests. Journal of Soils and Sediments 9: 400–410.
3. Hilscherova K, Machala M, Kannan K, Blankenship A, Giesy J (2000) Cell bioassays for detection of aryl hydrocarbon (AhR) and estrogen receptor (ER) mediated activity in environmental samples. Environmental Science and Pollution Research 7: 159–171.

4. Giltrap M, Macken A, McHugh B, McGovern E, Foley B, et al. (2011) In vitro screening of organotin compounds and sediment extracts for cytotoxicity to fish cells. Environmental Toxicology and Chemistry 30: 154–161.
5. Keiter S, Rastall A, Kosmehl T, Erdinger L, Braunbeck T, et al. (2006) Ecotoxicological Assessment of Sediment, Suspended Matter and Water Samples in the Upper Danube River. A pilot study in search for the causes for the decline of fish catches (12 pp). Environmental Science and Pollution Research 13: 308–319.
6. Lee LJ, Clemons J, Bechtel D, Caldwell S, Han K-B, et al. (1993) Development and characterization of a rainbow trout liver cell line expressing cytochrome P450-dependent monooxygenase activity. Cell Biology and Toxicology 9: 279–294.
7. Keiter S, Grund S, van Bavel B, Hagberg J, Engwall M, et al. (2008) Activities and identification of aryl hydrocarbon receptor agonists in sediments from the Danube river. Analytical and Bioanalytical Chemistry 390: 2009–2019.
8. Brack W (2003) Effect-directed analysis: a promising tool for the identification of organic toxicants in complex mixtures? Analytical and Bioanalytical Chemistry 377: 397–407.
9. Qu G, Shi J, Wang T, Fu J, Li Z, et al. (2011) Identification of tetrabromobisphenol A diallyl ether as an emerging neurotoxicant in environmental samples by bioassay-directed fractionation and HPLC-APCI-MS/MS. Environmental Science & Technology 45: 5009–5016.
10. Lübcke-von Varel U, Bataineh M, Lohrmann S, Löffler I, Schulze T, et al. (2012) Identification and quantitative confirmation of dinitropyrenes and 3-nitrobenzanthrone as major mutagens in contaminated sediments. Environment International 44: 31–39.
11. Brack W, Schirmer K, Erdinger L, Hollert H (2005) Effect-directed analysis of mutagens and ethoxyresorufin-O-deethylase inducers in aquatic sediments. Environmental Toxicology and Chemistry 24: 2445–2458.
12. Kaisarevic S, Varel UL-v, Orcic D, Streck G, Schulze T, et al. (2009) Effect-directed analysis of contaminated sediment from the wastewater canal in Pancevo industrial area, Serbia. Chemosphere 77: 907–913.
13. Chen G, White PA (2004) The mutagenic hazards of aquatic sediments: a review. Mutation Research/Reviews in Mutation Research 567: 151–225.
14. Hilscherova K, Kannan K, Kang Y-S, Holoubek I, Machala M, et al. (2001) Characterization of dioxin-like activity of sediments from a Czech River Basin. Environmental Toxicology and Chemistry 20: 2768–2777.
15. Liu LY, Wang JZ, Wei GL, Guan YF, Wong CS, et al. (2012) Sediment records of polycyclic aromatic hydrocarbons (PAHs) in the continental shelf of China: implications for evolving anthropogenic impacts. Environmental Science & Technology 46: 6497–6504.
16. Liu M, Hou L, Yang Y, Zou H, Lu J, et al. (2001) Distribution and sources of polycyclic aromatic hydrocarbons in intertidal flat surface sediments from the Yangtze estuary, China. Environmental Geology 41: 90–95.
17. Floehr T, Xiao H, Scholz-Starke B, Wu L, Hou J, et al. (2013) Solution by dilution?–A review on the pollution status of the Yangtze River. Environmental Science and Pollution Research: 1–38.
18. Müller B, Berg M, Yao ZP, Zhang XF, Wang D, et al. (2008) How polluted is the Yangtze river? Water quality downstream from the Three Gorges Dam. Science of The Total Environment 402: 232–247.
19. Wang C, Peng B (2002) analysis of micro organic compound pollution in major city river reaches of the main stem of the Changjiang river. Yangtze River 33: 4–9.
20. Sun YZ, Zhang B, Gao LR, Liu ZT, Zheng MH (2005) Polychlorinated Dibenzo-p-Dioxins and Dibenzofurans in Surface Sediments from the Estuary Area of Yangtze River, People's Republic of China. Bulletin of Environmental Contamination and Toxicology 75: 910–914.
21. Yang H, Zhuo S, Xue B, Zhang C, Liu W (2012) Distribution, historical trends and inventories of polychlorinated biphenyls in sediments from Yangtze River Estuary and adjacent East China Sea. Environmental Pollution 169: 20–26.
22. So M, Miyake Y, Yeung W, Ho Y, Taniyasu S, et al. (2007) Perfluorinated compounds in the Pearl River and Yangtze River of China. Chemosphere 68: 2085–2095.
23. Chen SJ, Gao XJ, Mai BX, Chen ZM, Luo XJ, et al. (2006) Polybrominated diphenyl ethers in surface sediments of the Yangtze River Delta: Levels, distribution and potential hydrodynamic influence. Environmental Pollution 144: 951–957.
24. Wu LL, Chen L, Hou JL, Zhang YL, Zhao JF, et al. (2010) Assessment of Sediment Quality of Yangtze River Estuary Using Zebrafish (Danio rerio) Embryos. Environmental Toxicology 25: 234–242.
25. Hollert H, Dürr M, Erdinger L, Braunbeck T (2000) Cytotoxicity of settling particulate matter and sediments of the Neckar River (Germany) during a winter flood. Environmental Toxicology and Chemistry 19: 528–534.
26. Luo J, Ma M, Liu C, Zha J, Wang Z (2009) Impacts of particulate organic carbon and dissolved organic carbon on removal of polycyclic aromatic hydrocarbons, organochlorine pesticides, and nonylphenols in a wetland. Journal of Soils and Sediments 9: 180–187.
27. Klee N, Gustavsson L, Kosmehl T, Engwall M, Erdinger L, et al. (2004) Changes in toxicity and genotoxicity of industrial sewage sludge samples containing nitro- and amino-aromatic compounds following treatment in bioreactors with different oxygen regimes. Environmental Science and Pollution Research 11: 313–320.
28. Borenfreund E, Puerner J (1985) A simple quantitative procedure using monolayer cultures for cytotoxicity assays (HTD/NR-90). Journal of tissue culture methods 9: 7–9.
29. Wölz J, Engwall M, Maletz S, Olsman Takner H, Bavel B, et al. (2008) Changes in toxicity and Ah receptor agonist activity of suspended particulate matter during flood events at the rivers Neckar and Rhine – a mass balance approach using in vitro methods and chemical analysis. Environmental Science and Pollution Research 15: 536–553.
30. Kennedy SW, Jones SP (1994) Simultaneous Measurement of Cytochrome P4501A Catalytic Activity and Total Protein Concentration with a Fluorescence Plate Reader. Analytical Biochemistry 222: 217–223.
31. Engwall M, Broman D, Brunström B, Ishaq R, Näf C, et al. (1996) Toxic potencies of lipophilic extracts from sediments and settling particulate matter (SPM) collected in a PCB-contaminated river system. Environmental Toxicology and Chemistry 15: 213–222.
32. Bols NC, Schirmer K, Joyce EM, Dixon DG, Greenberg BM, et al. (1999) Ability of Polycyclic Aromatic Hydrocarbons to Induce 7-Ethoxyresorufin-o-deethylase Activity in a Trout Liver Cell Line. Ecotoxicology and Environmental Safety 44: 118–128.
33. Keiter S, Braunbeck T, Heise S, Pudenz S, Manz W, et al. (2009) A fuzzy logic-classification of sediments based on data from in vitro biotests. Journal of Soils and Sediments 9: 168–179.
34. Rocha PS, Azab E, Schmidt B, Storch V, Hollert H, et al. (2010) Changes in toxicity and dioxin-like activity of sediments from the Tietê River (São Paulo, Brazil). Ecotoxicology and Environmental Safety 73: 550–558.
35. Yang G, Weng L, Li L (2008) Yangtze conservation and development report 2007: Science Press.
36. Liu M, Baugh PJ, Hutchinson SM, Yu L, Xu S (2000) Historical record and sources of polycyclic aromatic hydrocarbons in core sediments from the Yangtze Estuary, China. Environmental Pollution 110: 357–365.
37. Wen S, Hui Y, Yang F, Liu Z, Xu Y (2008) Polychlorinated dibenzo-p-dioxins (PCDDs) and dibenzofurans (PCDFs) in surface sediment and bivalve from the Changjiang Estuary, China. Chinese Journal of Oceanology and Limnology 26: 35.
38. Bian H, Li Z, Liu P, Pan J (2010) Spatial distribution and deposition history of nonylphenol and bisphenol A in sediments from the Changjiang River (Yangtze River) Estuary and its adjacent East China Sea. Acta Oceanologica Sinica 29: 44–51.
39. Schirmer K, Dixon DG, Greenberg BM, Bols NC (1998) Ability of 16 priority PAHs to be directly cytotoxic to a cell line from the rainbow trout gill. Toxicology 127: 129–141.
40. Fent K, Hunn J (1996) Cytotoxicity of organic environmental chemicals to fish liver cells (PLHC-1). Marine Environmental Research 42: 377–382.
41. Otte JC, Keiter S, Faβbender C, Higley EB, Rocha PS, et al. (2013) Contribution of Priority PAHs and POPs to Ah Receptor-Mediated Activities in Sediment Samples from the River Elbe Estuary, Germany. Plos One 8: e75596.
42. Giesy JP, Jude DJ, Tillitt DE, Gale RW, Meadows JC, et al. (1997) Polychlorinated dibenzo-p-dioxins, dibenzofurans, biphenyls and 2,3,7,8-tetrachlorodibenzo-p-dioxin equivalents in fishes from Saginaw Bay, Michigan. Environmental Toxicology and Chemistry 16: 713–724.
43. Yoo H, Khim JS, Giesy JP (2006) Receptor-mediated in vitro bioassay for characterization of Ah-R-active compounds and activities in sediment from Korea. Chemosphere 62: 1261–1271.
44. Xia J, Su G, Zhang X, Shi W, Giesy JP, et al. (2014) Dioxin-like activity in sediments from Tai Lake, China determined by use of the H4IIE-luc bioassay and quantification of individual AhR agonists. Environmental Science and Pollution Research 21: 1480–1488.
45. Hong S, Khim JS, Naile JE, Park J, Kwon BO, et al. (2012) AhR-mediated potency of sediments and soils in estuarine and coastal areas of the Yellow Sea region: A comparison between Korea and China. Environmental Pollution 171: 216–225.
46. Song M, Jiang Q, Xu Y, Liu H, Lam PK, et al. (2006) AhR-active compounds in sediments of the Haihe and Dagu Rivers, China. Chemosphere 63: 1222–1230.
47. Huisman P, De Jong J, Wieriks K (2000) Transboundary cooperation in shared river basins: experiences from the Rhine, Meuse and North Sea. Water Policy 2: 83–97.
48. Machala M, Vondráček J, Bláha L, Ciganek M, Neča J (2001) Aryl hydrocarbon receptor-mediated activity of mutagenic polycyclic aromatic hydrocarbons determined using in vitro reporter gene assay. Mutation Research/Genetic Toxicology and Environmental Mutagenesis 497: 49–62.
49. Clemons J, Myers C, Lee L, Dixon D, Bols N (1998) Induction of cytochrome P4501A by binary mixtures of polychlorinated biphenyls (PCBs) and 2, 3, 7, 8-tetrachlorodibenzo-p-dioxin (TCDD) in liver cell lines from rat and trout. Aquatic Toxicology 43: 179–194.
50. Hinger G, Brinkmann M, Bluhm K, Sagner A, Takner H, et al. (2011) Some heterocyclic aromatic compounds are Ah receptor agonists in the DR-CALUX assay and the EROD assay with RTL-W1 cells. Environmental Science and Pollution Research 18: 1297–1304.
51. Wölz J, Brack W, Moehlenkamp C, Claus E, Braunbeck T, et al. (2010) Effect-directed analysis of Ah receptor-mediated activities caused by PAHs in suspended particulate matter sampled in flood events. Science of The Total Environment 408: 3327–3333.

52. Engwall M, Broman D, Dencker L, Näf C, Zebühr Y, et al. (1997) Toxic potencies of extracts of sediment and settling particulate matter collected in the recipient of a bleached pulp mill effluent before and after abandoning chlorine bleaching. Environmental Toxicology and Chemistry 16: 1187–1194.
53. Brack W, Segner H, Möder M, Schüürmann G (2000) Fixed-effect-level toxicity equivalents–a suitable parameter for assessing ethoxyresorufin-O-deethylase induction potency in complex environmental samples. Environmental Toxicology and Chemistry 19: 2493–2501.
54. Gustavsson LK, Klee N, Olsman H, Hollert H, Engwall M (2004) Fate of Ah receptor agonists during biological treatment of an industrial sludge containing explosives and pharmaceutical residues. Environmental Science and Pollution Research 11: 379–387.
55. Grung M, Lichtenthaler R, Ahel M, Tollefsen K-E, Langford K, et al. (2007) Effects-directed analysis of organic toxicants in wastewater effluent from Zagreb, Croatia. Chemosphere 67: 108–120.
56. Thomas K, Langford K, Petersen K, Smith A, Tollefsen K (2009) Effect-directed identification of naphthenic acids as important in vitro xeno-estrogens and anti-androgens in North Sea offshore produced water discharges. Environmental Science & Technology 43: 8066–8071.
57. Tang Z, Liao H, Zhang L, Guo J, Wu F, et al. (2011) Distribution, source and risk assessment of polycyclic aromatic hydrocarbons in river sediment of Cheng-Yu economic zone. Huanjing Kexue (Environmental Science) 32: 2639–2644.
58. Feng C, Xia X, Shen Z, Zhou Z (2007) Distribution and sources of polycyclic aromatic hydrocarbons in Wuhan section of the Yangtze River, China. Environmental Monitoring and Assessment 133: 447–458.
59. Bouloubassi I, Fillaux J, Saliot A (2001) Hydrocarbons in Surface Sediments from the Changjiang (Yangtze River) Estuary, East China Sea. Marine Pollution Bulletin 42: 1335–1346.
60. Wang Y, Li X, Li BH, Shen ZY, Feng CH, et al. (2012) Characterization, sources, and potential risk assessment of PAHs in surface sediments from nearshore and farther shore zones of the Yangtze estuary, China. Environmental Science and Pollution Research 19: 4148–4158.
61. Whyte JJ, Jung RE, Schmitt CJ, Tillitt DE (2000) Ethoxyresorufin-O-deethylase (EROD) Activity in Fish as a Biomarker of Chemical Exposure. Critical Reviews in Toxicology 30: 347–570.
62. Hollert H, Dürr M, Olsman H, Halldin K, van Bavel B, et al. (2002) Biological and Chemical Determination of Dioxin-like Compounds in Sediments by Means of a Sediment Triad Approach in the Catchment Area of the River Neckar. Ecotoxicology 11: 323–336.
63. Brack W, Schirmer K (2003) Effect-directed identification of oxygen and sulfur heterocycles as major polycyclic aromatic cytochrome P4501A-inducers in a contaminated sediment. Environmental Science & Technology 37: 3062–3070.

11

A Novel Photocatalytic Material for Removing Microcystin-LR under Visible Light Irradiation: Degradation Characteristics and Mechanisms

Xin Sui[1], Xiangrong Wang[1], Honghui Huang[2], Guotao Peng[1], Shoubing Wang[1], Zhengqiu Fan[1]*

1 Department of Environmental Science & Engineering, Fudan University, Shanghai, China, **2** South China Sea Fisheries Research Institute, Chinese Academy of Fishery Sciences, Guangzhou, China

Abstract

Background and Purpose: Microcystin-LR (MC-LR), a common toxic species in contaminated aquatic systems, persists for long periods because of its cyclic structure. Ag_3PO_4 is an environment-friendly photocatalyst with relatively good degradation capacity for hazardous organic pollutants. This study aimed to investigate the degradation capacity of Ag_3PO_4 for MC-LR under visible light.

Methods: An Ag_3PO_4 photocatalyst was synthesized by the ion-exchange method and characterized by X-ray diffraction, field-emission scanning electron microscope, and UV-Vis spectrophotometer. MC-LR was quantified in each sample through high-performance liquid chromatograph. The degradation efficiency of MC-LR was affected by initial pH, initial Ag_3PO_4 concentration, initial MC-LR concentration, and recycle experiments. The degradation intermediates of MC-LR were examined by liquid chromatography-mass spectrometry (LC/MS).

Results: The degradation process can be well fitted with the pseudo-first-order kinetic model. The maximum MC-LR degradation rate of 99.98% can be obtained within 5 h under the following optimum conditions: pH of 5.01, Ag_3PO_4 concentration of 26.67 g/L, and MC-LR concentration of 9.06 mg/L. Nine intermediates were detected and analyzed by LC/MS. Three main degradation pathways were proposed based on the molecular weight of the intermediates and the reaction mechanism: (1) hydroxylation on the aromatic ring of Adda, (2) hydroxylation on the diene bonds of Adda, and (3) internal interactions on the cyclic structure of MC-LR.

Conclusion: Ag_3PO_4 is a highly efficient catalyst for MC-LR degradation in aqueous solutions.

Editor: Yinping Zhang, Tsinghua University, China

Funding: The project was supported by Country China Scholarship Council and Key Laboratory of Fishery Ecology and Environment, Guangdong Province, China. The funders had no role in study design, data collection and analysis, decision to publish, or preparation of the manuscript.

Competing Interests: The authors have declared that no competing interests exist.

* E-mail: zhqfan@fudan.edu.cn

Introduction

Eutrophication promoted by human activities causes the proliferation of *cyanobacteria* or blue–green algae [1]. *Microcystis* species are the most prevalent *Cyanobacteria* worldwide; these species produce hepatotoxin microcystins (MCs), a possible carcinogen. More than 90 types of MCs have been identified to date [2,3]. As a representative congener, microcystin-LR (MC-LR) is a common toxic species in contaminated aquatic systems and persists for long periods.

The general structure of the MCs is cyclo (*D*-Ala-*L*-X-*D*-erythro-*β*-Methyl-*D*-isoAsp-*L*-Y-Adda-isoGlu-N-methyldehydro-Ala), where X and Y represent variable amino acids and Adda represents 3-amino-9-methoxy-2,6,8-trimethyl-10-phenyldeca-4,6-dienoic-acid [4]. MCs are stable in natural aquatic systems because of their cyclic structure. The degradation of MC-LR is always slow by physical or biological methods. Therefore, developing a method for the fast removal of MC-LR is a significant research work.

Semiconductor materials used as environment-friendly photocatalysts for the removal of hazardous organic contaminants have drawn considerable attention over the last few decades. TiO_2-based heterogeneous photocatalytic oxidation leads to water splitting [5–7]. Methods that promote pollutant mineralization are based on the generation of highly reactive oxygen species (e.g., HO•, $O•_2^-$). Nevertheless, the photons in the visible region that occupy a major portion of the solar spectrum cannot be utilized by TiO_2 because of its large band gap ($E_g = 3.2$ eV), which reduces the degradation effect and increases the catalyst amount [6,8]. Yi et al. have recently reported that silver orthophosphate (Ag_3PO_4) was a highly efficient water oxidation visible light photocatalyst that could harness visible light to oxidize water and the oxygen evolution rate with Ag_3PO_4 was 2.5 times that with $BiVO_4$ [9,10]. Furthermore, many studies have also shown that Ag_3PO_4 has relatively good performance for the degradation of organic pollutants under visible light irradiation [11]. However, the

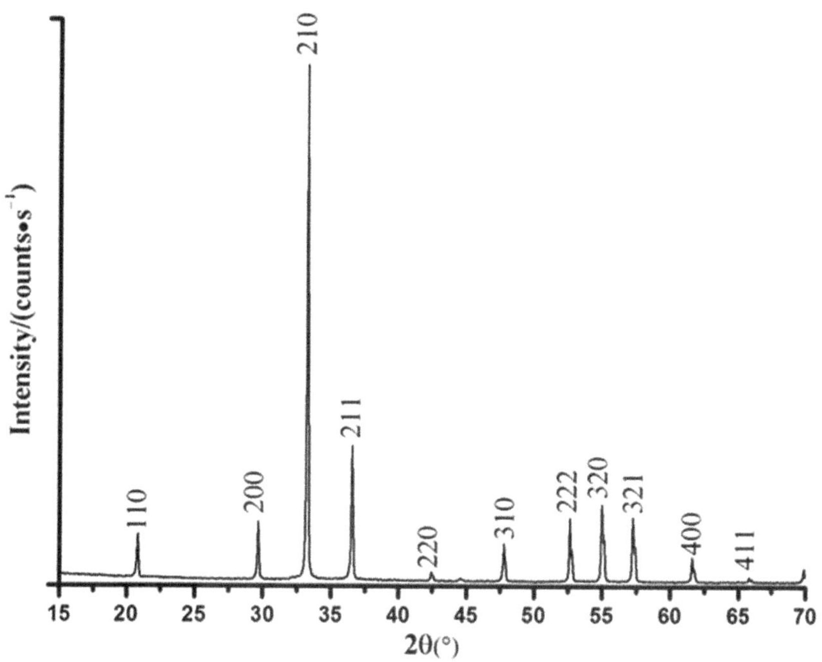

Figure 1. XRD patterns of the Ag_3PO_4 powders.

degradation capacity of Ag_3PO_4 for MC-LR under visible light irradiation has yet to be reported.

This study aims to investigate the degradation capacity of Ag_3PO_4 for MC-LR under visible light. The influencing parameters, including initial pH, initial Ag_3PO_4 concentration, initial MC-LR concentration, and recycle experiments, were studied. The degradation intermediates of MC-LR were also examined by liquid chromatography/mass spectrometry (LC/MS), and the degradation mechanism was analyzed.

Materials and Methods

Chemical and reagent

MC-LR standard sample was purchased from Express Technology Co., Ltd. (China) and kept at −20°C. The MC-LR agent used for removal was isolated and purified from cultured FACHB905 (Institute of Hydrobiology, Chinese Academy of Sciences, Wuhan, China) by using an improved method proposed by Ma et al. [12]. The purity of the MC-LR agent was over 95%, as determined by high-performance liquid chromatography (HPLC, Agilent 1200). $AgNO_3$ (99%) was purchased from Snn

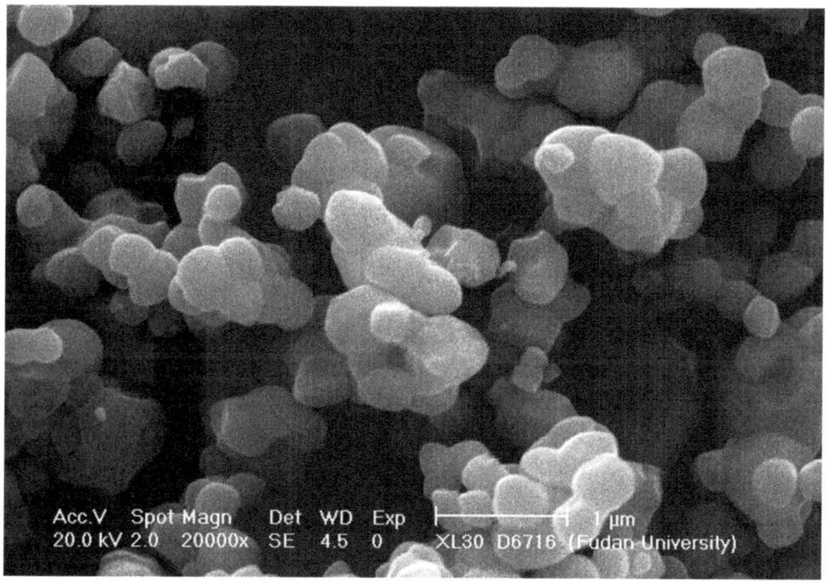

Figure 2. SEM image of the as-synthesized Ag_3PO_4 photocatalyst.

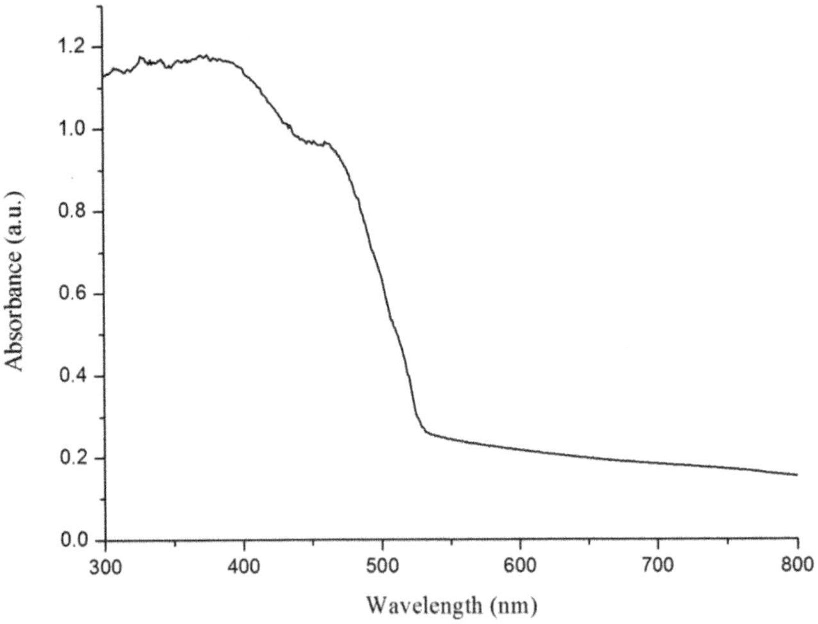

Figure 3. UV−vis diffuse reflectance spectra of the as-synthesized Ag_3PO_4 photocatalyst.

Chemical and Technology Co., Ltd. (Shanghai, China). Analytical-grade $Na_2HPO_4•12H_2O$, NaOH, and H_3PO_4 were purchased from Shanghai Chemical Reagent Co., Ltd. (China). Acetonitrile (HPLC grade) and trifluoroacetic acid (TFA, 99.5+%, HPLC grade) were purchased from Alfa Aesar, USA.

Synthesis and characteristics of Ag_3PO_4

Ag_3PO_4 was synthesized by the ion-exchange method as follows. $AgNO_3$ powder (0.075 mol) was dissolved in 500 mL of deionized H_2O. Then, 0.1 mol/L of Na_2HPO_4 (250 mL) aqueous solution was added in drops into the above solution under continuous stirring until yellow precipitates were formed. Finally, the mixture was washed with distilled water and dried at 60°C in air overnight.

The crystallographic properties of the products were obtained by X-ray diffraction (XRD, PANalytical/X' Pert PRO, Cu Kα radiation, Holland), $\lambda = 1.54056$ Å. The surface morphology and particle size of the photocatalyst were examined by a field-emission scanning electron microscope (SEM, Philips XL30, Holland). The

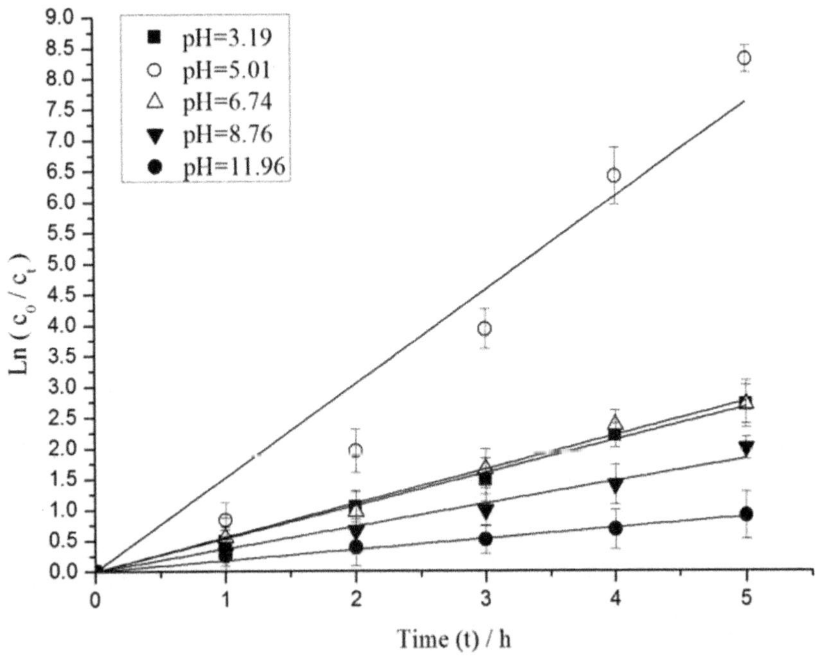

Figure 4. Pseudo-first-order kinetics of MC-LR degradation under different initial pH values.

ultraviolet-visible diffuse reflectance spectra were recorded with a UV-vis spectrophotometer using $BaSO_4$ as the reference (PerkinElmer, Lambda 35, USA).

Photocatalytic degradation of MC-LR

The effects of parameters such as initial pH, initial Ag_3PO_4 concentration, initial MC-LR concentration, and recycle experiments on the degradation capacity of Ag_3PO_4 for MC-LR were evaluated. The different initial pH values were 3.19, 5.01, 6.74, 8.76, and 11.96, by using 0.1 mol/L NaOH or 0.1 mol/L H_3PO_4 as buffer. The different initial Ag_3PO_4 concentrations used were 0, 7.07, 13.67, 20.40, 26.67, and 33.47 g/L. MC-LR solution was extracted from cultured algal cells. A 20.34 mg/L MC-LR stock solution was diluted with distilled water to obtain the concentration gradient (2.61, 4.05, 6.57, 8.30, 9.06, 9.76, and 10.09 mg/L). The stability of the as-prepared Ag_3PO_4 photocatalyst was investigated by four successive cycling runs of experiments with pH 5.06, and the catalyst amount and the initial MC-LR concentration used in each run were 26.67 and 9.06 mg/L, respectively. After 5 h of visible light irradiation, the suspension was extracted from the beaker and centrifuged at 5000 rpm for 20 min to remove the impurities. The solid catalyst was washed thrice with deionized water and then further centrifuged. Finally, the solid catalyst was dried at 60 °C for 12 h for the next run [13]. Moderate MC-LR solution was added into a 15 mL quartz glass vial, in which Ag_3PO_4 was added. A preferable test interval of 1 h was adopted to further analyze the degradation process, with each initial parameter having three parallel samples.

Photocatalytic degradation was conducted at 26±1°C and atmospheric pressure under visible light irradiation. The mixture was magnetically stirred in the dark for 60 min to ensure the adsorption–desorption equilibrium of MC-LR on the catalysts and the adsorption of approximately 5% MC-LR. The reactor was then irradiated with visible light emitted by a 500 W xenon lamp with a 420 nm cutoff filter. A sample of 0.8 mL was taken and filtered to remove the photocatalyst particles. The quantification of MC-LR in each sample was determined by HPLC. For HPLC analysis, the injection volume to a C-18 discovery column (Supelco) was 20 μL at 40 °C. The mobile phase in isocratic mode with a flow rate of 1 mL/min was a mixture of 40% acetonitrile and 60% water, both containing 0.05% TFA. MC-LR was measured with a photodiode array detector at 238 nm.

LC-MS Analysis

A Bruker micrOTOF II mass spectrometer was utilized for the MS identification of the reaction intermediates. Positive ions were analyzed using electrospray ionization, and MS data were acquired in full-scan mode (400 Da/scan to 1200 Da/scan). The voltage of electrospray process was 4 kv. The MS consisted of a Dionex Ultimate 3000 HPLC Pump and a Dionex Ultimate 3000 autosampler. The HPLC column was ZORBAX SB-C18 column (4.6 mm×250 mm, 5 μm particle size, Agilent, USA), and the temperature was maintained at 40 °C. The mobile phase consisted of a mixture of 0.05% TFA in acetonitrile and 0.1% TFA in Milli-Q water. Gradient elution method was programmed according to the study of Liu et al. [14]. The flow rate was controlled at 1.0 mL/min. The injection volume of the treated sample was 20 μL. A 1 mg standard of MC-LR dry powder was prepared by adding 20 mL Milli-Q water.

Degradation kinetic analysis

As widely used of the kinetic model in the scope of chemical reaction or photo catalytic degradation process [13,15,16], the influence of each factor on MC-LR degradation was described by the pseudo-first-order kinetic model (Equation 1):

$$\ln(c_0/c_t) = kt \quad (1)$$

Where c_0 and c_t were the concentrations of reactant at time 0 and t, respectively, and k was the pseudo-first-order rate constant. k was obtained from the gradient of the graph of $\ln(c_0/c_t)$ versus t.

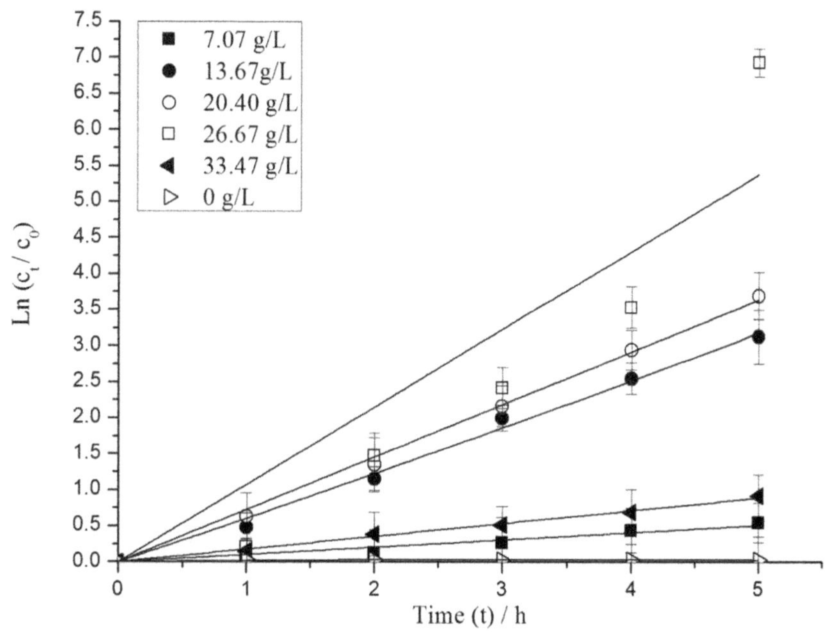

Figure 5. Pseudo-first-order kinetics of MC-LR degradation under different initial Ag_3PO_4 concentration.

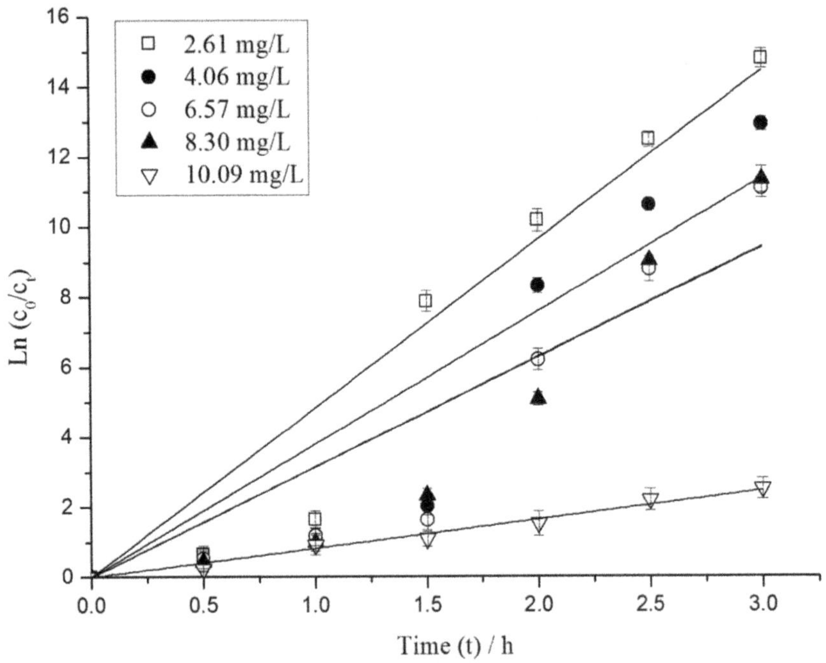

Figure 6. Pseudo-first-order kinetics of MC-LR degradation under different initial MC-LR concentration.

Results

Characteristics of the Ag_3PO_4 photocatalyst

The powder was subjected to XRD to determine the crystal phase composition and crystallite size of Ag_3PO_4. As shown in Figure 1, all diffraction peaks of the as-prepared Ag_3PO_4 samples were in line with the standard spectrum (JCPDS card No. 06-0505). In addition, no diffraction peaks resulting from impurities were detected [17]. The eleven distinctive peaks at 20.83°, 29.66°, 33.27°, 36.55°, 42.47°, 47.78°, 52.68°, 54.98°, 57.24°, 61.60°, and 65.87° matched with the (110), (200), (210), (211), (220), (310), (222), (320), (321), (400), and (411) crystal planes of Ag_3PO_4, respectively.

The morphologies and particle of the as-synthesized Ag_3PO_4 were demonstrated in the SEM images. As shown in Figure 2,

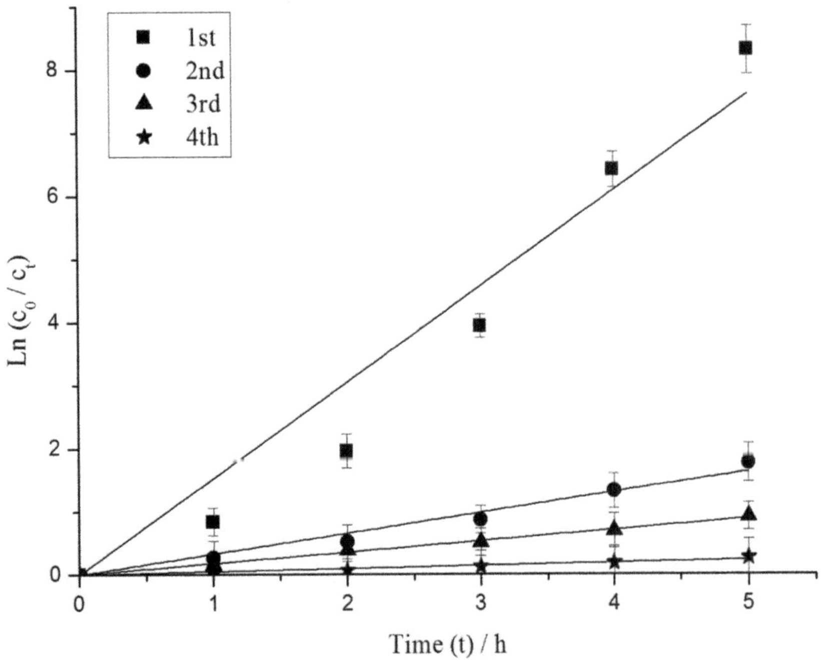

Figure 7. Pseudo-first-order kinetics of MC-LR degradation under 4 cycling runs.

Table 1. Degradations kinetics of MC-LR in different conditions.

Factors	Conditions	Pseudo-first-order kinetics ln (c_0/c_t) = kt	
		k (h^{-1})	R^2
pH	3.19	0.53	0.99813
	5.01	1.52	0.97555
	6.74	0.55	0.99649
	8.76	0.36	0.99100
	11.96	0.18	0.99435
Ag_3PO_4 (g/L)	0.00	0.01	0.79063
	7.07	0.10	0.97653
	13.67	0.63	0.99735
	20.40	0.73	0.99895
	26.67	1.08	0.91456
	33.47	0.18	0.99769
MC-LR (mg/L)	2.61	4.82	0.9696
	4.06	3.79	0.91146
	6.57	3.15	0.91345
	8.30	3.14	0.91289
	10.09	0.82	0.99248
cycling runs	1st Run	1.52	0.97555
	2nd Run	0.33	0.98872
	3rd Run	0.11	0.99254
	4th Run	0.05	0.97268

Ag_3PO_4 consisted of agglomerated smooth spherical particles with an average particle size of approximately 200 nm to 500 nm.

The UV–vis diffuse reflectance spectra showed the optical properties of the Ag_3PO_4 photocatalyst. As shown in Figure 3, the absorption band edge of the prepared Ag_3PO_4 was at about 530 nm in the spectrum. The optical absorption near the band edge of a crystal obeys the following Equation 2 [18]:

$$\alpha h\nu = A\left(h\nu - E_g\right)^{n/2} \tag{2}$$

Where α, ν, and A were absorption coefficient, light frequency, and proportionality constant, respectively; and n decided the transition type in a semiconductor ($n = 1$, direct absorption; $n = 4$, indirect absorption; n for Ag_3PO_4 was 1) [13]. By applying this equation, the band gap of Ag_3PO_4 was approximated as 2.35 eV, which was significantly lower than that of TiO_2.

Degradation characterization of MC-LR using Ag_3PO_4

Effect of initial pH. Figure 4 shows the degradation profile of MC-LR under visible light irradiation at different initial pH with the same initial Ag_3PO_4 concentration of 26.67 g/L and initial MC-LR concentration of 9.06 mg/L. The maximum MC-LR degradation rate occurred at pH 5.01, with the pseudo-first-order kinetic constant k of 1.52 h^{-1} and a MC-LR removal rate of 99.98% in 5 h. The degradation rate of MC-LR increased as the pH was increased from 3.19 to 5.01. By contrast, the degradation rate of MC-LR reduced when the pH was higher than 5.01. The pseudo-first-order kinetic constant k dropped to 0.18 h^{-1} and only 59.19% MC-LR was removed when the pH was 11.96.

The photocatalytic degradation of MCs was pH dependent [19]. Photocatalytic oxidation occurred on or near the surface of the catalyst in the Ag_3PO_4 catalysts system. The change in pH influenced hydrophobicity of MC-LR, which was increased as pH decreases. Through the hydrophobic effect, hydrophobic compounds preferentially moved to surfaces from the bulk of aqueous solvents [19]. The MC-LR showed very low absorption under basic conditions (pH 8.76 and 11.96). These findings demonstrated that hydrophobicity was a contributory factor influencing the initial adsorption of MC-LR to the surface of Ag_3PO_4. The change in pH also influenced solubility and catalytic activity of Ag_3PO_4 catalysts. On one hand, Ag_3PO_4 catalysts could be dissolved in the strong acidic solution which would reduce the catalyst dosage and then led the decrease of absorption [18]. On the other hand, the concentration of hydroxide ion (OH^-) might affect catalytic activity of Ag_3PO_4 in aqueous solution. In the basic solution, excess OH^- would promote the formation of Ag^+ to metallic Ag. Ag deposited onto the surface of Ag_3PO_4 particles in the form of Ag nanoparticles to cover the active sites, which influenced the absorption of visible light [18]. The results suggested that Ag_3PO_4 catalysts exhibited the best photocatalytic performance for MC-LR degradation at pH 5.01.

Effect of initial Ag_3PO_4 concentration. Figure 5 shows the degradation of MC-LR ($c_0 = 9.76$ mg/L) for various initial concentrations of Ag_3PO_4 at pH 5.06. The control experiment showed that the removal of MC-LR without Ag_3PO_4 under visible light irradiation was almost negligible. The removal rate for MC-LR increased from 41.90% to 99.90% when the initial Ag_3PO_4 concentration was increased from 7.07 g/L to 26.67 g/L. The pseudo-first-order kinetic constant k reached 1.08 h^{-1} when the Ag_3PO_4 concentration was 26.67 g/L. Nevertheless, the degradation rate of MC-LR was reduced to 59.91% when the Ag_3PO_4 concentration was increased to 33.47 g/L. Therefore, the optimal concentration of Ag_3PO_4 was 26.67 g/L. A maximum value of the effective area of receiving light was detected in the photocatalytic system. The photocatalytic system reached a saturation condition as the initial concentration of Ag_3PO_4 was further increased from 26.67 g/L to 33.47 g/L. Excessive dosage of catalyst led to the scattering effect, which affected the light absorption of the photocatalyst and then decreased the pseudo-first-order kinetic constant [20]. Ag_3PO_4 displayed a high photocatalytic activity under weak acidic condition ($k = 1.08$ h^{-1}). The photocatalytic activity of Ag_3PO_4 was 0.8 times that of mesoporous N-TiO_2 ($k = 0.60$ h^{-1}), the well-known TiO_2 system photocatalyst with the best photocatalytic performance [21].

Effect of initial MC-LR concentration. Figure 6 shows the degradation rate at different initial MC-LR concentrations when the initial Ag_3PO_4 concentration was 26.67 g/L, initial pH was approximately 5.07, and the test interval was 0.5 h. The degradation of MC-LR was significantly influenced by the concentration of MC-LR. When the initial MC-LR concentration was increased from 2.61 mg/L to 10.09 mg/L, the degradation rate of MC-LR decreased from nearly 100% to 91.92% and the pseudo-first-order kinetic constant k decreased from 4.82 h^{-1} to 0.82 h^{-1} after 3 h of reaction. These results indicated that the degradation efficiency was decreasing with the MC-LR concentration. The yield rate of HO• was remained as a constant with the same Ag_3PO_4 concentration. The increasing concentration of MC-LR resulted in a decrease in the ratio of decomposed amount of MC-LR to the total amount of MC-LR.

Effect of recycle experiments of Ag_3PO_4 photocatalyst. As shown in Figure 7, the degradation rate of MC-LR gradually reduced with increasing cycling runs of Ag_3PO_4. The pseudo-first-order kinetic constant k decreased from

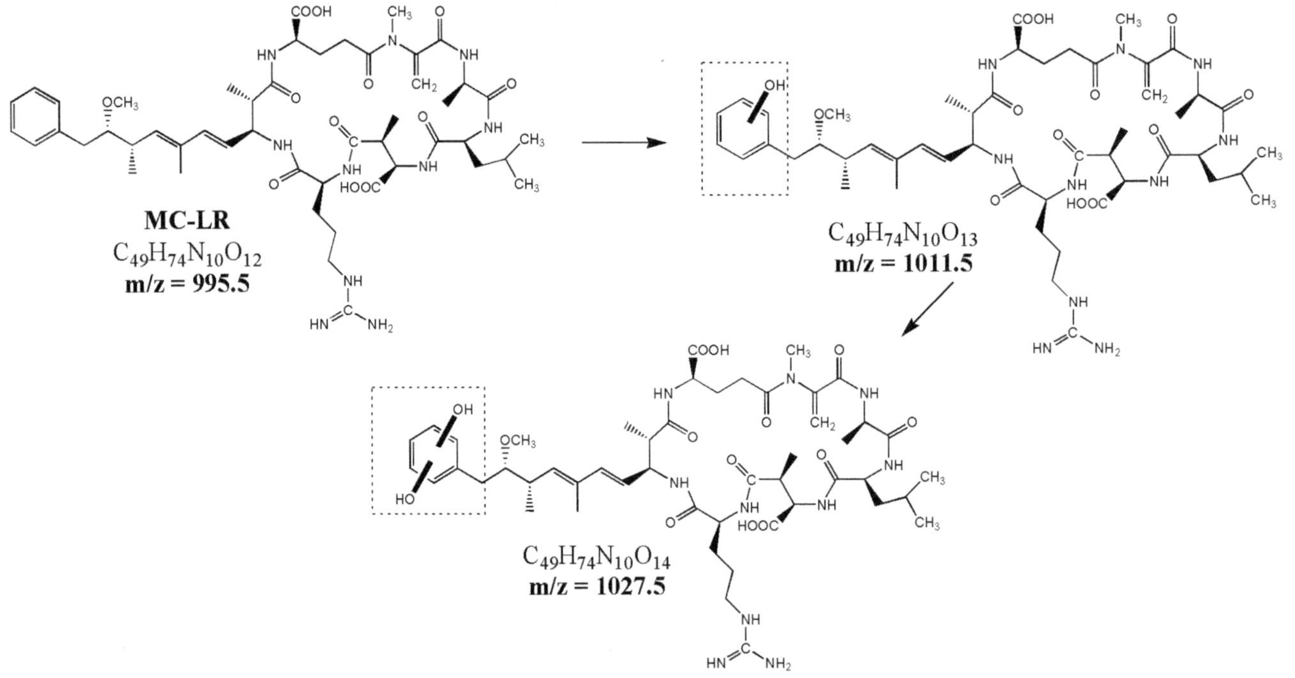

Figure 8. Proposed MC-LR's degradation pathways: hydroxylation on the aromatic ring of Adda.

1.52 h^{-1} to 0.05 h^{-1} when the cycling runs of Ag_3PO_4 were increased from 1 to 4. The photocatalytic activity of the Ag_3PO_4 product had a significantly loss. Possibly, the increased number of cycling runs promoted the deposition of the Ag nanoparticles on the surfaces of the Ag_3PO_4 particles during the photocatalytic process. The deposited nanoparticles covered the active sites [22]. Hence, Ag_3PO_4 exhibited the best photocatalytic performance for MC-LR degradation in the first run.

Error caused by logarithmic computation and automatic integration of peak area made some data were not well fitted the pseudo-first-order kinetic model. Nevertheless, parallel experiment analyzed the error of data was within 10%, which revealed that the data were reliable.

Therefore, the degradation parameters including initial pH, initial Ag_3PO_4 concentration, initial MC-LR concentration, and recycle experiments were in line with the pseudo-first-order kinetic model. All factors are listed in Table 1. The maximum MC-LR degradation rate of 99.98% can be obtained within 5 h under the following optimum conditions: pH of 5.01, Ag_3PO_4 concentration of 26.67 g/L, MC-LR concentration of 9.06 mg/L, and cycling run of one time.

Degradation mechanisms of MC-LR

Photocatalytic degradation depends on the electron–hole pairs that are produced from the inter-band excitation. When the Ag_3PO_4 photocatalyst absorbs radiation with energy equal to or larger than its band gap, the electrons and holes that migrate to the surface of catalyst serve as redox sources and react with adsorbed water to generate highly reactive oxygen species (e.g., HO•). According to previously described degradation mechanisms with HO•, this reactive oxygen species can change the structure of MC-LR or trigger some internal reactions on the cyclic structure of MC-LR [23]. The production of HO• by Ag_3PO_4 photocatalyst was shown in the following Equations 3–6:

$$Ag_3PO_4 + h\nu \rightarrow Ag_3PO_4(e^- + h^+) \quad (3)$$

$$e^- + O_2 \rightarrow O_2^- \quad (4)$$

$$h^+ + H_2O \rightarrow HO\bullet + H^+ \quad (5)$$

$$h^+ + OH^- \rightarrow HO\bullet \quad (6)$$

Where e^- represents electrons in the conduction band and h^+ represents holes in the valence band.

The degradation products of MC-LR were observed in the total ion chromatogram of the LC/MS spectrogram. The peaks in the initial sample must have a signal-to-noise ratio of 3 and the difference between the peak areas of the initial and treated samples must be at least two times to eliminate the interference of impurities [5]. Ultimately, nine intermediates fitted these requirements: m/z 1011.5, 1027.5, 1029.5, 781.3, 835.4, 1009.5, 877.4, 764.3, and 681.4. The following three main degradation pathways were proposed according to the molecular weight of the products and the reaction mechanism between HO• and peptides or proteins.

Hydroxylation on the aromatic ring of Adda. The intermediates m/z 1011.5 with retention time (R.T.) = 20.7 min and 27.9 min probably belonged to the aromatic ring of Adda [24]. The radical HO• combined with the double bonds of the aromatic ring, formed a carbon-centered radical, and then formed a peroxy radical by reacting with oxygen [25]. The alkyl group directly connected to the aromatic ring of Adda donated the

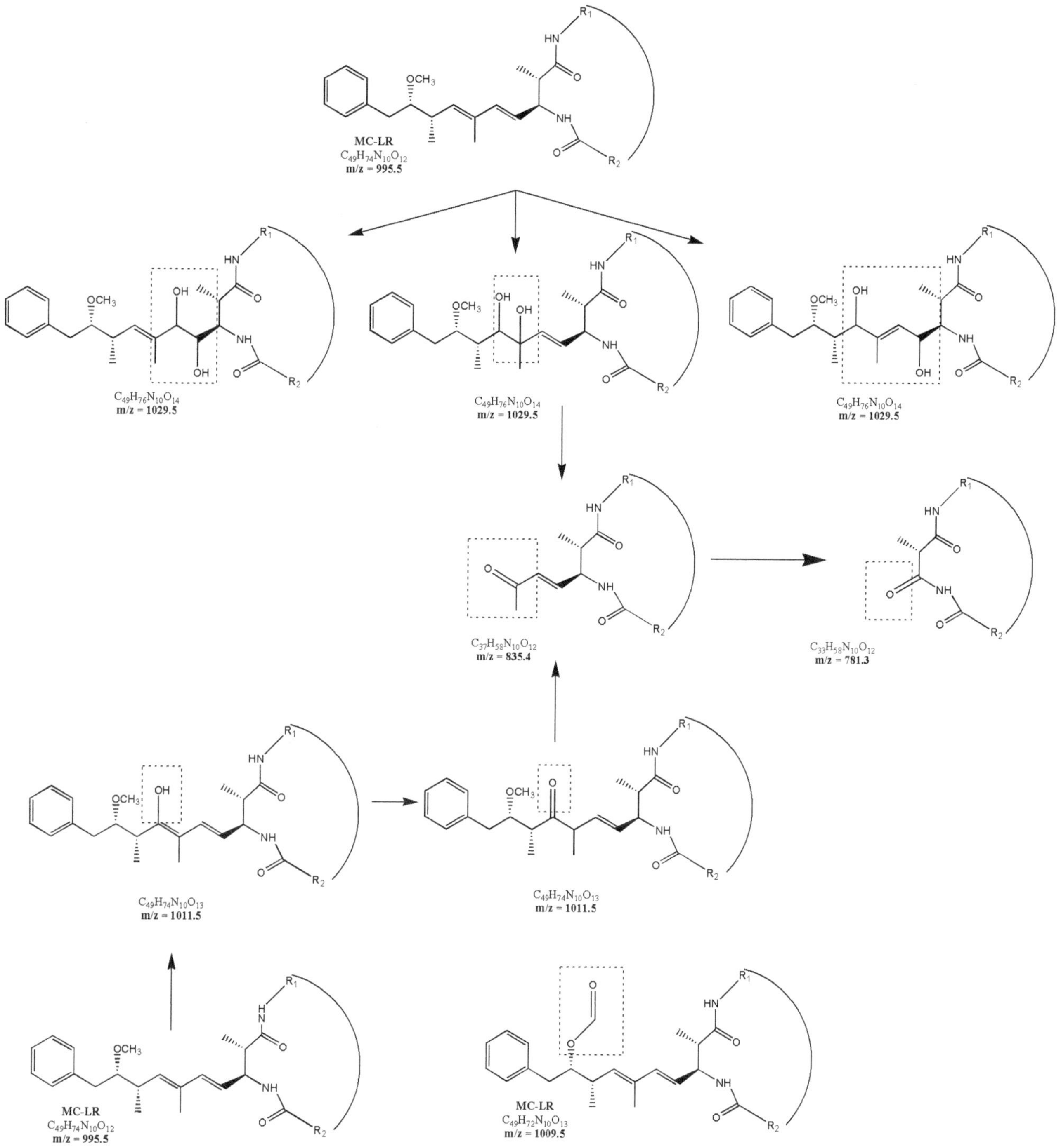

Figure 9. Proposed MC-LR's degradation pathways: hydroxylation on the diene bonds of Adda.

electrons. Electrophilic substitution with HO• radicals was believed to occur at the *ortho* and *para* positions.

The first aromatic hydroxylation increased the electron density on the aromatic ring and induced the second hydroxylation, which was *m/z* 1027.5 (Figure 8).

Hydroxylation on the diene bonds of Adda. The diene bonds of the Adda chain was likely another attacked site. Hydroxyl addition possibly occurred between C_4–C_5, C_6–C_7, or to C_4 and C_7, forming the *m/z* 1029.5 intermediates (Figure 9) [25]. In the second oxidation route, the Adda chain was completely removed and formed the ketone-derivative *m/z* 781.3, which had never been reported yet. One probable pathway was the double hydroxyl addition to the C_6-C_7 of the Adda chain; this mechanism formed the *m/z* 1029.5 intermediate products. Then, the C_6-C_7 of the Adda chain was split to form the intermediate *m/z* 835.4. The intermediate *m/z* 835.4 was subjected to a series of oxidation processes, forming the *m/z* 781.3 ketone derivative with an intact cyclic structure. Another possible route was through the substitution of an HO group to the hydrogen of C_7 of the Adda chain. This mechanism formed an enol-MC-LR (*m/z* 1011.5) that

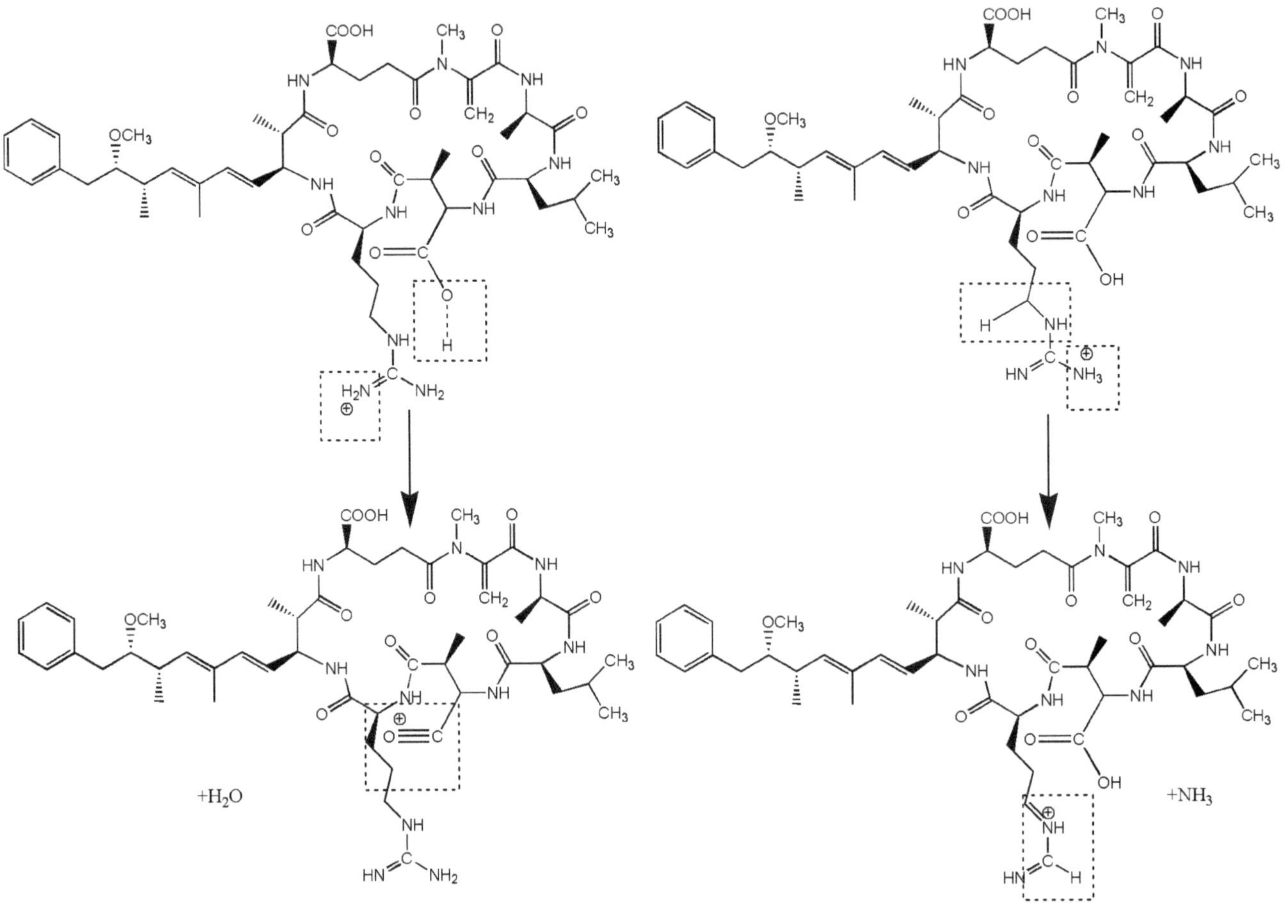

Figure 10. Possible interactions between Arg and MeAsp on the cyclic structure of MC-LR.

was isomerized to the ketone–MC-LR [25]. After a series of oxidative-induced bond cleavage steps, the ketone derivative *m/z* 781.3 was formed. The ketone derivative was more stable than hydroxyl derivative.

The detected *m/z* 1009.5 was probably formed by the hydrogen abstraction reaction on the methoxy group of Adda by HO•.

Internal interactions on the cyclic structure of MC-LR. In the photocatalytic process, the excited Ag_3PO_4 photocatalyst that produced a large amount of HO• could form a strong oxidizing condition. This condition could trigger internal reactions on the cyclic structure of MC-LR; meanwhile, some neutral molecules (such as H_2O, NH_3, CO_2, CO, and $C_9H_{10}N$) would be lost [26]. Dehydration occurred on the free carboxylic groups of amino acids, hydroxyl-substituted bonds, and the backbone of the peptides, which could form the linearization structure of MC-LR [6]. Studies showed that the dissociation energy of amino acids adjacent to Arg is evidently higher than those without Arg. Other works reported that the loss of an ammonia molecule lead to the cyclic structure open in non-Arg MC derivatives [27,28]. In the case of MC-LR, MeAsp with a free carboxylic group adjoined to Arg was more likely to interact with the guanidine group of Arg than the peptide bond, followed by the loss of neutral molecules (Figure 10).

As shown in Figure 10, another intermediate *m/z* 1011.5 probably occurred because of the removal of ammonia from the Arg moiety and the double hydroxylation on the C_6-C_7 of the Adda chain. The peptide bond between MeAsp and Leu was fractured because Arg transferred to the nitrogen of MeAsp. Then, the Adda moiety was broken to form *m/z* 877.4. Then, the peptide bond between Leu and Mdha was fractured to form *m/z* 764.3. Ultimately, *m/z* 681.4 was formed by the removal of Mdha through the same mechanism. The pathways were in accordance with the study of Antoniou et al. [6] (Figure 11).

The diene bonds of Mdha were other sites for the double hydroxylation and oxidation of aldehydes, forming the *m/z* 1029.5 and 1011.5 fragments, respectively [5].

Discussion

The Ag_3PO_4 photocatalyst was synthesized and used to remove MC-LR from aqueous solutions for the first time in this study. The degradation efficiency of MC-LR was influenced by several parameters, including initial pH, initial Ag_3PO_4 concentration, initial MC-LR concentration, and recycle experiments. In addition, the degradation process was well fitted with the pseudo-first-order kinetic model. Three main degradation pathways were proposed based on the molecular weight of the intermediates and the reaction mechanism: (1) hydroxylation on the aromatic ring of Adda, (2) hydroxylation on the diene bonds of Adda, and (3) internal interactions on the cyclic structure of MC-LR.

In conclusion, Ag_3PO_4 is a highly efficient catalyst for MC-LR degradation in aqueous solutions. However, further studies must be conducted to determine the entire degradation process. Furthermore, the property of the intermediates formed during

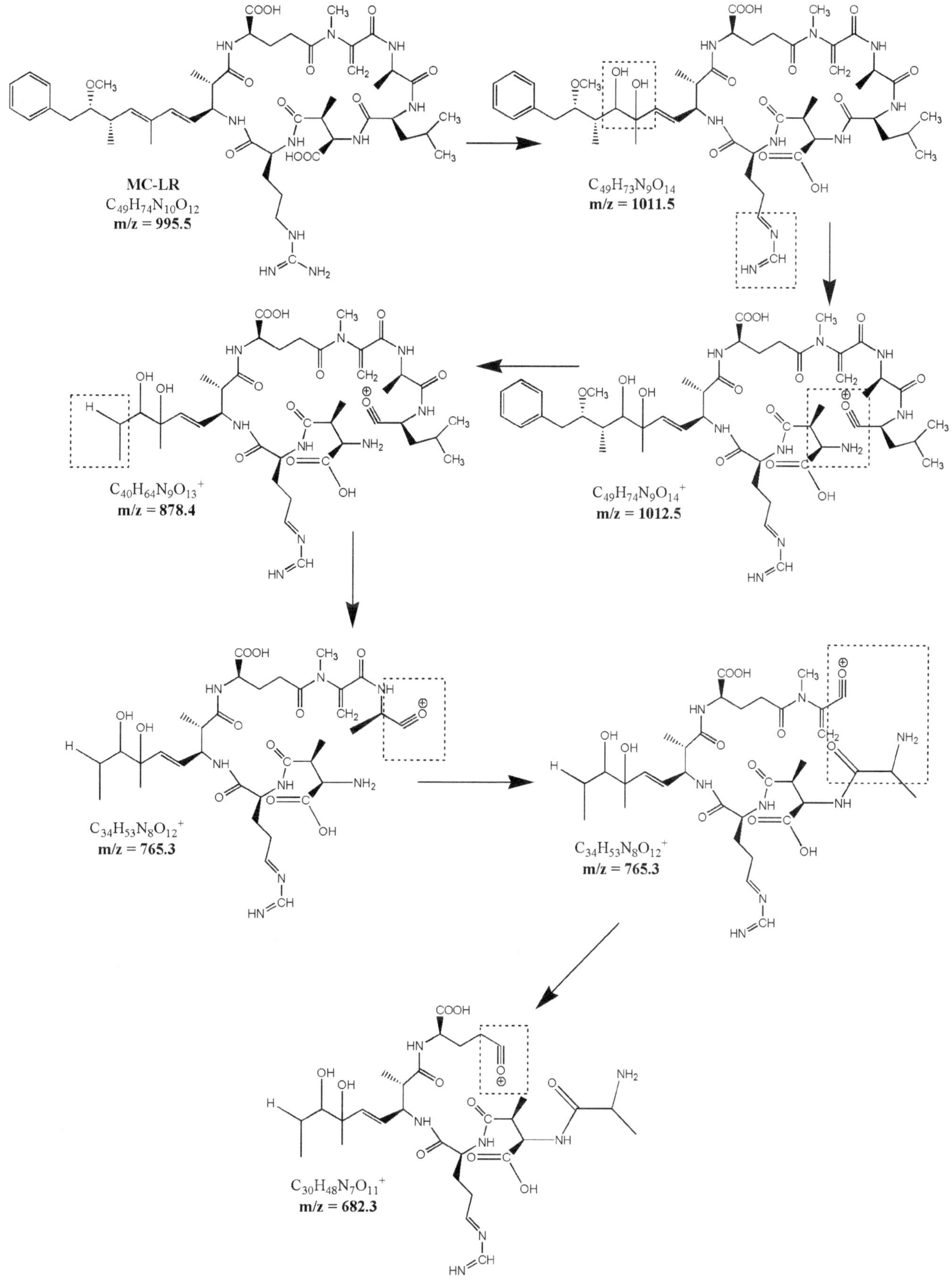

Figure 11. Proposed MC-LR's degradation pathways: internal interactions on the cyclic structure of MC-LR.

the MC-LR degradation should be investigated to promote the practical application of the Ag_3PO_4 photocatalyst.

Acknowledgments

The project was supported by Country China Scholarship Council and Key Laboratory of Fishery Ecology and Environment, Guangdong Province, China.

Author Contributions

Conceived and designed the experiments: ZQF XS. Performed the experiments: XS. Analyzed the data: XS. Contributed reagents/materials/analysis tools: ZQF XRW XS HHH SBW GTP. Wrote the paper: XS ZQF.

References

1. Jacobs L, Peralta-Zamora P, Campos FR, Pontarolo R (2013) Photocatalytic degradation of microcystin-LR in aqueous solutions. Chemosphere 90:1552–1557.
2. Ziegmann M, Abert M, Muller M, Frimmel FH (2010) Use of fluorescence fingerprints for the estimation of bloom formation and toxin production of *Microcystis aeruginosa*. Water Res 44:195–204.
3. Jiang JL, Shan ZJ, Xu WL, Wang XR, Zhou JY, et al. (2013) Microcystin-LR Induced reactive oxygen species mediate cytoskeletal disruption and apoptosis of hepatocytes in *Cyprinus carpio* L. PLOS ONE 8.
4. Dai GF, Quan CY, Zhang XZ, Liu J, Song LR, et al. (2012) Fast removal of cyanobacterial toxin microcystin-LR by a low-cytotoxic microgel-Fe(III) complex. Water Res 46:1482–1489.
5. Antoniou MG, Nicolaou PA, Shoemaker JA, de la Cruz AA, Dionysiou DD (2009) Impact of the morphological properties of thin TiO_2 photocatalytic films on the detoxification of water contaminated with the cyanotoxin, microcystin-LR. Appl Catal B-Environ 91:165–173.
6. Antoniou MG, Shoemaker JA, de la Cruz AA, Dionysiou DD (2008) LC/MS/MS structure elucidation of reaction intermediates formed during the TiO_2 photocatalysis of microcystin-LR. Toxicon 51:1103–1118.
7. Khan SB, Rahman MM, Akhtar K, Asiri AM, Rub MA (2014) Nitrophenol chemi-sensor and active solar photocatalyst based on spinel hetaerolite nanoparticles. PLOS ONE 9.
8. Zhou DX, Ji ZX, Jiang XM, Dunphy DR, Brinker J, et al. (2013) Influence of material properties on TiO_2 nanoparticle agglomeration. PLOS ONE 8.
9. Yi ZG, Ye JH, Kikugawa N, Kako T, Ouyang SX, et al. (2010) An orthophosphate semiconductor with photooxidation properties under visible-light irradiation. Nature Materials 9:559–564.
10. Rawal SB, Sung SD, Lee WI (2012) Novel Ag_3PO_4/TiO_2 composites for efficient decomposition of gaseous 2-propanol under visible-light irradiation. Catalysis Communications 17:131–135.
11. Wang WG, Cheng B, Yu JG, Liu G, Fan WH (2012) Visible-light photocatalytic activity and deactivation mechanism of Ag_3PO_4 spherical particles. Chemistry-An Asian Journal 7:1902–1908.
12. Ma QW, Ren J, Huang HH, Wang XR, Wang SB, et al. (2011) A novel degradation method of microcystin-LR induced by nitrous acid. Fresen Environ Bull 20:1007–1012.
13. Liu W, Wang ML, Xu CX, Chen SF, Fu XL (2013) Ag_3PO_4/ZnO: An efficient visible-light-sensitized composite with its application in photocatalytic degradation of Rhodamine B. Mater Res Bull 48:106–113.
14. Liu I, Lawton LA, Robertson P (2003) Mechanistic studies of the photocatalytic oxidation of microcystin-LR: An investigation of byproducts of the decomposition process. Environ Sci Technol 37:3214–3219.
15. Chen P, Zhu L, Fang S, Wang C, Shan G (2012) Photocatalytic degradation efficiency and mechanism of microcystin-RR by mesoporous Bi_2WO_6 under near ultraviolet light. Environ Sci Technol 46:2345–2351.
16. Ma QW, Ren J, Huang HH, Wang SB, Wang XR, et al. (2012) Kinetic and mechanistic study of microcystin-LR degradation by nitrous acid under ultraviolet irradiation. J Hazard Mater 215:75–82.
17. Thomas M, Ghosh SK, George KC (2002) Characterization of nanostructured silver orthophosphate. Mater Lett 56:386–392.
18. Ge M, Zhu N, Zhao YP, Li J, Liu L (2012) Sunlight-assisted degradation of dye pollutants in Ag_3PO_4 suspension. Ind Eng Chem Res 51:5167–5173.
19. Lawton LA, Robertson P, Cornish B, Marr IL, Jaspars M (2003) Processes influencing surface interaction and photocatalytic destruction of microcystins on titanium dioxide photocatalysts. J Catal 213:109–113.
20. Teng W, Li XY, Zhao QD, Zhao JJ, Zhang DK (2012) In situ capture of active species and oxidation mechanism of RhB and MB dyes over sunlight-driven Ag/Ag_3PO_4 plasmonic nanocatalyst. Appl Catal B-Environ 125:538–545.
21. Choi H, Antoniou MG, Pelaez M, De la Cruz AA, Shoemaker JA, et al. (2007) Mesoporous nitrogen-doped TiO_2 for the photocatalytic destruction of the cyanobacterial toxin Microcystin-LR under visible light irradiation. Environ Sci Technol 41:7530–7535.
22. Bi YP, Ouyang SX, Cao JY, Ye JH (2011) Facile synthesis of rhombic dodecahedral AgX/Ag_3PO_4 (X = Cl, Br, I) heterocrystals with enhanced photocatalytic properties and stabilities. Phys Chem Chem Phys 13:10071–10075.
23. Qiao RP, Li N, Qi XH, Wang QS, Zhuang YY (2005) Degradation of microcystin-RR by UV radiation in the presence of hydrogen peroxide. Toxicon 45:745–752.
24. Buxton GV, Greenstock CL, Helman WP, Ross AB (1988) Critical-review of rate constants for reactions of hydrated electrons, hydrogen-atoms and hydroxyl radicals ($\bullet OH/\bullet O^-$) in aqueous-solution. J Phys Chem Ref Data 17:513–886.
25. Antoniou MG, Shoemaker JA, De La Cruz AA, Dionysiou DD (2008) Unveiling new degradation intermediates/pathways from the photocatalytic degradation of microcystin-LR. Environ Sci Technol 42:8877–8883.
26. Papayannopoulos IA (1995) The interpretation of collision-induced dissociation tandem mass-spectra of peptides. Mass Spectrom Rev 14:49–73.
27. Diehnelt CW, Dugan NR, Peterman SM, Budde WL (2006) Identification of microcystin toxins from a strain of *Microcystis aeruginosa* by liquid chromatography introduction into a hybrid linear ion trap-fourier transform ion cyclotron resonance mass spectrometer. Anal Chem 78:501–512.
28. Vachet RW, Asam MR, Glish GL (1996) Secondary interactions affecting the dissociation patterns of arginine-containing peptide ions. J Am Chem Soc 118:6252–6256.

12

Novel Microbiological and Spatial Statistical Methods to Improve Strength of Epidemiological Evidence in a Community-Wide Waterborne Outbreak

Katri Jalava[1*¤], Hanna Rintala[2], Jukka Ollgren[1], Leena Maunula[3], Vicente Gomez-Alvarez[4], Joana Revez[3], Marja Palander[1], Jenni Antikainen[5], Ari Kauppinen[6], Pia Räsänen[1], Sallamaari Siponen[1], Outi Nyholm[1], Aino Kyyhkynen[1], Sirpa Hakkarainen[2], Juhani Merentie[2], Martti Pärnänen[2], Raisa Loginov[5], Hodon Ryu[4], Markku Kuusi[1], Anja Siitonen[1], Ilkka Miettinen[6], Jorge W. Santo Domingo[4], Marja-Liisa Hänninen[3], Tarja Pitkänen[6]

1 Department of Infectious Disease Surveillance and Control, National Institute for Health and Welfare, Helsinki, Finland, **2** Siilinjärvi municipality, Finland, **3** Department of Food Hygiene and Environmental Health, University of Helsinki, Helskinki, Finland, **4** Office of Research and Development, United States Environmental Protection Agency, Cincinnati, Ohio, United States of America, **5** Laboratory HUSLAB, Helsinki University Hospital, Helsinki, Finland, **6** Department of Environmental Health, National Institute for Health and Welfare, Kuopio, Finland

Abstract

Failures in the drinking water distribution system cause gastrointestinal outbreaks with multiple pathogens. A water distribution pipe breakage caused a community-wide waterborne outbreak in Vuorela, Finland, July 2012. We investigated this outbreak with advanced epidemiological and microbiological methods. A total of 473/2931 inhabitants (16%) responded to a web-based questionnaire. Water and patient samples were subjected to analysis of multiple microbial targets, molecular typing and microbial community analysis. Spatial analysis on the water distribution network was done and we applied a spatial logistic regression model. The course of the illness was mild. Drinking untreated tap water from the defined outbreak area was significantly associated with illness (RR 5.6, 95% CI 1.9–16.4) increasing in a dose response manner. The closer a person lived to the water distribution breakage point, the higher the risk of becoming ill. Sapovirus, enterovirus, single *Campylobacter jejuni* and EHEC O157:H7 findings as well as virulence genes for EPEC, EAEC and EHEC pathogroups were detected by molecular or culture methods from the faecal samples of the patients. EPEC, EAEC and EHEC virulence genes and faecal indicator bacteria were also detected in water samples. Microbial community sequencing of contaminated tap water revealed abundance of *Arcobacter* species. The polyphasic approach improved the understanding of the source of the infections, and aided to define the extent and magnitude of this outbreak.

Editor: Martyn Kirk, The Australian National University, Australia

Funding: No external funding execpt routine governmental work funding was obtained for this work. The funders had no role in study design, data collection and analysis, decision to publish, or preparation of the manuscript.

Competing Interests: The authors have declared that no competing interests exist.

* Email: katri.jalava@thl.fi

¤ Current address: Environmental Health Department, City of Tampere, Finland

Introduction

Community-wide waterborne outbreaks are characterized by a large number of exposed people with high attack rates [1–6]. Waterborne outbreaks are frequently associated with large number of symptomatic cases in a point source manner. Such outbreaks may be caused by a failure in the drinking water distribution system [1,3,4] or water treatment breakthrough of contaminating agents due to heavy rainfall or other excess weather conditions [2,5]. The water distribution system can be potentially contaminated with multiple pathogens during a relatively short period of time as the result of intrusion from surface or waste water [7]. Indeed, waterborne outbreaks with multiple causative organisms, e.g. *Campylobacter* spp., norovirus-like organisms, *Shigella* and enterohaemorrhagic *Escherichia coli* (EHEC) have been described [1,3,4,6,8,9]. In particular, when norovirus and sapovirus types are implicated in large scale waterborne outbreaks, this strongly indicates drinking water distribution system contamination by a human faecal sources [10,11]. Sapovirus usually causes sporadic infections [12] but has been isolated from cases of waterborne outbreaks [12,13].

Waterborne outbreaks may be classified according to the level of evidence indicating that the drinking water was the cause of the outbreak. Evidence may be found by microbiological and/or epidemiological studies and the level of evidence can be assessed according to standardized criteria [14]. During the years 1998–2009 there have been 3–10 waterborne outbreaks in Finland annually and the outbreaks have typically been detected in small community groundwater plants with fewer than 500 consumers [7]. The implicated technical failures for the groundwater contamination in Finland have been flooding and surface run-off caused by heavy rains or rapid melting of snow. Also intrusion of contaminated water and cross-connections in the water distribution system play important role as a cause of Finnish waterborne outbreaks [15,16]. Most common causative microbes have been norovirus and *Campylobacter* [7].

In waterborne outbreak investigations, the delay between environmental investigation and the original presence of the pathogen within a water body has often hampered the detection of causative microbiological agents [15]. Special methods suitable for concentration of large water volumes, such as the use of large diameter membrane filters or ultrafiltration apparatus have been developed to increase the sensitivity of microbial detection in dilute water samples [17–19]. For the identification of a faecal contamination source, various detection methods targeting specific pathogens and microbial water quality indicators [15] and faecal source host specific molecular assays are available [20]. The use of RNA-based methods may not only help detecting but also identify active and thus potentially infective pathogens, providing a better estimate of public health risks than current DNA-based methods [21]. Additionally, next generation sequencing (NGS) can provide a path towards detecting multiple microbial taxa within complex microbial communities including rare members [17]. The high number of reads per sample in NGS applications also enables the use of these techniques for source tracking and identification of the transmission route of an outbreak [22].

Epidemiological cohort or case control studies may be applied in drinking water outbreaks due to the point source nature of the contamination and a well-defined population [3,11,23]. Web-based questionnaires are now increasingly used in outbreak investigations [24], especially in large and/or widely spread outbreaks [25]. Additionally, novel spatial methods have shown potential to define the source and location of the outbreak [26–28]. Modelling may be used to describe the person-to-person transmission or infection spreading from the environment [28]. Spatial variation in disease incidence has been studied in detail [27]. Also distance from water tanks has been evaluated by statistical methods with respect to diarrhoeal incidence [26].

The outbreak

The main water pipe was accidentally broken on 4th July, 2012 during road construction work in Vuorela, a community of 3000 inhabitants within the municipality of Siilinjärvi in Eastern Finland. The pipe breakage caused the contents of the upper drinking water storage reservoir to leak into the road construction pit. The pipe breakage was fixed within 14 hours, flushed and quality of the water was shown to fulfill the hygienic quality criteria. On the 16th July, the local environmental health authorities were informed by the health care centre of an excess number of patients with gastrointestinal symptoms. Given the recent pipe break in the area, a waterborne outbreak was suspected. The local outbreak control team of health and environmental authorities and waterworks personnel was activated, national outbreak awareness team was informed and consulted. On the following day, the results from tap water samples revealed faecal contamination of the water further confirming the waterborne nature of the outbreak. An immediate boil water notice was issued on 17th July and collection of patient and drinking water samples was initiated. The water distribution pipeline was subsequently flushed, the water storage reservoir was cleaned and the whole distribution system was disinfected with chlorine. The boil water notice was cancelled when the outbreak was declared over on 3rd August 2012.

The aim of our study was to reveal the role of contaminated water as a cause of a community wide outbreak detected in a small Finnish municipality during July 2012. We used a polyphasic approach carrying out advanced epidemiological, microbiological and environmental investigation to verify the source and scale of the outbreak. We included advanced and novel statistical, epidemiological and molecular microbiology methods, not previously applied to outbreak investigations according to our knowledge.

Materials and Methods

This study was part of public health response. According to Finnish legislation, no ethical approval is needed in this type of response. This study and related sampling and modelling were part of an official waterborne outbreak investigation, which is on the responsibility of municipal health and public health authorities (many of the coauthors are representatives of respective authorities).

Epidemiological investigation

Data collection. The contaminated part of the water distribution system provides drinking water for the area of Vuorela and Toivala in the municipality of Siilinjärvi. The total population of this area was 5934 and the exposed population who were served by the contaminated water distribution system in the defined outbreak area was 2931 persons (source population). The age, sex and living coordinates for the population were obtained from the National Population Register. A case was defined as a person staying or living in the Vuorela area during July 2012 with diarrhoea or two of the following symptoms: nausea, vomiting, stomach ache or fever. We excluded persons who were absent from the outbreak area during the whole study period and those who travelled abroad during July 2012. Based on the geographical coordinates we also determined the number of households or blocks of houses with unique water delivery points in the affected area.

Study design. A retrospective cohort study was conducted. A web based questionnaire was designed to define the extent and cause of the outbreak. The exposed population was informed by the local newspaper and press releases on the municipality website to participate in the study. All inhabitants of the Vuorela and Toivala area were invited as well as those visiting or working in this area. Data was collected between 19th July and 1st August, 2012, and any person living or spending time during the study period in the defined outbreak area was eligible to participate in the study. Study participants were asked about their basic demographic characteristics, clinical symptoms (from 4th to 20th July, 2012) and habits of consumption of tap water in the defined outbreak area (from 1st to 30th July, 2012).

Data analysis. We used a commercial web based questionnaire from Webropol (www.webropol.fi). The data was analysed as univariate factors calculating the risk ratios for all risk factors asked and frequencies of the illness as a retrospective cohort sample using R [29]. Subsequently, a binary log and logistic regression models with case status as the outcome variable and those explanatory variables that were significant in the univariate analysis were included, the analysis were performed in R. Furthermore, as we had the information of all the inhabitants, we compared the age and sex between the non-cases and the source population using a standard Wald's statistic for calculating the confidence intervals for the observed difference in percentages between the groups.

Spatial analysis and regression. For the spatial analysis, only those persons with address information available from the National Population Register and who replied to the cohort study were included. We obtained information on name, address, date of birth and living location co-ordinates. Sampling locations in the water distribution pipeline were plotted on a schematic map obtained from the local construction office (Figure 1). The shortest distance via the water pipe to the pipe breakage was determined from the obtained digital map using an R package gdistance. The distance data was subsequently allocated to each person living in the area. We also included a spatial correlation variable for the model to explain the possible transmission of the infection within households and closely living contacts and due to the other possibly unmeasured spatially correlated factors. We evaluated the model by residual diagnostics, residuals were plotted against the predicted values and Cook's diagnostic values were calculated [30]. We subsequently categorised the distance to four equal groups and compared the proportion of ill persons in each.The statistical calculations were done in R using both routine and special spatial packages [29]. The code for calculating the spatial distance via the water pipe in R which is presented in (File S1).

Clinical microbiology investigation

Microbiological analysis of patient samples. A total of 25 patients were sampled between 17th July and 2nd August, 2012. The age of the sampled patients had a median of 43 years (range 5–85 years). The potential causative agents were tested broadly, each from 5–21 samples aiming of qualitative detection of possible pathogens (Table 1). The initial samples were analysed in the local clinical laboratory with routine tests for common pathogens. As no major pathogens were detected, more samples were collected and analysed in national reference or specialized laboratories. A part of the patient samples were tested for enteric bacterial pathogens using standard clinical microbiology culture methods and PCR for the target genes of *Campylobacter jejuni/coli* (*rimR*, *gyrB*), *Salmonella* spp. (*invA*), *Shigella* spp./enteroinvasive *E. coli* (*ipaH*, *invE*) and *Yersinia* spp. (*virF*, *rumB*) [31]. The presence of diarrheagenic *E. coli* (i.e. EHEC, EPEC, ETEC, EIEC/*Shigella* spp. and EAEC) virulence genes (*pic*, *bfpB*, *invE*, *hlyA*, *elt*, *ent*, *escV*, *eae*, *ipaH*, *aggR*, stx_1, stx_2, *estlb*, *estla* and *astA*) were determined using multiplex PCR and qPCR techniques from mixed preliminary cultures grown on CLED (cystine lactose electrolyte deficient) agar medium plates [31,32]. The *ipaH* and *invG* genes are specific for both *Shigella* spp. and EIEC and the tests do not distinguish between these two organisms. If PCR for EHEC genes was positive, the specific colony was picked out when possible from the mixed culture plate for a single PCR of the *stx* genes. Electron microscopy was performed to detect enteric viruses, such as noro-, rota-, adeno-, entero-, sapo- and astroviruses. In addition, for norovirus reverse transcription (RT)-qPCR [33,34] was performed. For sapovirus analysis, two different PCR protocols were used [35,36] and nucleic acid sequences were determined from the amplicons of the polymerase region [37]. Samples were also tested for enteric parasites *Giardia* and *Cryptosporidium* using enzyme immunoassay method for antigen detection [38]. Seven frozen faecal samples were tested in a retrospective testing scenario for the presence of *Arcobacter* spp. by culture method as described previously [39] and the plates were inspected regularly for up to 3 weeks [40]. A species-specific multiplex-PCR was performed for detection of *A. buzleri*, *A. cryaerophilus* and *A. skirrowii* as described previously [41]. The samples were also tested using an additional *Arcobacter* genus specific PCR method [42].

Environmental investigation

Environmental sampling and analysis. The drinking water in the distribution network of Vuorela and Toivala area is UV-disinfected groundwater produced in the nearby Jäläniemi waterworks. The Siilinjärvi municipality owns and operates this public drinking waterworks and its distribution network. The employees of the Siilinjärvi municipality, including health protection authorities and personnel of the waterworks were responsible of environmental sampling and delivery of samples to expert laboratories. The water quality management routines include testing for coliform bacteria, heterotrophic plate counts and *Clostridium perfringens* at the waterworks and around the

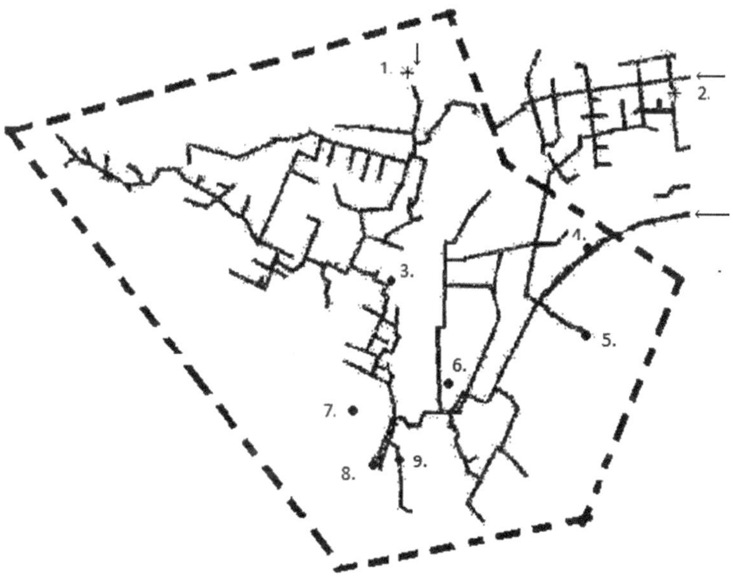

Figure 1. Schematic map of the water pipe of a defined outbreak area in Vuorela, July 2012. The outbreak (boil water notice) area is indicated by dashed line (- - -), the water sampling points (1–9) are coded as (•) with a positive culture finding and (*) with a negative finding. Arrows indicate the inflow points of the water from the water plant (outside the figure). Points 1,2,3,6,7 and 9 are tap water sampling locations, point 4 is the water pipe line breakage point (surface water), point 5 represent drinking water from the upper water storage reservoir and point 8 is the municipal effluent sampling location.

Table 1. Microbiological results of faecal samples from symptomatic patients of a waterborne outbreak in Vuorela, July 2012.

Microbial pathogen and methods used	Number of patients tested positive/number of tested
Campylobacter jejuni/coli[1,2]	1[5]/21
Salmonella spp.[1,2]	0/21
Shigella spp.[1,2]	0/21
Yersinia spp.[1,2]	0/21
E. coli	
- EHEC[1,2]	2[6]/12
- EPEC[2]	6/12
- ETEC[2]	0/12
- EIEC/*Shigella*[1,2]	0/12
- EAEC[2]	2/12
Arcobacter spp.[1,2]	0/7
Norovirus[2,3]	0[2]/12, 0[3]/17
Adenovirus[3]	0/17
Enterovirus[3]	1/17
Sapovirus[2,3]	5[2,7]/12, 3[3]/17
Astrovirus[2,3]	0[2]/7, 0[3]/17
Giardia[4]	0/19
Cryptosporidium[4]	0/19

Methods used were cultivation[1], PCR/RT-PCR[2] (polymerase chain reaction/reverse transcripition-PCR), EM[3] (electron microscopy) and/or EIA[4] (Enzyme Immunoassay). [5]*Campylobacter jejuni* and [6]EHEC O157:H7 (from one sample) and [7]sapovirus GII.P3 were isolated from the samples.

water distribution network according to European regulation [43]. The water quality results were acceptable prior to the outbreak with the last results dating back to 7th June and 2nd July, 2012. After accidental pipe breakage at 4th July, the pipeline was fixed within 14 hours, then flushed and sampled from two points (locations before the breakage point outside the schematic map and after the breakage point, point 6 in Figure 1) in 5th July, 2012 for indicator bacteria testing. The water quality was shown to comply the legal microbiological quality criteria (i.e. absence of coliform bacteria, *E. coli* and intestinal enterococci with no abnormal change in heterotrophic plate counts, Table 2).

A total of 65 water samples were collected between 5th July and 29th August (Figure 1). The majority of samples (n = 54) were tap water samples collected from various parts of the distribution system (1–10 sampling events/location) (Table 2). Also a raw water sample from a water production well, two biofilm samples from water meters and a sample from the community wastewater influent were collected. Surface water from a rainwater collection well and ditches at the pipeline breakage location (n = 5) were sampled as potential contamination sources. Both small (500–2000 mL) and large (10 L) scale water samples were analysed for pathogens. On-site large volume sampling (more than 100 L) using tangential flow ultrafiltration was conducted 17th July at one of the distribution system sites where high counts of faecal bacteria were detected at the previous day small scale sampling (point 7 represented in Figure 1). The ultrafiltration was conducted without sodium polyphosphate using a semi-automated system [17,18] and the bovine-serum pre-treated ultrafilter was eluted after sampling with 9:1 retentate/elution solution and 500 ml elution solution [17,44].

Culture methods were used to analyse *E. coli* and coliform bacteria [45,46], intestinal enterococci and heterotrophic plate counts [46], and vegetative cells and spores of *Clostridium perfringens* [47] from water samples. The presence/absence of faecal bacterial pathogens, thermotolerant *Campylobacter* spp. and *Salmonella* spp., was determined using culture-based selective enrichment methods [46,48], and EHEC was analysed using serotype O157 specific immunomagnetic separation coupled with the selective culture enrichment [49]. Enteric viruses were concentrated as previously described [50], except using Sartolon polyamide membranes (diameter 47 mm, pore size 45 µm; Sartorius, Goettingen, Germany). Norovirus, rotavirus and adenovirus were analysed using previously described RT-qPCR and qPCR methods [48,50,51]. For sapovirus analysis, the faecal protocol was applied to water as described earlier. Protozoan parasites *Giardia* and *Cryptosporidium* were determined from a tap water concentrate using immunomagnetic separation and epifluorescence microscopy according to the international standard method [52] and from water meter biofilm samples using molecular techniques [53]. The concentration of free and total chlorine in the drinking water distribution was measured on-site using the Palintest Micro 1000 photometer and in the laboratory using ISO 7393 standard method [54].

Further identification of pathogens. *E. coli* isolates from seven tap water samples (n = 19) and DNA extracts (n = 4) were screened for presence of virulence genes of EHEC, EPEC, ETEC, EIEC and EAEC [31,32]. In addition, the same virulence genes were tested from mixed cultures (tap water and rainwater sample) collected from membranes during the *E. coli*/coliform analysis. PFGE-profiles of *C. jejuni* isolates (n = 6) from patient and environmental samples were produced using *Kpn*I and *Sma*I restriction enzymes [55].

Extracted nucleic acids from total (DNA) and active bacterial fraction (RNA) in water samples were used as template to amplify faecal source identifiers and the bacterial 16S rRNA gene for microbial source tracking (MST) and NGS applications, respectively. MST assays were performed as previously described [21]. The samples were obtained from the upper storage reservoir water

Table 2. The counts of water quality indicator bacteria (range of MPN or CFU/100 mL), occurrence of faecal pathogens and chlorine concentrations in the water samples taken in July and August, 2012 in Vuorela, Finland.

Date (number of samples)	*E. coli*	Coliform bacteria	Entero-coccus	Faecal pathogens	Chlorine (mg/l)
Drinking water distribution system samples from the clean area					
5–19 Jul (12)	0	0–1	0	Not detected/3–6 samples[1]	ND, 0.4–1.8[2]
Drinking water distribution system samples from the boil water notice area					
5 Jul (1)	0	0	ND	ND	ND
16 Jul (4)	0–150	0–150	0–17	*Arcobacter* spp./1–4 samples[3]	ND
17 Jul (4)	0–21	0–34	0–2	EHEC, EPEC and EAEC virulence genes[4]	ND
18–20 Jul (10)	0[5]	0	0	Not detected/2–4 samples[1]	<0.1–1.6
23–30 Jul (13)	ND[6]	ND	ND	Not detected/3 samples[7]	<0.1–2.0
1–29 Aug (9)	ND[6]	ND	ND	ND	0.4–1.9
Water storage reservoir					
17 July (1)	110	190	15	Norovirus and adenovirus[8]	ND
21 July (1)	0	0	ND	Not detected[9]	ND
Biofilms from the water meters removed from the boil water notice area					
1 Aug (2)	ND	ND	ND	*Arcobacter* spp.[10]	ND
Raw water at the groundwater abstraction plant					
9 August (1)	0	0	ND	Not detected[11]	ND
Surface water samples (contaminant source)					
23 Jul (1)	86	450	44	*Campylobacter jejuni* 10–100 cfu/l, EPEC virulence genes[12]	ND
29 Aug (5)	0–94	ND	5–80	ND	ND
Community wastewater					
25 Jul (1)	ND	ND	ND	Sapovirus	ND

ND; not determined.
[1]A portion of the samples were selected for *Salmonella*, *Campylobacter*, enterohaemorragic *E. coli* (EHEC) culture analyses and for norovirus analysis.
[2]Measured from three locations at 19 July.
[3]Samples were tested for noro-, adeno-, rota- and sapoviruses, *Campylobacter* and *E. coli* virulence genes. *Arcobacter* was tested from DNA extracts and genus specific PCR was positive in one sample (point 9 in Figure 1).
[4]*E. coli* virulence genes were detected after ultrafiltration from one sampling location (point 5 in Figure 1). *Salmonella*, *Campylobacter*, EHEC (culture method), noro-, rota- and sapovirus, *Giardia* and *Cryptosporidium* were not detected (1–4 samples tested/method).
[5]One colony of *Clostridium perfringens* was found from 1 000 mL of tap water sample taken from the most contaminated area.
[6]*Clostridium perfringens* was analyzed and not detected from 5 000 mL samples.
[7]Samples were tested for sapovirus.
[8]Sample was tested for noro-, adeno-, rota- and sapoviruses, *Salmonella*, *Campylobacter* and *E. coli* virulence genes. *Clostridium perfringens* was detected (10 CFU/L).
[9]Sample was tested for *Campylobacter* and noroviruses. *Clostridium perfringens* was detected (2 CFU/L).
[10]Sample tested for *Campylobacter*, *Arcobacter*, *Giardia* and *Cryptosporidium*.
[11]Sample tested for *Campylobacter*.
[12]Sample tested for *Campylobacter*, *E. coli* virulence genes, noro- and adenoviruses, *Giardia* and *Cryptosporidium*. *Clostridium perfringens* was detected (40 CFU/L).

before and after cleaning (point 5 in Figure 1) and from a tap water sample collected during the contamination episode (point 7 in Figure 1). The MST assays included the analysis of faecal bacterial groups *Bacteroidales* spp. (GenBac3 assay [56]) and human-specific *Bacteroidales* (HF183 assay [57]).

For Illumina MiSeq NGS, we utilized barcoded primers 515F and 806R to produce 250 bp pair-ended sequences [58]. Reads were processed and analyzed using the software MOTHUR v1.30.2 ([59]; http://www.mothur.org) as described previously [60]. Briefly, fastq files with forward and reverse reads were used to form contigs. The reads that met the following criteria were excluded from further analysis: the length was no greater than 255 bp; contained ambiguous bases (N) or homopolymers greater than 7 bases; identified as chimera; or classified as Chloroplasts or Mitochondria. Reads were aligned and sorted with >97% similarity into operational taxonomic units (OTUs). Taxonomic classification was obtained using the tool Classifier in the Ribosomal Database Project II release 10.28 [61]. The raw reads were deposited in the NCBI Sequence Read Archive (SRA) under accession number SRP041117. Using MEGA v5.2 [62], a phylogenetic tree based on the aligned 16S rRNA gene sequences (~255 bp) was constructed with the Maximum Likelihood (ML) method using Tamura-Nei model [63] with 1,000 bootstrap replicates. The tree was used to infer the phylogenetic relationship among sequences classified as *Arcobacter* obtained in this study. *Sulfurospirillum deleyianum* (NR_074378) and *Campylobacter fetus* (L04314) were used as outgroup. Three water samples and two water meter samples were subsequently analysed for *Arcobacter* spp. using genus specific PCR [42] and species specific multiplex-PCR [41] from the DNA extracted as previously described [50].

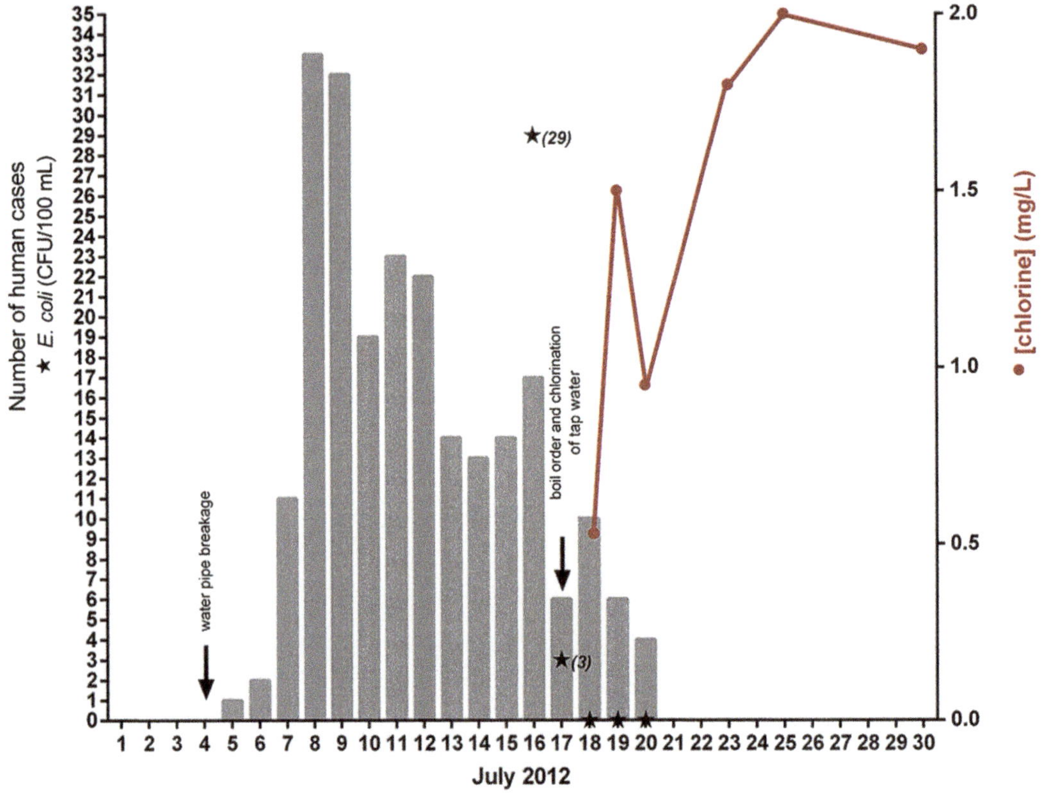

Figure 2. Epidemic curve of a waterborne outbreak in Vuorela, July 2012 based on the reported onset date of illness of the cases, and *E. coli* bacteria counts and chlorine levels in the point 7 (See Fig. 1) of the water distribution network.

Results

Epidemiological investigation

Descriptive analysis. Of all 2931 inhabitants (source population) of the defined outbreak area, 473 (16%) persons participated in the study (study population). We excluded 19 persons absent from the outbreak area during the whole study period and 23 due to travelling abroad. In total, we identified 225 cases and 206 healthy persons from the cohort study according to the case definition. However, for the individual risk factors data contained a few missing values. During the four week period after the water distribution breakage, an estimated 800 excess persons visited either a nurse or GP in the local health care centre. The outbreak curve presented in Figure 2 implicates a point source outbreak.

Demographic characteristics. Of the 431 respondents, 33% (135/408) were male. The mean age was 41.2 years (range 1–80 years). As there were few children among the respondent (10 in total, 3 among non-cases), we excluded the children from comparison analysis. The proportion of male (33%, 60/182) among the non-cases in the study population was lower to that of the source population (48%, 1074/2252). This difference for the male proportions of 14.5% (7.4%–21.6%) was statistically significant. Persons were slightly older among the source population compared to the study population, Table 3. Also, there were 50.2% (232/462) households (unique water delivery points) with at least one person under 18 years of age in the source population compared to 68.5% (126/184) among the study population. This difference was significant, CI for the difference was 18.3% (10.2%–26.4%). Therefore, households with children were more likely to respond to the study.

The course of the illness was mild, only one person was admitted to hospital. The main symptoms were stomach ache 88% (199/225), nausea 85% (191/225) and diarrhoea 82% (185/225). The length of the illness had a median of 3 days (range 1–30), 16% (35/225) of the cases sought for medical assistance and one patient was hospitalized (37-year old female with no underlying medical condition). Absence from work due to outbreak illness was

Table 3. Comparison of the study population to source population with respect to age groups in a waterborne outbreak in Vuorela, July 2012.

Age group	Study population (non-cases)% of population (number of persons)	Source population % of population (number of persons)	% difference (95% confidence intervals)
20–39 years	40% (78/195)	29% (655/2252)	11% (3.9%–18.1%)
≥40 years	60% (117/195)	71% (1597/2252)	−11% (−18.1%–−3.9%)

Table 4. The Univariate and multivariate results for individual risk factors and the generalized additive model risk ratios with the spatial term of a waterborne outbreak in Vuorela, July 2012.

	Explanatory variable	Risk ratio or univariate log regression exp (β-values), (95% confidence intervals) for individual risk factors	Multivariable generalized additive logit model, exp (β-values), (p-value)
Personal characteristic	Age (continuous in years)	0.99 (0.98–0.99)	0.975 (0.0061)
Drinking at home	Tap water	2.2 (1.2–4.1)	5.90 (0.0037)
	Water from own well	1.0 (0.57–1.87)	n/a
	Bottled water	0.86 (0.69–1.07)	n/a
	Boiled water	0.69 (0.53–0.90)	n/a
Drinking water in Vuorela (outside home)	Tap water	1.6 (1.2–2.0)	n/a
	Water from own well	0.83 (0.56–1.22)	n/a
	Bottled water	0.90 (0.72–1.12)	n/a
	Boiled water	0.80 (0.56–1.15)	n/a
Spatial variables	Distance from the breakage by waterpipe (metres)	0.99950 (0.99930–0.99969)	0.998 (0.060)
	Spatial variable (coordinates)	n/a	n/a (0.002)

reported by 31% (133/431) of the study participants, the total number of working days lost because of illness was 398. Taking care of a sick child was not included.

Univariate analysis of the cohort. Of all possible cases occurring in the area, 225 cases responded to the study in an area with a population of 2931 persons. Of the water related risk factors, drinking untreated tap water in the outbreak area had a risk ratio of 5.6 (95% CI 1.9–16.4), also drinking untreated tap water at home RR 2.2 (95% CI 1.2–4.1) and outside home RR 1.6 (95% CI 1.2–2.0) in the outbreak area were associated with illness, Table 4. Drinking boiled water at home was a protective factor, RR 0.69 (95% CI 0.53–0.90), drinking well water or bottled water were not significant either at home or outside home. It was observed that the risk ratio increased as the number of glasses of water consumed increased. This further strengthened the role of the contaminated tap water as a cause of the outbreak in a dose response manner, table 5.

Multivariate and spatial analysis of the cohort. Of those fulfilling eligibility criteria, 20 participants had no address information, 81 lived outside the defined boil water notice area and 17 could not be found in the national population register, leaving 313 persons for the spatial analysis. After these exclusions, 154 cases and 159 non-cases were used for the spatial analysis. Age, drinking tap water at home and distance from the breakage point were significant in the multivariable model, Table 4. Younger persons were more likely to become ill ($p = 0.0061$), risk also increased if tap water was consumed at home ($p = 0.0037$). Distance from the leakage point was inversely associated with becoming a case ($p = 0.060$). Also spatial variable was significant in the generalized additive model ($p = 0.002$), this is likely to reflect the person to person transmission within households or neighbourhoods (Table 4). The distance via the water pipe was shorter for the cases compared with the non-cases (Table 4). This indicates that the closer one lived to the water breakage point, the more contaminated the drinking water was and therefore the likelihood of the illness was higher. The diagnostics of the model for randomized quantile residuals were normally distributed to suggest correct specification of the model. Between the categorised groups, the proportion of ill persons was higher among those living closer to water breakage point, Table 6.

Clinical microbiology investigation

Patient faecal samples. The microbiological analysis of faecal samples from the patients identified several pathogens (Table 1). The pathogens identified included sapovirus (detected from 5 patients of 12 tested, 2 sapoviruses were genotyped as

Table 5. The dose-response between the illness and the amount of water consumed at home in a waterborne outbreak in Vuorela, July 2012.

Number of glasses of water consumed at home per day	Cases/total number of persons in the implicated group (%)	Risk ratio (95% confidence intervals)	p-value
0	6/28	reference	
1–3	49/114 (43%)	2.01 (1.06–4.81)	0.065
4–6	112/193 (58%)	2.71 (1.48–6.40)	0.0066
7–9	37/48 (77%)	3.60 (1.94–8.54)	0.00054
10 or more	9/19 (47%)	2.21 (0.96–5.68)	0.068

Those not drinking water at home served as a control group.

Table 6. Categorized distance and proportion of cases within those groups in a waterborne outbreak in Vuorela, July 2012.

Distance categories	% (Cases/total)
Distance 1 (<2332)	72.4% (63/87)
Distance 2 (2332–2713)	45.5% (35/77)
Distance 3 (2713–3202)	43.5% (37/85)
Distance 4 (>3202)	29.7% (19/64)

GII.P3), enterovirus (detected from one patient out of 17 tested) and *C. jejuni* (isolated from one patient out of 21 tested). Two sapovirus nucleic acid sequences were submitted to Genbank with accession numbers KJ200380 and KJ200381. In addition, specific virulence genes of pathogenic *E. coli* were detected, including genes from EHEC (2/12), EPEC (6/12) and EAEC (2/12). In subsequent culturing of the virulence gene positive samples, one *E. coli* O157:H7 positive culture was found. No suspect *Arcobacter* spp. were detected by culture methods or by PCR.

Environmental investigation

Faecal microbes in the drinking water distribution. Faecal indicator bacteria were detected in four out of 11 tap water samples taken from different sampling points of the distribution on 16th and 17th July, 2012 (Table 2). The bacterial counts from the drinking water reservoir sample (point 5 in Figure 1) taken at the same time were high for total coliform bacteria, *E. coli* and enterococci. In addition, low numbers of adenoviruses and noroviruses (genogroup II) were detected in the storage reservoir sample. Water samples tested negative for sapovirus, although the community waste water influent was shown to contain sapovirus. Specific virulence genes of EHEC, EPEC and EAEC were detected from the large volume tap water sample (point 7 in Figure 1).

Cause of the water contamination. Technical investigations revealed that the water storage reservoir had been rapidly filled with contaminated water after the pipe breakage repair. The microbial contaminants unintentionally funnelled to the storage reservoir could remain viable as no disinfection was used in the distribution system (Table 2). Subsequently the contaminated water was introduced to the Vuorela and Toivala distribution system due to the water usage. The water level in the storage reservoir varies depending on water usage and pumping of the fresh water from the groundwater abstraction plant into the reservoir. The water quality at the groundwater abstraction plant was tested acceptable and free of microbial contaminants shortly after the breakage (Table 2). De-contamination of the reservoir was ensured by emptying and then mechanical cleaning, washing and chlorination of the inner surfaces of the storage reservoir. Indicator bacteria were no longer detected from the water distribution system after flushing and successful chlorination (Table 2). However, it took approximately a week before *C. perfringens* spores were absent in the water samples from the contaminated distribution system and chlorine levels reached the minimum target concentration (0.5 mg/l) at the most distant parts of the distribution.

Microbial source tracking. GenBac3 and HF183 assays showed the presence of *Bacteroides* spp. and human-specific *Bacteroidales*, respectively, in the contaminated tap water and in the storage reservoir (Table S1). There were more rRNA markers present compared to the rRNA gene (DNA) markers (rRNA:rDNA ratios were higher) in the storage reservoir sample than in the contaminated tap water sample. After cleaning of the storage reservoir, the GenBac3 and HF183 marker concentrations declined to below the quantification and detection limits, respectively (Table S1). A reactive surface water sample from the rainwater well was collected one week after the outbreak notification to evaluate the potential environmental sources near the pipe breakage location (Table 2). *C. jejuni* and EPEC virulence genes were identified from the surface water sample. Also indicator bacteria; coliforms, *E. coli* and intestinal enterococci were abundant. Five *C. jejuni* isolates from this suspected contamination source had identical *Kpn*I and *Sma*I profiles in PFGE, but different from those of the *C. jejuni* patient isolate (data not shown).

Bacterial communities in water. Community analysis based on the bacterial 16S rRNA region showed a higher diversity for contaminated sampling sites, while upper storage reservoir after the cleaning showed a lower diversity (Table S2). It is noteworthy that the community diversity measured as a total number of OTUs in the community was higher within the DNA reads than within the RNA reads (Table S2). Taxonomic analysis of contaminated water samples indicated a high presence of the class Beta-proteobacteria including the family *Comamonadaceae* and genus *Zoogloea*, and Gamma-proteobacteria including affiliated family *Methylococcaceae* and genera *Methylobacter* and *Pseudomonas* (Figure 3, A and B; Table S3).

Interestingly, also a significant share of reads generated from the contaminated water samples affiliated with Epsilon-proteobacteria genus *Arcobacter* (Table S3 and Table S4). Further sequence analysis revealed close relationship to *A. cryaerophilus* group 1A LMG9865 (OTU00012 in Figure 4). However, the corresponding tap water sample was negative for the species-specific *Arcobacter* multiplex-PCR (for *A. butzleri*, *A. cryaerophilus*, *A. skirrowii*) and the storage reservoir samples were no longer available for species-specific testing. *Arcobacter* genus specific PCR was positive for a tap water sample and a water meter sample collected from another location (point 9 in Figure 1) but these samples remained negative by culture. Although the sequences from both samples (Genbank with accession number KJ196910) showed high sequence similarity (>98%) with *A. cryaerophilus* (uncultured), *A. skirrowii* and *A. cibarius*, no species attribution can be made.

After the cleaning and chlorination of the upper storage reservoir, the relative abundance of Beta- and Gamma-proteobacteria decreased and the abundance of *Sphingobacteria* including genus *Pedobacter* increased (Figure 3, C; Table S3). The reads associated *Bacteroides*, *Escherichia* and *Clostridia* in the contaminated water samples were absent from the storage reservoir sample after cleaning and chlorination (Table S4).

Discussion

This study describes a large-scale municipality area wide outbreak due to breakage of drinking water distribution pipeline during a road construction work. The illness in the community became apparent only two weeks after the incident but the cause of the illness was obvious. This was confirmed using a polyphasic approach applying microbiological, epidemiological and statistical methods. Multiple causative microorganisms were isolated from patient samples. No major definite causative pathogen was identified but sapovirus was most frequently detected in patient samples. Also various types of virulent *E. coli*, *C. jejuni* and *Arcobacter* spp. might have played a role in the onset of gastrointestinal symptoms. All these organisms have been previously associated with waterborne transmission [5,9,12,13,23]. By

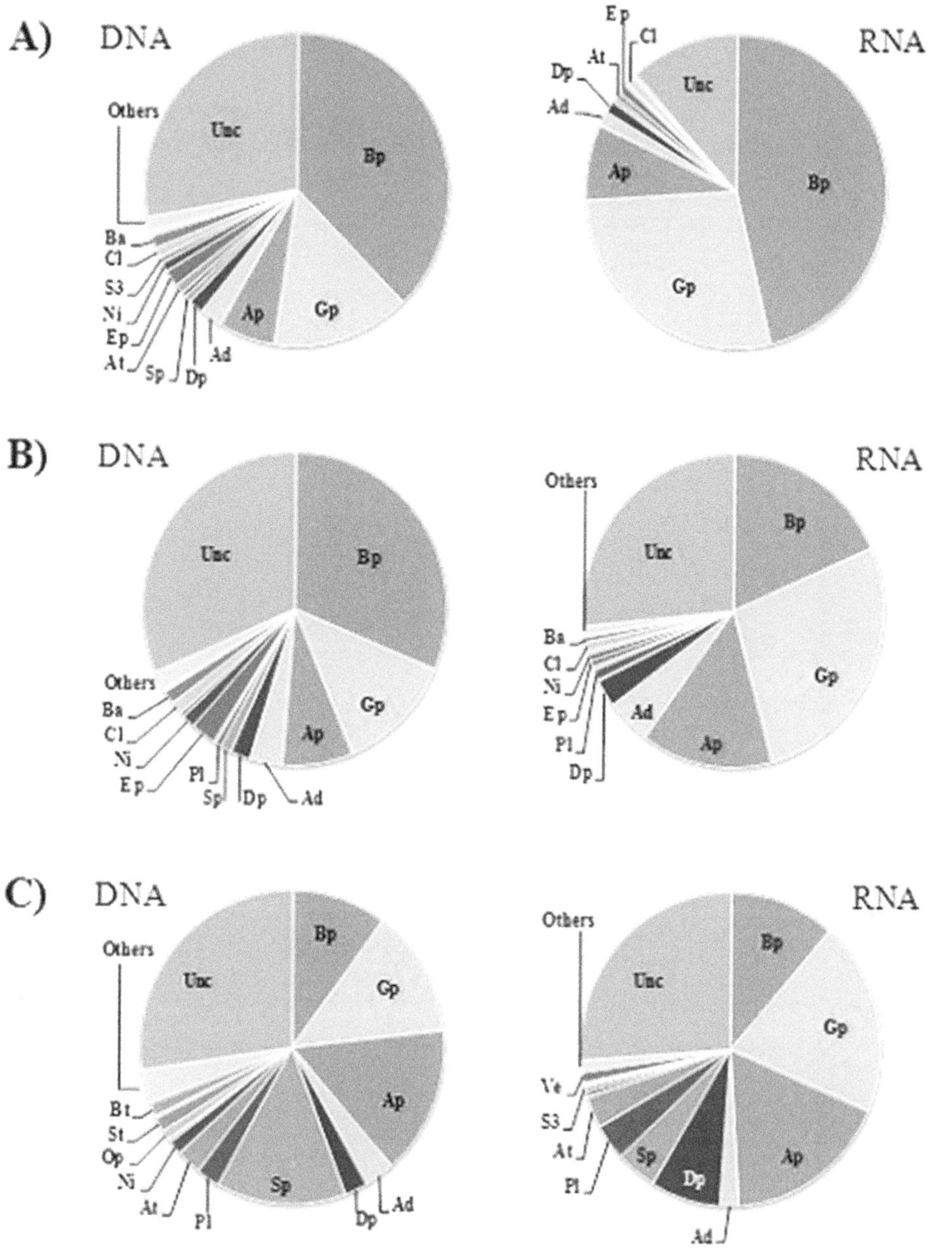

Figure 3. Distribution of the Bacteria domain as determined by taxonomic identification of partial 16S rRNA gene sequencing (at class level) in a waterborne outbreak in Vuorela, July 2012. Charts and tables represent the cumulative distribution of total DNA and RNA extracted from samples of A) the upper storage reservoir before cleaning (point 5 in Figure 1), B) tap water during contamination (point 7 in Figure 1) and C) the upper storage reservoir after cleaning. Legend: Beta-proteobacteria (Bp); Gamma-proteobacteria (Gp); Alpha-proteobacteria (Ap); Acidobacteria (Ad); Delta-proteobacteria (Dp); Sphingobacteria (Sp); Planctomycetacia (Pl); Actinobacteria (At); Epsilon-proteobacteria (Ep); Nitrospira (Ni); Verrucomicrobia Subdivision3 (S3); Clostridia (Cl); Bacteroidia (Ba); Opitutae (Op); Spartobacteria (St); Verrucomicrobiae (Ve); Bacteroidetes (Bt); Others (classes each representing <1%); unclassified (Unc).

applying the spatial and microbiological analysis, we could identify the contamination route for this outbreak. The distance to the water breakage point was inversely associated with illness both by epidemiological and spatial methods. We further confirmed the role of the contaminated water as a vehicle by novel spatial analysis. The spatial method provided here could be used to compare the likelihoods between possible candidate point source locations. Effective control measures and rapid and continuous communication with the media were essential to ensure appropriate public health measures.

The overall clinical illness in most patients was mild with only 16% accessing medical assistance and only one patient being hospitalized. Majority of the cases had gastrointestinal symptoms with mild course and relatively short length. Overall, the number of cases was likely to be an underestimate for the whole population. The case definition used in the present study was relatively sensitive due to the mild overall nature of the illness. The

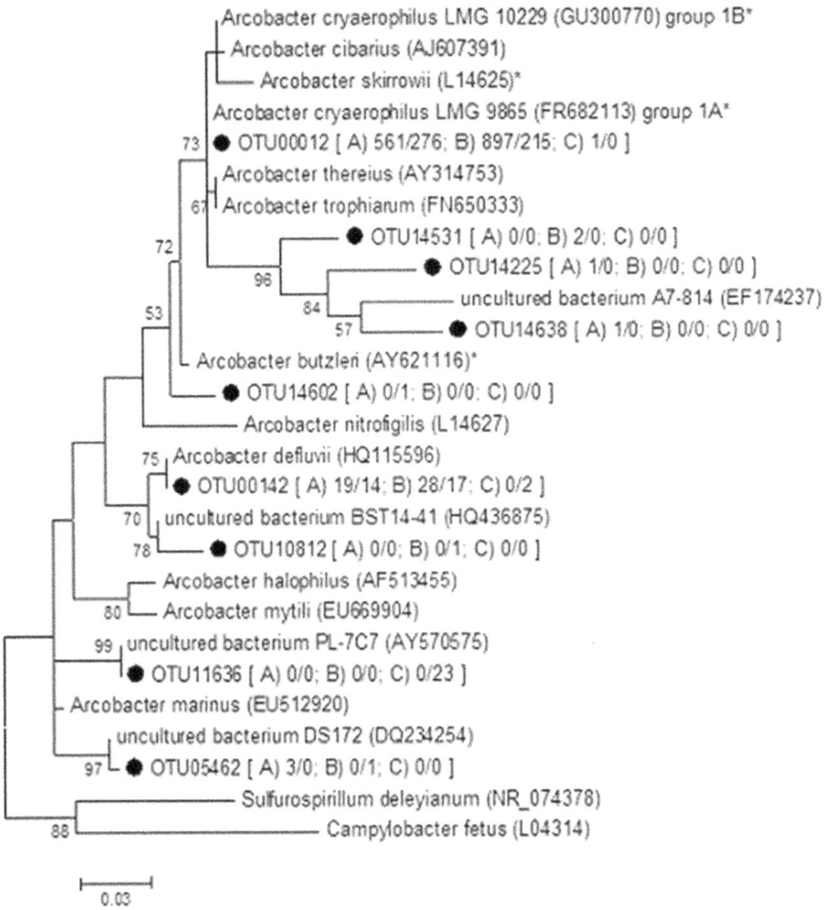

Figure 4. Phylogenetic relationships among OTUs (•) of the genus *Arcobacter* in a waterborne outbreak in Vuorela, July 2012. The tree was inferred from a maximum likelihood analysis of aligned 16S rRNA gene sequence (≈255 bp) and nodes with a bootstrap value ≥50% of 1 000 replicates are identified. *Sulfurospirillum deleyianum* (NR_074378) and *Campylobacter fetus* (L04314) were used as outgroup. Number in bracket represents the total amount of DNA/RNA reads identified in samples from A) the upper storage reservoir before cleaning, B) tap water during contamination and C) the upper storage reservoir after cleaning. **A. butzleri, A. cryaerophilus* and *A. skirrowii* have been associated with gastrointestinal diseases [76].

contaminated water consumed whether at or outside home in the defined outbreak area was associated with illness. We further showed by epidemiological methods that drinking contaminated tap water was associated with illness in a dose-response manner. The effect levelled off at the highest dose, often observed in food or waterborne outbreaks [64,65], this has been observed also in previous studies [1,3,9]. Furthermore, drinking boiled water was a protective factor indicating the effectiveness of the implemented control measures.

It is challenging to estimate the true attack rate in the present outbreak. It is quite likely that people showing symptoms responded more actively in the study compared to non-cases. We had a unique opportunity to compare the study population to the source population and found that women responded more actively to the study. This has been found also previously [66–68]. As we also found slightly more younger people among the non-cases compared to the source population, therefore, the observed effect of age should be interpreted with caution. The crude attack rate obtained from the questionnaire study (225/431, 52%) is within the same range to what has been observed in similar outbreaks, namely between 31–88% [2–4]. The true number of ill persons is difficult to estimate, but the educated evaluation of the local general practitioner (800 cases) is a fair estimate.

By applying the spatial and microbiological analysis, we could identify the contamination route for this outbreak. The contaminated water from the breakage initially filled the water storage reservoir and subsequently this water was distributed to the community over a number of days. The drinking water storage reservoirs have contributed to the transmission of waterborne infections also elsewhere, usually due to the improper maintenance and structure of the reservoirs [46,69,70]. In our case, the exact cause of the contamination was pipeline breakage at the road construction site. The storage reservoir prolonged the contaminant transport time to the water consumers. The likelihood of illness was higher closer to the water breakage point as measured by water pipeline length. Also the higher amount of water consumed in the boil water notice area increased the likelihood of illness. Additionally, younger persons (i.e. parents of small children) might have been more likely to become infected presumably due to the fact that they had more contact to young infants.

The source of the illness was obvious from the beginning of the outbreak, but we initially searched for one major pathogen as a causative agent yet aware that this type of outbreak is often caused by multiple pathogens. The sapovirus finding was novel and has rarely been detected in previous Finnish waterborne outbreaks [13]. According to recent reports, sapoviruses can be commonly found in environmental waters [10,71,72] highlighting the importance of including sapovirus detection into the outbreak investigations. Sapoviruses cause acute gastroenteritis primarily among young children [36], and indeed, young age was suspected to be one risk factor in this study. This result, together with the presence of sapoviruses in communal wastewater influent, support the assumption that sapovirus was commonly circulating in the community. We did not detect sapoviruses in the drinking water samples, presumably due to delayed sampling and low viral concentration in the drinking water. We also found that adults had symptomatic sapovirus infections. Adults have been found to have sapovirus infections in other studies [12,73,74]. The occurrence of human infecting sapoviruses among the patient cases together with the detection of human-specific genetic marker HF183, adenoviruses and noroviruses from the contaminated drinking water suggests that human faecal material was present at the pipe breakage site. No leakage was identified in a waste-water pipe line located in the same construction pit as the drinking water pipeline. Instead, it was concluded that surface water runoff from the nearby recreational area was the potential source of the faecal microbes. The abundant pathogenic *E. coli* findings from patient and water samples were not surprising bearing in mind the nature of the contamination. One positive *C. jejuni* finding in a patient may have had sporadic origin as the environmental isolate was of different type.

We used a novel NGS approach to study the microbial communities in the tap water distributed in the contaminated water pipe network. To our knowledge, this was the first time that high-throughput sequencing was employed for detection of potential causative bacterial agents in a waterborne outbreak case. By using the novel microbial community analysis for the water samples, we could show an abundance of *Arcobacter* spp. in the drinking water distribution during the contamination. Further analysis found high sequence similarity with potential pathogenic *Arcobacter* spp. (*A. cryaerophilus*, *A. skirrowii* and *A. cibarius*). Although the abundance of reads of *Arcobacter* correlated with the quality of water (increase in contaminated waters), the resolution of *Arcobacter* species lineage was not possible due to ambigious nature of the 16S rRNA sequences [75]. Since *Arcobacter* spp. have been suspected as a cause of gastroenteritis in humans [5,76], we aimed to isolate these from the remaining patient samples. No *Arcobacter* spp. were isolated perhaps due to the small quantity, long storage time (more than one year) or storage conditions of the samples. Nevertheless, considering that there is no gold standard for the isolation of these putative pathogens, their presence cannot be ruled out.

The NGS approach also provided information on the bacterial communities in the non-chlorinated and faecally contaminated drinking water. Until now, the analysis of indigeneous microbial community composition in non-chlorinated drinking water distribution systems has been scarce [77]. The detected change in the storage reservoir bacterial community structure after the cleaning and chlorination proves the effectiveness of these measures. Moreover, our result showing higher community diversity within the DNA reads than within the RNA reads suggest that DNA-based methods may not effectively discriminate between active and dormant populations. However, more work is needed to better characterize the bacterial community changes in the different drinking water distribution systems. For example, understanding the role of drinking water retention time in the network [77] and the effect of drinking water treatment processes such as filtration on the bacterial communities in the distribution [78] would facilitate the sustainable management of the microbial water quality in the distribution networks.

We further confirmed the role of the contaminated water as a vehicle by novel spatial analysis. By calculating the distance for each household to the water breakage point via the water pipe, we could show that the probability of illness decreased by the increasing length of the water pipe. However, it should be noted that non-symptomatic persons in the same households and neighbourhoods participated more actively in the cohort study. We used all cases in the model, including potential secondary cases. As the spatial term describes just secondary spread and was significant in the spatial logistic regression model, it suggests that there were indeed a number of secondary cases among the case patients. In addition to distance from the breakage point, age was inversely associated with illness and consumption of water in the Vuorela area was positively associated in a spatial logistic regression.

The study was limited by a slow response due to notification delay as it took a relatively long time before the outbreak was detected and confirmed. Therefore, the causative agents could not be isolated from the water samples. The microbial community analysis also warrants further studies. We used 16S rRNA as a target of the NGS studies. In the future, with careful design of pathogen specific primers for sequencing purposes, it might be possible to abandon the requirement of culture isolation for genotyping purposes. In successful NGS applications, the huge number of reads per sample potentially enables the identification phylogenetic relatedness between the causative pathogenic strains in different samples [79]. Subsequently, this might allow the use of NGS approach for source tracking and understanding of the transmission route of an outbreak.

Conclusions

We used novel and existing statistical, microbiological and spatial methods to characterize a community wide waterborne outbreak. These methods may be applied to wide range of food and waterborne outbreak investigations and beyond. In particular, microbial community analysis in combination with traditional culture and PCR based methods aided to clarify the potential causative agents. In addition, the cohort study was carried out as a rapid web based application added with a novel spatial method to show the statistical association to the suspected point source leakage of the water pipe network. We also confirmed previous observations that women and younger people were more likely to respond to this type of study. In the present outbreak the event associated with the outbreak was fairly obvious, but in many outbreak situations the source cannot be easily identified [3]. Failures in the water distribution networks are common causes of waterborne outbreaks [2]. The method provided here could be used to compare the likelihoods between possible candidate point source locations.

Supporting Information

Table S1 Numbers of the GenBac3 and HF183 markers ($\log_{10}$ copies 100 mL^{-1}) in TaqMan rRNA-targeted RT-qPCR (rRNA) and rRNA gene-targeted qPCR (rDNA) assays.

Table S2 Community diversity estimates (±CI) of the domain *Bacteria*.

Table S3 Taxonomic affiliation of the most abundant *Bacteria* domain representatives.

Table S4 Abundance of reads associated with faecal bacteria and/or disease agents detected in the Vuorela and Toivala drinking water distribution system.

File S1 R code for calculating the shortest direct distance and the distance via the water pipe between each inhabitant location and water pipe breakage point.

Acknowledgments

We thank Anne-Mari Rissanen, Paula Muona and Mirja Rissanen and all the other personnel at the water works, in the local analytical laboratories and municipality of Siilinjärvi for their help in the outbreak investigation. We thank Pekka Pere from the University of Helsinki in helping with the statistical analysis. The opinions expressed in the paper are those of the authors and do not necessarily reflect the official positions and policies of the US Environmental Protection Agency or any other public authorities mentioned in this study. Any mention of product or trade names does not constitute recommendation for use by the US Environmental Protection Agency or any other public authorities mentioned in this study.

Author Contributions

Conceived and designed the experiments: KJ JO LM VGA JR H. Rintala AS JWSD MLH TP. Performed the experiments: KJ H. Rintala LM VGA JR M. Palander JA A. Kauppinen PR SS ON A. Kyyhkynen SH JM M. Pärnänen RL H. Ryu TP. Analyzed the data: KJ JO LM VGA JR A. Kauppinen A. Kyyhkynen SH M. Pärnänen MK AS JWSD MLH TP. Contributed reagents/materials/analysis tools: KJ JO LM VGA JR M. Palander JA A. Kyyhkynen SH JM M. Pärnänen RL H. Ryu MK AS IM JWSD MLH TP. Wrote the paper: KJ JO LM VGA JR A. Kauppinen MK AS JWSD MLH TP.

References

1. Maurer AM, Sturchler D (2000) A waterborne outbreak of small round structured virus, *Campylobacter* and *Shigella* co-infections in la Neuveville, Switzerland, 1998. Epidemiol Infect 125: 325–332.
2. Laursen E, Mygind O, Rasmussen B, Ronne T (1994) Gastroenteritis: A waterborne outbreak affecting 1600 people in a small Danish town. J Epidemiol Community Health 48: 453–458.
3. Jakopanec I, Borgen K, Vold L, Lund H, Forseth T, et al. (2008) A large waterborne outbreak of campylobacteriosis in Norway: The need to focus on distribution system safety. BMC Infect Dis 8: 128-2334-8-128.
4. Laine J, Huovinen E, Virtanen MJ, Snellman M, Lumio J, et al. (2011) An extensive gastroenteritis outbreak after drinking-water contamination by sewage effluent, Finland. Epidemiol Infect: 1–9.
5. Fong TT, Mansfield LS, Wilson DL, Schwab DJ, Molloy SL, et al. (2007) Massive microbiological groundwater contamination associated with a waterborne outbreak in Lake Erie, South Bass Island, Ohio. Environ Health Perspect 115: 856–864. 10.1289/ehp.9430.
6. Bopp DJ, Sauders BD, Waring AL, Ackelsberg J, Dumas N, et al. (2003) Detection, isolation, and molecular subtyping of *Escherichia coli* O157:H7 and *Campylobacter jejuni* associated with a large waterborne outbreak. J Clin Microbiol 41: 174–180.
7. Zacheus O, Miettinen IT (2011) Increased information on waterborne outbreaks through efficient notification system enforces actions towards safe drinking water. J Water Health 9: 763–772. 10.2166/wh.2011.021.
8. Dev VJ, Main M, Gould I (1991) Waterborne outbreak of *Escherichia coli* O157. Lancet 337: 1412.
9. Gubbels SM, Kuhn KG, Larsson JT, Adelhardt M, Engberg J, et al. (2012) A waterborne outbreak with a single clone of *Campylobacter jejuni* in the Danish town of Koge in May 2010. Scand J Infect Dis 44: 586–594. 10.3109/00365548.2012.655773.
10. Nenonen NP, Hannoun C, Larsson CU, Bergstrom T (2012) Marked genomic diversity of norovirus genogroup I strains in a waterborne outbreak. Appl Environ Microbiol 78: 1846–1852. 10.1128/AEM.07350-11.
11. Parshionikar SU, Willian-True S, Fout GS, Robbins DE, Seys SA, et al. (2003) Waterborne outbreak of gastroenteritis associated with a norovirus. Appl Environ Microbiol 69: 5263–5268.
12. Svraka S, Vennema H, van der Veer B, Hedlund KO, Thorhagen M, et al. (2010) Epidemiology and genotype analysis of emerging sapovirus-associated infections across Europe. J Clin Microbiol 48: 2191–2198. 10.1128/JCM.02427-09.
13. Rasanen S, Lappalainen S, Kaikkonen S, Hamalainen M, Salminen M, et al. (2010) Mixed viral infections causing acute gastroenteritis in children in a waterborne outbreak. Epidemiol Infect 138: 1227–1234. 10.1017/S0950268809991671.
14. Tillett HE, de Louvois J, Wall PG (1998) Surveillance of outbreaks of waterborne infectious disease: Categorizing levels of evidence. Epidemiol Infect 120: 37–42.
15. Pitkänen T (2013) Review of *Campylobacter* spp. in drinking and environmental waters. J Microbiol Methods 95: 39–47. 10.1016/j.mimet.2013.06.008.
16. Miettinen IT, Zacheus O, von Bonsdorff CH, Vartiainen T (2001) Waterborne epidemics in Finland in 1998–1999. Water Sci Technol 43: 67–71.
17. Hill VR, Kahler AM, Jothikumar N, Johnson TB, Hahn D, et al. (2007) Multistate evaluation of an ultrafiltration-based procedure for simultaneous recovery of enteric microbes in 100-liter tap water samples. Appl Environ Microbiol 73: 4218–4225. 10.1128/AEM.02713-06.
18. Rhodes ER, Hamilton DW, See MJ, Wymer L (2011) Evaluation of hollow-fiber ultrafiltration primary concentration of pathogens and secondary concentration of viruses from water. J Virol Methods 176: 38–45. 10.1016/j.jviromet.2011.05.031.
19. Hijnen WA, Biraud D, Cornelissen ER, van der Kooij D (2009) Threshold concentration of easily assimilable organic carton in feedwater for biofouling of spiral-wound membranes. Environ Sci Technol 43: 4890–4895.
20. Toledo-Hernandez C, Ryu H, Gonzalez-Nieves J, Huertas E, Toranzos GA, et al. (2013) Tracking the primary sources of fecal pollution in a tropical watershed in a one-year study. Appl Environ Microbiol 79: 1689–1696. 10.1128/AEM.03070-12.
21. Pitkänen T, Ryu H, Elk M, Hokajarvi AM, Siponen S, et al. (2013) Detection of fecal bacteria and source tracking identifiers in environmental waters using rRNA-based RT-qPCR and rDNA-based qPCR assays. Environ Sci Technol 47: 13611–13620. 10.1021/es403489b.
22. Di Bella JM, Bao Y, Gloor GB, Burton JP, Reid G (2013) High throughput sequencing methods and analysis for microbiome research. J Microbiol Methods. 10.1016/j.mimet.2013.08.011.
23. Riera-Montes M, Brus Sjolander K, Allestam G, Hallin E, Hedlund KO, et al. (2011) Waterborne norovirus outbreak in a municipal drinking-water supply in Sweden. Epidemiol Infect 139: 1928–1935. 10.1017/S0950268810003146.
24. de Jong B, Ancker C (2008) Web-based questionnaires - a tool used in a *Campylobacter* outbreak investigation in Stockholm, Sweden, October 2007. Euro Surveill 13: 18847.
25. Oxenford CJ, Black AP, Bell RJ, Munnoch SA, Irwin MJ, et al. (2005) Investigation of a multi-state outbreak of *Salmonella hvittingfoss* using a web-based case reporting form. Commun Dis Intell Q Rep 29: 379–381.
26. Bessong PO, Odiyo JO, Musekene JN, Tessema A (2009) Spatial distribution of diarrhoea and microbial quality of domestic water during an outbreak of diarrhoea in the Tshikuwi community in Venda, South Africa. J Health Popul Nutr 27: 652–659.
27. Dangendorf F, Herbst S, Reintjes R, Kistemann T (2002) Spatial patterns of diarrhoeal illnesses with regard to water supply structures–a GIS analysis. Int J Hyg Environ Health 205: 183–191. 10.1078/1438-4639-00151.
28. Tuite AR, Tien J, Eisenberg M, Earn DJ, Ma J, et al. (2011) Cholera epidemic in haiti, 2010: Using a transmission model to explain spatial spread of disease and identify optimal control interventions. Ann Intern Med 154: 593–601. 10.1059/0003-4819-154-9-201105030-00334.
29. R_Core_Team R (2013) R: A language and environment for statistical computing. Vienna, Austria.
30. Hosmer DWJ, Lemeshow S, Strudivant RX (2013) Applied logistic regression. 3rd edition, Wiley.
31. Antikainen J, Kantele A, Pakkanen SH, Laaveri T, Riutta J, et al. (2013) A quantitative polymerase chain reaction assay for rapid detection of 9 pathogens directly from stools of travelers with diarrhea. Clin Gastroenterol Hepatol 11: 1300–1307.e3. 10.1016/j.cgh.2013.03.037.
32. Antikainen J, Tarkka E, Haukka K, Siitonen A, Vaara M, et al. (2009) New 16-plex PCR method for rapid detection of diarrheagenic *Escherichia coli* directly from stool samples. Eur J Clin Microbiol Infect Dis 28: 899–908. 10.1007/s10096-009-0720-x.
33. Summa M, von Bonsdorff CH, Maunula L (2012) Pet dogs–a transmission route for human noroviruses? J Clin Virol 53: 244–247. 10.1016/j.jcv.2011.12.014.
34. Ronnqvist M, Ratto M, Tuominen P, Salo S, Maunula L (2013) Swabs as a tool for monitoring the presence of norovirus on environmental surfaces in the food industry. J Food Prot 76: 1421–1428. 10.4315/0362-028X.JFP-12-371.
35. Oka T, Katayama K, Hansman GS, Kageyama T, Ogawa S, et al. (2006) Detection of human sapovirus by real-time reverse transcription-polymerase chain reaction. J Med Virol 78: 1347–1353. 10.1002/jmv.20699.
36. van Maarseveen NM, Wessels E, de Brouwer CS, Vossen AC, Claas EC (2010) Diagnosis of viral gastroenteritis by simultaneous detection of adenovirus group F, astrovirus, rotavirus group A, norovirus genogroups I and II, and sapovirus in two internally controlled multiplex real-time PCR assays. J Clin Virol 49: 205–210. 10.1016/j.jcv.2010.07.019.

37. Reuter G, Zimsek-Mijovski J, Poljsak-Prijatelj M, Di Bartolo I, Ruggeri FM, et al. (2010) Incidence, diversity, and molecular epidemiology of sapoviruses in swine across Europe. J Clin Microbiol 48: 363–368. 10.1128/JCM.01279-09.
38. Rimhanen-Finne R, Hanninen ML, Vuento R, Laine J, Jokiranta TS, et al. (2010) Contaminated water caused the first outbreak of giardiasis in Finland, 2007: A descriptive study. Scand J Infect Dis 42: 613–619. 10.3109/00365541003774608.
39. Revez J, Huuskonen M, Ruusunen M, Lindstrom M, Hanninen ML (2013) Arcobacter species and their pulsed-field gel electrophoresis genotypes in Finnish raw milk during summer 2011. J Food Prot 76: 1630–1632. 10.4315/0362-028X.JFP-13-083.
40. Merga JY, Leatherbarrow AJ, Winstanley C, Bennett M, Hart CA, et al. (2011) Comparison of *Arcobacter* isolation methods, and diversity of *Arcobacter* spp. in Cheshire, United Kingdom. Appl Environ Microbiol 77: 1646–1650. 10.1128/AEM.01964-10.
41. Houf K, Tutenel A, De Zutter L, Van Hoof J, Vandamme P (2000) Development of a multiplex PCR assay for the simultaneous detection and identification of *Arcobacter butzleri*, *Arcobacter cryaerophilus* and *Arcobacter skirrowii*. FEMS Microbiol Lett 193: 89–94.
42. Harmon KM, Wesley IV (1996) Identification of *Arcobacter* isolates by PCR. Lett Appl Microbiol 23: 241–244.
43. European U (1998) Council directive 98/83/EC of 3rd November on the quality of water intended for human consumption. Official Journal of European Communities L330: 32–54.
44. Polaczyk AL, Narayanan J, Cromeans TL, Hahn D, Roberts JM, et al. (2008) Ultrafiltration-based techniques for rapid and simultaneous concentration of multiple microbe classes from 100-L tap water samples. J Microbiol Methods 73: 92–99. 10.1016/j.mimet.2008.02.014.
45. ISO_9308-2 A (2012) Water quality - enumeration of Escherichia coli and coliform bacteria – part 2: Most probable number method. International Organization for Standardization, Geneva, Switzerland.
46. Pitkänen T, Miettinen IT, Nakari UM, Takkinen J, Nieminen K, et al. (2008) Faecal contamination of a municipal drinking water distribution system in association with *Campylobacter jejuni* infections. J Water Health 6: 365–376.
47. ISO/CD 6461-2 (2002) Water quality – Detection and enumeration of the spores of sulfite-reducing anaerobes (clostridia) – part 2: Method by membrane filtration. Revised committee draft. International Organization for Standardization, Geneva, Switzerland.
48. Pradhan SK, Kauppinen A, Martikainen K, Pitkänen T, Kusnetsov J, et al. (2013) Microbial reduction in waste water treatment using Fe^{3+} and Al^{3+} coagulants and PAA disinfectant. J Water Health 11: 581–589.
49. NMKL 164 A (2005) *Escherichia coli* O157. Detection in food and feeding stuffs, Nordic committee on food analysis.
50. Kauppinen A, Ikonen J, Pursiainen A, Pitkänen T, Miettinen IT (2012) Decontamination of a drinking water pipeline system contaminated with adenovirus and *Escherichia coli* utilizing peracetic acid and chlorine. J Water Health 10: 406–418. 10.2166/wh.2012.003.
51. Kauppinen A, Martikainen K, Matikka V, Veijalainen AM, Pitkänen T, et al. (2014) Sand filters for removal of microbes and nutrients from wastewater during a one-year pilot study in a cold temperate climate. J Environ Manage 133: 206–213. 10.1016/j.jenvman.2013.12.008.
52. ISO_15553 (2006) Water quality - isolation and identification of *Cryptosporidium* oocysts and *Giardia* cysts from water. International Organization for Standardization, Geneva, Switzerland.
53. Guy RA, Payment P, Krull UJ, Horgen PA (2003) Real-time PCR for quantification of *Giardia* and *Cryptosporidium* in environmental water samples and sewage. Appl Environ Microbiol 69: 5178–5185.
54. ISO 7393 (1985) Water quality – Determination of free chlorine and total chlorine. International Organization for Standardization, Geneva, Switzerland.
55. Nakari UM, Hakkinen M, Siitonen A (2011) Identification of persistent subtypes of *Campylobacter jejuni* by pulsed-field gel electrophoresis in Finland. Foodborne Pathog Dis 8: 1143–1145. 10.1089/fpd.2011.0882.
56. Siefring S, Varma M, Atikovic E, Wymer L, Haugland RA (2008) Improved real-time PCR assays for the detection of fecal indicator bacteria in surface waters with different instrument and reagent systems. J Water Health 6: 225–237. 10.2166/wh.2008.022.
57. Kildare BJ, Leutenegger CM, McSwain BS, Bambic DG, Rajal VB, et al. (2007) 16S rRNA-based assays for quantitative detection of universal, human-, cow-, and dog-specific fecal *Bacteroidales*: A bayesian approach. Water Res 41: 3701–3715. 10.1016/j.watres.2007.06.037.
58. Caporaso JG, Lauber CL, Walters WA, Berg-Lyons D, Lozupone CA, et al. (2011) Global patterns of 16S rRNA diversity at a depth of millions of sequences per sample. Proc Natl Acad Sci U S A 108 Suppl 1: 4516–4522. 10.1073/pnas.1000080107.
59. Schloss PD, Westcott SL, Ryabin T, Hall JR, Hartmann M, et al. (2009) Introducing mothur: Open-source, platform-independent, community-supported software for describing and comparing microbial communities. Appl Environ Microbiol 75: 7537–7541. 10.1128/AEM.01541-09.
60. Kozich JJ, Westcott SL, Baxter NT, Highlander SK, Schloss PD (2013) Development of a dual-index sequencing strategy and curation pipeline for analyzing amplicon sequence data on the MiSeq illumina sequencing platform. Appl Environ Microbiol 79: 5112–5120. 10.1128/AEM.01043-13.
61. Cole JR, Wang Q, Cardenas E, Fish J, Chai B, et al. (2009) The ribosomal database project: Improved alignments and new tools for rRNA analysis. Nucleic Acids Res 37: D141–5. 10.1093/nar/gkn879.
62. Tamura K, Peterson D, Peterson N, Stecher G, Nei M, et al. (2011) MEGA5: Molecular evolutionary genetics analysis using maximum likelihood, evolutionary distance, and maximum parsimony methods. Mol Biol Evol 28: 2731–2739. 10.1093/molbev/msr121.
63. Tamura K, Nei M, Kumar S (2004) Prospects for inferring very large phylogenies by using the neighbor-joining method. Proc Natl Acad Sci U S A 101: 11030–11035. 10.1073/pnas.0404206101.
64. ter Waarbeek HL, Dukers-Muijrers NH, Vennema H, Hoebe CJ (2010) Waterborne gastroenteritis outbreak at a scouting camp caused by two norovirus genogroups: GI and GII. J Clin Virol 47: 268–272. 10.1016/j.jcv.2009.12.002 [doi].
65. Teunis P, Van den Brandhof W, Nauta M, Wagenaar J, Van den Kerkhof H, et al. (2005) A reconsideration of the *Campylobacter* dose-response relation. Epidemiol Infect 133: 583–592.
66. Volken T (2013) Second-stage non-response in the swiss health survey: Determinants and bias in outcomes. BMC Public Health 13: 167-2458-13-167. 10.1186/1471-2458-13-167 [doi].
67. Johnson TP, Wislar JS (2012) Response rates and nonresponse errors in surveys. JAMA 307: 1805–1806. 10.1001/jama.2012.3532 [doi].
68. Martikainen P, Laaksonen M, Piha K, Lallukka T (2007) Does survey non-response bias the association between occupational social class and health? Scand J Public Health 35: 212–215. 10.1080/14034940600996563.
69. Miettinen IT, Pitkänen T, Nakari UM, Hakkinen M, Wermudsen K, et al. (2006) 5th nordic drinking water conference, 8.-10. June, 2006.
70. Richardson G, Thomas DR, Smith RM, Nehaul L, Ribeiro CD, et al. (2007) A community outbreak of *Campylobacter jejuni* infection from a chlorinated public water supply. Epidemiol Infect 135: 1151–1158. 10.1017/S0950268807007960.
71. Kitajima M, Oka T, Haramoto E, Katayama H, Takeda N, et al. (2010) Detection and genetic analysis of human sapoviruses in river water in Japan. Appl Environ Microbiol 76: 2461–2467. 10.1128/AEM.02739-09.
72. Murray TY, Mans J, Taylor MB (2013) First detection of human sapoviruses in river water in South Africa. Water Sci Technol 67: 2776–2783. 10.2166/wst.2013.203.
73. Mikula C, Springer B, Reichart S, Bierbacher K, Lichtenschopf A, et al. (2010) Sapovirus in adults in rehabilitation center, upper austria. Emerg Infect Dis 16: 1186–1187. 10.3201/eid1607.091789.
74. Lee LE, Cebelinski EA, Fuller C, Keene WE, Smith K, et al. (2012) Sapovirus outbreaks in long-term care facilities, Oregon and Minnesota, USA, 2002–2009. Emerg Infect Dis 18: 873–876. 10.3201/eid1805.111843.
75. Vandamme P, Pot B, Gillis M, de Vos P, Kersters K, et al. (1996) Polyphasic taxonomy, a consensus approach to bacterial systematics. Microbiol Rev 60: 407–438.
76. Collado L, Figueras MJ (2011) Taxonomy, epidemiology, and clinical relevance of the genus *Arcobacter*. Clin Microbiol Rev 24: 174–192. 10.1128/CMR.00034-10.
77. Lautenschlager K, Hwang C, Liu WT, Boon N, Koster O, et al. (2013) A microbiology-based multi-parametric approach towards assessing biological stability in drinking water distribution networks. Water Res 47: 3015–3025. 10.1016/j.watres.2013.03.002.
78. Pinto AJ, Xi C, Raskin L (2012) Bacterial community structure in the drinking water microbiome is governed by filtration processes. Environ Sci Technol 46: 8851–8859. 10.1021/es302042t.
79. Cox MJ, Cookson WO, Moffatt MF (2013) Sequencing the human microbiome in health and disease. Hum Mol Genet 22: R88–94. 10.1093/hmg/ddt398.

13

Abundances of Tetracycline, Sulphonamide and Beta-Lactam Antibiotic Resistance Genes in Conventional Wastewater Treatment Plants (WWTPs) with Different Waste Load

Mailis Laht[1], Antti Karkman[2], Veiko Voolaid[1], Christian Ritz[3], Tanel Tenson[1], Marko Virta[2], Veljo Kisand[1]*

1 Institute of Technology, University of Tartu, Tartu, Estonia, **2** Department of Food and Environmental Sciences, University of Helsinki, Helsinki, Finland, **3** Department of Nutrition, Exercise and Sports, Faculty of Science, University of Copenhagen, Copenhagen, Denmark

Abstract

Antibiotics and antibiotic resistant bacteria enter wastewater treatment plants (WWTPs), an environment where resistance genes can potentially spread and exchange between microbes. Several antibiotic resistance genes (ARGs) were quantified using qPCR in three WWTPs of decreasing capacity located in Helsinki, Tallinn, and Tartu, respectively: sulphonamide resistance genes (*sul1* and *sul2*), tetracycline resistance genes (*tetM* and *tetC*), and resistance genes for extended spectrum beta-lactams (bla_{oxa-58}, bla_{shv-34}, and $bla_{ctx-m-32}$). To avoid inconsistencies among qPCR assays we normalised the ARG abundances with 16S rRNA gene abundances while assessing if the respective genes increased or decreased during treatment. ARGs were detected in most samples; *sul1*, *sul2*, and *tetM* were detected in all samples. Statistically significant differences (adjusted $p<0.01$) between the inflow and effluent were detected in only four cases. Effluent values for bla_{oxa-58} and *tetC* decreased in the two larger plants while *tetM* decreased in the medium-sized plant. Only bla_{shv-34} increased in the effluent from the medium-sized plant. In all other cases the purification process caused no significant change in the relative abundance of resistance genes, while the raw abundances fell by several orders of magnitude. Standard water quality variables (biological oxygen demand, total phosphorus and nitrogen, etc.) were weakly related or unrelated to the relative abundance of resistance genes. Based on our results we conclude that there is neither considerable enrichment nor purification of antibiotic resistance genes in studied conventional WWTPs.

Editor: Jorge M. O. Fernandes, University of Nordland, Norway

Funding: The research was funded by the Academy of Finland and EnSTe graduate school, and the European Regional Development Fund through the Centre of Excellence in Chemical Biology in Estonia. Additional funding was provided by COST Action TD0803. The funders had no role in study design, data collection and analysis, decision to publish, or preparation of the manuscript.

Competing Interests: The authors have declared that no competing interests exist.

* Email: kisand@ut.ee

These authors contributed equally to this work.

Introduction

Antibiotic resistance (AR) has become a worldwide problem, making infectious diseases more resilient thus making treatment more difficult and costly [1]. AR is not confined to hospital environments, and is able to spread between both human dominated and natural environments. Increased amounts of antibiotic residues [2–4] and antibiotic resistant bacteria (ARB) are found in human-related environments such as agricultural settings (e.g. farms, soil etc.) [5,6] and surface-, drinking- and wastewaters [7–11]. Wastewater treatment plants (WWTPs) receive sewage from various sources, including hospitals and households which are both important sources of antibiotics and their residues [12–14]. and antibiotic resistant bacteria (ARB) [15–17]. The presence of antibiotics and antibiotic residues [18]. ARB, and antibiotic resistance genes (ARG) have been confirmed in many WWTPs [19–23]. Bacteria from various environments, including human, soil, and activated sludge, are mixed in WWTPs and therefore these facilities are considered to be important "hot-spots" for AR and spread of resistance genes [18,24–26]. The presence of antibiotics, ARB, and ARG in the same setting creates an environment that selects for AR and provides an opportunity for genetic material housing ARGs to transfer between bacterial species via horizontal gene transfer [13,21,24,25,27,28]. In a metagenomic study of plasmids it was shown that numerous medically relevant ARG can be found in WWTPs (140 ARGs in a single WWTP and 123 in the effluent water) [29]. Therefore, there is a concern that resistance genes will spread in the bacterial population and further into more natural environments less impacted by human activity [30–33].

Many studies of antibiotic resistant bacteria and resistance genes have used culture-based assays, which are biased towards specific cultivable pathogenic or environmental species [17,19,21,23,34]. Culture-dependent data does not reflect the real variability or the actual amount of resistance genes present in a given WWTP, so these studies normally characterize only a small subset of the total population [35,36]. Surprisingly, only a limited number of studies have used quantitative methods to investigate resistance genes in the total communities in effluent waters [21,33,37–40]. The small number of quantitative studies could be one reason why the overall impact of WWTPs in spreading resistance to the environment has not yet been properly evaluated.

The current study was designed to investigate the role of three conventional WWTPs in the distribution of ARGs using culture-independent quantitative methods. We thus focus on quantitative whole community level measurements on the water-phase of the inflow and effluent of WWTPs which has not been studied using quantitative methods in sufficient detail to assess the impact of WWTPs on the distribution of ARGs. To quantify the number of ARGs we analysed the total DNA from wastewater and effluent samples using quantitative real-time PCR (qPCR). Our original hypothesis was that conventional WWTPs increase the relative abundance of ARGs during processing because they do not employ technologies that target the removal of genetic elements.

Materials and Methods

1.1. Sample collection

Both raw wastewater and final effluent water samples were collected from three city WWTPs of decreasing capacity located in the Baltic Sea catchment area: large (Helsinki, Finland), medium (Tallinn, Estonia), and small (Tartu, Estonia). No specific permissions were required for sampling these locations, and the sampling was carried out in collaboration with each WWTP staff. The wastewater treatment technology employed in these three WWTPs is similar and typical of other facilities located in both Nordic (Finland, Sweden, Norway) and Eastern European (Estonia, Latvia, Lithuania, Poland) countries (EEA 2013). The WWTP in Helsinki is both the largest wastewater treatment plant in Finland and all Nordic countries; with about 0.8 million residents in the Helsinki metropolitan area. The majority of wastewater in Estonia is produced in Tallinn (~350 000 residents) and Tartu (~100 000). To abide by European law (91/271/EEC), the main steps of treatment are: primary treatment - mechanical treatment steps (sand, grit, fat and grease removal, pre-sedimentation); secondary treatment - biological treatment (activated sludge); followed by tertiary treatment - final deep purification using a combination of methods (chemical, mechanical and biological as in secondary sedimentation, bio-filters etc.). The WWTP effluent in both Helsinki and Tallinn is directed into the Baltic Sea while the WWTP effluent in Tartu flows into the Emajõgi River which forms part of the Baltic Sea catchment area. The main steps in the WWTP process and water-phase treatment are presented in Figure S1 in File S1.

Samples were collected over a one year period from December 2010 to December 2011 at five different time points, each representing a different season (four seasons; winter was sampled twice, Table S1A in File S1). In each sampling period three consecutive samples were taken on separate days at 1–3 day intervals; the exact sampling dates and monitored variables are given in Table S1 in File S1. Composite samples were taken, collected over 24 h periods, except in Tartu in the 2011 winter season when a grab sample was taken owing to technical problems (the automatic sampler was frozen because of extremely low temperatures).

1.2. Collection of total microbial community

The water samples were stored at 4°C pending filtration (within a couple of hours). Ten ml of influent water and 100 ml of effluent water were filtered through polycarbonate filters (pore size 0.22 µm, diameter 47 mm, GE Water & Process Technologies). For the last two time points at the Tallinn WWTP 200 ml of effluent water was filtered because a new treatment step (bio-filter) was added in September 2011.

1.3. DNA extraction

For the first three time points (Dec 2010; March 2011; June 2011) from the Helsinki WWTP, DNA was extracted from the samples using a MoBio PowerWater DNA isolation kit (MoBio Laboratories, Inc., CA, USA). For all other samples, the DNA was extracted using the modified bead beating and silica-membrane method (nucleic acid binding on to silica particles [41]. Method in brief: for lysis: 400 µl TE+50 µl lysozyme (50 mg/ml from egg yolk) was added to the filter and incubated at 37°C for 15 min. Fifty µl of 10% SDS and 500 µl lysis buffer BQ1 (NucleoSpin, Macherey-Nagel)+zirconium beads (0.1 mm diameter; burned at 500°C) were added and the samples were processed by a 5 min beating on a bead beater (Biospec products, Minibead beater). Thereafter, proteinase K was added followed by 15 min at 65°C with constant shaking. Five hundred µl of 96% ethanol was added and the whole volume was applied to commercial silica-membrane columns (NucleoSpin Macherey-Nagel). Finally, the total DNA was recovered in 50 µl of elution buffer BE (NucleoSpin Macherey-Nagel, 5 mM Tris/HCl pH 8.5). The concentration of extracted DNA was measured using a NanoDrop Spectrophotometer ND-1000 (absorption readings at 260 nm). The extracted DNA was stored at −20°C pending further analysis.

1.4. Detection and quantification of ARG copy number by qPCR

Seven resistance genes were surveyed: *sul1, sul2, tetM, tetC,* $bla_{shv\text{-}34}$, $bla_{ctx\text{-}m\text{-}32}$, and $bla_{oxa\text{-}58}$. For the first three time point samples (Dec 2010; March 2011; June 2011) from the Helsinki WWTP (Assay 1), qPCR was performed using a Dynamo Flash SYBR Green qPCR kit (Thermo Scientific, Lithuania) and a 7300 Real-Time PCR system (Applied Biosystems, Foster City, CA, USA). The thermal cycling conditions were as follows: 95°C for 7 min, 40 cycles at 95°C for 10 s and Tm for 30 s. A melting curve was obtained to confirm specificity of amplification. Reactions were conducted in 10 µl volumes on 96-well plates containing 1×Dynamo Flash SYBR Green master mix, 0.3 µM of each primer and 1×ROX passive reference dye. Template DNA was used in qPCR reactions in the range 2–12 ng DNA per reaction; a fixed dilution of raw DNA exctract was used. In parallel with the ARGs, the16S rRNA gene copy numbers were quantified.

The qPCR for detecting 16S rRNA gene and ARGs for all the other samples (Assay 2) used the 7900HT Fast Real-Time PCR System (Applied Biosystems). Reactions were conducted in 10 µl volumes on 384-well plates containing 1×Maxima SYBR Green/ROX qPCR Master Mix (Thermo Scientific), 0.4 µM of each primer. The two-step thermal cycling conditions for detecting 16S rRNA gene were as follows: 95°C for 10 min, 40 cycles of 95°C 15 s, 60°C 1 min. For ARGs the thermal cycling conditions were as follows: 95°C for 10 min, 40 cycles at 95°C for 15 s, Tm °C for 30 s and 72°C for 30 s. Melting curves were obtained to confirm specificity of amplification. Template DNA was used in the qPCR reaction in the range 3×10^{-4} to 22.4 ng DNA per reaction. This was obtained by a wider range of dilutions (to avoid inhibition) of the raw DNA extraction product, three levels of 10 fold dilutions

Table 1. Primers used for detecting the target genes and the melting temperatures (Tm) used for primers.

Gene name	Primers					
	Forward	Reverse	Tm °C	Ref.	Amplification efficiency	
					Assay 1	Assay 2
16S rRNA gene	5′-AGA GTT TGA TCC TGG CTC AG-3′			[60]		
		5′-CTG CTG CSY CCC GTA GGA-3′	60	[61] modif.		82–100
		5′-CTG CTG CCT CCC GTA GG-3′	60	[61]	87	
tetC	5′-TGC GTT GAT GCA ATT TCT ATG C-3′	5′-GGC GCC TAC AAT CCA TG-3′	64	[62]	80	93–101
tetM	5′-GCA ATT CTA CTG ATT TCT GC-3′	5′-CTG TTT GAT TAC AAT TTC CGC-3′	60	[61]	90–93	89–108
sul1	5′-CGG CGT GGG CTA CCT GAA CG-3′	5′-GCC GAT CGC GTG AAG TTC CG-3′	64	[63]	90–93	81–88
sul2	5′-GCG CTC AAG GCA GAT GGC ATT-3′	5′-GCG TTT GAT ACC GGC ACC CGT-3′	64	[62]	102	91–103
blactx-m-32	5′-CGT CAC GCT GTT GTT AGG AA-3′	5′-CGC TCA TCA GCA CGA TAA AG-3′	64	[59]	87	89–101
blashv-34	5′-GCG TTA TTT TCG CCT GTG TA-3′	5′-AGG TGC TCA TCA TGG GAA AG-3′	60	[63]	92–94	97–108
blaoxa-58	5′-GCA ATT GCC TTT TAA ACC TGA-3′	5′-CTG CCT TTT CAA CAA AAC CC-3′	60	[63]	90	97–111

qPCR amplification efficiency is given for 16S RNA gene and for ARGs, R^2 of the linear range of standards was always >0.99.

were used. Primers and annealing temperature, Tm (°C), are given in Table 1. Number of technical replicates in both qPCR assays was 3.

1.5. Standards used for quantification

A plasmid vector and fragments of ARGs were constructed and used as standards for quantifying the raw qPCR results. The standard plasmids were checked for the correct inserts by sequencing. In Assay 1 for the three Helsinki time points (Dec 2010; March 2011 and June 2011) the 16S rRNA gene quantification standard was genomic DNA from *E. coli* K12 (genome size 4.6 Mbp with seven copies of the rRNA operon). In Assay 2, used for all other samples, the standard for quantifying 16S rRNA gene was constructed from a 16S rRNA gene fragment from the natural aquatic *Chryseobacterium* strain isolated from Emajõgi River, which receives WWTP effluent water, and was cloned into a plasmid and validated by sequencing. Information about the plasmids used and Genbank accession numbers are given in Table 2.

1.6. Quantification and normalisation of ARGs

Standard curves (Ct per log copy number) for 16S rRNA gene and ARG quantification were obtained for each run using the plasmid constructs (or genomic *E. coli* DNA for 16S rRNA gene in Assay 1) (Table 2) in ten-fold serial dilutions. The gene copy number of a standard was determined from the plasmid/genomic DNA concentration (measured using a NanoDrop in Assay 1 or by fluorescent staining with PicoGreen (Invitrogen) and using VICTOR X3 Multilabel Plate Readers (Perkin-Elmer) in Assay 2). The ARG levels in the sample were calculated using the standard curve equation and measured Ct value, the quality control of raw Ct values for standard curve and unknown samples was done before further analysis. The limit of quantification (LOQ) was defined as the lowest point of the linear part of standard curve: Assay 1, ARGs 100 and 16S rRNA gene 1000 gene copy number per reaction; Assay 2, ARGs and 16S rRNA gene 100, except *sul1* with 10 copy numbers per reaction. The negative control was the reaction mix with nuclease-free water instead of the template DNA. The negative control had always a Ct value at least 3.3 cycles lower than the smallest standard used

Table 2. Plasmids and PCR fragments used as standards.

Gene name	Standard constructs	Accession number	Reference
16S rRNA gene	**Assay 1**		
	Genomic DNA from *E. coli* K12 genome size 4.6 Mbp with 7 copies of rRNA operon		
	Assay 2		
	PCR product cloned in plasmid PGEM-T Easy Vector System (Promega)	KF737394	present study
tetC	pDrive (Qiagen)		[62]
tetM	pDrive (Qiagen)		[62]
sul1	R388		[64]
sul2	RSF1010		[64]
$bla_{ctx\text{-}m\text{-}32}$	pUC19	KF737395	present study
$bla_{shv\text{-}34}$	PGEM-T Easy Vector System (Promega)	KF737397	present study
$bla_{oxa\text{-}58}$	pUC19	KF737396	present study

for calculation of LOQ. Technical replicates were incorporated in the statistical analysis to reflect differences in quantification (see below for details).

1.7. Water quality variables

WWTPs in Europe analyse water quality parameters according to the EU directive for urban wastewater treatment 91/271/ECC. These parameters are monitored regularly in accredited laboratories according to standard methods. The parameters used in this study were: biochemical oxygen demand (BOD_7) - standard method EN 1899-2; total suspended solids (SS) - EN 872; total phosphorus (Ptot) – EN ISO 6878; total nitrogen (Ntot) – EN ISO 11905. In addition, the automated measurements of flow rate and water temperature in the process were recorded. We received the data for water variables from the staff of each WWTP.

1.8. Statistics

Linear mixed models were fitted to the data using the functionality of the package lme4 [42]. the statistical software used was R version 3.0.1 [43]. Average levels of gene copy numbers (16S rRNA gene or ARGs) were modelled using the location of the WWTP (Helsinki, Tallinn, Tartu), sample source (IF/EF) and method protocol (Assay1/Assay2) as fixed effects and technical and biological replicates as random effects. Significance of fixed effects was assessed by an F-test using a (pre-specified) significance level of 1%. In addition, the combined effect of each of the environmental variables BOD_7, SS, Ptot, Ntot, flow rate (i.e water discharge - WD) and temperature in inflowing versus effluent waste on gene copy numbers was described by a linear mixed model version of an analysis of covariance model. Model checking was based on residual plots and normal probability plots using the raw residuals. Models were reduced using the likelihood ratio test. A 1% or 5% significance level was used. Pairwise comparisons were evaluated based on adjusted p-values obtained using the single-step method [44]. Values below LOQ were not included in the analyses.

Principal component analysis (PCA) and its extensions to between groups (BGA) and within groups (WGA) analyses (ade4 package in R) was used to analyse the grouping of inflowing/effluent samples by monitored water quality variables (nutrients, suspended solids, biological oxygen demand, water discharge and temperature).

Results and Discussion

2.1. Selection of ARGs for the study

Initially, we made a screening for 25 selected genes that could spread with higher probability from any WWTPs into environment. Initial selection based on four major criteria: (i) clinically relevant genes (risk to human health) previously detected in WWTPs (e.g. [29]); (ii) genes found in various mobile elements, demonstrating their potential for transfer between bacteria [45–48]. (iii) high consumption antibiotics–sulphonamides, tetracyclines, and beta-lactams; (iv) incorporation of long-used antibiotics (tetracycline, sulfonamides) and the newer extended spectrum beta-lactams (carbapenems and 3rd and 4th generation cephalosporins) and their resistance genes. Altogether 12 *bla*, 6 *tet*, 3 *sul*, 3 *qnr* (floroquinolones) and 1 vancomycin resistant gene were tested using traditional PCR. Finally, two resistance genes for sulphonamide (*sul1* and *sul2*), two for tetracycline (*tetM* and *tetC*), and three for extended spectrum beta-lactams (bla_{oxa-58}, bla_{shv-34}, *and* $bla_{ctx-m-32}$) remained in the study based on their abundance and frequency in screening study. It was necessary to choose a small number of relevant target genes because the variety of different ARGs is large. In the Comprehensive Antibiotic Resistance Database [49] (http://arpcard.mcmaster.ca/accessed Sept. 23, 2013), the number of ARGs is 2153 and testing all these genes quantitatively in one study would be prohibitively difficult.

2.2. Abundance of genes (16S rRNA gene and ARGs)

To evaluate the abundance of the total bacterial community, we quantified 16S rRNA gene in the samples using qPCR. The amplification efficiency is given in Table 1. Initially, the raw gene copy numbers were used to estimate the general changes of bacterial levels during wastewater purification. The copy number of 16S rRNA gene was several orders of magnitude lower in the effluent (EF) than the inflow (IF) (Figure 1A, Table S2 in File S1). The differences between the IF and EF samples were statistically significant in all cities (adjusted $p<0.01$) (Figure 1A).

The raw gene copy numbers of ARGs/ml decreased during processing in the WWTP water phase. The levels of ARGs detected in the EF were lower than IF in all three plants (Figure 1B, Table S2 in File S1). The decrease of abundance from IF to EF was statistically significant ($p<0.01$) for all ARGs in the large (Helsinki) WWTP. In the medium (Tallinn) WWTP, for all ARGs except bla_{shv-34} ($p>0.05$), the decrease was statistically significant ($p<0.01$). In the small WWTP (Tartu), the raw abundance of ARGs decreased after purification but the decrease was statistically significant only for $bla_{ctx-m-32}$ and *tetM* ($p<0.01$), and weakly significant for bla_{oxa-58} ($p = 0.03$). Earlier studies have demonstrated the large variation of treatment plant efficiency in removing microorganisms [50] and micropollutants [51]. which also depends on the capacity of the WWTP plant. Low capacity treatment facilities are more vulnerable to changes in inflowing wastewater composition and flow rates. In addition, the WWTP in Tartu did not have biological post-filtration at the time of sampling.

ARGs are surprisingly rarely quantified directly using community DNA in both IF and EF water samples from WWTPs. As in our study (Figure 1B), a few other studies have found from 100 to 1000 fold reductions of raw ARG copy numbers during the purification process; e.g. *sul1, tetW*; [52]; *tetC, tetA*; [53]. *tetG, tetQ* [54]. In one study, the relative abundance of *sul1* increased, while *sul2* decreased slightly in WWTP effluent [31]. In our study, resistance genes for "older" antibiotics (with exception of *tetC*) were more commonly detected. *sul1*, *sul2,* and *tetM* were present above LOQ in all sites and samples (Table 3 and Figure S2 in File S1). High abundances of various tetracycline resistance genes and sulphonamide resistance genes were also demonstrated in other studies [21,23]. Quantitative studies that target resistance to the newer beta-lactams in community DNA in WWTPs are almost completely absent–only bla_{TEM} by Lachmayr et al. [38]. in addition, *bla* genes were quantified in the river water under the influences of wastewater but not directly in the WWTP effluent [55].

2.3. Normalised/Relative abundances of ARGs

To avoid inconsistencies among qPCR assays, including suboptimal efficiency in some cases, we used 16S rRNA gene-normalised values, and the different protocol (Assay1/Assay2) was added as an additional fixed effect into the statistical models. This type of data analysis allows one to quantify the relative changes in ARG abundances, whether more or fewer ARGs appear per microbial genome. When relative abundances were compared, statistically significant ($p<0.01$) differences between IF and EF were detected in only four cases (Figure 2). For bla_{oxa-58} and *tet*C in the Helsinki plant we observed a relative decrease after purification. In addition, *tet*M decreased in the Tallinn WWTP EF

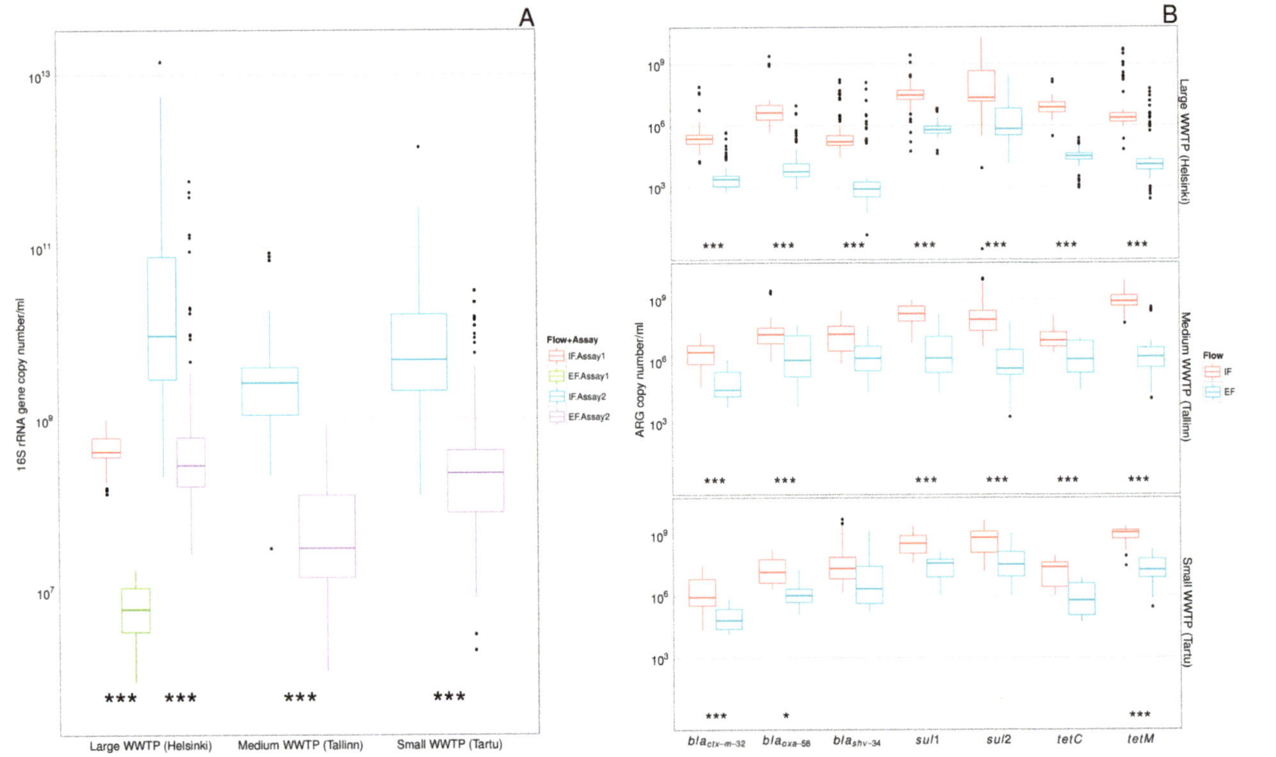

Figure 1. Raw gene copy numbers detected in a WWTP sample (copy number/ml). A - 16S rRNA gene in inflow (IF) and effluent (EF). Assay1 was used only for samples from large WWTP (Helsinki) from Winter 2010 to Autumn 2011; B–Antibiotic resistance genes (ARGs). Statistical significance between inflow wastewater and effluent samples: *** - $p<0.01$; *$0.03>p>0.01$. For the pairs not marked the statistical difference between inflow and outflow was statistically insignificant. The line in each box marks the median and boxes: 25th and 75th percentiles; whiskers: 5th and 95th percentiles and outliers ±1.5 * IQR. See Figure S2 in File S1 for abundances of same genes presented by each sampling event.

samples. The only increase in EF was observed for bla_{shv-34} in the Tallinn WWTP. In all other cases the purification process had no significant effect on the relative abundances of resistance genes.

We conclude from our study that there is neither considerable enrichment (selection) nor purification of ARGs during processes in WWTPs (Figure 2) at the whole community level. Effective selection would be assumed when there are appropriate conditions, i.e. increased concentration of ABs occur. In two WWTPs studied, the measured levels of AB concentrations of several compounds were very low but measurable compared to highly labile beta-lactams [56]. This suggests that conditions could favour enrichment of at least *tet* and *sul* genes within studied WWTPs. Although, quantitative enrichment of ARGs (*sul*) responsible for resistance against refractory ABs with longer history of usage has

Table 3. Detection of ARGs in different WWTPs (total of all analyses per gene, n = 15).

ARG	City	IF %(number) detected	EF %(number) detected
tetC	Helsinki	93 (14)	80 (12)
	Tallinn	67 (10)	27 (4)
	Tartu	67 (10)	73 (11)
***bla*$_{oxa-58}$**	Helsinki	100 (15)	87 (13)
	Tallinn	100 (15)	47 (7)
	Tartu	87 (13)	80 (12)
***bla*$_{shv-34}$**	Helsinki	100 (15)	100 (15)
	Tallinn	87 (13)	87 (13)
	Tartu	100 (15)	100 (15)
***bla*$_{ctx-m-32}$**	Helsinki	100 (15)	100 (15)
	Tallinn	87 (13)	40 (6)
	Tartu	80 (12)	47 (7)

Only genes that were sometimes not detected are given. sul1, sul2 and tetM were detected 100% in all IF and EF samples from the WWTPs.

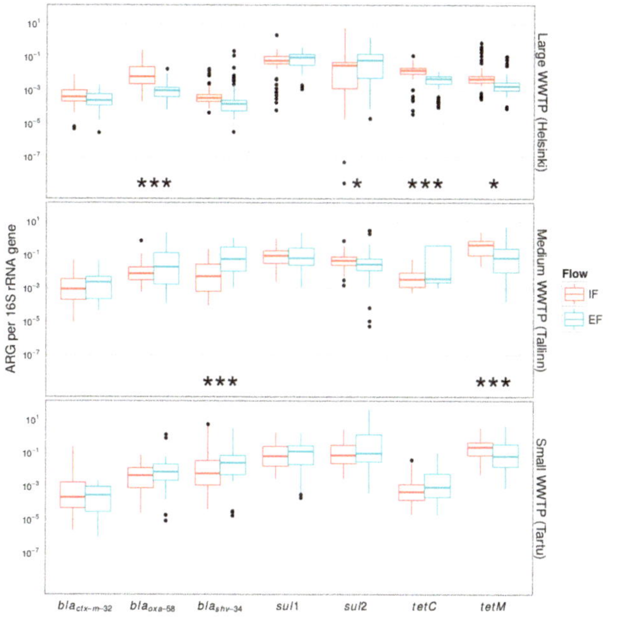

Figure 2. Normalised ARG abundances. Antibiotic resistance gene copy numbers normalised to 16S rRNA gene copy numbers. The results are given for all samples for one gene for a WWTP, no seasonal comparison. Statistically significant comparison results are marked with *** at $p<0.01$, $*0.03>p>0.01$. The line in each box marks the median and boxes: 25th and 75th percentiles; whiskers: 5th and 95th percentiles and outliers ±1.5 * IQR.

been demonstrated in some studies [31]. reduction or no change has been observed in most studies [21,23,40]. To date, only one study demonstrated similar case of positive selection for newer ARG possessing organisms in a WWTP, which suggests that bacteria harbouring bla_{TEM} are released more from effluent water compared to wastewater [38].

2.4. Abundance of ARGs and treatment efficiency of wastewater

In EU countries, treatment efficiency of WWTPs is estimated by monitoring a few water quality measures, according to European directive 271/1991/EC. The compulsory parameters monitored are total nitrogen (Ntot) and phosphorus (Ptot), Biological Oxygen Demand (BOD_7), and suspended solids (SS). In addition, a few generic background parameters are measured in all WWTPs e.g. Water Discharge (WD) and temperature. Such water quality parameters are good for evaluating wastewater purification in the traditional sense–removal of excess nutrients, labile organic compounds etc. Obviously, the wastewater was purified of excess nutrients and organic compounds in the WWTPs studied because all parameters were up to an order of magnitude lower in EF than IF. We observed a change of between 6 to 45 fold depending on the parameter. Averages in IF: BOD_7 - 231; P_{tot} 8; N_{tot} 51 and SS 306 mg/l and in EF: BOD_7 5; P_{tot} 0.6; N_{tot} 8 and SS 8 mg/l (Table S1B in File S1). Volumetric concentrations of nutrients, SS and labile organic compounds were higher in the IF of the small (Tartu) than in medium (Tallinn) and large (Helsinki) WWTP (Figure S4 in File S1; IF samples are strongly associated with these variables, permutation test, 1000 replicates, $p<0.01$). This could be caused by shorter solid retention time in smaller plants [57]. At the same time, the efficiency of purification in the traditional sense did not differ dramatically among plants (Figure S3 in FileS1; residual differences among WWTPs disappear after decomposing the IF/EF level differences, permutation test, 1000 replicates, $p>0.05$). A relationship between the change of ARG abundances and the efficiency of nutrient removal and temperature has been reported previously (e.g. [58,59]). These studies suggest that changes in ARG abundance could depend on processes and conditions in the WWTP. However, none of these parameters are designed to estimate threats associated with the spread of either ARB or ARGs. Indeed, in our study, temperature, water discharge and concentrations of nutrients did not help in estimating the efficiency of ARG removal. This was demonstrated by the absence of the combined effect of monitored environmental variables and abundances of ARGs (Figure S4 in File S1). Moreover, a new treatment step (installation of biological post-filtration for final effluent treatment mainly for nitrogen removal) was added during the study period in the Tallinn WWTP (July 2011). However, no statistically significant effect was observed on ARG removal after this event. The Helsinki WWTP had the biological post-filtration installed throughout the study period.

In conclusion, these results force us to reject our original hypothesis. All ARGs were detected in most wastewater and effluent samples, however, the conventional WWTPs under study seem not to be important sites for changes in the relative abundance of ARGs at the whole community level: no enrichment in relative abundance was observed. Furthermore, no additional reduction of ARGs occurred; raw abundance changed in proportion to the decrease of bacterial abundance. We conclude that many unknown factors may influence the biological purification processes in conventional WWTPs and the evaluation of their relationship to ARG removal or selection requires more complex case studies.

Acknowledgments

We acknowledge the staff from HSY at Viikinmäki, Tartu Veevärk AS and Tallinna Vesi AS for providing samples, and Dr. David Schryer for language revision.

Author Contributions

Conceived and designed the experiments: TT MV VK. Performed the experiments: ML AK VV. Analyzed the data: ML AK VV CR VK. Contributed reagents/materials/analysis tools: TT MV. Contributed to the writing of the manuscript: ML AK VV CR TT MV VK.

References

1. Levy SB, Marshall B (2004) Antibacterial resistance worldwide: causes, challenges and responses. Nat Med 10: S122–9.
2. Lindberg RH, Wennberg P, Johansson MI, Tysklind M, Andersson BA V (2005) Screening of Human Antibiotic Substances and Determination of Weekly Mass Flows in Five Sewage Treatment Plants in Sweden. Environ Sci Technol 39: 3421–3429.
3. Le-Minh N, Khan SJ, Drewes JE, Stuetz RM (2010) Fate of antibiotics during municipal water recycling treatment processes. Water Res 44: 4295–4323.
4. Novo A, André S, Viana P, Nunes OC, Manaia CM (2013) Antibiotic resistance, antimicrobial residues and bacterial community composition in urban wastewater. Water Res 47: 1875–1887. Available: http://www.sciencedirect.com/science/article/pii/S0043135413000274. Accessed 23 January 2013.

5. Khachatourians GG (1998) Agricultural use of antibiotics and the evolution and transfer of antibiotic-resistant bacteria. Can Med Assoc J 159: 1129–1136.
6. Durso LM, Miller DN, Wienhold BJ (2012) Distribution and Quantification of Antibiotic Resistant Genes and Bacteria across Agricultural and Non-Agricultural Metagenomes. PLoS One 7: e48325.
7. Schwartz T, Kohnen W, Jansen B, Obst U (2003) Detection of antibiotic-resistant bacteria and their resistance genes in wastewater, surface water, and drinking water biofilms. FEMS Microbiol Ecol 43: 325–335.
8. Kümmerer K (2004) Resistance in the environment. J Antimicrob Chemother 54: 311–320.
9. Baquero F, Martínez J-L, Cantón R (2008) Antibiotics and antibiotic resistance in water environments. Curr Opin Biotechnol 19: 260–265.
10. Allen HK, Donato J, Wang HH, Cloud-Hansen KA, Davies J, et al (2010) Call of the wild: antibiotic resistance genes in natural environments. Nat Rev Microbiol 8: 251–259. Available: http://dx.doi.org/10.1038/nrmicro2312. Accessed 21 March 2014.
11. Stoll C, Sidhu JPS, Tiehm A, Toze S (2012) Prevalence of clinically relevant antibiotic resistance genes in surface water samples collected from Germany and Australia. Environ Sci Technol 46: 9716–9726.
12. Brown KD, Kulis J, Thomson B, Chapman TH, Mawhinney DB (2006) Occurrence of antibiotics in hospital, residential, and dairy effluent, municipal wastewater, and the Rio Grande in New Mexico. Sci Total Environ 366: 772–783.
13. Duong HA, Pham NH, Nguyen HT, Hoang TT, Pham HV, et al (2008) Occurrence, fate and antibiotic resistance of fluoroquinolone antibacterials in hospital wastewaters in Hanoi, Vietnam. Chemosphere 72: 968–973.
14. Chang X, Meyer MT, Liu X, Zhao Q, Chen H, et al (2010) Determination of antibiotics in sewage from hospitals, nursery and slaughter house, wastewater treatment plant and source water in Chongqing region of Three Gorge Reservoir in China. Environ Pollut 158: 1444–1450.
15. Reinthaler F., Posch J, Feierl G, Wüst G, Haas D, et al (2003) Antibiotic resistance of E. coli in sewage and sludge. Water Res 37: 1685–1690.
16. Kim S, Aga DS (2007) Potential Ecological and Human Health Impacts of Antibiotics and Antibiotic-Resistant Bacteria from Wastewater Treatment Plants. J Toxicol Environ Heal Part B 10: 559–573.
17. Figueira V, Vaz-Moreira I, Silva M, Manaia CM (2011) Diversity and antibiotic resistance of Aeromonas spp. in drinking and waste water treatment plants. Water Res 45: 5599–5611.
18. Michael I, Rizzo L, McArdell CS, Manaia CM, Merlin C, et al (2013) Urban wastewater treatment plants as hotspots for the release of antibiotics in the environment: A review. Water Res 47: 957–995.
19. Łuczkiewicz A, Jankowska K, Fudala-Książek S, Olańczuk-Neyman K, Łuczkiewicz A, et al (2010) Antimicrobial resistance of fecal indicators in municipal wastewater treatment plant. Water Res 44: 5089–5097. Available: http://www.sciencedirect.com/science/article/pii/S0043135410005713. Accessed 23 January 2013.
20. Li D, Qi R, Yang M, Zhang Y, Yu T (2011) Bacterial community characteristics under long-term antibiotic selection pressures. Water Res 45: 6063–6073.
21. Munir M, Wong K, Xagoraraki I (2011) Release of antibiotic resistant bacteria and genes in the effluent and biosolids of five wastewater utilities in Michigan. Water Res 45: 681–693.
22. Zhang X-X, Zhang T (2011) Occurrence, Abundance, and Diversity of Tetracycline Resistance Genes in 15 Sewage Treatment Plants across China and Other Global Locations. Environ Sci Technol 45: 2598–2604.
23. Gao P, Munir M, Xagoraraki I (2012) Correlation of tetracycline and sulfonamide antibiotics with corresponding resistance genes and resistant bacteria in a conventional municipal wastewater treatment plant. Sci Total Environ 421–422: 173–183.
24. Guardabassi L, Lo Fo Wong DM., Dalsgaard A (2002) The effects of tertiary wastewater treatment on the prevalence of antimicrobial resistant bacteria. Water Res 36: 1955–1964.
25. Moura A, Henriques I, Smalla K, Correia A (2010) Wastewater bacterial communities bring together broad-host range plasmids, integrons and a wide diversity of uncharacterized gene cassettes. Res Microbiol 161: 58–66.
26. Jury KL, Khan SJ, Vancov T, Stuetz RM, Ashbolt NJ (2011) Are Sewage Treatment Plants Promoting Antibiotic Resistance? Crit Rev Environ Sci Technol 41: 243–270.
27. Courvalin P (1994) Transfer of antibiotic resistance genes between gram-positive and gram-negative bacteria. Antimicrob Agents Chemother 38: 1447–1451.
28. Schlüter A, Szczepanowski R, Pühler A, Top EM, Schlüter A, et al (2007) Genomics of IncP-1 antibiotic resistance plasmids isolated from wastewater treatment plants provides evidence for a widely accessible drug resistance gene pool. FEMS Microbiol Rev 31: 449–477.
29. Szczepanowski R, Linke B, Krahn I, Gartemann K, Guetzkow T, et al (2009) Detection of 140 clinically relevant antibiotic-resistance genes in the plasmid metagenome of wastewater treatment plant bacteria showing reduced susceptibility to selected antibiotics. Microbiology-Sgm 155: 2306–2319.
30. West BM, Liggit P, Clemans DL, Francoeur SN (2010) Antibiotic Resistance, Gene Transfer, and Water Quality Patterns Observed in Waterways near CAFO Farms and Wastewater Treatment Facilities. Water, Air, Soil Pollut 217: 473–489.
31. Czekalski N, Berthold T, Caucci S, Egli A, Bürgmann H (2012) Increased Levels of Multiresistant Bacteria and Resistance Genes after Wastewater Treatment and Their Dissemination into Lake Geneva, Switzerland. Front Microbiol 3.
32. Pruden A, Arabi M, Storteboom HN (2012) Correlation between upstream human activities and riverine antibiotic resistance genes. Environ Sci Technol 46: 11541–11549.
33. Rizzo L, Manaia C, Merlin C, Schwartz T, Dagot C, et al (2013) Urban wastewater treatment plants as hotspots for antibiotic resistant bacteria and genes spread into the environment: A review. Sci Total Environ 447: 345–360.
34. Da Silva MF, Tiago I, Verssimo A, Boaventura RAR, Nunes OC, et al (2006) Antibiotic resistance of enterococci and related bacteria in an urban wastewater treatment plant. FEMS Microbiol Ecol 55: 322–329.
35. Levy SB (2002) Factors impacting on the problem of antibiotic resistance. J Antimicrob Chemother 49: 25–30. Available: http://jac.oxfordjournals.org/content/49/1/25.
36. Jury KL, Khan SJ, Vancov T, Stuetz RM, Ashbolt NJ (2011) Are Sewage Treatment Plants Promoting Antibiotic Resistance? Crit Rev Environ Sci Technol 41: 243–270.
37. Auerbach EA, Seyfried EE, McMahon KD (2007) Tetracycline resistance genes in activated sludge wastewater treatment plants. Water Res 41: 1143–1151.
38. Lachmayr KL, Kerkhof LJ, DiRienzo AG, Cavanaugh CM, Ford TE (2009) Quantifying Nonspecific TEM β-Lactamase (blaTEM) Genes in a Wastewater Stream. Appl Environ Microbiol 75: 203–211.
39. Zhang T, Zhang M, Zhang X, Fang HH (2009) Tetracycline Resistance Genes and Tetracycline Resistant Lactose-Fermenting Enterobacteriaceae in Activated Sludge of Sewage Treatment Plants. Environ Sci Technol 43: 3455–3460. Available: http://dx.doi.org/10.1021/es803309m. Accessed 27 March 2013.
40. Börjesson S, Mattsson A, Lindgren P-E (2010) Genes encoding tetracycline resistance in a full-scale municipal wastewater treatment plant investigated during one year. J Water Health 08: 247.
41. Boom R, Sol CJ, Salimans MM, Jansen CL, Dillen PMW, et al (1990) Rapid and simple method for purification of nucleic acids. J Clin Microbiol 28: 495–503.
42. Bates D, Bolker MM and Ben (2013) lme4: Linear mixed-effects models using S4 classes.
43. R Development Core Team (2013). R: A language and environment for statistical computing. R Foundation for Statistical Computing, Vienna, Austria. URL http://www.R-project.org/.
44. Hothorn T, Bretz F, Westfall P (2008) Simultaneous inference in general parametric models. Biom J 50: 346–363.
45. Toleman MA, Bennett PM, Walsh TR (2006) Common regions e.g. orf513 and antibiotic resistance: IS91-like elements evolving complex class 1 integrons. J Antimicrob Chemother 58: 1–6.
46. Bennett PM (2008) Plasmid encoded antibiotic resistance: acquisition and transfer of antibiotic resistance genes in bacteria. Br J Pharmacol 153: S347–S357.
47. Roberts AP, Mullany P (2009) A modular master on the move: the Tn916 family of mobile genetic elements. Trends Microbiol 17: 251–258.
48. Liu M, Zhang Y, Yang M, Tian Z, Ren L, et al (2012) Abundance and Distribution of Tetracycline Resistance Genes and Mobile Elements in an Oxytetracycline Production Wastewater Treatment System. Environ Sci Technol 46: 7551–7557.
49. McArthur AG, Waglechner N, Nizam F, Yan A, Azad MA, et al (2013) The comprehensive antibiotic resistance database. Antimicrob Agents Chemother 57: 3348–3357.
50. Koivunen J, Siitonen A, Heinonen-Tanski H (2003) Elimination of enteric bacteria in biological/chemical wastewater treatment and tertiary filtration units. Water Res 37: 690–698. doi:10.1016/S0043-1354(02)00305-6.
51. Luo Y, Guo W, Ngo HH, Nghiem LD, Hai FI, et al (2014) A review on the occurrence of micropollutants in the aquatic environment and their fate and removal during wastewater treatment. Sci Total Environ 473<unicode: 619–641. doi:10.1016/j.scitotenv.2013.12.065.
52. Gao P, Munir M, Xagoraraki I (2012) Correlation of tetracycline and sulfonamide antibiotics with corresponding resistance genes and resistant bacteria in a conventional municipal wastewater treatment plant. Sci Total Environ 421–422: 173–183.
53. Zhang T, Zhang M, Zhang X, Fang HH (2009) Tetracycline Resistance Genes and Tetracycline Resistant Lactose-Fermenting Enterobacteriaceae in Activated Sludge of Sewage Treatment Plants. Environ Sci Technol 43: 3455–3460.
54. Auerbach EA, Seyfried EE, McMahon KD (2007) Tetracycline resistance genes in activated sludge wastewater treatment plants. Water Res 41: 1143–1151.
55. Graham DW, Olivares-Rieumont S, Knapp CW, Lima L, Werner D, et al (2011) Antibiotic Resistance Gene Abundances Associated with Waste Discharges to the Almendares River near Havana, Cuba. Environ Sci Technol 45: 418–424.
56. Lillenberg M (2011) Residues of some pharmaceuticals in sewage sludge in estonia, their stability in the environment and accumulation into food plants via fertilizing residues of some pharmaceuticals in sewage sludge in estonia, their stability in the environment and accumul Estonian University of Life Sciences. Available: http://dspace.emu.ee/xmlui/bitstream/handle/10492/146/Thesis_Lillenberg_2011.pdf.
57. Clara M, Kreuzinger N, Strenn B, Gans O, Kroiss H (2005) The solids retention time-a suitable design parameter to evaluate the capacity of wastewater treatment plants to remove micropollutants. Water Res 39: 97–106.
58. Novo A, Manaia CM (2010) Factors influencing antibiotic resistance burden in municipal wastewater treatment plants. Appl Microbiol Biotechnol 87: 1157–1166. Available: http://www.ncbi.nlm.nih.gov/pubmed/20396880.

59. Nõlvak H, Truu M, Tiirik K, Oopkaup K, Sildvee T, et al (2013) Dynamics of antibiotic resistance genes and their relationships with system treatment efficiency in a horizontal subsurface flow constructed wetland. Sci Total Environ 461–462: 636–644.
60. Edwards U, Rogall T, Blöcker H, Emde M, Böttger EC (1989) Isolation and direct complete nucleotide determination of entire genes. Characterization of a gene coding for 16S ribosomal RNA. Nucleic Acids Res 17: 7843–7853.
61. López-Gutiérrez JC, Henry S, Hallet S, Martin-Laurent F, Catroux G, et al (2004) Quantification of a novel group of nitrate-reducing bacteria in the environment by real-time PCR. J Microbiol Methods 57: 399–407.
62. Tamminen M, Karkman A, Lohmus A, Muziasari WI, Takasu H, et al (2010) Tetracycline Resistance Genes Persist at Aquaculture Farms in the Absence of Selection Pressure. Environ Sci Technol 45: 386–391.
63. Pei R, Kim S-C, Carlson KH, Pruden A (2006) Effect of River Landscape on the sediment concentrations of antibiotics and corresponding antibiotic resistance genes (ARG). Water Res 40: 2427–2435.
64. Heuer H, Smalla K (2007) Manure and sulfadiazine synergistically increased bacterial antibiotic resistance in soil over at least two months. Environ Microbiol 9: 657–666.

14

Unexpected Attraction of Polarotactic Water-Leaving Insects to Matt Black Car Surfaces: Mattness of Paintwork Cannot Eliminate the Polarized Light Pollution of Black Cars

Miklos Blaho[1], Tamas Herczeg[1], Gyorgy Kriska[2,3], Adam Egri[1], Denes Szaz[1], Alexandra Farkas[1], Nikolett Tarjanyi[3], Laszlo Czinke[1,2], Andras Barta[1], Gabor Horvath[1]*

1 Environmental Optics Laboratory, Department of Biological Physics, Physical Institute, Eötvös University, Budapest, Hungary, **2** Group for Methodology in Biology Teaching, Biological Institute, Eötvös University, Budapest, Hungary, **3** Danube Research Institute, Centre for Ecological Research, Hungarian Academy of Sciences, Budapest, Hungary

Abstract

The horizontally polarizing surface parts of shiny black cars (the reflection-polarization characteristics of which are similar to those of water surfaces) attract water-leaving polarotactic insects. Thus, shiny black cars are typical sources of polarized light pollution endangering water-leaving insects. A new fashion fad is to make car-bodies matt black or grey. Since rough (matt) surfaces depolarize the reflected light, one of the ways of reducing polarized light pollution is to make matt the concerned surface. Consequently, matt black/grey cars may not induce polarized light pollution, which would be an advantageous feature for environmental protection. To test this idea, we performed field experiments with horizontal shiny and matt black car-body surfaces laid on the ground. Using imaging polarimetry, in multiple-choice field experiments we investigated the attractiveness of these test surfaces to various water-leaving polarotactic insects and obtained the following results: (i) The attractiveness of black car-bodies to polarotactic insects depends in complex manner on the surface roughness (shiny, matt) and species (mayflies, dolichopodids, tabanids). (ii) Non-expectedly, the matt dark grey car finish is much more attractive to mayflies (being endangered and protected in many countries) than matt black finish. (iii) The polarized light pollution of shiny black cars usually cannot be reduced with the use of matt painting. On the basis of these, our two novel findings are that (a) matt car-paints are highly polarization reflecting, and (b) these matt paints are not suitable to repel polarotactic insects. Hence, the recent technology used to make matt the car-bodies cannot eliminate or even can enhance the attractiveness of black/grey cars to water-leaving insects. Thus, changing shiny black car painting to matt one is a disadvantageous fashion fad concerning the reduction of polarized light pollution of black vehicles.

Editor: Mark G. Kuzyk, Washington State University, United States of America

Funding: This work was supported by the grant TabaNOid 232366 (Trap for the Novel Control of Horse-flies on Open-air Fields) funded by the European Commission under the 7th Framework Programme received by G. Horvath and G. Kriska. Gabor Horváth thanks the German Alexander von Humboldt Foundation for an equipment donation and a three-month research fellowship (3.3-UNG/1073032 STP, between 1 June and 31 August 2013) in Regensburg. The funders had no role in study design, data collection and analysis, decision to publish, or preparation of the manuscript.

Competing Interests: The authors have declared that no competing interests exist.

* Email: gh@arago.elte.hu

Introduction

Nowadays cars have frequently a matt black or dark grey finish. Matt car surfaces can be produced via paint or foil. Such a finishing gives an eye striking, unusual matt colour to the car-body. Since this painting or covering is rather expensive, predominantly the luxurious cars are produced to be matt.

Shiny car-bodies attract water-leaving insects [1–3], because the hood, roof and boot reflect horizontally polarized light [4,5], and these insects are lured to this optical signal since they detect water by means of the horizontal polarization of water-reflected light [6–19]. This positive polarotaxis induced by the reflection polarization of artificial surfaces is the main reason for polarized light pollution [20]. Thus, shiny black cars are typical sources of polarized light pollution [4,21], a spectacular consequence of which can be seen in Fig. 1 showing mass-swarming mayflies attracted to shiny black cars. The mayflies in Fig. 1 laid their egg batches (each containing 6000-9000 eggs) onto car-bodies, and these eggs perish quickly due to dehydration. *Ephoron virgo* (Ephemeroptera: Polymitarcyidae; Fig. 1D) is not only endangered [22], but also a highly protected mayfly species in Europe [23].

A new fashion fad is that cars are painted matt black or matt dark grey (row 1 in Fig. 2, Fig. S1A-E). Alternatively, the whole car-body or its fragments (e.g., the roof, hood or boot) are covered with carbon foil resulting in a similarly matt black/grey appearance (Fig. S1F). Since rough (matt) surfaces depolarize the reflected light [4,5,8,11,14,16–18,20], one of the possible ways of elimination of polarized light pollution of shiny black artificial

Figure 1. Mayflies attracted *en masse* to shiny black cars due to the highly and horizontally polarized light reflected from the car-body. (A, B) Mass-swarming *Ephemerella hendrickson*. (C) Egg-laying *Ephemera danica*. (D) Thousands of mass-swarming female *Ephoron virgo* mayflies landed on a windscreen, onto which they laid their yellow egg batches. Photos A and B were taken by Dr. Rebecca Allen (Michigan State University, USA), while photos C and D originate from Dr. György Kriska (Eötvös University, Budapest, Hungary).

surfaces is to make them matt [15,20]. Thus, one could hypothesize that matt black/grey cars may less induce polarized light pollution, because they may reflect weakly polarized light, that does not attract polarotactic insects. If this were the case, the spreading of matt black cars would be a fashion fad that could be welcomed, because they would not attract polarotactic insects, the egg batches of which would inevitably perish due to dehydration when laid onto car-bodies.

To compare the attractivenes of matt and shiny black car surfaces to polarotactic insects, we performed two field experiments with horizontal shiny and matt black/grey car-body fragments. We checked the attractiveness of these test surfaces to polarotactic mayflies, dolichopodid flies and tabanid flies. The studied mayfly species are endangered in Europe. Mayflies and dolichopodids were abundant in the site of our first experiment. Since tabanid flies are serious pests, they are neither rare nor protected species. The investigated mayflies, dolichopodids and tabanids were used simply as indicators of strongly and horizontally polarized reflected light, because they are positively polarotactic insects [15,16,18,24–31]. We showed that changing shiny black car paintwork to matt one can be a disadvantageous fashion fad concerning the polarized light pollution of black vehicles.

Results

Reflection-polarization characteristics of matt black cars

Figure 2 shows the reflection-polarization characteristics of a matt black car measured from five different directions of view in the blue (450 nm) part of the spectrum with imaging polarimetry. The degree of polarization d of light reflected from the skylit part of the tilted windscreen was very high ($85\%<d<100\%$), while the light reflected from the vertical windows was only weakly polarized ($d<15\%$). The roof, hood and boot reflected moderately polarized light ($25\%<d<55\%$), and the vertical parts of the car-body reflected weakly polarized light ($d<15\%$). The spatial distribution of d of light reflected from the roof, hood and boot was rather homogeneous. The tilted windscreen and the horizontal roof, hood and boot reflected horizontally polarized light, while the other (vertical and tilted) parts of the car-body reflected light with vertical or oblique directions of polarization. Areas of the car-body that reflect light with degree of polarization high enough ($d>15\%$) and with nearly horizontal direction of polarization ($80^{o}<\alpha<100^{o}$) are sensed as water by polarotactic insects (Horváth and Varjú 2004; Kriska *et al.* 2009). According to Fig. 2, these areas of the matt black car-body occured on the windscreen, roof, hood and boot. The reflection-polarization characteristics of car-bodies covered by matt black/grey carbone foils are quite similar (Fig. S5).

Figure 2. Photograph, patterns of the degree of linear polarization d and the angle of polarization α (clockwise from the vertical), and areas detected as water by polarotactic insects (for which the reflected light has the following characteristics: $d>15\%$, $80°<\alpha<100°$) of a matt black car measured with imaging polarimetry from five different directions of view in the blue (450 nm) part of the spectrum. The angle of elevation of the optical axis of the polarimeter was −20° from the horizontal. The number-plate of the car is screened by a white rectangle.

Reflection-polarization characteristics of test surfaces

Figure 3 shows and Table 1 contains the average and standard deviation of the degree of polarization d of light reflected from the three horizontal test surfaces (shiny black, matt black, matt grey) used in experiments 1 and 2 measured with imaging polarimetry in the red, green and blue parts of the spectrum when the surfaces reflected sun- and skylight, or light from trees and bushes (Fig. S4). The major part of the test surfaces reflected horizontally polarized light, a small part reflected vertically or obliquely polarized light, while from a neutral point (at the border of the horizontally and vertically polarizing surface regions) unpolarized light was reflected. The shiny black and matt black test surfaces reflected light with the highest d, while the matt grey surface was the weakest polarizing (Table 1, Fig. 3). The polarization characteristics of the black and grey test surfaces depended only slightly on the wavelength. When the test surfaces reflected sun- and skylight, their degree of polarization was higher in comparison to the situation when they reflected light from trees and bushes. In the blue spectral range the least standard deviation of d occurred on the matt grey test surface, and in the green part of the spectrum the least standard deviation of d possessed the matt grey test surface (Table 1). Comparing the reflection-polarization characteristics of shiny/matt, black/grey car-bodies (Fig. 2, Fig. S5) with those of our test surfaces (Fig. 3, Fig. S4, Table 1), it was evident that they were practically the same. Thus, our test surfaces imitated well the polarization characteristics of matt/shiny and black/grey car-bodies.

According to Table 1 (see also Fig. 3), depending on the direction of view, in the blue, green and red spectral ranges the standard deviation Δd of d of the shiny black test surface (blue: ±16.4–19.6%, green: ±15.7–18.0%, red: ±15.7–18.8%) was 1.9–2.8, 1.8–2.5 and 1.5–2.8 times higher, respectively, than that of the matt grey surface (blue: ±5.8–10.5%, green: ±6.4–10.2%, red: ±6.7–10.3%). Similarly, Δd of the matt black test surface (blue: ±11.9–14.6%, green: ±11.8–13.9%, red: ±12.0–13.7%) was 1.4–2.1, 1.4–1.8 and 1.3–1.8 times higher, respectively, than that of the matt grey surface. These results are important in the explanation of our finding that the attractiveness of the horizontally polarizing matt grey car-body fragment to mayflies was significantly larger than that of the two black test surfaces (see Discussion).

Attractiveness of test surfaces to mayflies, dolichopodids and tabanids

Figure 4 shows the numbers of mayflies landed on the three test surfaces in experiment 1 (Table S1). There was no statistically significant difference between the attractiveness of the shiny black and matt black test surfaces to mayflies (Table 2). Interestingly, the matt grey test surface attracted about 10.7–15.7 times more

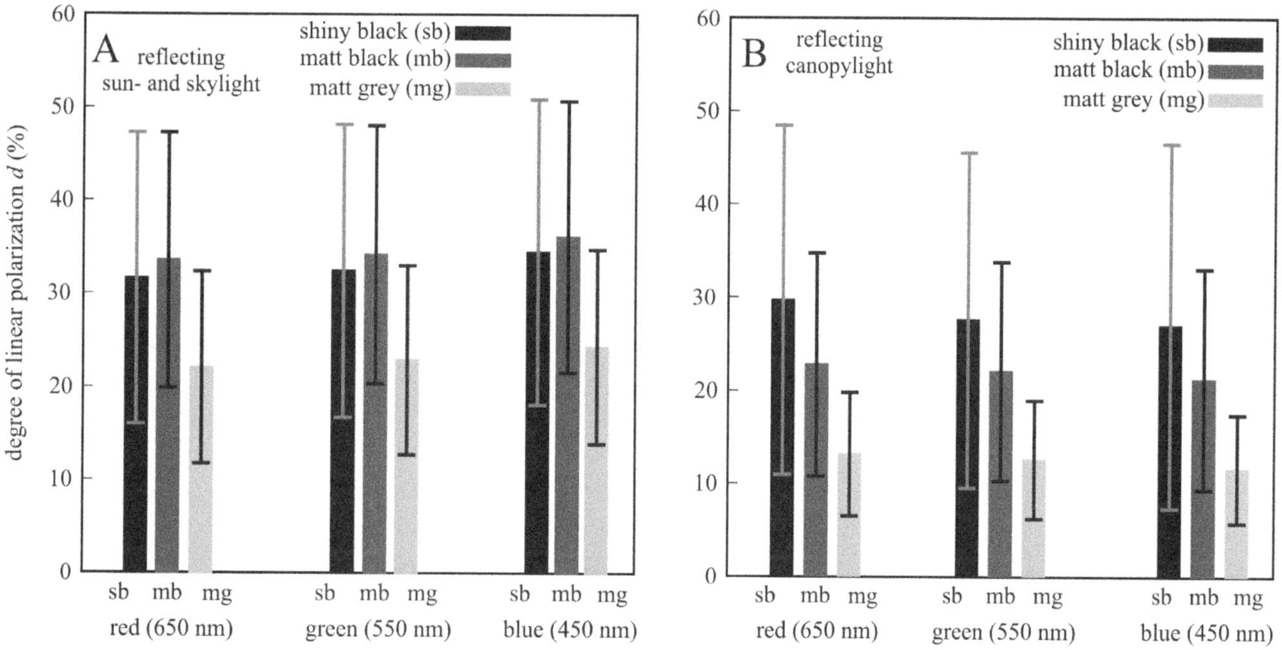

Figure 3. Degree of linear polarization *d* of the shiny black (sb), matt black (mb) and matt grey (mg) horizontal test surfaces used in experiments 1 and 2 measured with imaging polarimetry in the red (650 nm), green (550 nm) and blue (450 nm) parts of the spectrum when sun- and skylight (A, Supplementary Fig. S4I) or canopylight originating from trees and bushes (B, Supplementary Fig. S4II) was reflected by the test surfaces. Columns: averages. Vertical bars: standard deviations. The average is calculated for the whole area of each test surface (corresponding to 2500×4000 = 10 000 000 pixels in the pictures and polarization patterns).

mayflies than the two black test surfaces, which differences were statistically significant (Table 2). This result was very surprising, and we propose an explanation for it in the Discussion.

Figure 5 shows the numbers of dolichopodids landed on the three test surfaces in experiment 1 (Table S1). The shiny black test surface was the most attractive to dolichopodids and the matt grey surface was the least attractive. The differences between the attractiveness of the shiny black and matt grey as well as between the matt black and matt grey test surfaces were significant, while there was no significant difference between the numbers of dolichopodids landed on the shiny black and matt black surfaces (Table 3).

Figure 6 shows the numbers of landing, touching and looping of tabanids at the three test surfaces in experiment 2 (Table S2). Considering these three reactions, the shiny black test surface was significantly the most attractive to tabanids, while considering landing and touching, the matt grey surface was significantly the least attractive. The attractiveness of the matt black test surface was between that of the shiny black and the matt grey surfaces (Fig. 6, Table 4).

The results of our experiments show the following: (i) The attractiveness of black car-bodies to polarotactic insects depends in complex manner on the surface roughness (shiny, matt) and species (mayflies, dolichopodids, tabanids). (ii) Depending on species, matt black car-bodies can be less or equally attractive to polarotactic insects than shiny black ones. (iii) The polarized light pollution of shiny black cars usually cannot be reduced with the use of matt painting. (iv) Non-expectedly, the matt dark grey car

Table 1. Degree of linear polarization *d* (average ± standard deviation, averaged to the whole surface area) of the three horizontal test surfaces (shiny black, matt black, matt grey) used in experiments 1 and 2 measured with imaging polarimetry in the red (650 nm), green (550 nm) and blue (450 nm) parts of the spectrum when sun- and skylight (Fig. S4I) or canopylight originating from trees and bushes (Fig. S4II) was reflected by the test surfaces.

light incident to the test surfaces	test surface	degree of linear polarization *d* (%)		
		red	green	blue
sun- skylight	shiny black	31.6±15.7	32.4±15.7	34.5±16.4
	matt black	33.6±13.7	34.2±13.9	36.1±14.6
	matt grey	22.0±10.3	22.9±10.2	24.3±10.5
canopylight from trees and bushes	shiny black	29.6±18.8	27.4±18.0	26.8±19.6
	matt black	22.6±12.0	21.9±11.8	21.1±11.9
	matt grey	13.0±6.7	12.4±6.4	11.4±5.8

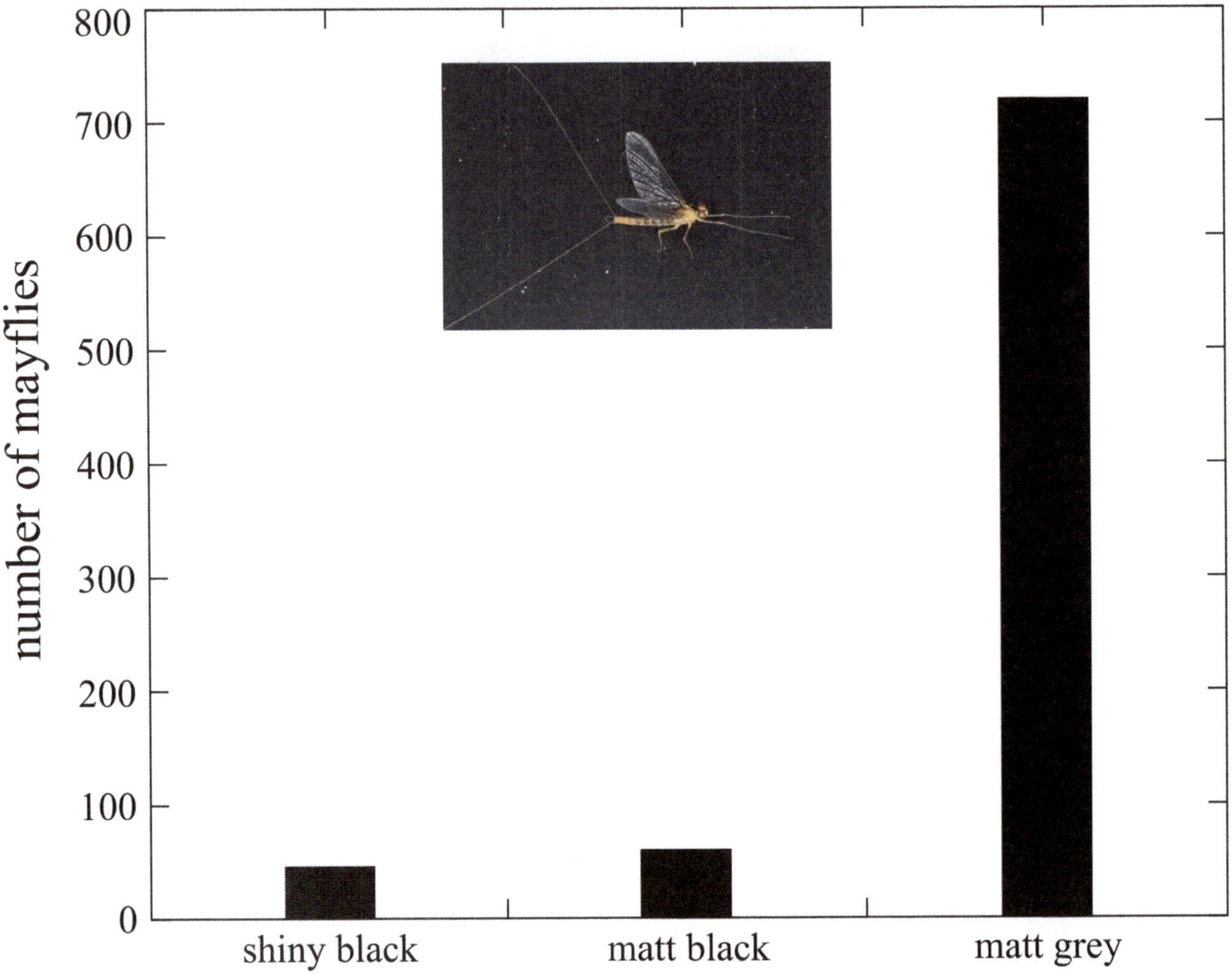

Figure 4. Total numbers of mayflies landed on the shiny black, matt black and matt grey horizontal test surfaces in experiment 1. The inset is a photograph of a mayfly landed on the shiny black test surface. The number of repetition is 6 (see Materials and methods, and Discussion).

finish is much more attractive to mayflies (being endangered and protected in many countries) than matt black finish.

Discussion

One could assume that changing the shiny paintwork of black car-bodies to a matt one, the polarized light pollution of the horizontally polarizing roof, hood and boot can be reduced, since roughness (mattness) depolarizes the reflected light. In our field experiments we obtained that this is not what always happens, because the investigated horizontally polarizing matt black and matt grey car-body fragments were similarly attractive, or even more attractive to polarotactic mayflies than the studied horizontally polarizing shiny black car-body fragment. The reason for this is that although (i) the matt grey car-body surface reflected light with lower degrees of polarization *d* than the shiny black surface, and (ii) the matt black car finish was less polarizing than the shiny black finish when reflecting canopylight (Table 1, Fig. 3B), this

Table 2. Statistical comparisons (Kruskal Wallis and Mann-Whitney U test) between the numbers of mayflies landed on the shiny black, matt black and matt grey horizontal test surfaces in experiment 1 (Fig. 4, Table S1).

comparison	test type	test result	significancy
shiny black *versus* matt black *versus* matt grey	Kruskal-Wallis	$H = 176.2$, $df = 2$, $p < 0.0001$	significant
shiny black *versus* matt grey	Mann-Whitney U with Bonferroni correction	$U = 172.5$, $p < 0.0001$	significant
shiny black *versus* matt black	Mann-Whitney U with Bonferroni correction	$U = 4372.5$, $p = 0.38$	not significant
matt black *versus* matt grey	Mann-Whitney U with Bonferroni correction	$U = 210$, $p < 0.0001$	significant

Figure 5. Total numbers of dolichopodids landed on the shiny black, matt black and matt grey horizontal test surfaces in experiment 1. The inset is a photograph of a dolichopodid fly landed on the matt black test surface. The number of repetition is 6 (see Materials and methods, and Discussion).

reduction of *d* was not large enough to make them unattractive to polarotactic insects. On the other hand, (iii) the matt black finish was even more polarizing than the shiny black finish when reflecting sun- and skylight (Table 1, Fig. 3A). One of the prerequisites of attraction of water-leaving insects by polarizing reflectors is that *d* of reflected light must be higher than the threshold d^* of insect polarization sensitivity. Species-specific values of d^* are known for certain dragonfly, mayfly and tabanid fly species [25]. Thus, the attractiveness of a car paintwork to polarotactic insects can be ensured only if *d* of reflected light is higher than d^*.

More remarkably, the attractiveness of the horizontal matt grey car-body fragment to polarotactic mayflies was significantly larger than that of the two black test surfaces. One of the reasons for this could be that the standard deviation Δd of *d* of the shiny black test surface was about 2–3 times higher than that of the matt grey surface, and Δd of the matt black test surface was 1.3–2.1 times higher than that of the matt grey surface (Table 1, Fig. 3). Calm

Table 3. Statistical comparisons (Kruskal Wallis and Mann-Whitney U test) between the numbers of dolichopodids landed on the shiny black, matt black and matt grey horizontal test surfaces in experiment 1 (Fig. 4, Table S1).

comparison	test type	test result	significancy
shiny black *versus* matt black *versus* matt grey	Kruskal-Wallis	H = 16.8, df = 2, p = 0.0002	significant
shiny black *versus* matt grey	Mann-Whitney U with Bonferroni correction	U = 3111.5, p = 0.0001	significant
shiny black *versus* matt black	Mann-Whitney U with Bonferroni correction	U = 4166, p = 0.25	not significant
matt black *versus* matt grey	Mann-Whitney U with Bonferroni correction	U = 3468, p = 0.003	significant

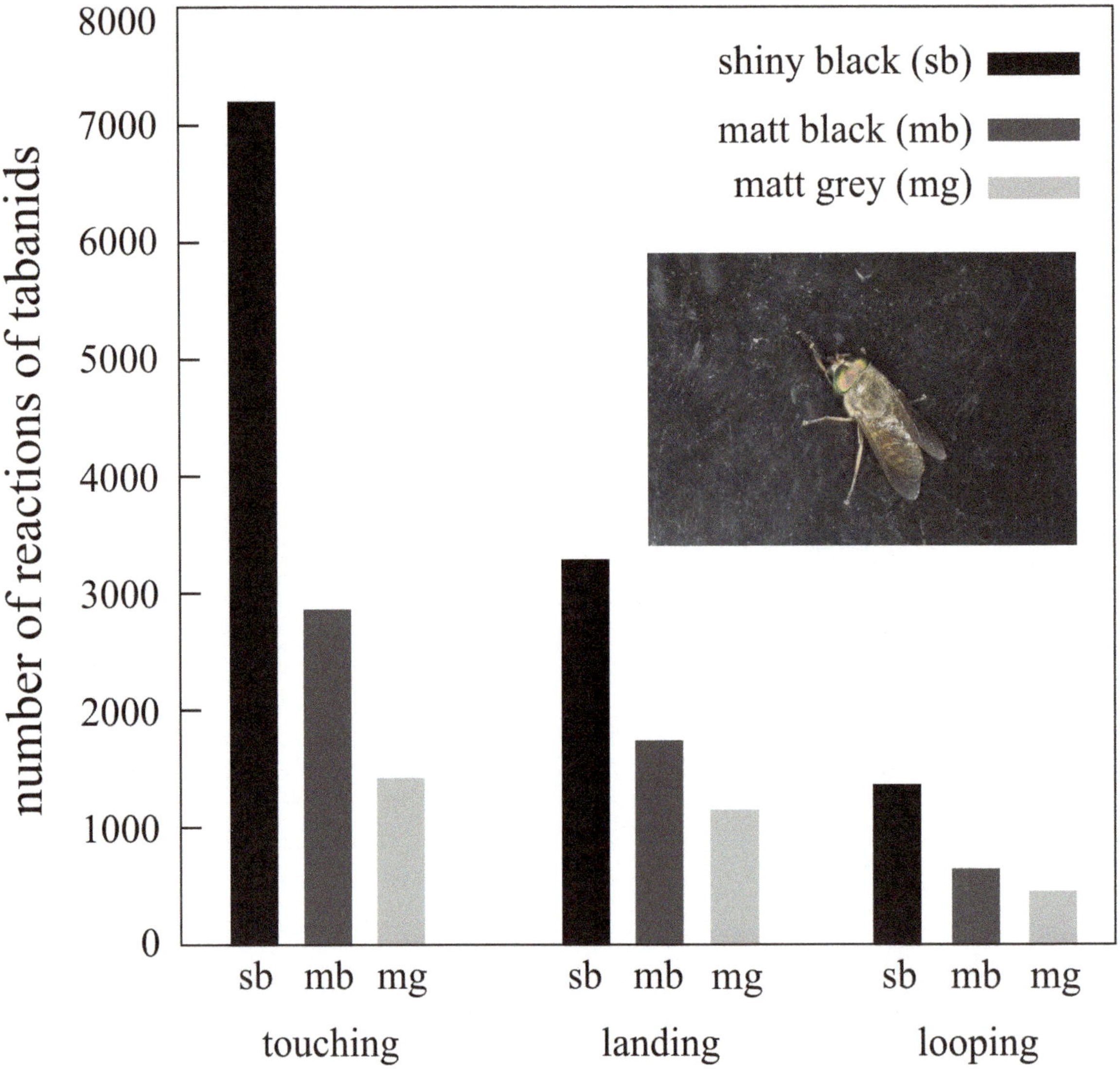

Figure 6. Total numbers of reactions (touching, landing and aerial looping) of tabanids to the shiny black (sb), matt black (mb) and matt grey (mg) horizontal test surfaces in experiment 2. The inset is a photograph of a tabanid fly landed on the matt grey test surface. The number of repetition is 20 (see Materials and methods, and Discussion).

water bodies with a smooth surface are characterized by small Δd of reflected light due to the lack of ripples [11,32,33]. Turbulent waters with rough, ripplening/undulating surfaces reflect light with large Δd, because d of reflected light depends strongly on the direction of incidence relative to the normal vector of the surface, which changes randomly spatio-temporally if the water is turbulently flowing. Since the mayflies investigated in experiment 1 prefer quiescent water bodies [34], the matt grey test surface with small Δd of reflected light might have imitated a calm water surface to them in the visible spectral range, while the two black test surfaces with approximately 1.5–3 times larger Δd might have been sensed by mayflies as turbulent waters. Therefore, the matt grey test surface was much more attractive to mayflies than the two black test surfaces.

Another reason for the greater attractiveness of the matt grey car-body surface to mayflies can be that the matt grey test surface (with a greyness of 90%) was by 10% brighter than the black test surfaces (with a greyness of 100%), and thus the former was more attractive to mayflies due to their possible positive phototaxis. However, it has been shown earlier that the studied polarotactic mayfly species (*Baetis rhodani*, *Epeorus sylvicola*, *Rhithrogena semicolorata*) are not phototactic [24]: their attractiveness to shiny/matt white/grey/black horizontal surfaces is governed by the linear polarization of reflected light, rather than by the light intensity (brighness/darkness).

Table 1 (see also Fig. 3) provides the values of the degree of polarization d in three (red, green, blue) spectral ranges to show that the reflection-polarization characteristics of our test surfaces used in experiments 1 and 2 were practically independent of the wavelength in the visible part of the spectrum: at a given test surface under a given illumination condition the spectral differences in d are less than 5%, which is below the polarization sensitivity threshold of any known animal [35].

In our experiments mayflies, dolichopodids and tabanids attracted by test surfaces with different reflection-polarization characteristics were counted. Experiment 1 was repeated 6 times on 6 days during the extremely short (a few days) swarming period of the investigated mayflies. This meant $6 \times 2 = 12$ h observation

Table 4. Statistical comparisons (Kruskal Wallis and Mann-Whitney U test) between the numbers of the three reactions (landing, touching, looping) of tabanids to the shiny black, matt black and matt grey horizontal test surfaces in experiment 2 (Fig. 6, Table S2).

comparison		test type	test result	significancy
landing	shiny black *versus* matt black *versus* matt grey	Kruskal-Wallis	H = 38.6, df = 2, p<0.0001	significant
	shiny black *versus* matt grey	Mann-Whitney U with Bonferroni correction	U = 11, p<0.0001	significant
	shiny black *versus* matt black	Mann-Whitney U with Bonferroni correction	U = 27.5, p<0.0001	significant
	matt black *versus* matt grey	Mann-Whitney U with Bonferroni correction	U = 67.5, p = 0.0003	significant
touching	shiny black *versus* matt black *versus* matt grey	Kruskal-Wallis	H = 34.2, df = 2, p<0.0001	significant
	shiny black *versus* matt grey	Mann-Whitney U with Bonferroni correction	U = 20, p<0.0001	significant
	shiny black *versus* matt black	Mann-Whitney U with Bonferroni correction	U = 44, p<0.0001	significant
	matt black *versus* matt grey	Mann-Whitney U with Bonferroni correction	U = 71.5, p = 0.0005	significant
looping	shiny black *versus* matt black *versus* matt grey	Kruskal-Wallis	H = 11.2, df = 2, P = 0.004	significant
	shiny black *versus* matt grey	Mann-Whitney U with Bonferroni correction	U = 82, P = 0.001	significant
	shiny black *versus* matt black	Mann-Whitney U with Bonferroni correction	U = 127.5, p = 0.049	not significant
	matt black *versus* matt grey	Mann-Whitney U with Bonferroni correction	U = 139.5, p = 0.10	not significant

time, during which 96 photographs were taken about each test surface (Table S1). After each photography the three test surfaces were reordered with cyclical permutation to eliminate site effects. The studied mayflies live as adult only for one day. Thus, during a given 2-hour session of experiment 1 (on a given day) a given mayfly might react to certain test surfaces more than once, meaning pseudo-replication of this experiment after the reordering of the test surfaces every 5 minutes (24 times in the 2 h of a given session). Such a possible pseudo-replication could not be avoided during the 2 h of a given session, because the test surfaces must have been dry. Pseudo-replication could have been avoided only with the use of sticky test surfaces. However, the matt black test surfaces must not be covered with any glue, since the glue layer would make them shiny. On the other hand, each day of the 6-day experiment 1 always new mayflies might have occurred due to their one-day life-time. Thus, pseudo-replication might have been minimal in experiment 1. The adults of dolichopodids live more than one day, thus, considering the numbers of dolichopodids, some degree of pseudo-replication may not be excluded in experiment 1. However, it is highly probable that every day of the 6-day experiment 1 not the same dolichopodid individuals reacted to our test surfaces all the time. Similar was true for the number of tabanids in the 20-day experiment 2 repeated 20 times with hourly cyclically permutated re-ordering of the three test surfaces. According to our results and experiences gained in our earlier similar field experiments with mayflies, dolichopodids and tabanids [14,16,18,24–31], the number of repetitions and the duration of experiments 1 and 2 were large enough to detect significant differences in the attractiveness of test surfaces to polarotactic insects.

In our research the exact species names are practically irrelevant, since the studied water-leaving insects were only indicators of the polarized light pollution of the investigated test surfaces. The only important aspect was that positively polarotactic insects should be involved into our field experiments, the aim of which was to test the attractiveness of matt/shiny and black/grey car-bodies. All the involved species are positively polarotactic as shown earlier.

In every year more and more cars are running on the roads. They are predominantly shiny, and if their paint strongly absorbs light in a given spectral range, they induce strong polarized light pollution [4,21] like the black or dark grey asphalt roads themselves [20,24,36]. We showed here that, unfortunately, this kind of light pollution cannot be eliminated by the use of matt black/grey car paintworks available presently in the market. The technology of matt paintworks has been developed to provide car-owners with a striking visual appearance of their cars and/or the protection of vulnerable paintworks against scratches. It would be worth improving this technology to ensure a much greater reduction of reflection polarization in order to eliminate polarized light pollution of car-bodies. If this were realized, the mattness of black cars could be an advantageous fashion fad considering the protection of endangered populations of polarotactic water-leaving insects.

The physical reason for the high polarization reflection (eliciting attractiveness to polarotactic insects) of black car surfaces is the following: A shiny (smooth) surface of a given dielectric medium reflects two components. The first component, the light reflected from the air-medium interface, is partially linearly polarized with a direction of polarization parallel to the surface. The second component, the light backscattered from the medium and refracted at the medium-air interface, is also partially linearly polarized but with direction of polarization perpendicular to the surface. The superposition of these orthogonally polarized components reduces the net degree of polarization of surface-reflected light. If the first or the second component dominates, the degree of polarization is high with a direction of polarization parallel or perpendicular to the surface, respectively. In the case of the smooth (shiny) surface of a black medium the first component dominates, because the second component is strongly absorbed by the black medium, thus, the degree of polarization of reflected light is high. If the surface of a black medium is rough (matt), in a microscopic scale it is composed of countless tiny surface fragments (facets), the surfaces of which are smooth on their own but oriented randomly in all possible directions. Due to these random facet orientation the surface reflects light diffusely (in all possible directions). A given facet reflects light with high degrees of polarization due to the absorption of the above-mentioned second component, but the direction of polarization of facet-reflected light is random because of the random facet orientation. Furthermore, incident light can also be reflected more than once from different facets. All these result in that the net degree of polarization of light reflected by a matt surface is reduced because of the superposition of the numerous individual facet-reflected components with random directions of polarization. More details of the physics/optics of absorbing and reflecting materials can be read e.g. in [37,38].

Materials and Methods

Experiment 1

Experiment 1 was performed between 15 and 29 May 2013 on six warm days in the Hungarian Duna-Ipoly National Park at Dömörkapu (47° 40' N, 19° 03' E), where an asphalt road is running parallel (at a distance not more than 7 m) to a mountain creek. The creek is the emergence site of different mayfly (Ephemeroptera: Baetidae, Heptageniidae) and dolichopodid fly (Diptera: Dolichopodidae) species. At dusk from May to July every year adult mayflies and dolichopodids emerge from the creek and swarm in large numbers near or above the asphalt road, which is thus an ideal place for choice experiments. As the studied mayfly species are endangered in Europe, we obtained a permission from the Central Danube Environmental Protection and Water Management Inspectorate to perform our field experiment at this site. In this experiment our aim was to test the attractiveness of horizontal car-body parts, such as the hood, roof and boot of car-bodies to mayflies and dolichopodids, which are positively polarotactic insects [14,24–26]. There were three different test surfaces: (i) shiny black (greyness = 100%), (ii) matt black (greyness = 100%) and (iii) matt grey (greyness = 90%). They were composed of a metal plate (80 cm×80 cm, 10 kg) painted or covered with the same paints or carbon foil as used presently in the car industry. These test surfaces were produced by a professional Hungarian firm (Lakk-Mix Ltd., Budapest) dealing with car-bodies, car-body painting and covering, and using strict standardized methods. Only one of the test surfaces was covered by a matt grey carbone foil (Avery 502), the other two surfaces were painted by RAL (shiny black, matt black) paint.

These car-body fragments were laid on the asphalt surface (of a small bridge running above the creek) along a straight line 1 m apart from each other. On a given day the experiment began at 19 h (= local summer time = UTC+2 h) and stopped at 21 h. The order of the test surfaces was cyclically permutated every 5 minutes to eliminate site effects. After such a reordering, we photographed all three test surfaces with a digital camera (Nikon D90) to document the insects landed on or flying immediately (a few dm) above them (Figs. S2, S3). Later, in the laboratory we counted the number of mayflies and dolichopodids on these photographs. Although these mayflies and dolichopodids could not be taxonomically identified, they surely belonged to the order Ephemoreptera, family Baetidae, Heptageniidae and order Diptera, family Dolichopodidae, respectively, as was visually determined by one of the authors (G. K.), who is an expert of these insects. According to our earlier field experiments at the same site [24–26], we know that the following mayfly and dolichopodid species occurred in the air during experiment 1: *Baetis rhodani*, *Epeorus sylvicola*, *Rhithrogena semicolorata* (mayflies), *Dolichopus ungulatus*, *Dolichopus acuticornis*, *Dolichopus agilis* (dolichopodids).

Experiment 2

Experiment 2 was performed between 24 June and 27 July 2013 twenty times on sunny, warm days on a Hungarian horse farm in Szokolya (47° 52' N, 19° 00' E), where tabanid flies are abundant in summer [18,25–31]. Our aim was to test the attractiveness of the three (shiny black, matt black, matt grey) test surfaces used in experiment 1 to tabanids, which are positively polarotactic insects [16,28,31,39]. The test surfaces were laid on the ground along a straight line 1 m apart from each other in a meadow near the horse farm, 5 m from a row of trees and bushes. Two persons sat 2 m from the row of the three test surfaces and continuously counted the reactions of tabanids to them. On a given day the experiment began at 9 h (UTC+2 h) and stopped at 14 h. The order of the test surfaces was cyclically permutated hourly to eliminate site effects. Three different tabanid reactions were distinguished: (1) aerial looping (a flying tabanid approached the test surface and performed at least one loop in the air above it at a height of a few decimeters), (2) touch-down (a tabanid touched at least once the test surface then flew away), and (3) landing (a tabanid landed on the test surface and remained on it at least for 3 seconds). These reactions are typical to tabanid flies at horizontally polarizing reflecting surfaces on the ground [16,40]. Although the observed tabanids could not be taxonomically identified, they surely belonged to the family Tabanidae as was visually determined by the observers (authors of this work), who are experience in tabanid field experiments. In the horse farm two other experiments using different tabanid traps ran simultaneously with experiment 2. Since the tabanids captured in these experiments were later identified (by Mónika Gyurkovszky and Róbert Farkas, Department of Parasitology and Zoology, Faculty of Veterinary Science, Szent István University, Budapest), we know that the following tabanid species occurred in the air during experiment 2: *Tabanus tergestinus*, *T. bromius*, *T. bovinus*, *T. autumnalis*, *Atylotus fulvus*, *A. loewianus*, *A. rusticus*, *Haematopota italica*.

Imaging polarimetry

The reflection-polarization characteristics of cars and the test surfaces used in experiments 1 and 2 were measured by imaging polarimetry in the red (650±40 nm = wavelength of maximal sensitivity ± half bandwidth of the CCD detectors of the polarimeter), green (550±40 nm) and blue (450±40 nm) spectral ranges. Mayflies, dolichopodids and tabanids possess ultraviolet-, blue- and green-sensitive photoreceptors [41]. It is, however, still unknown in which spectral range these insects sense the polarization of light. The method of imaging polarimetry has been described in detail by Horváth and Varjú [11,32]. Here we present only the polarization patterns measured in the blue part of the spectrum. In the case of the black and grey cars and test surfaces similar patterns were obtained in the red and green spectral ranges, because due to their colourless feature their reflection-polarization characteristics depend only slightly on the wavelength of light due to the blueness of incident skylight.

Statistical analysis

Non-parametric Kruskal-Wallis tests [42] were used to compare the reactions of attracted insects (Tables S1 and S2) to the three different test surfaces in experiments 1 and 2. Since the Kruskal-Wallis tests were significant for all three insect groups (dolichopodids, mayflies and tabanid flies), we have done separate Mann Whitney U tests with Bonferroni correction as post-hoc comparisons [42] to find out which groups differ significantly. All statistical tests were performed with the use of the program Statistica 7.0.

Field study permits/approvals

Many thanks to Csaba Viski (Szokolya, Hungary), who allowed our experiments on his horse farm. We are grateful to Mónika Gyurkovszky and Prof. Róbert Farkas (Department of Parasitology and Zoology, Faculty of Veterinary Science, Szent István

University, Budapest) for taxonomically identifying the tabanid species occurring in the air during our experiment 2. The logistic help of István Gubek (Eötvös University, Budapest) is also acknowledged. We thank Györgyi Antoni (Center for Innovation and Grant Affairs, Eötvös University, Budapest) and Emese Kovács (Vörösmarty Tourist House, Mátraháza, Hungary) for providing us with the matt black cars in Figures 2 and S5. Thanks are also to Rebecca Allen (Michigan State University, USA) for the photos in Fig. 1A,B. We thank the permission from the Central Danube Environmental Protection and Water Management Inspectorate to investigate mayflies.

Conclusions

From the results of our field experiments presented here we conclude that making matt the car-body cannot reduce the polarized light pollution of black cars. Matt car surfaces can even attract more individuals of certain polarotactic insect species (e.g. mayflies) than shiny black cars. Thus, changing shiny black paintwork to matt one can be a disadvantageous fashion fad concerning environmental protection.

Supporting Information

Figure S1 Cars with different matt black/grey painting (A-E), or carbon foil on the hood and roof (F) (photographs taken by Gábor Horváth). The number-plates are screened by white rectangles.

Figure S2 Photographs of the shiny black, matt black and matt grey horizontal car-body fragments used in experiment 1 with some mayflies and dolichopodids above or on the test surfaces. On such photographs were counted the attracted insects.

Figure S3 Photographs of mayflies (A, B), dolichopodids (C) and tabanids (D) landed on the car-body fragments used in experiments 1 and 2. (A) An egg laying female (down) and a male (up) *Rhithrogena semicolorata* mayfly on the matt black test surface. (B) Male *R. semicolorata* on the shiny black test surface. (C) A dolichopodid fly on the matt black test surface. (D) Tabanid flies on the shiny black test surface.

Figure S4 Photograph, patterns of the degree of linear polarization d and the angle of polarization α (clockwise from the vertical), and areas detected as water by polarotactic insects (for which the reflected light has the following characteristics: $d>15\%$, $80^\circ<\alpha<100^\circ$) of the shiny black, matt black and matt grey horizontal test surfaces used in experiments 1 and 2 measured with imaging polarimetry from two different directions of view in the blue (450 nm) part of the spectrum. The polarimeter saw: (I) toward an open field (the surfaces reflected sun- and skylight), (II) toward trees and bushes (the surfaces reflected light from a tree canopy). The angle of elevation of the optical axis of the polarimeter was -45° from the horizontal.

Figure S5 Photograph, patterns of the degree of linear polarization d and the angle of polarization α (clockwise from the vertical), and areas detected as water by polarotactic insects (for which the reflected light has the following characteristics: $d>15\%$, $80^\circ<\alpha<100^\circ$) of a shiny red car, the hood and roof of which are covered with matt black carbon foil. The patterns were measured in the blue (450 nm) part of the spectrum with imaging polarimetry from two different directions of view under a cloudy sky when the sun was shining from behind a large thin cloud layer. The polarimeter saw toward the antisolar half of the sky. The angle of elevation of the optical axis of the polarimeter was -20° from the horizontal. In the α-pattern double-headed arrows show the local direction of polarization of light reflected from the car-body. The number-plate of the car and two persons are screened by white rectangles.

Table S1 Numbers of mayflies (M) and dolichopodids (D) landed on the shiny black, matt black and matt grey horizontal test surfaces in experiment 1 counted on the photographs taken after each permutation of the order of the surfaces. No.: number of repetition of experiment, %: percentage of mayflies/dolichopodids relative to their total number counted on all three test surfaces, AV: average, SD: standard deviation. The number of repetition is 6 (see Materials and methods, and Discussion).

Table S2 Numbers of three reactions (LA: landing, TO: touching, LO: looping) of tabanids to the shiny black, matt black and matt grey horizontal test surfaces in experiment 2 as a function of time in 2013 (6: June, 7: July). The number of repetition is 20 (see Materials and methods, and Discussion).

Author Contributions

Conceived and designed the experiments: MB TH GK AE GH. Performed the experiments: MB TH GK AE DS AF NT LC AB GH. Analyzed the data: MB GK AE GH. Contributed reagents/materials/analysis tools: MB TH GK AE GH. Contributed to the writing of the manuscript: MB TH GK GH.

References

1. Jäch MA (1997) Daytime swarming of rheophilic water beetles in Austria (Coleoptera: Elmidae, Hydraenidae, Haliplidae). Latissimus 9: 10–11.
2. Nilsson AN (1997) On flying *Hydroporus* and the attraction of *H. incognitus* to red car roofs. Latissimus 9: 12–16.
3. Vondel BJ van (1998) Another case of water beetles landing on a red car roof. Latissimus 10: 29.
4. Kriska G, Csabai Z, Boda P, Malik P, Horváth G (2006) Why do red and dark-coloured cars lure aquatic insects? The attraction of water insects to car paintwork explained by reflection-polarization signals. Proceedings of the Royal Society B 273: 1667–1671.
5. Horváth G, Móra A, Bernáth B, Kriska G (2011) Polarotaxis in non-biting midges: female chironomids are attracted to horizontally polarized light. Physiology and Behavior 104: 1010–1015.
6. Schwind R (1991) Polarization vision in water insects and insects living on a moist substrate. Journal of Comparative Physiology A 169: 531–540.
7. Schwind R (1995) Spectral regions in which aquatic insects see reflected polarized light. Journal of Comparative Physiology A 177: 439–448.
8. Wildermuth H (1998) Dragonflies recognize the water of rendezvous and oviposition sites by horizontally polarized light: a behavioural field test. Naturwissenschaften 85: 297–302.
9. Wildermuth H (2007) Polarotaktische Reaktionen von *Coenagrion puella* und *Libellula quadrimaculata* auf Erdbeerkulturen als ökologische Falle (Odonata: Coenagrionidae, Libellulidae). Libellula 26: 143–150.
10. Bernáth B, Gál J, Horváth G (2004) Why is it worth flying at dusk for aquatic insects? Polarotactic water detection is easiest at low solar elevations. Journal of Experimental Biology 207: 755–765.

11. Horváth G, Varjú D (2004) *Polarized Light in Animal Vision- Polarization Patterns in Nature*. Springer-Verlag, Heidelberg - Berlin - New York.
12. Csabai Z, Boda P, Bernáth B, Kriska G, Horváth G (2006) A 'polarisation sundial' dictates the optimal time of day for dispersal by flying aquatic insects. Freshwater Biology 51: 1341–1350.
13. Csabai Z, Kálmán Z, Szivák I, Boda P (2012) Diel flight behaviour and dispersal patterns of aquatic Coleoptera and Heteroptera species with special emphasis on the importance of seasons. Naturwissenschaften 99: 751–765.
14. Kriska G, Bernáth B, Horváth G (2007) Positive polarotaxis in a mayfly that never leaves the water surface: polarotactic water detection in *Palingenia longicauda* (Ephemeroptera). Naturwissenschaften 94: 148–154 + cover picture.
15. Horváth G, Kriska G (2008) Polarization vision in aquatic insects and ecological traps for polarotactic insects. In: Lancaster, J., Briers, R. A. (eds.) *Aquatic Insects: Challenges to Populations*. Wallingford, CAB International Publishing, Oxon, UK, Chapter 11, pp. 204–229.
16. Horváth G, Majer J, Horváth L, Szivák I, Kriska G (2008) Ventral polarization vision in tabanids: horseflies and deerflies (Diptera: Tabanidae) are attracted to horizontally polarized light. Naturwissenschaften 95: 1093–1100.
17. Lerner A, Meltser N, Sapir N, Erlick C, Shashar N, et al. (2008) Reflected polarization guides chironomid females to oviposition sites. Journal of Experimental Biology 211: 3536–3543.
18. Egri Á, Blahó M, Sándor A, Kriska G, Gyurkovszky M, et al. (2012) New kind of polarotaxis governed by degree of polarization: attraction of tabanid flies to differently polarizing host animals and water surfaces. Naturwissenschaften 99: 407–416 + electronic supplement.
19. Boda P, Csabai Z (2013) When do beetles and bugs fly? A unified scheme for describing seasonal flight behaviour of highly dispersing primary aquatic insects. Hydrobiologia 703: 133–147.
20. Horváth G, Kriska G, Malik P, Robertson B (2009) Polarized light pollution: a new kind of ecological photopollution. Frontiers in Ecology and the Environment 7: 317–325.
21. Wildermuth H, Horváth G (2005) Visual deception of a male *Libellula depressa* by the shiny surface of a parked car (Odonata: Libellulidae). International Journal of Odonatology 8: 97–105.
22. Russev BK (1992) Threatened species of Ephemeroptera (Insecta) from Bulgaria. Lauterbornia H. 9: 13–17.
23. 13/2001 (V. 9.) Decree of Hungarian Ministry of Environment and Water (HMEW). Website of HMEW. Available: http://net.jogtar.hu/jr/gen/hjegy_doc.cgi?docid=A0100013.KOM and http://natura.2000.hu/doc/13_2001._V.9._KoM_rendelet_2.pdf. Accessed 1 July 2014.
24. Kriska G, Horváth G, Andrikovics S (1998) Why do mayflies lay their eggs *en masse* on dry asphalt roads? Water-imitating polarized light reflected from asphalt attracts Ephemeroptera. Journal of Experimental Biology 201: 2273–2286.
25. Kriska G, Bernáth B, Farkas R, Horváth G (2009) Degrees of polarization of reflected light eliciting polarotaxis in dragonflies (Odonata), mayflies (Ephemeroptera) and tabanid flies (Tabanidae). Journal of Insect Physiology 55: 1167–1173.
26. Horváth G, Blahó M, Egri Á, Kriska G, Seres I, et al. (2010) Reducing the maladaptive attractiveness of solar panels to polarotactic insects. Conservation Biology 24: 1644–1653 + electronic supplement.
27. Horváth G, Blahó M, Kriska G, Hegedüs R, Gerics B, et al. (2010) An unexpected advantage of whiteness in horses: the most horsefly-proof horse has a depolarizing white coat. Proceedings of the Royal Society B 277: 1643–1650.
28. Blahó M, Egri Á, Barta A, Antoni G, Kriska G, et al. (2012) How can horseflies be captured by solar panels? A new concept of tabanid traps using light polarization and electricity produced by photovoltaics. Veterinary Parasitology 189: 353–365.
29. Blahó M, Egri Á, Báhidszki L, Kriska G, Hegedüs R, et al. (2012) Spottier targets are less attractive to tabanid flies: on the tabanid-repellency of spotty fur patterns. *Public Library of Science ONE* (*PLoS ONE*) 7(8): e41138. doi: 10.1371/journal.pone.0041138 + supporting information.
30. Egri Á, Blahó M, Kriska G, Farkas R, Gyurkovszky M, et al. (2012) Polarotactic tabanids find striped patterns with brightness and/or polarization modulation least attractive: An advantage of zebra stripes. Journal of Experimental Biology 215: 736–745 + electronic supplement.
31. Egri Á, Blahó M, Száz D, Barta A, Kriska G, et al. (2013) A new tabanid trap applying a modified concept of the old flypaper: Linearly polarising sticky black surfaces as an effective tool to catch polarotactic horseflies. International Journal for Parasitology 43: 555–563.
32. Horváth G, Varjú D (1997) Polarization pattern of freshwater habitats recorded by video polarimetry in red, green and blue spectral ranges and its relevance for water detection by aquatic insects. Journal of Experimental Biology 200: 1155–1163.
33. Bernáth B, Szedenics G, Wildermuth H., Horváth G (2002) How can dragonflies discern bright and dark waters from a distance? The degree of polarization of reflected light as a possible cue for dragonfly habitat selection. Freshwater Biology 47: 1707–1719.
34. Encalada AC, Peckarsky BL (2007) A comparative study of the cost of alternative mayfly oviposition behaviors. Behavioral Ecology and Sociobiology 61: 1437–1448.
35. Henze MJ, Labhart T (2007) Haze, clouds and limited sky visibility: Polarotactic orientation of crickets under difficult stimulus conditions. Journal of Experimental Biology 210: 3266–3276.
36. Horváth G, Kriska G, Malik P, Hegedüs R, Neumann L, et al. (2010) Asphalt Surfaces as Ecological Traps for Water-Seeking Polarotactic Insects: How can the Polarized Light Pollution of Asphalt Surfaces be Reduced? Series: Environmental Remediation Technologies, Regulations and Safety. Nova Science Publishers, Inc., Hauppauge, New York, USA, p. 47.
37. Umow N (1905) Chromatische Depolarisation durch Lichtzerstreuung. Physikaliche Zeitschrift 6: 674–676.
38. Lerner A, Shashar N, Haspel C (2012) Sensitivity study on the effects of hydrosol size and composition on linear polarization in absorbing and nonabsorbing clear and semi-turbid waters. Jounal of the Optical Society of America A 29: 2394–2405.
39. Krcmar S (2013) Comparison of the efficiency of the olfactory and visual traps in the collection of horseflies (Diptera: Tabanidae). Entomol. Gener. 34: 261–267
40. Krcmar S, Lajos P (2011) Response of horse flies to aged equine urine (Diptera: Tabanidae). Entomol. Gener. 33: 245–250.
41. Briscoe AD, Chittka L (2001) The evolution of color vision in insects. Annual Review of Entomology 46: 471–510.
42. Zar JH (2010) *Biostatistical Analysis*. Pearson Prentice Hall, New Jersey, USA.

15

Transgenic *Bacillus thuringiensis* (Bt) Rice Is Safer to Aquatic Ecosystems than Its Non-Transgenic Counterpart

Guangsheng Li[1], Yongmo Wang[1]*, Biao Liu[2], Guoan Zhang[1]

1 Hubei Insect Resources Utilization and Sustainable Pest Management Key Laboratory, College of Plant Science and Technology, Huazhong Agricultural University, Wuhan, Hubei, P.R. CHINA, 2 Nanjing Institute of Environmental Science, Ministry of Environmental Protection of the People's Republic of China, Nanjing, P.R. CHINA

Abstract

Rice lines genetically modified with the crystal toxin genes from *Bacillus thuringiensis* (Bt) have experienced rapid development, with biosafety certificates for two Bt rice lines issued in 2009. There has still been no commercial release of these lines yet due to public concerns about human health and environmental risks. Some studies confirmed that Bt rice was as safe as conventional rice to non-target organisms when pesticides were not applied, however, pesticides are still required in Bt rice to control non-lepidopteran pests. In this study, we assessed the environmental effects of two Bt rice lines expressing either the *cry1Ab/1Ac* or *cry2A* genes, respectively, by using zooplanktons as indicator species under normal field management practices using pesticides when required. In the whole rice growing season, non-Bt rice was sprayed 5 times while Bt rice was sprayed 2 times, which ensured both rice achieved a normal yield. Field investigations showed that rice type (Bt and non-Bt) significantly influenced zooplankton abundance and diversity, which were up to 95% and 80% lower in non-Bt rice fields than Bt rice fields. Laboratory rearing showed that water from non-Bt rice fields was significantly less suitable for the survival and reproduction of *Daphnia magna* and *Paramecium caudatum* in comparison with water from Bt rice fields. Higher pesticide residues were detected in the water from non-Bt than Bt rice fields, accounting for the bad performance of zooplankton in non-Bt field water. Our results demonstrate that Bt rice is safer to aquatic ecosystems than non-Bt rice, and its commercialization will be beneficial for biodiversity restoration in rice-based ecosystems.

Editor: Juan Luis Jurat-Fuentes, University of Tennessee, United States of America

Funding: This study was supported by the Major Projects of National Genetically Modified Organisms Breeding (NO. 2014ZX08012005) and the National Natural Science Funds (31370544). The funders had no role in study design, data collection and analysis, decision to publish, or preparation of the manuscript.

Competing Interests: The authors have declared that no competing interests exist.

* Email: ymwang@mail.hzau.edu.cn

Introduction

Rice (*Oryza sativa* L.) is a staple food for more than three billion people [1], and more than 90% of total rice is produced and consumed in Asia [2]. Over 100 species of insects attack and damage rice, and lepidopteran pests such as rice stem borers are the most serious with yield losses ranging from 5% to 25% [3]. Rice has been successfully modified to express the cry genes derived from the *Bacillus thuringiensis* (Bt) bacterium, and several Bt rice lines with high resistance to lepidopteran pests have been developed over the past 15 years in China [4]. The China Ministry of Agriculture (CMOA) has confirmed that safety certificates were issued in November 2009 for two Bt rice lines, Huahui no. 1 and Bt Shanyou 63 [5], both of which express a fusion *cry1Ab/1Ac* gene. However, the safety certificates did not lead to final commercial release because there were fierce public debates on the health and environmental risks of Bt rice [6]. There is therefore an urgent need to provide the public with scientifically based knowledge of the actual environmental effects of genetically modified rice.

Many field and laboratory experiments have assessed the impact of Bt rice on paddy arthropods, including parasitoids, predators and non-target herbivores. In general, Bt rice has no negative impact (in comparison to non-Bt rice) to the individual fitness, population abundance and community diversity of non-target arthropods [7–13]. However, almost all of the field studies have excluded pesticides from both the Bt and non-Bt rice fields. Although Bt rice is effective in reducing pesticide sprays, some pesticide application is still required to control fungal diseases and non-lepidopteran pests if full yield is to be achieved [14,15]. To accurately assess the environmental impact of Bt rice it is therefore necessary to conduct assessments under normal agricultural practices, including pesticide sprays when required.

Previous environmental assessments of Bt rice have mainly focused on terrestrial arthropods, and have not considered the aquatic ecosystem [16]. In most growth stages of rice, a layer of water is sustained in the rice field. In tropic or subtropical areas, this paddy water layer teems with zooplankton such as rotifers, cladocerans and copepods, most of which are shared by adjacent ponds and lakes, and those organisms constitute an important food source for fish and predatory insects [17]. These organisms also carry out active functions in nutrient cycling in agroecosystems [18]. Further, pesticides readily dissolve in water, and the environmental effects of pesticide use are likely to be most acute in these ecosystems [19].

In the present study, we evaluated the environmental effects of two Bt rice lines, one of which has been granted a safety certificate by the CMOA, using paddy zooplanktons as indicator species. Bt rice and non-transgenic rice were planted in plots in a complete randomized block design. Experimental fields were inspected weekly for rice pest occurrence and sprayed with pesticides when required to guarantee that non-Bt and Bt rice achieved equal yields (Fig. 1). The diversity and abundance of three types of zooplankton were compared between Bt and non-Bt treatments. Field water was collected to conduct laboratory rearing experiments of water quality using *Daphnia magna* and *Paramecium caudatum*, two species commonly used to evaluate water pollutants [20,21]. Pesticide residues in field-collected water were also compared between Bt and non-Bt rice fields.

Materials and Methods

Ethics Statement

All field studies were carried out at an experimental field which is maintained by Huazhong Agricultural University (HZAU) and located in the campus (latitude 30°34′ N, longitude 114°47′ E). Special permits for the field studies were obtained from Science and Technology Department of HZAU which executes the function of GMO management under the supervision of China National Agricultural GMO Safety Committee (CNAGSC). All materials used during rice planting (pesticides and fertilizer etc.) were registered for legal use, and the products of Bt rice (seed and straw) were disposed according to the regulations formulated by CNAGSC. The field studies did not involve endangered or protected species.

Bt rice materials and field design

Two Bt rice lines, Bt rice 1 and Bt rice 2, and their common non-transgenic isogenic counterpart, were used as plant materials. Both the Bt rice lines are variety of Minghui 63, an elite indica rice restorer line. Bt Rice 1 was transformed with a synthetic *cry1Ab/1Ac* gene and Bt rice 2 was transformed with the *cry2A* gene [22,23]. Bt rice 1 is also called Huahui no. 1, which was issued with a safety certificate by the CMOA in 2009 [5]. Field experiments were carried out at an experimental field at the HZAU campus in Wuhan (China) in 2013. The experimental field was constructed from a small hill in 2008, and planted with rice for 3 years without pesticide spray. According to a randomized complete block design (RCBD), the field was divided into 3 blocks, and each block divided into 3 plots which were planted with Bt rice 1, Bt rice 2 and non-Bt rice randomly. Each plot covered an area of 220 m^2. Solid ridges and drainage channels were built around plots to avoid water flow between the plots. Both Bt and non-Bt rice were sowed in early May and transplanted by hand one month later. Spacing between rows was 25 cm, and hills within rows were 20 cm apart. The field was irrigated with water from a lake nearby. The soil in the experimental field was typically a yellow-brown soil consisting of 25.6% clay, 68.8% silt, 5.6% sand, and 11.2% organic matter. Field management was performed following conventional agronomic practices. Pesticides including avermectins, triazophos, difenoconazole, propiconazole and buprofezin were used when any of the rice pests exceeded action threshold (see Table S1). Action threshold refers to a point at which pest population or environmental conditions indicate that pest control action must be taken to guarantee no economic loss. A windshield was held downwind between plots when spraying.

Field investigation of zooplanktons

Five litres of water were collected from each plot in the morning before 10 a.m., and were concentrated to 50 ml using plankton net (64 μm mesh size). The concentrated samples were fixed with Lugol's solution, and stored in amber bottles. Three kinds of

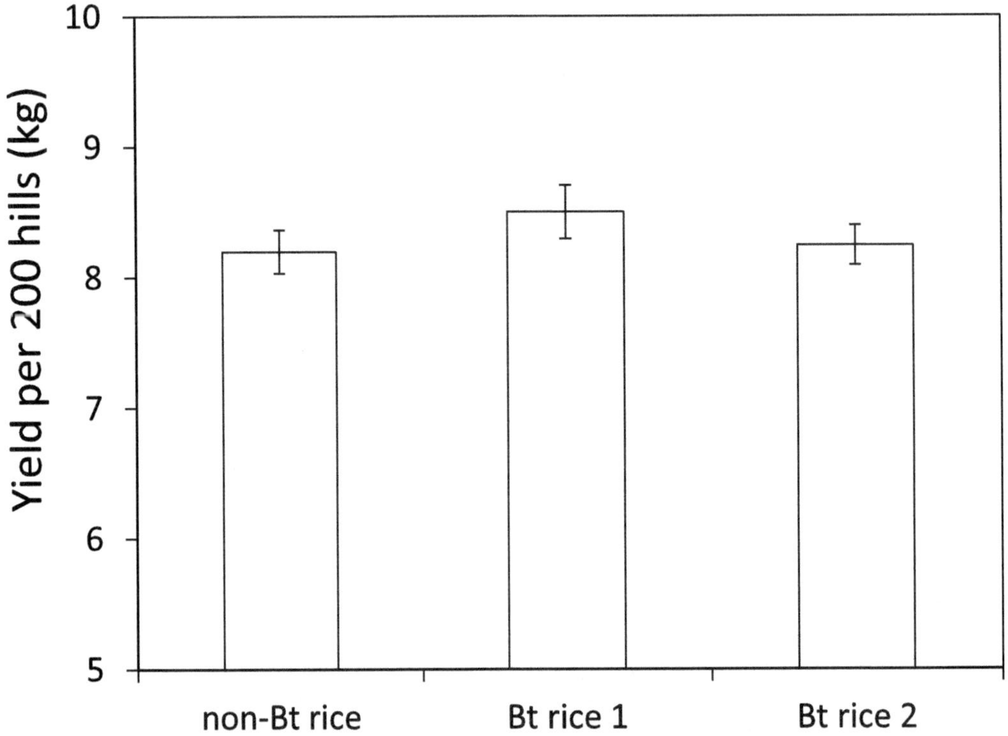

Figure 1. Yield of Bt and non-Bt rice. Non-Bt rice = Minghui 63, Bt rice 1 = Minghui 63 *cry1Ab/1Ac*, Bt rice 2 = Minghui 63 *cry2A*. Non-Bt rice as well as Bt rice were sprayed with pesticides when insect or fungal pests reached the action threshold. Error bars indicate the standard error. Planned comparison of means showed that the yield of non-Bt rice and Bt rice did not differ significantly (d.f. = 6; t = 1.303; p = 0.240).

zooplanktons, Rotifera, Cladocera and Copepoda, were identified down to the lowest possible taxonomic level, and were counted with the drop count method [24]. Each sample was counted twice and the mean was recorded. The zooplankton investigation was conducted twice, 5 and 12 days after the final pesticide spray on August 1st (see table S1).

Daphnia magna culturing in field-collected water

Daphnia magna is highly sensitive to toxic substances in aquatic ecosystem, and has a short generation time [25]. *Daphnia magna* used in the experiments were from a single clonal population cultured in laboratory. Field water was collected from each plot and was filtered by 64 μm plankton net to remove suspended particles and plankton. The filtered water was used as culture medium in the water quality tests. A newly hatched *D. magna* (< 24 h old) was transferred to a 100-ml beaker filled with 80 ml filtered field-collected water using a broad-tipped pipette. This was replicated 30 times per plot (9 plots in total). Three millilitres of *Scenedesmus obliquus* solution ($4\sim5\times10^5$ cells/ml), cultivated according to Stein 1973 [26], was added to each beaker as food source for the *D. magna* [27]. Every 3 days, each *D. magna* was transferred to a new beaker with new filtered field-collected water and food source of *S. obliquus*. Each beaker was inspected daily to record survival and presence of newly born *D. magna*, and removal of offspring. All cultures were maintained under static conditions at 23±1°C and a controlled photoperiod at 12 h light and 12 h darkness.

Field-collected water quality test using *Paramecium caudatum*

Paramecium are unicellular organisms belonging to Ciliophora that are commonly found in ponds and lakes. They are often used as bioindicators of aquatic environmental pollutants [28]. Stock cultures of *P. caudatum* were cultured in laboratory according to Ishikawa and Hota's procedure [29]. Twenty grams of dry rice straw (from a field without pesticide application) was boiled in 2 L tap water for 5 min and the supernatant was used as liquid food of *P. caudatum*. The experimental unit consisted of a 50 ml centrifuge tube with 10 ml culture medium, 3 ml *P. caudatum* stock culture and 37 ml filtered field-collected water. The initial density in the tubes was 54.3 cells ml. Each plot had 5 experimental replicates. The cultures were maintained in a climate chamber at 23±1°C, with a photoperiod of 12 h light and 12 h darkness. The density in each tube was determined every day in the morning using a stereo microscope and counting chamber. This experiment lasted for 8 days.

Pesticide residue test in field-collected water

Field water samples were collected from each plot on August 11, 2013 (see Table S1 for pesticide application dates). All samples were filtered through a plankton net (64 μm mesh size) and stored into 100 ml amber glass bottles at 4°C. The samples were sent to Technology Service Centre Co., Ltd (http://www.labfaster.com/) the next day, and subjected to liquid chromatography-mass spectrometry (LC-MS/MS) analysis for residues of 5 pesticides. Those pesticides had been used in the experimental field before the sampling date. Each sample was analyzed twice and the average residue was recorded.

Rice yield test

After rice grain fully ripened in September, 20 rows of rice each containing 10 hills were selected randomly in each plot. Panicles were threshed and rice grain was dried in sunlight and weighed.

Data analysis

Yield data were analyzed by one-way RCBD ANOVA, and means were compared between non-Bt rice and Bt rice by a planned comparison. Data of zooplankton in field investigation were recorded as each group abundance and total abundance, and those abundances were transformed by ln (x+1) to meet the assumption of homogeneity of variances and then were subjected to one-way RCBD ANOVA. The means of abundance were separated by Fisher's LSD test. In order to explore the effect of overall rice type on abundance of zooplankton, multivariate analysis of variances (MANOVA) combining each group of zooplankton were also conducted. The survival rates of *D. magna* were arcsine-transformed and subjected to repeated measures ANOVA to estimate rice type and culturing time effect and their interaction; data of total reproduction were square root transformed and analysed by one-way RCBD ANOVA followed by a LSD test. The population growth of *P. caudatum* were also ln (x+1) transformed and subjected to repeated measures ANOVA. All analyses were carried out using general linear model (GLM) in SPSS package (version 16.0, SPSS Inc).

Results

Zooplankton abundance and diversity in Bt and non-Bt rice field

Thirty-three species were identified, 17 species in Rotifera, 14 in Cladocera and 2 in Copepoda (see Table S2). In the first investigation, 4, 21 and 18 species were found in non-Bt rice, Bt rice 1 and Bt rice 2 treatment, respectively, and 16, 24 and 19 species in the second investigation (see Table S2). One-way randomized complete block design (RCBD) ANOVAs detected significant rice type effect (d.f. = 2,4; $F>10.659$; $p<0.035$) and non-significant block effect (d.f. = 2,4; $F<4.515$; $p>0.094$) in both investigations on abundance of each zooplankton group (Table 1). The lowest abundance was always found in non-Bt rice plots. In the first investigation, total abundance in Bt rice treatments was around 1,500 individuals L^{-1}, which was significantly higher than that in non-Bt rice treatment (below 50 individuals L^{-1}) (LSD test, $p<0.01$); the differences between Bt rice 1 and Bt rice 2 were not significant (LSD test, $p=0.974$). The second investigation found rebound in zooplankton abundance to around 1,600 and 640 individuals L^{-1} in Bt rice and non-Bt rice respectively, but the trend of differences between Bt and non-Bt did not change (Table 1). Multivariate analyses of variance (MANOVA) were used to assess overall rice type effect on abundance of 3 groups of zooplankton, and significant effects were detected in the first investigation (d.f. = 6,4; $F=9.049$; $p=0.026$), as well as in the second (d.f. = 6,4; $F=7.678$; $p=0.034$). Therefore the water layer in Bt rice fields harbored more abundant and diverse zooplankton than that of non-Bt rice.

Laboratory rearing of *D. magna* and *P. caudatum*

We tested the survival of *D. magna* and *P. caudatum* in water collected from Bt and non-Bt rice plots in the laboratory. Daphnids (<24 hrs old) were reared in 100 ml field-collected water with *Scenedesmus obliquus* as feed, and every 3 days the culture medium was refreshed with field-collected water. Differences on survival rate were found between Bt and non-Bt field-collected water one day after the onset of this experiment (Fig. 2A). In following days, survival rate in the non-Bt treatment decreased gradually to zero by the end of this 13-day experiment, while survival rates in the two Bt treatments remained relatively stable at 52% and 60% by day 13. Repeated-measures ANOVA within RCBD found significant rice type effect (d.f. = 2,4;

Table 1. Abundance (individuals L^{-1}) and diversity (number of species) of Rotifera, Cladocera and Copepoda in Bt and non-Bt rice plots.

Treatment	Rotifera	Cladocera	Copepoda	Total
1st investigation (abundance)				
non-Bt rice	10.33±5.17b	0±0b	35.33±25.56b	45.66±29.55b
Bt rice 1	172.66±74.99a	29.00±7.57a	1501.00±651.00a	1702.66±688.26a
Bt rice 2	254.67±119.32a	56.26±12.33a	1031.67±335.56a	1342.50±348.50a
1st investigation (number of species)				
non-Bt rice	1.33±0.67a	0±0b	1.33±0.67b	2.67±1.33b
Bt rice 1	4.67±1.20a	7.00±1.00a	3.67±0.33a	15.33±1.20a
Bt rice 2	3.67±0.33a	6.33±2.08a	3.63±0.33a	15.70±3.00a
2nd investigation (abundance)				
non-Bt rice	425.33±107.33b	4.66±2.33b	148.00±43.39b	638.00±69.51b
Bt rice 1	1291.33±201.17a	34.33±8.76a	598.00±150.89a	1923.67±325.42a
Bt rice 2	829.67±25.69a	38.33±10.84a	521.33±50.11a	1317.33±63.89a
2nd investigation (number of species)				
non-Bt rice	4.33±0.67a	2.00±0.58b	3.00±0.58a	9.33±0.88b
Bt rice 1	5.67±0.33a	7.00±1.53a	3.67±0.33a	16.33±1.86a
Bt rice 2	5.00±0.00a	6.00±0.00a	4.00±0.00a	15.00±0.00a

Non-Bt and Bt rice plots were sprayed with pesticides when rice pests exceeded the action threshold (Table S1). The first and the second investigation were conducted 5 days and 12 days after the final spray, respectively. The values presented in the table are untransformed means ± SE of each treatment. Means followed by the same letter within the same column in each investigation do not differ significantly at $p = 0.05$ (LSD post hoc test after ANOVA, n = 3).

$F = 34.393$, $p = 0.003$) and measuring time effect (d.f. = 12,52; $F = 59.489$; $p < 0.001$), and no significant block effect (d.f. = 2,4; $F = 1.587$; $p = 0.311$). The survival rate did not differ between the two Bt rice treatments (LSD test, $p = 0.549$), but did between non-Bt rice and Bt rice 1 or Bt rice 2 (LSD test, $p = 0.002$ and $p = 0.003$). Reproduction was observed in the culture medium on the 7th day of the experiment in all treatments, which indicated rice type had no effect on sexual maturity age of *D. magna*. A significant effect of rice type on average reproduction rate was observed (d.f. = 2,4; $F = 8.963$; $p = 0.033$), and LSD tests revealed significant differences between non-Bt rice and Bt rice 1 ($p = 0.018$) or Bt rice 2 ($p = 0.026$) (Fig. 2B). The growth (body length) of *D. magna* was also recorded, but no significant differences were found.

A stock of *P. caudatum* was cultured in the laboratory by traditional methods [29]. The stock culture was diluted to 54.3 cells ml^{-1} in 50 ml culture medium with 10 ml rice straw infusion and approx. 37 ml filtered field water collected from Bt and non-Bt plots. The population growth was inspected daily until the 8th day. The *P. caudatum* expressed lower growth rate in non-Bt rice compared to the two Bt rice treatments (Fig. 3). Repeated-measures ANOVA within RCBD detected significant effects of rice type (d.f. = 2,4; $F = 195.707$; $p < 0.001$), measuring time (d.f. = 8,32; $F = 1790.116$; $p < 0.001$) and the interaction of rice type × measuring time (d.f. = 16,32; $F = 13.109$; $p < 0.001$). The growth rates did not differ between Bt rice 1 and Bt rice 2 (LSD test, $p = 0.072$). The mean density was about 300 cells/ml in non-Bt culture, and 450 and 500 in Bt rice 1 and Bt rice 2 cultures by the end of this experiment.

Pesticide residue in field-collected water

In our previous study, we revealed that Bt rice (with gene *cry1Ab/1Ac*) in itself has no negative impact on paddy zooplanktons [30]. In order to determine what caused the differences in survival found earlier, we tested pesticide residues in the water collected from the field 10 days after the final spray (see Table S1 for pesticide application dates). We found that residues of pesticides targeting Homopteran and fungal pests were in the same order of magnitude in Bt and non-Bt rice plots, since those pesticides were used both in Bt rice and non-Bt rice (Table 2). The residues of chlorantraniliprole and triazophos, targeting lepidopteran pests, were100 times higher in non-Bt plots than Bt plots (Table 2). Theoretically, pesticides targeting lepidopteran pests would not be detected in Bt plots because they were not used in Bt rice. However, they were detected in Bt plots probably because pesticide spray drifted by wind or/and water although we had made efforts to avoid this. In short, the pesticide residues in non-Bt plots were far higher than Bt plots, which likely accounts for the poor fitness performance of zooplanktons in the non-Bt treatment.

Discussion

Research on genetic modification has developed rapidly in China. In November 2009, COMA announced that biosafety licenses were issued to two lepidopteran-resistant Bt rice lines in designated areas. However, the decision was the subject of heated scientific and public debates in the following years. The core of these debates is the uncertainty regarding food safety and potential environmental risks.

In this study, we evaluated the environmental risk of two Bt rice lines from a new angle. Firstly, we focused on paddy aquatic organisms as the non-target organisms in the environmental assessment of Bt rice. Previous non-target effect studies of Bt crops mainly focuses on terrestrial invertebrates, such as earthworms, collembolans, mites, woodlice and nematodes [31–35]. Like those terrestrial invertebrates, aquatic organisms also carry out important ecological functions, such as nutrient cycling, decomposition of organic matter and maintaining aquatic food webs [36]. In

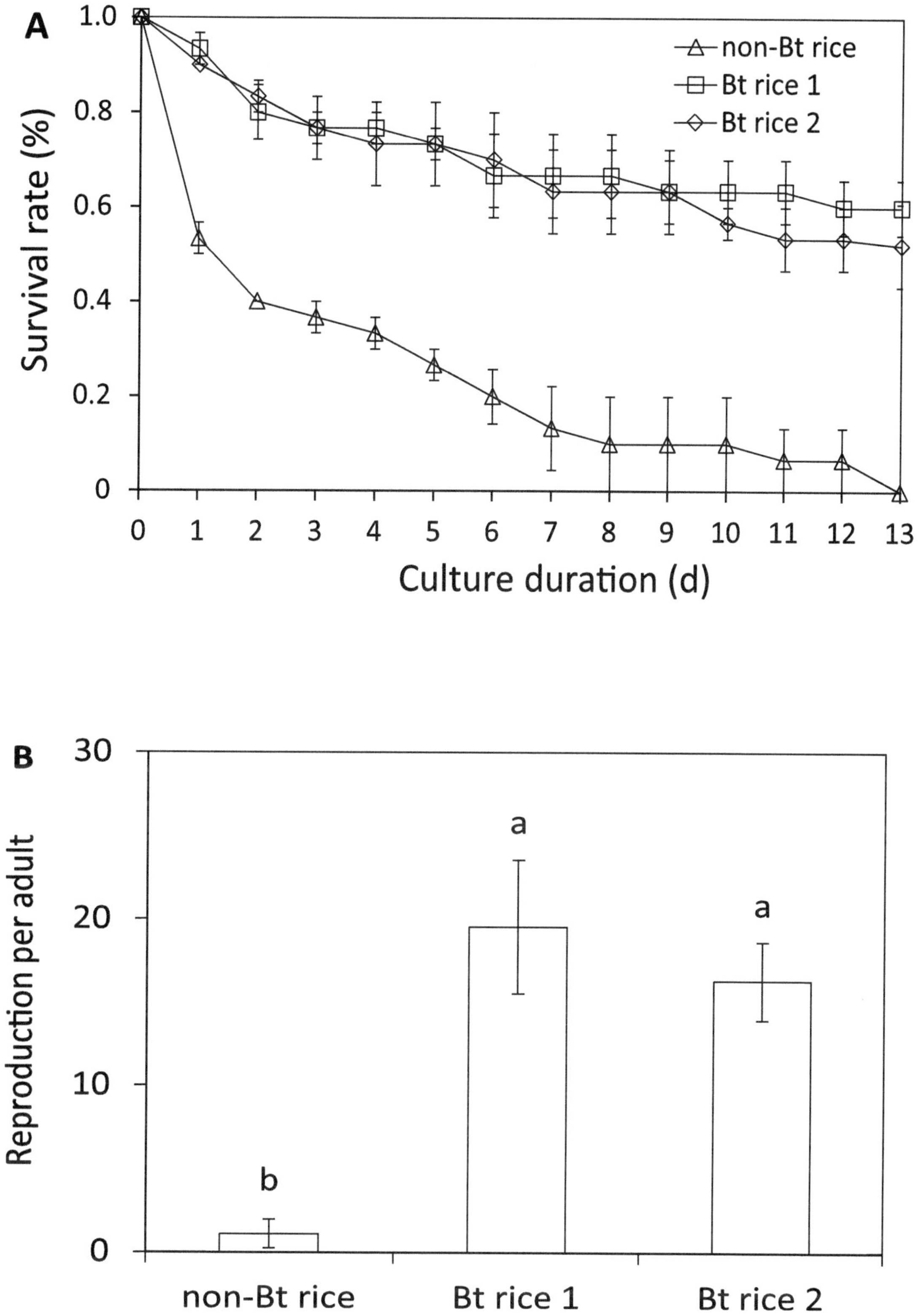

Figure 2. Survival (A) and reproductive rate (B) of *D. magna* reared in water collected from non-Bt rice and Bt rice plots. Water was collected 6 days after the final pesticide sprays and the culture medium was renewed every 3 days with field water collected on that very day. Non-Bt and Bt rice plots were sprayed with pesticides when rice pests exceeded the action threshold (see Table S1). Error bars indicated standard error. Different letters capped on the bars indicate significant difference at $p = 0.05$ (LSD post hoc test after ANOVA).

many areas, the water matrix of rice fields is connected to nearby ponds and lakes which are important resources for rice farming farmers for additional income and food by fish production [37]. Secondly, we assessed Bt rice's environmental effects in compar-

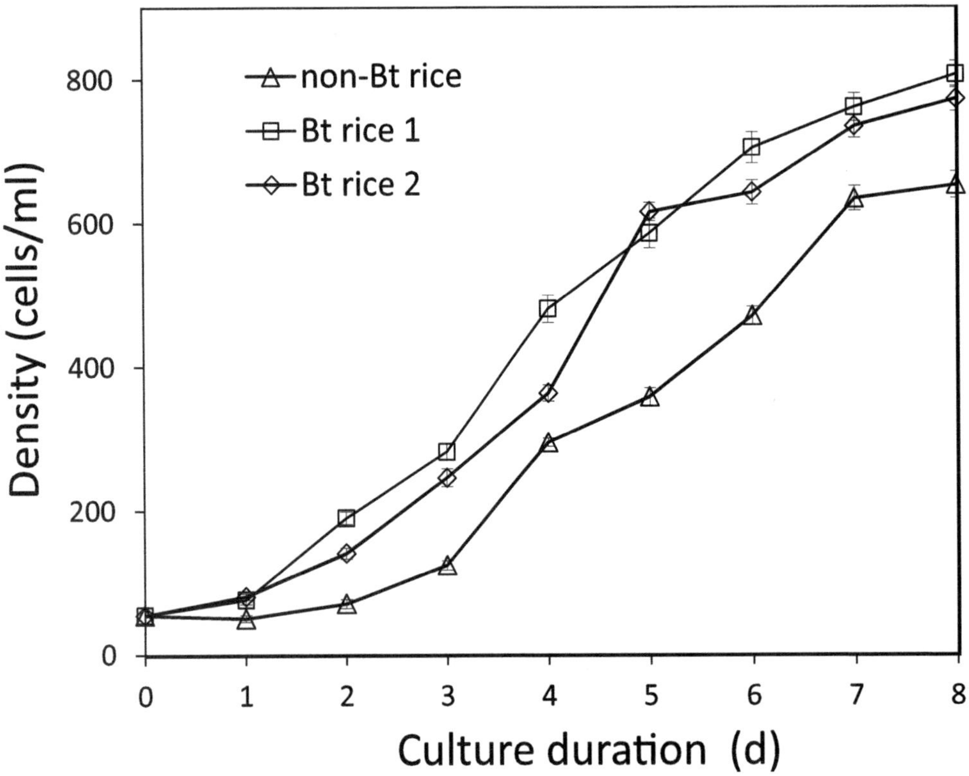

Figure 3. Population growth of *P. caudatum* in culture medium made of field water. Field water was collected from non-Bt rice and Bt rice plots 11 days after the final pesticide spray (see Table S1). Error bars indicate the standard error.

ison with non-Bt rice under realistic agricultural conditions in which pesticides were used in both Bt and non-Bt fields as required. Former field studies demonstrated that although Bt rice (*cry1Ab/Ac*) reduced pesticide use by up to 60%, they were still required against fungal and non-lepidopteran pests in order to guarantee no yield loss [14,15]. However, almost all previous studies on Bt rice's non-target effect were conducted without considering the factor of pesticide application [4].

In this study, Bt rice was sprayed twice and non-Bt rice was sprayed 5 times throughout the whole growing season (Table S1), which resulted in equal yield between Bt and non-Bt rice (Fig. 1). The Bt rice expressed good resistance to lepidopteran pests, more specifically rice leaf folder and rice stem borers, and the two sprays only targeted a fungal disease and rice planthopper. The same results were reported in Bt cotton and Bt maize which also greatly reduced insecticide applications [38].

Our first field investigation found that the total abundance of three groups of zooplankton Rotifera, Cladocera and Copepoda in non-Bt field was about 95% lower than that in Bt rice field, and the total number of species was 80% lower (Table 1). The second sampling was conducted 7 days after the first one, and a rise in abundance and diversity was observed, however zooplanktons were still significantly more abundant and diverse in the Bt rice field. In contrast to this result, our previous two-year field study, in which we did not use pesticides, found the total population density of rotifera, cladocera and copepods did not differ significantly between Bt (*cry1Ab/1Ac*) and non-Bt rice fields [30]. The differences we report here are therefore not due to Bt rice itself but from the differences in pesticide application. We tested pesticide residues using LC-MS/MS in water samples collected from rice fields and found that pesticide (targeting lepidopteran pests) residues were around 100 times higher in non-Bt than Bt

Table 2. Pesticide residue ± SE ($\mu g\ L^{-1}$) in water collected from non-Bt and Bt rice plots.

Pesticide	Non-Bt rice	Bt rice 1	Bt rice 2
chlorantraniliprole	15.19±1.41	0.17±0.04	0.17±0.04
difenoconazole	5.44±1.37	1.47±0.89	2.57±1.38
propiconazole	35.21±1.15	42.67±14.92	26.93±8.89
triazophos	80.45±11.91	0.78±0.23	1.19±0.52
buprofezin	0.73±0.29	0.16±0.08	0.28±0.21

Field water was collected 10 days after the final spray (see table S1 for the spraying date). Pesticide residues were quantified by liquid chromatography-tandem mass spectrometry (LC-MS/MS).

plots (Table 2). It is widely documented that pesticides enter into aquatic environments and accumulate to impact certain aquatic organisms. Some zooplanktons are frequently used in ecotoxicological tests of anthropogenic chemicals [39]. The qualitative and quantitative characteristics of zooplankton population depend on the degree of water pollution [40]. The high abundance of zooplankton in Bt rice fields corresponds with low water pollution from reduced pesticide application, and can be considered a positive effect of Bt rice on aquatic environment.

In addition to field investigation, we conducted laboratory experiment to rear *D. magna* and *P. caudatum* in field-collected water. The survival rate of *D. magna* declined to zero in non-Bt rice field-collected water by the end of the 13-days rearing experiment, while the survival rate was up to 60% in Bt rice field water (Fig. 2A). The amount of reproduction in Bt rice field water was more than 15 times higher than in non-Bt rice field water (Fig. 2B). The population of *P. caudatum* also developed faster in Bt rice field-collected water, and the final population density was about 20% higher in Bt rice field-collected water than non-Bt rice water (Fig. 3). Up to now, only a few studies have assessed non-target effects of Bt crops on aquatic organisms. Rosi-Marshall and colleagues reported that consumption of Bt corn (*cry1Ab*) debris reduced growth and increased mortality of caddisfly larvae in laboratory feeding trials [41]. Bøhn's study demonstrated that *D. magna* fed on Bt maize (*cry1Ab*) showed significantly reduced fitness performance compared to the non-Bt control [42]. Rosi-Marshall's study concluded that Bt δ-endotoxin in Bt maize byproducts harmed aquatic trichopterans insects which are phylogenetically close to Lepidoptera, and Bøhn's study concluded a toxic effect rather than a lower nutritional value of Bt maize accounting for the lower fitness of *D. magna*. Their findings were quite different from the results of our present study. Although Bt protein is sometimes detected in Bt rice field water [43], we did not consider Bt proteins to have negative impact on paddy *Daphnia* because a another water flea species, *Daphnia hyalina*, was shown to perform as well in culture medium with high concentrations of Bt toxin as in the control medium [30]. In this study, we cultured *D. magna* and *P. caudatum* just in water collected from rice field and did not feed with Bt rice materials because water flea and paramecium typically do not feed directly on rice products or by-products [44]. In their studies, the non-target organisms were fed with high doses of Bt maize materials for a longer time than ours. Nevertheless, those aquatic organisms would not use a single food source for that long in the natural environment. Furthermore, they did not consider agricultural chemicals (such as pesticides) that probably existed in their materials, making their conclusions questionable [45]. We found that field-collected water from Bt rice fields was more suitable for survive survival and reproduction of paddy zooplanktons in short-term rearing trials, and we also detected higher pesticide residues in the water from non-Bt rice fields than Bt rice fields. Our results therefore indicate that pesticide residues in field-collected water play a more important role in zooplankton fitness than Bt toxins and Bt rice, which highlight the importance of considering pesticide application in non-target environmental assessments of Bt crops.

Conclusions

In order to settle the debate on Bt rice and realize its potential benefit as early as possible [46], scientific experiments are needed to reveal the real environmental effects of Bt rice. We assessed the influence of two Bt rice lines on aquatic organisms under normal field management which used pesticide when pests exceeded the action threshold. The results indicated that Bt rice required reduced pesticide application therefore produced lower pesticide residues in the field water body. The abundance and diversity of zooplankton were significantly higher in Bt rice fields than non-Bt rice fields, and *D. magna* and *P. caudatum* displayed far higher survival and fecundity in Bt rice field water than non-Bt rice field water. These results demonstrate that the Bt rice lines are more friendly to the aquatic ecosystem than conventional rice. Pesticide application in rice fields has long been blamed for biodiversity loss in rice-based ecosystems [47] and it is expected that biodiversity will gradually be restored following the release of Bt rice.

Acknowledgments

We thank Hui Jing for identifying and counting zooplanktons and Mingyuan He for providing *D. magna* samples and Yongjun Lin for providing Bt rice seeds. We also thank James Hereward for valuable comments on grammar and syntax of the manuscript.

Author Contributions

Conceived and designed the experiments: YW. Performed the experiments: GL. Analyzed the data: YW GL. Contributed to the writing of the manuscript: YW. Provided some advice in experimental design: BL. Performed field management: GZ.

References

1. Redona ED (2004) Rice Biotechnology for Developing Countries in Asia. In: A Eaglesham editors. Agricultural Biotechnology: Finding Common International Goals. National Agricultural Biotechnology Council (NABC) Report No. 16. Ithaca (New York): National Agricultural Biotechnology Council. 201–232.
2. Khush GS (1997) Origin, dispersal, cultivation and variation of rice. Oryza: From Molecule to Plant, (Springer), 25–34.
3. Pathak MD, Khan ZR (1994) Insect pests of rice Insect pests of rice. International Rice Research Institute, Los Banos, Philippines. 89 p.
4. Chen M, Shelton A, Ye GY (2011) Insect-resistant genetically modified rice in China: from research to commercialization. Annu Rev Entomol 56: 81–101.
5. Lu C (2010) The first approved transgenic rice in China. GM Crops 1(3): 113–115.
6. Niu S, Stanway D (8 Mar 2013) China delays GMO corn and rice to woo the public. The China Post. Available: https://www.chinapost.com.tw/china/national-news/2013/03/08/372455/China-delays.htm Accessed 15 Mar 2014.
7. Tian JC, Liu ZC, Yao HW, Ye GY, Peng YF (2008) Impact of transgenic rice with a cry1Ab gene on parasitoid subcommunity structure and the dominant population dynamics of parasitoid wasps in rice paddy. J Environ Entomol 1: 003.
8. Liu Z, Ye G, Hu C, Swapan KD (2001) Effects of Bt transgenic rice on population dynamics of main non-target insect pests and dominant spider species in rice paddies. Acta Phytophylacica Sinica 29(2): 138–144.
9. Jiang Y, He J, Dong B, Xu H, Fu Q, et al. (2003) Effects of transgenic Bt rice on the biological characteristics of *Apanteles chilonis* (Munakata) (Hymenoptera: Braconidae). Acta Entomol Sinica 47(1): 124–129.
10. Chen M, Zhao JZ, Ye GY, Fu Q, Shelton AM (2006) Impact of insect-resistant transgenic rice on target insect pests and non-target arthropods in China. Insect Sci 13(6): 409–420.
11. Chen M, Ye G, Liu Z, Fang Q, Hu C, et al. (2009) Analysis of Cry1Ab toxin bioaccumulation in a food chain of Bt rice, an herbivore and a predator. Ecotoxicology 18(2): 230–238.
12. Bai Y, Jiang M, Cheng J, Wang D (2006) Effects of Cry1Ab toxin on *Propylea japonica* (Thunberg) (Coleoptera: Coccinellidae) through its prey, *Nilaparvata lugens* Stål (Homoptera: Delphacidae), feeding on transgenic Bt rice. Environ Entomol 35(4): 1130–1136.

13. Bai Y, Jiang M, Cheng J (2005) Effects of transgenic cry1Ab rice pollen on fitness of Propylea japonica (Thunberg). J Pest Sci 78(3): 123–128.
14. Wang Y, Zhang G, Du J, Liu B, Wang M (2010) Influence of transgenic hybrid rice expressing a fused gene derived from cry1Ab and cry1Ac on primary insect pests and rice yield. Crop Prot 29(2): 128–133.
15. Huang J, Hu R, Rozelle S, Pray C (2005) Insect-resistant GM rice in farmers' fields: assessing productivity and health effects in China. Science 308(5722): 688–690.
16. Carstens K, Anderson J, Bachman P, De Schrijver A, Dively G, et al. (2012) Genetically modified crops and aquatic ecosystems: considerations for environmental risk assessment and non-target organism testing. Transgenic Res 21(4): 813–842.
17. Ali AB (1990) Seasonal dynamics of microcrustacean and rotifer communities in Malaysian rice fields used for rice-fish farming. Hydrobiologia 206(2): 139–148.
18. Vanni MJ (2002) Nutrient cycling by animals in freshwater ecosystems. Annu Rev Ecol Syst 33(1): 341–370.
19. Liess M, Ohe PCVD (2005) Analyzing effects of pesticides on invertebrate communities in streams. Environ Toxicol Chem 24(4): 954–965.
20. Biesinger KE, Christensen GM (1972) Effects of various metals on survival, growth, reproduction, and metabolism of Daphnia magna. J Fish Board of Can 29(12): 1691–1700.
21. Shiny K, Remani K, Nirmala E, Jalaja T, Sasidharan V (2005) Biotreatment of wastewater using aquatic invertebrates, *Daphnia magna* and *Paramecium caudatum*. Bioresour Technol 96(1): 55–58.
22. Tu J, Zhang G, Datta K, Xu C, He Y, et al. (2000) Field performance of transgenic elite commercial hybrid rice expressing *Bacillus thuringiensis* δ-endotoxin. Nat Biotechnol 18(10): 1101–1104.
23. Chen H, Tang W, Xu C, Li X, Lin Y, et al. (2005) Transgenic indica rice plants harboring a synthetic cry2A* gene of *Bacillus thuringiensis* exhibit enhanced resistance against lepidopteran rice pests. Theor Appl Genet 111(7): 1330–1337.
24. Emmanuel B, Onyema I (2007) The Plankton and fishes of a tropical creek in South-western Nigeria. Turk. J. Fish. Aquat. Sci 7(2): 105–113.
25. Lovern SB, Strickler JR, Klaper R (2007) Behavioral and physiological changes in *Daphnia magna* when exposed to nanoparticle suspensions (titanium dioxide, nano-C60, and C60HxC70Hx). Environ Sci Technol 41(12): 4465–4470.
26. Stein JR (1973) Handbook of phycological methods: Culture methods and growth measurements (Cambridge University Press, New York). 472 p.
27. Villegas-Navarro A, Gonzalez M, Lopez ER, Aguilar RD, Marcal WS (1999) Evaluation of *Daphnia magna* as an indicator of toxicity and treatment efficacy of textile wastewaters. Environ Int 25(5): 619–624.
28. Miyoshi N, Kawano T, Tanaka M, Kadono T, Kosaka T, et al. (2003) Use of Paramecium Species in Bioassays for Environmental Risk Management: Determination of IC50 Values for Water Pollutants. J Health Sci 49(6): 429–435.
29. Ishikawa T, Hota M (2006) Interaction of two swimming Paramecia. J Exp Biol 209(22): 4452–4463.
30. Wang Y, Huang J, Hu H, Li J, Liu B, et al. (2013) Field and laboratory studies on the impact of two Bt rice lines expressing a fusion protein Cry1Ab/1Ac on aquatic organisms. Ecotox Environ Safe 92: 87–93.
31. Zwahlen C, Hilbeck A, Howald R, Nentwig W (2003) Effects of transgenic Bt corn litter on the earthworm *Lumbricus terrestris*. Mol Ecol 12(4): 1077–1086.
32. Heckmann LH, Griffiths BS, Caul S, Thompson J, Pusztai-Carey M, et al. (2006) Consequences for Protaphorura armata (Collembola: Onychiuridae) following exposure to genetically modified *Bacillus thuringiensis* (Bt) maize and non-Bt maize. Environ Pollut 142(2): 212–216.
33. Griffiths BS, Caul S, Thompson J, Birch ANE, Scrimgeour C, et al. (2006) Soil Microbial and Faunal Community Responses to Maize and Insecticide in Two Soils. J Environ Qual 35(3): 734–741.
34. Clark BW, Prihoda KR, Coats JR (2006) Subacute effects of transgenic Cry1Ab *Bacillus thuringiensis* corn litter on the isopods *Trachelipus rathkii* and *Armadillidium nasatum*. Environ Toxicol Chem 25(10): 2653–2661.
35. Saxena D, Stotzky G (2001) *Bacillus thuringiensis* (Bt) toxin released from root exudates and biomass of Bt corn has no apparent effect on earthworms, nematodes, protozoa, bacteria, and fungi in soil. Soil Biol Biochem 33(9): 1225–1230.
36. Simpson I, Roger PA (1995) The impact of pesticides on nontarget aquatic invertebrates in wetland ricefields: a review. Impact of pesticides on farmer health and the rice environment, (Springer), 249–270.
37. Kangmin L (1988) Rice-fish culture in China: A review. Aquaculture 71(3): 173–186.
38. Benbrook CM (2012) Impacts of genetically engineered crops on pesticide use in the US-the first sixteen years. Environ Sci Eur 24(1): 1–13.
39. Hanazato T (2001) Pesticide effects on freshwater zooplankton: an ecological perspective. Environ Pollut 112(1): 1–10.
40. Vandysh O (2004) Zooplankton as an indicator of the state of lake ecosystems polluted with mining wastewater in the Kola Peninsula. Russ J Ecol 35(2): 110–116.
41. Rosi-Marshall EJ, Tank JL, Royer TV, Whiles MR, Evans-White M, et al. (2007) Toxins in transgenic crop byproducts may affect headwater stream ecosystems. Proc Natl Acad Sci USA 104(41): 16204–16208.
42. Bøhn T, Primicerio R, Hessen DO, Traavik T (2008) Reduced fitness of *Daphnia magna* fed a Bt-transgenic maize variety. Arch Environ Contam Toxicol 55(4): 584–592.
43. Wang Y, Hu H, Huang J, Li J, Liu B, et al. (2013) Determination of the movement and persistence of Cry1Ab/1Ac protein released from Bt transgenic rice under field and hydroponic conditions. Soil Biol Biochem 58: 107–114.
44. O'Sullivan P, Reynolds CS (2008) The lakes handbook: limnology and limnetic ecology, (John Wiley & Sons). 667 p.
45. Beachy RN, Fedoroff NV, Goldberg RB, McHughen A (2008) The burden of proof: A response to Rosi-Marshall, et al. Proc. Natl. Acad. Sci. USA 105: E9.
46. James C (2009) Global Status of Commercialized Biotech/GM Crops, 2009 (ISAAA Ithaca, NY).
47. Halwart M (2008) Biodiversity, nutrition and livelihoods in aquatic rice-based ecosystems. Biodiversity 9(1–2): 36–40.

16

Differential Hepatic Metal and Metallothionein Levels in Three Feral Fish Species along a Metal Pollution Gradient

Lieven Bervoets[1]*, Dries Knapen[1,2], Maarten De Jonge[1], Karen Van Campenhout[3], Ronny Blust[1]

1 Department of Biology, Systemic Physiological & Ecotoxicological Research, University of Antwerp, Antwerp, Belgium, 2 Laboratory for Veterinary Physiology, Department of Veterinary Sciences, University of Antwerp, Antwerp, Belgium, 3 Department of Environment and Health, Flemish Government, Brussels, Belgium

Abstract

The accumulation of cadmium, copper and zinc and the induction of metallothioneins (MT) in liver of three freshwater fish species was studied. Gudgeon (*Gobio gobio*), roach (*Rutilus rutilus*) and perch (*Perca fluviatilis*) were captured at 6 sampling sites along a cadmium and zinc gradient and one reference site in a tributary of the Scheldt River in Flanders (Belgium). At each site up to 10 individuals per species were collected and analyzed on their general condition factor (K), hepatosomatic index (HSI) and gonadosomatic index (GSI). From each individual fish the liver was dissected and analyzed on Cd, Cu and Zn and MT-content. Although not all species were present at each site, hepatic Cd and Zn levels generally followed the pollution gradient and highest levels were measured in perch, followed by roach and gudgeon. Nevertheless also an effect of site was observed on this order. MT-levels appeared to be the highest in gudgeon although differences with the other species were not very pronounced and depended on the site. Significant relationships were found between hepatic zinc accumulation and MT levels. For each species the ratio $MT_{theoretical}/MT_{measured}$ was calculated, which gives an indication of the relative capacity to induce MTs and thus immobilize the metals. Perch had the lowest capacity in inducing MTs (highest ratio). Relationships between hepatic metal levels and fish condition indices were absent or very weak.

Editor: Sandra Maria Feliciano de Oliveira Azevedo, Federal University of Rio de Janeiro, Brazil

Funding: This work was sponsored by two projects of the Flemish Government (VLINA/9903 and VLINA/0004). MDJ is supported by the Agency for Innovation by Science and Technology. The funders had no role in study design, data collection and analysis, decision to publish, or preparation of the manuscript.

Competing Interests: The authors have declared that no competing interests exist.

* E-mail: lieven.bervoets@ua.ac.be

Introduction

Despite a decrease in discharges of metals in most West-European countries, including Belgium, several water courses are still historically contaminated with metals [1–3], potentially affecting different stages of the aquatic food chain. Fish are interesting species for the evaluation of biological effects of metal pollution under natural circumstances because they might accumulate metals from water, sediment as well as food [4–6]. However, accumulation of metals does not necessarily indicates deleterious effects since organisms have possibilities to protect themselves from metal toxicity by increased excretion, differential allocation among organs and by binding the metals intracellularly [7–9]. Metallothioneins (MTs) are are low-molecular weight, heat-stable and cysteine-rich proteins involved in the binding and regulation of essential metals such as copperand zinc, and the detoxification of non-essential metals such as cadmium and mercury [10]. The induction of MTs as a response to elevated levels of waterborne and dietary metal exposure, has been frequently used as a biomarker for metal exposure, both under laboratory and field conditions [7,11–20]. Fish species that differ in their feeding strategies and/or detoxification capacities might accumulate metals to a different extent [21–24]. Bottom dwelling species such as the gudgeon (*Gobio gobio*) will be exposed to water, sediment and food whereas pelagic species are mainly exposed via water and food.

As a consequence, given the different way of exposure, both metal accumulation and MT-induction might differ among different species [13,25], which might be responsible for differences in sensitivity to metals [26,27].

In order to relate possible differences in MT levels to differences in sensitivity multiple effects on fish can be studied. Integrative measures such as condition factor and hepatosomatic index can provide valuable information concerning the overall effect of pollutants on individual fish [28–31] and can be related to both metal and MT levels in fish tissues.

The aim of this study was to assess whether different fish species living along a Cd and Zn pollution gradient differentially accumulate metals and induce MTs. Three species were investigated; the bottom dwelling gudgeon (*Gobio gobio*), roach (*Rutilus rutilus*) an invertivorous fish and perch (*Perca fluviatilis*) a piscivorous fish. Metal and MT levels in liver were compared among species and environmental metal concentrations were related to accumulated levels. Furthermore it was investigated whether differences in accumulated metals and induced MTs resulted in different condition. Condition indices that were measured are hepatosomatic index, gonadosomatic index and condition factor.

Materials and Methods

Ethics Statement

The study was conducted in accordance to national and international guidelines (directive 2007/526/EC of the European Commission) for the protection of animal welfare. All necessary permits were obtained for the described field studies. Permission to catch fish with electrofishing and to sacrifice a limited number of

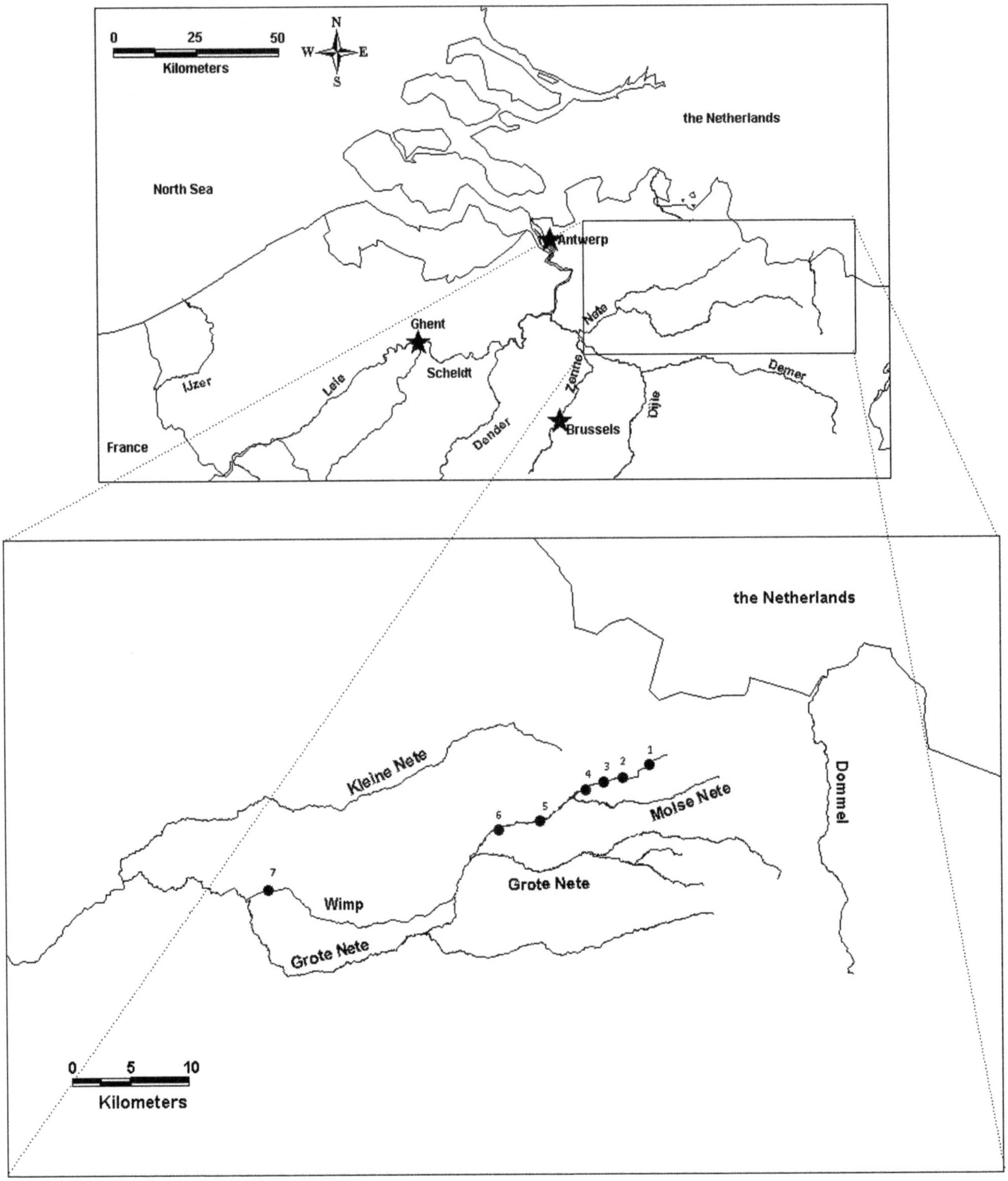

Figure 1. Map of the sampling sites.

each species at each location was obtained from the Flemish Government (LNE/Department of Environment, Nature and Energy).

Study area and sampling design

Six sampling sites were selected along the River Molse Nete, located in Flanders (Belgium) and belonging to the basin of the River Scheldt. In addition a reference site was sampled (site 7) at the River Wimp, belonging to the same basin (Fig 1). Sampling sites 1–6 are situated along an existing cadmium and zinc

Table 1. Average water quality characteristics at the different sampling sites.

Site	O_2 mg/l	pH	SO_4^{2-} mg/l	Cl^- mg/l	NO_3^-+ NO_2^--N	NH_4^+-N mg/l	PO_4^{3-}-P mg/l	Cond. μS/cm	Cd μg/l	Cu μg/l	Zn μg/l
1	8.2	7.2	82	51	2.2	0.26	0.05	461	3.9	4.1	712
2	9.3	7.2	85	42	2.0	0.23	0.05	457	62	8.3	5539
3	8.5	7.1	77	39	1.7	0.41	0.08	409	38	6.1	3853
4	8.6	7.2	95	38	1.6	0.35	0.09	409	58	8.1	4864
5	9.3	7.3	61	37	1.5	0.68	0.11	376	17	5.7	1762
6	8.7	7.3	57	36	2.6	0.88	0.16	380	8.4	5.4	922
7	7.6	6.8	60	53	1.6	2.2	0.13	391	< 0.10	3.5	62
CQC	5.5-9.5	6.5-9.0	500	250	NA	*	NA		0.02	24	30
FQC	≥ 5.0	6.5–8.0	250	200	10.0	1.0	0.30	1000	1.0	50	200

CQC: Canadian Quality Criteria for aquatic freshwater life [65] FQC: Flemish Quality Criteria [66] NA: Not Available; * pH-dependent; ND: Not Determined

concentration gradient with the highest metal concentrations measured at site 2 [28,31,32]. Water characteristics such as oxygen level, pH, water hardness and water temperature (4.6 – 6 °C) were within the same range for all sites (table 1).

Water samples were collected in duplicate monthly between November and March 2001–2002 at all sampling stations. To measure the total metal concentrations water samples were acidified with nitric acid (HNO_3; 69%) to a pH of 2.0 and filtered through a 0.45-μm Millipore (Bedford, MA, USA) membrane filter. All samples were stored in 20-ml polypropylene vials at 4 °C until analysis. Sediments were collected twice, i.e. in December and in February. Samples were taken with a 'Petit Ponar' grab sampler (Wildco cat.no. 1728; 235 cm^2). At each site a mixed sample was taken, composed of 5 grab samples [33]. Samples were sieved with the site-water using a 500 μm-mesh sieve and stored in 500 ml polyethylene beakers at 4°C. Subsequently, supernatant was decanted carefully to prevent loss of sediment. After decantation the sediment sample was homogenised with a plastic spatula. Prior to extraction the sediment of each sampling station was centrifuged (10,000 *g*) to collect the pore water [34]. The remaining wet sediment was analysed on total metal content: sediments were dried at 60 °C during 48 hours and a mixture of concentrated HNO_3 and HCl (4:1) was added. Eventually, samples were put in Teflon bombs and digested in a microwave oven [28,35].

At all sampling sites fish were caught within one month between December and January by electrofishing, using an Electra catch WFC7 generator producing 150 V. Three successive samplings were conducted at each site (length: ± 100 m). All fish were identified to the species level and counted. Of all specimens, forklength was measured (± 1 mm) and weight determined using a Kern 442.43 balance (± 0.1 g).

From each sampling site up to 8 specimens (if present) of three fish species, i.e. gudgeon (*Gobio gobio*), perch (*Perca fluviatilis*), and roach (*Rutilus rutilus*), were sacrificed using an overdose of the anesthetic ethyl meta-aminobenzoate methanesulfonic acid (MS 222) and liver tissues were collected and weighed (0.001 g) in the field and immediately stored in liquid nitrogen and transported to the lab. In the lab, the liver was homogenized and separated in two parts, one part for determination of metal concentrations stored at –20 °C and one part for metallothionein determination stored at –80 °C. Tissue samples for metal determination were put in acid-washed polypropylene pre-weighted vials and dried for 24 h at 60 °C. Subsequently, the biological material was digested with a mixture of concentrated ultrapure HNO_3 and hydrogen peroxide (H_2O_2; 29%) (3:1; v/v) in a microwave oven and stored until measurement [28].

Chemical measurements

Water quality characteristics (pH; temperature (°C); conductivity (μS/cm); oxygen (mg/l)) were measured on-site and at each sampling moment with a WTW multiline F/SET-3 field kit. Chloride, sulphate, phosphate, ammonium, nitrate and nitrite were determined using an automated analyser (Skalar: FAS, SA 20/40).

Cadmium, copper, and zinc were measured in the surface water, sediment pore water, bulk sediment and fish tissue samples with either an Inductively Coupled Plasma-Atomic Emission Spectrometer (ICP-AES, Varian Liberty Series II, Victoria, Australia) equipped with a micro concentric groove nebulizer [36] or an Inductively Coupled Plasma-Mass Spectrophotometer (ICP-MS, Varian UltraMass 700, Victoria, Australia).

Concentrations of the metals in the water are expressed in μmol l^{-1}, the liver metal concentrations are expressed on a wet weight (wet/wet) basis in nmol g^{-1}. The accuracy of the metal determination was verified using certified reference material of the Community Bureau of Reference (EU), i.e. standard for trace elements in river sediment (CRM 320) and mussel tissue (CRM 278). Recoveries were within 10 % of the certified values.

Metallothionein determination

After thawing, the tissue samples were homogenized using an Ultra-turrax T8 (IKA, Labortechnik, Staufen, Germany) in 3 volumes (v/w) of 10 mM Tris-HCl buffer (pH 7.4) containing 5 mM β-mercaptoethanol (to prevent oxidation), and 0.1 mM phenylmethanesulfonylfluoride (PMSF, protease inhibitor). Tris-(hydroxylmethyl)-aminomethane (Tris), PMSF, and β-mercapto-ethanol were obtained from Sigma (Sigma-Aldrich, St. Louis, MO, USA). The samples were centrifuged at 9000 *g* and 4 °C for 10 min. (Eppendorf Centrifuge 5804R, Hamburg, Germany) and ultra centrifuged at 100 000 *g* for 60 min. at 4 °C (Sorval Discovery TM 90 Ultra speed centrifuge, Newton, Connecticut, USA). Aliquots of supernatants (cytosolic fractions) were stored at – 80 °C for a period not longer than 24 hrs before use. Total cytosolic MT concentrations were measured using the cadmium thiomolybdate saturation assay from Klein *et al.* [37].

The general concept of this assay is to remove all MT-bound Cd, Cu, and Zn ions and, subsequently, to saturate the MT

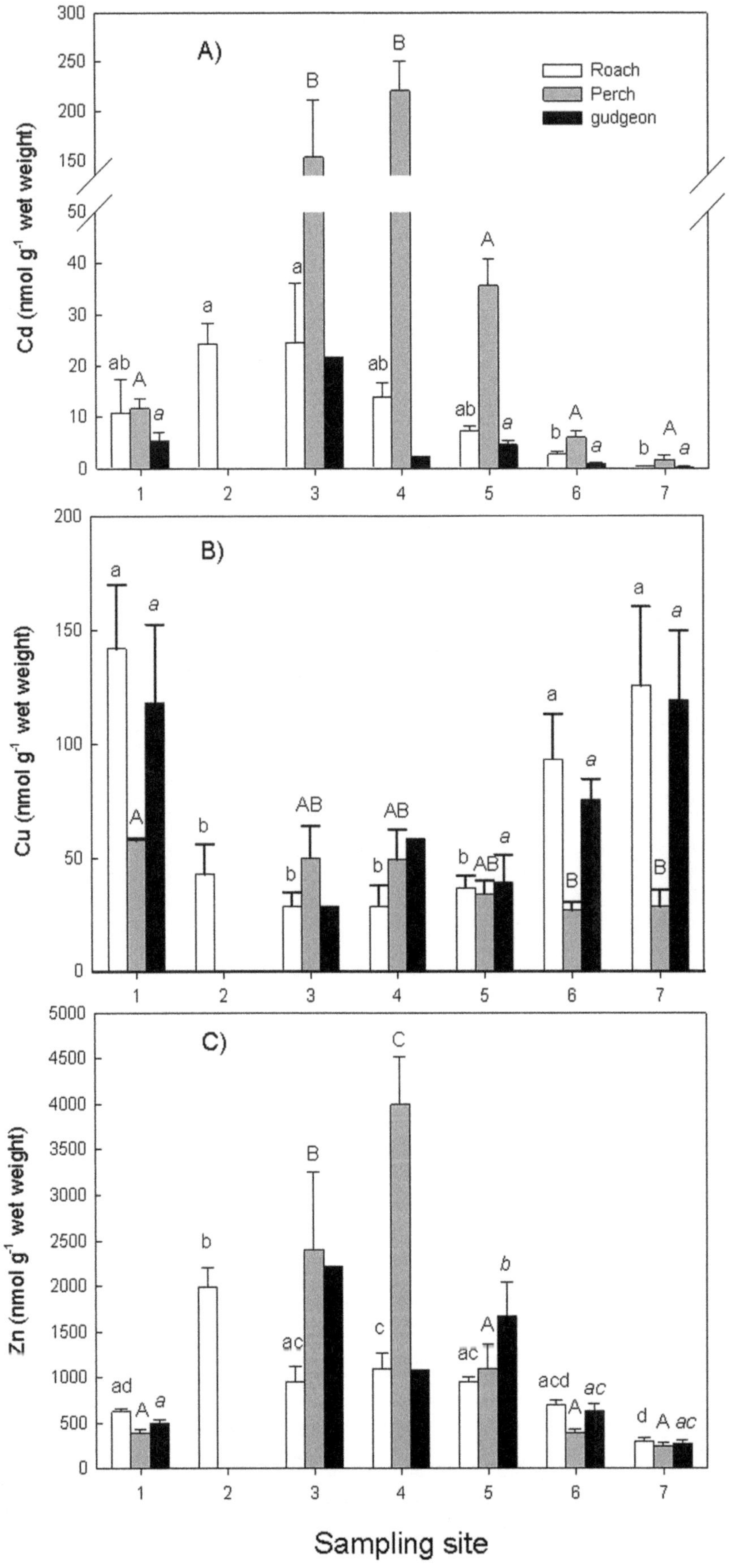
A)
Roach
Perch
gudgeon
Cd (nmol g^{-1} wet weight)
B)
Cu (nmol g^{-1} wet weight)
C)
Zn (nmol g^{-1} wet weight)
Sampling site

Figure 2. Mean metal concentration (± standard deviation) in liver of the three fish species in μg g^{-1} dry weight. (a) Cd, (b) Cu, (c) Zn. Different letters (A, a or *a* for respectively Roach, Perch and Gudgeon) mean significant differences in hepatic tissue within a species ($p < 0.05$).

molecules with the ^{109}Cd isotope, which was quantified using a Minaxi-Autogamma 5530 counter (Canberra Packard, Boston, MA, USA). The excessive ^{109}Cd was complexed by Chelex-100 (Merck, Darmstadt, Germany). The MT concentrations were calculated on the basis of a molar ratio of Cd/MT of 7 [38]. It has been shown that this method compares very well with other established techniques such as ELISA or the Ag-saturation method [39].

Measuring MT concentrations using the cadmium thiomolybdate saturation assay of Klein et al. [37] allows the determination of MT binding places including the ones initially occupied by Cu, Cd and Zn. Measured MT concentrations in liver were compared to Cu, Cd and Zn concentrations measured in those tissues. In order to investigate differences in detoxification capacities among the three fish species, the theoretical level of MT needed to bind all the present Cd, Cu and Zn in the liver was calculated for the three species. Based on the specific binding capacity of the thiol groups of MT for the three metals (7 mol Cd, 12 mol Cu and 7 mol Zn/mol MT, Kito et al. [38]) the following equation was used: $MT_{theoretical} = [Cd]tissue/7 + [Cu]tissue/12+[Zn]tissue/7$. The ratio $MT_{theoretical}/MT_{measured}$ (further referred to as MT_t/MT_m) provides an indication of how efficient an organism is inducing MTs after exposure to metals.

Condition indices

The relative condition factor (K) was calculated according to the formula $K = W/W'$ [40,41], where the observed weight of an individual fish (W in g) is compared to its expected weight (W') based on its observed length (L). This expected weight is derived from the length-weight regressions $W = a L^b$, with *a* and *b* as fitting constants from reference populations of each studied species. K indicates whether an individual of a species is in a better (>1) or worse (<1) condition compared to an average individual of the same length in a reference population.

The Hepatosomatic Index (HIS) and Gonadosomatic Index (GSI) were calculated by comparing the respective organ weights to the total body weight (organ weight/total body weight x 100).

Statistical analysis

ANOVA (Analysis of Variance) and Duncan's multiple range tests were used to determine the significance of the differences between the different sampling sites. If the data were not normally distributed or showed heterogeneous variances, they were log transformed to get homogeneous variances and a normal distribution. Multiple linear regressions were used to construct empirical models to relate exposure concentrations to the tissue metal concentrations and the tissue MT concentrations. All statistical analyses were performed using the software package STATISTICA (StatSoft Inc., Tulsa).

Results

Table 1 summarizes the results of the water quality characteristics from the different sampling sites. General water quality characteristics all met the Flemish and international water quality standards. Total Cd and Zn levels clearly followed a gradient between site 2 and site 7. For copper no pattern was observed. Cd and Zn levels exceeded quality standards at all sites, including site 1 upstream of the pollution source but except the reference site,

Table 2 gives the results of the sediment and pore water analysis. Cadmium and Zn concentrations in the sediment did not follow the same gradient as for the dissolved metals but also exceeded sediment quality standards. Highest levels of both metals were measured more downstream of the pollution source. At the downstream sites increased copper levels were measured although nowhere the quality standards were exceeded.

For each species fish of comparable size were selected for analysis (total length: roach: 13.9 ± 2.8 cm; gudgeon: 11.1 ± 1.6; perch: 14.0 ± 5.6 cm). Roach could be captured in sufficient numbers at all sites. Gudgeon however, was absent at site 2 and only one specimen was captured at sites 3 and 4. Perch was absent at site 2. In total, metal and MT levels in the liver could be measured in 45 roach, 23 perch and 33 gudgeons captured in the pollution gradient.

Figure 2 shows the results of the Cd, Cu and Zn concentrations in the liver of the three fish species caught at the different sampling. The average wet weight/dry weight ratios for the liver were 3.63 ± 0.39, 3.85 ± 0.31, 3.05 ± 0.60 for Roach, Perch and

Table 2. Total metal levels in the sediments and the pore water at the different sampling sites.

Site	Cd-sed μg/g dw	Cu-sed μg/g dw	Zn-sed μg/g dw	Cd-pw μg/l	Cu-pw μg/l	Zn-pw μg/l
1	1.88	1.64	89.7	1.23	0.43	26.4
2	19.3	2.56	313	nd	nd	nd
3	35.3	3.15	1431	4.95	1.14	1275
4	51.6	6.78	1625	2.05	1.55	341
5	40.8	18.6	2306	1.43	0.42	95.0
6	51.8	35.7	2634	1.19	1.29	555
7	0.27	9.17	177	0.12	2.80	17.0
CCC				2.2	9.0	120
PEC	3.5	197	315			

Sed: sediment; pw: pore water
CCC: Criterion Continuous Concentration [43] aquatic freshwater life; ND: Not Determined
PEC: Probable Effect Concentrations [42].

Table 3. Two-way analysis of variance for the effect of species and site on hepatic metal levels.

Source of variation	df	Mean of Squares	F_s
Cadmium			
Species	2	17932	90.7^{***}
site	4	8017	40.5^{***}
Interaction	8	10695	54.1^{***}
Zinc			
Species	2	689305	2.23^{ns}
site	4	4287589	13.9^{***}
Interaction	8	893124	2.89^{**}
Copper			
Species	2	22891	9.10^{***}
site	4	15777	6.27^{***}
Interaction	8	2744	1.09^{ns}

Gudgeon respectively. The effect of site and species was analyzed with a two-way ANOVA. Because not all species were present at all sites this two-way ANOVA could be applied only on fish from sites 1, 5, 6 and 7. One-way ANOVA, followed by a Duncan's multiple range test, was used to compare the levels among the sites per species. For Cd there was a significant effect of site and species and a significant interaction (table 3). Highest cadmium levels were measured in liver of perch at site 5. Considering the species separately, in roach highest hepatic cadmium levels were measured at site 2 and 3, in perch at site 3 and 5 and in gudgeon at site 3 (no perch or gudgeon were caught at site 2). Also for Cu there was a significant effect of site and species but no significant interaction (table 3) with highest levels in roach and gudgeon at the sites 1, 6 and 7. For gudgeon however, no significant differences among sites were observed and also for perch differences were less pronounced with significant lower levels at site 6 and 7 compared to site 1 (Figure 2B). For zinc again there was a significant effect of species and site and a significant interaction (table 3). Significant highest zinc levels were measured in liver of perch at site 3 and 4. Looking at the individual species the pattern found was comparable to cadmium with for roach highest levels at site 2, for perch highest levels at site 4 and for gudgeon highest levels at site 5.

Table 4. Two-way analysis of variance for the effect of species and site on hepatic Metallothioneine levels.

Source of variation	df	Mean of Squares	F_s
Species	2	2214	16.5^{***}
site	4	3852	28.8^{***}
Interaction	8	688	5.14^{***}

Figure 3 shows the hepatic MT levels in the 3 species at the different sites. The effect of site and species on MT levels was analyzed with a two-way ANOVA. Because no sufficient numbers of all species could be captured at all sites, this two-way ANOVA could be applied only on fish from sites 1, 5, 6 and 7. Significant effects of site and species and a significant interaction were found (table 4). Significant highest hepatic MT levels were found in gudgeon from site 5. When the MT levels in the individual species were compared among sites, in roach highest levels were found at sites 2 and 3, in perch at 3 and 4 and in gudgeon at site 3, but here only one individual was captured.

Multiple regression analysis linking measured MT concentrations to Cu, Cd and Zn concentrations in liver showed strong relationships for all three species with R^2 of 0.63; 0.92 and 0.61 for roach, perch and gudgeon respectively. However, the only metal

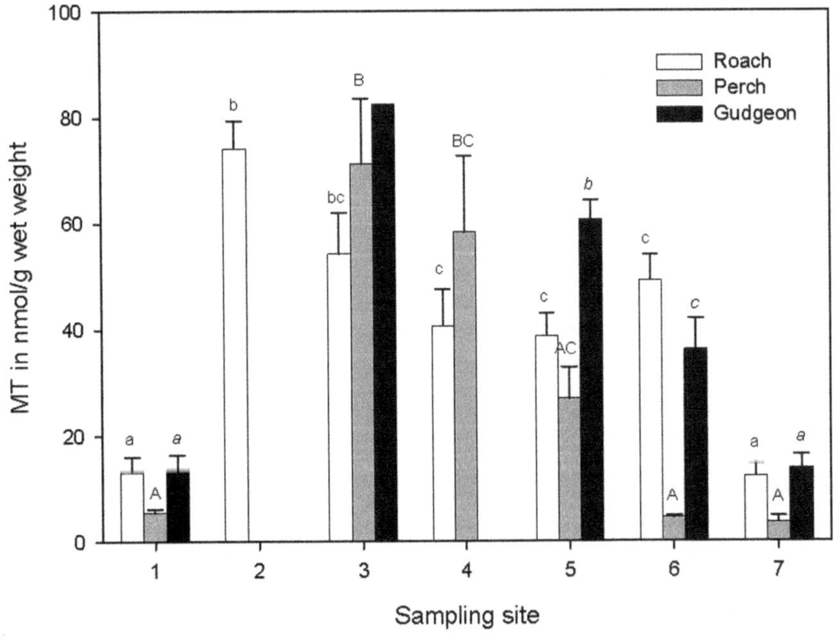

Figure 3. Metallothionein level (MT) in liver of the three fish species (nmol MT g-1 wet weight). Different letters (A, a or *a* for respectively Roach, Perch and Gudgeon) mean significant differences in hepatic tissue within species ($p < 0.05$).

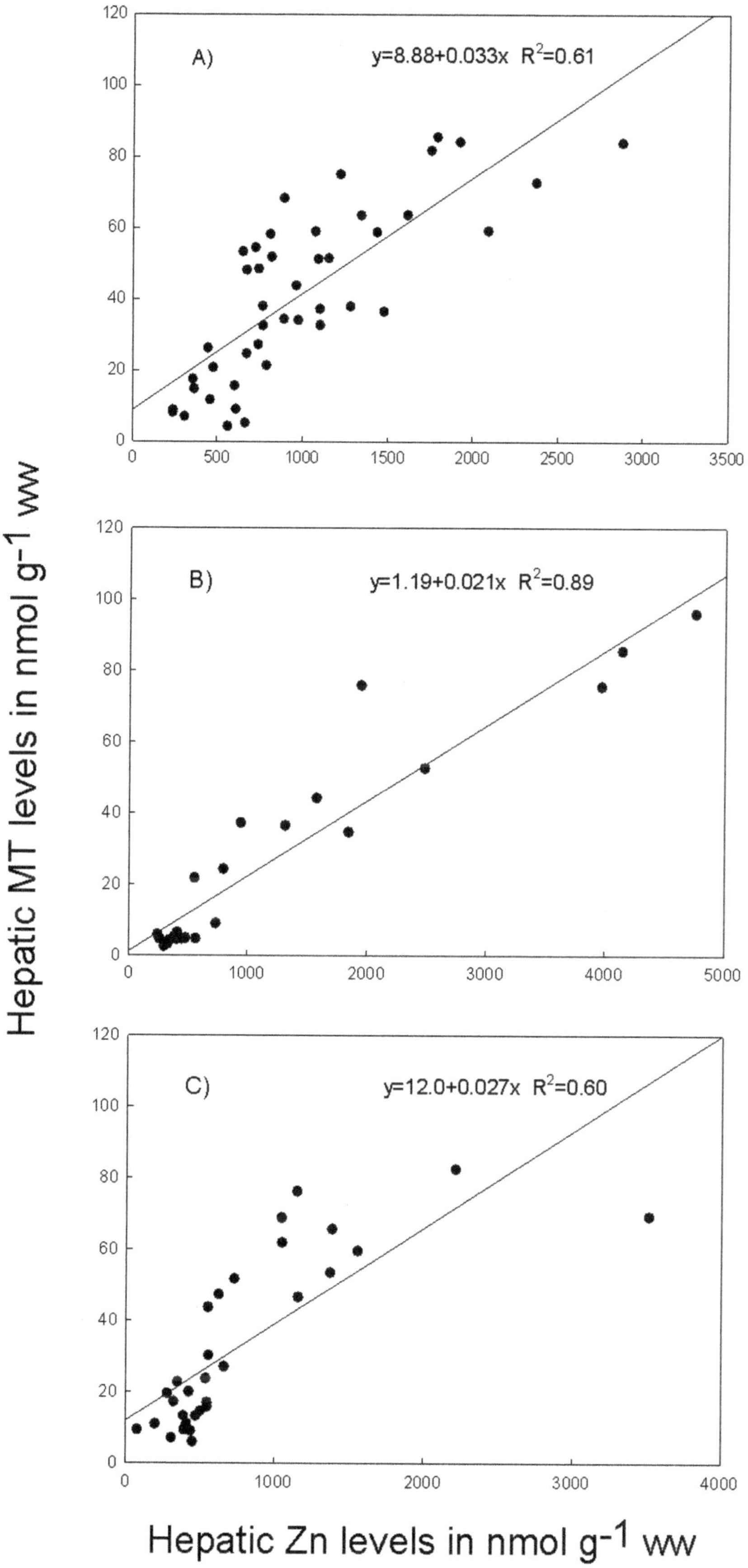

Figure 4. Relationship between the hepatic Zn and MT level in the three fish species. (a) Roach; (b) Perch; (c) Gudgeon.

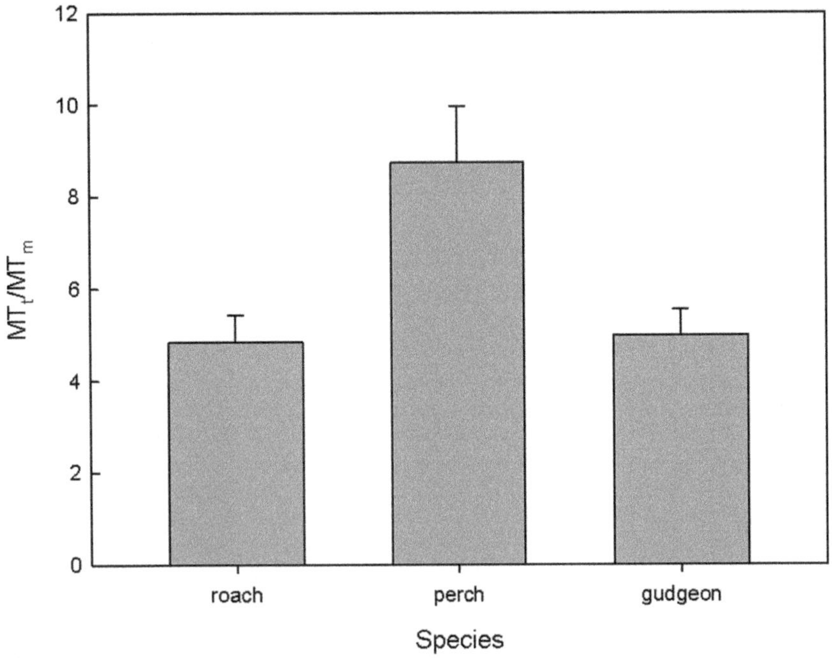

Figure 5. Ratio of theoretical hepatic MT/measured hepatic MT for the three fish species.

that was significantly contributing in the linear regression to the MT levels was Zn. However, Zn was strongly correlated with Cd and including only Cd in the model also revealed significant relationships with R^2 between 0.60 and 0.87. Figure 4 illustrates the relationship between hepatic zinc and MT levels for the three species. Figure 5 presents the MT_t/MT_m ratio for the three species. The higher the ratio the lower is the efficiency. Perch proved to induce lower MT levels compared to the two other species (ANOVA, figure 5).

Figure 6 summarizes the results of the different condition indices measured for the three fish species. Values were tested on significant differences with an ANOVA, followed by a Duncan's multiple range test. For Roach only significant differences were observed for HSI with lowest levels at site 3 and highest at site 4. From site 4 on HSI decreased more downstream. For Perch only K differed significantly among the sites with highest levels at site 5. For gudgeon no significant differences were found for none of the measured indices among the sites.

With regression analysis the individual K, HSI and GSI were related for each species to the individual hepatic metal concentration. For Roach a significant negative relationship was found between hepatic Cd and K although the R^2 was only 0.16. For Perch significant but also rather week, negative relationships were found between hepatic Cd and GSI and between hepatic Zn and K (R^2 of respectively 0.14 and 0.17). Finally for gudgeon a significant negative relationship was found between hepatic Cd and HSI ($R^2 = 0.19$).

Relating the MT_t/MT_m-ratio to the condition indices revealed only a significant negative relationship with K for perch ($R^2 = 0.17$) and with HSI for gudgeon ($R^2 = 0.25$).

Discussion

Cd and Zn in surface water and sediment exceeded the Flemish and International quality standards [42,43] at all investigated sites, except the reference site. Also compared to other polluted sites the measured environmental levels can be considered as very high [34,44,45] and among the highest levels reported before [46]. This was reflected in the hepatic Cd and Zn levels of the three fish species. Hepatic levels in roach ranged from 0.01 to 9.17 µg/g dw for Cd, from 0.62 to 18.4 µg/g dw for Cu and from 15.5 to 185 µg/g dw for Zn. Cadmium and zinc levels measured in fish from the present study were comparable or slightly to much higher than the levels measured in liver from roach and perch from metal polluted sites in literature whereas copper levels were comparable to levels in fish from uncontaminated sites [19,31,32,47–50]. For gudgeon only data were found from the same pollution gradient but from another period and levels in the present study were slightly lower [31,32,51].

In the present study hepatic Cd and Zn levels were higher in perch than in gudgeon and roach at most of the sites. This was opposite to what was found by Blanchard et al [49], Teil et al. [52] and Szarek-Gwiazda et al [51] who all found equal or lower Cd and Zn levels in perch compared to roach but in their studies environmental levels were lower compared to the present study. Andres et al. [48] however, measured higher levels in roach than in perch at environmental levels comparable to the present study. We cannot find any explanation for these differences in literature. Although in most studies no biomagnification is observed for metals other than Hg [53], in the present study highest levels were measured in the piscivorous fish. A possible reason for the highest levels in perch is that this species could be less capable in eliminating the metals than the other two species.

Differences in hepatic MT levels among the three fish species were site-specific although at most sites MT levels were comparable between roach and gudgeon but much lower in perch. At the reference site average MT levels were 12.4, 3.65 and 13.9 nmol/g ww for roach, perch and gudgeon respectively.

MT levels in roach from the reference site were comparable to levels found in unexposed roach [54,55] or from a reference site in the same River basin [19]. Studies of hepatic MT levels in gudgeon could only be found from the same river basin [16,26].

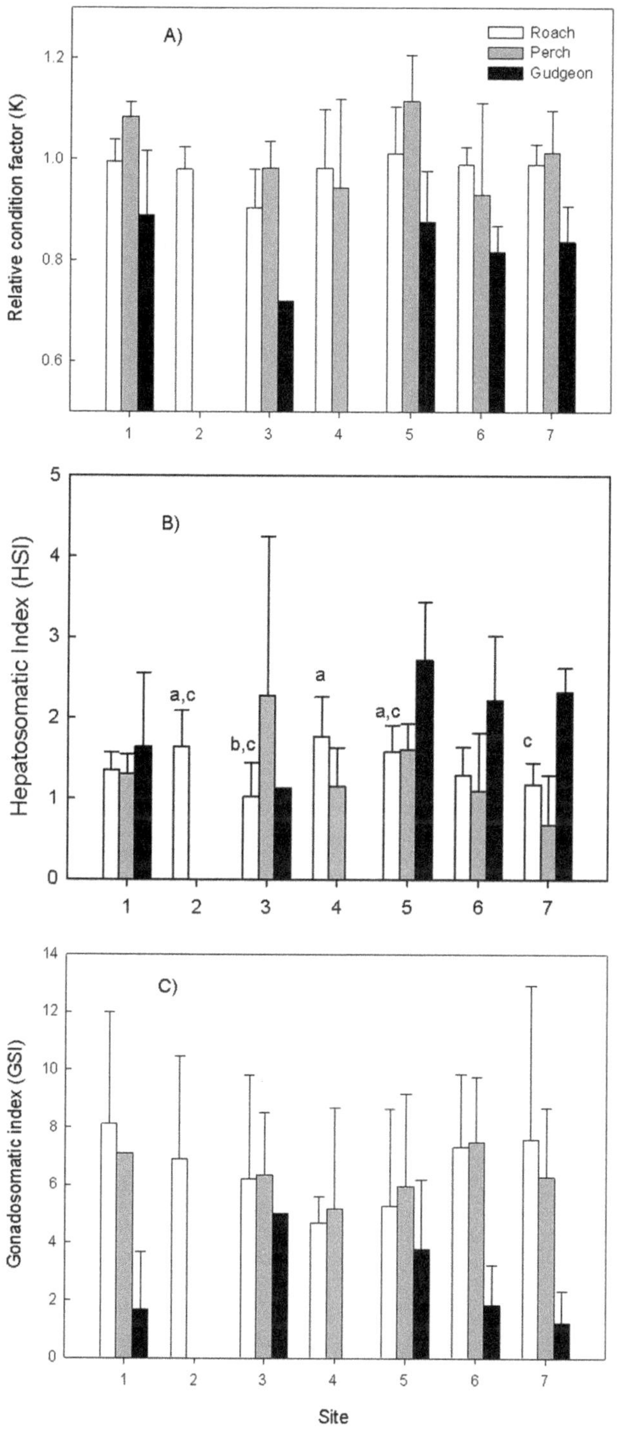

Figure 6. Condition measures for the three fish species (mean ± standard dev.). A) Relative condition factor; B) Hepatosomatic Index; C) Gonadosomatic Index. Different letters mean significant differences within species ($p < 0.05$)

Levels in the present study from the reference site were comparable or slightly lower than the levels found by Van Campenhout et al [16] and Knapen et al. [26] at reference sites. At the contaminated sites MT levels in gudgeon in the present study were slightly to much lower compared to the levels found by Van Campenhout et al [16], which measured MT levels in gudgeon sampled in 1998 from the same contaminated sites. This is probably due to the decreased metal levels between the two study periods [16]. Levels in perch from our study were comparable to the levels found by Hogstrand et al. [56].

For all three species good to very good relationships were found between hepatic zinc and hepatic MT levels (figure 4). Several studies have investigated the effect of pollutants on the condition factor, HSI or GSI [e.g.19,57–59]. In the present study few differences in K, HSI or GSI were found among the sites for the different species.

Few significant and rather weak relationships were found between hepatic metal levels and condition indices. Ozmen et al. [60] found a reduced condition factor (CF) in carp captured at sites polluted with various contaminants including metals. A strong significant inverse relationship between CF and Cd concentration in liver of brown trout was found in a study of Clements and Rees [57]. Reynders et al. [19] found significant differences in CF of caged carp among sites but no significant relationship with accumulated metal. In a study of Bervoets et al. [61] condition factor was significantly related to metal load in the kidney but not in the other tissues. Knapen et al. [31] however, found significant differences in K for gudgeon along the same pollution gradient and reference site. However, lowest K was observed at the reference site and the least contaminated site of the gradient. Other studies did not find any relationships between environmental pollution and condition factor in fish [59,62]. In the present study the described variation in CF by hepatic metal levels was rather low, which can be attributed to other factors than (metal) pollution such as habitat quality and food availability [63].

A negative but weak relationship was found for gudgeon between hepatic Cd and the hepatosomatic index. This is in contrast to what was found by Ozmen et al. [60] and Bervoets et al. [61] who found a positive relationship between hepatic Cd levels and HSI in carp. However, Pereira et al. [64] measured a decreased HSI in winter flounder (*Pleuronectes americanus*) exposed to high cadmium concentrations for 71 days.

Since MTs are also induced by Hg and Ag, the approach of the MTt/MTm-ratio is only valid in a study area with low levels of those metals. In the present study we Hg and Ag were measured in the livers (data not shown) but levels were not significant among sites and species and generally very low or often even below the detection limit. Moreover, from the database of Flemish Environment Agency (www.vmm.be) it was obvious that concentrations of both metals were low in the pollution gradient.

If we compare the ratios of the theoretical over the actual MT levels perch proved to induce less MTs compared to the two other species. From this observation we would expect that perch is more sensitive to the metal pollution than the other two species. This was however not tested in the present study and relationships between the MTt/MTm-ratio and condition indices were non-significant or only weak relationships were found such as for perch with K and for gudgeon with the HSI. The absence of clear relationships between the MTt/MTm-ratio, which can be considered as a measure for sensitivity could be due to other factors influencing those indicators. Possibly other sub-lethal endpoints such as growth or reproduction should be selected in order to assess the protective effect of MT on fish condition or health.

From this study we can conclude that different fish species exposed within a same pollution gradient differentially accumulate metals and differentially induce metal binding proteins. Perch accumulated highest levels of Cd and Zn but showed the lowest detoxification capacity. Only few and weak relationships between detoxification capacity and measured endpoints could be found.

Probably more sensitive chronic effects have to be studied that are less affected by other environmental factors.

Acknowledgments

We thank Tine Moermans, Tom De Bie, Liesbeth, Bruyndoncx, Veerle Mees, and Judith Voets for assistance in the field and/or the laboratory.

Author Contributions

Conceived and designed the experiments: LB RB. Performed the experiments: LB KVC DK. Analyzed the data: LB DK MDJ. Contributed reagents/materials/analysis tools: LB RB . Wrote the paper: LB DK MDJ.

References

1. De Jonge M, Blust R, Bervoets L (2010) The relation between Acid Volatile Sulfides (AVS) and metal accumulation in aquatic invertebrates: Implications of feeding behavior and ecology. Environ Pollut 158: 1381-1391.
2. Hutchinson SM, Rothwell JJ (2008) Mobilisation of sediment-associated metals from historical Pb working sites on the River Sheaf, Sheffield, UK. Environ Pollut 155, 61-71.
3. Nguyen HL, Braun M, Szaloki I, Baeyens W, Van Grieken R, et al. (2009) Tracing the Metal Pollution History of the Tisza River Through the Analysis of a Sediment Depth Profile. Water Air Soil Poll 200: 119-132.
4. Dallinger R, Kautzky H (1985) The importance of contaminated food for the uptake of heavy metals by rainbow trout (*Salmo gairdneri*): a field study. Oecologia 67: 82-89.
5. Chowdhury MJ, Bucking C, Wood C (2008) Pre-exposure to waterborne nickel downregulates gastrointestinal nickel uptake in rainbow trout: Indirect evidence for nickel essentiality. Environ Sci Technol 42: 1359-1364.
6. Van Campenhout K, Bervoets L, Redeker ES, Blust R (2009) A kinetic model for the relative contribution of waterborne and dietary cadmium and zinc in the common carp (*Cyprinus carpio*). Environ Toxicol Chem 28: 209-219.
7. Dallinger R, Egg M, Kock G, Hofer R (1997) The role of metallothionein in cadmium accumulation of Arctic char (*Salvelinus alpinus*) from high alpine lakes. Aquat Toxicol 38: 47-66
8. Rainbow PS (2007) Trace metal bioaccumulation: models, metabolic availability and toxicity. Environ Int 33:576-582.
9. Van Campenhout K, Infante HE, Hoff PT, Moens L, Goemans G, et al. (2010) Cytosolic distribution of Cd, Cu and Zn, and metallothionein levels in relation to physiological changes in gibel carp (*Carassius auratus gibelio*) from metal-impacted habitats. Ecotox Environ Safe 73: 296-305.
10. Coyle P, Philcox JC, Carey LC, Rofe AM (2002) Metallothionein: The multipurpose protein. Cell Mol Life Sci 59: 627-647.
11. Cosson RP (1994) Heavy metal intracellular balance and relationship with metallothionein induction in the liver of carp after contamination by silver, cadmium and mercury following or not pretreatment by zinc. Biometals 7: 9-19.
12. Chowdhury MJ, Baldisserotto B, Wood CM (2005) Tissue-specific cadmium and metallothionein levels in rainbow trout chronically acclimated to waterborne or dietary cadmium. Arch Environ Con Tox 48: 381-390.
13. De Boeck G, Ngo TTH, Van Campenhout K, Blust R (2003) Differential metallothionein induction patterns in three freshwater fish during sublethal copper exposure. Aquat Toxicol 65: 413-424.
14. Costa PM, Repolho T, Caeiro S, Diniz ME, Moura I, et al. (2008) Modelling metallothionein induction in the liver of *Sparus aurata* exposed to metal-contaminated sediments. Ecotox Environ Safe 71: 117-124
15. Olsvik PA, Hindar K, Zachariassen KE, Andersen RA (2001) Brown trout (*Salmo trutta*) metallothioneins as biomarkers for metal exposure in two Norwegian rivers. Biomarkers 6: 274-288.
16. Van Campenhout K, Bervoets L, Blust R (2003) Metallothionein concentrations in natural populations of gudgeon (*Gobio gobio*): relationship with metal concentrations in tissues and environment. Environ Toxicol Chem 22:1548-1555.
17. Van Campenhout K, Infante HG, Goemans G, Belpaire C, Adams F, et al. (2008) A field survey of metal binding to metallothionein and other cytosolic ligands in liver of eels using an on-line isotope dilution method in combination with size exclusion (SE) high pressure liquid chromatography (HPLC) coupled to Inductively Coupled Plasma time-of-flight Mass Spectrometry (ICP-TOFMS). Sci Total Environ 394: 379-389.
18. Bird DJ, Rotchell JM, Hesp SA, Newton LC, Hall NG, et al. (2008) To what extent are hepatic concentrations of heavy metals in Anguilla anguilla at a site in a contaminated estuary related to body size and age and reflected in the metallothionein concentrations? Environ Pollut 151: 641-651.
19. Reynders H, Bervoets L, Gelders M, De Coen WM, Blust R (2008) Accumulation and effects of metals in caged carp and resident roach along a metal pollution gradient Sci Total Environ 391: 82-95.
20. Mieiro CL, Bervoets L, Joosen S, Blust R, Duarte AC, et al. (2011) Metallothioneins failed to reflect mercury external levels of exposure and bioaccumulation in marine fish – Considerations on tissue and species specific responses. Chemosphere 85:114-121.
21. AllenGil SM, Martinov VG (1995) Heavy-metal burdens in 9 species of freshwater and anadromous fish from the Pechora River, Northern Russia. Sci Total Environ 160-61: 653-659.
22. Terra BF, Araujo FG, Calza CF, Lopes RT, Teixeira TP (2008) Heavy metal in tissues of three fish species from different trophic levels in a Tropical Brazilian River. Water Air and Soil Poll 187: 275-284.
23. Wang YM, Cheng P, Cui RN, Si WT, Zhang YM, et al. (2010) Heavy metal concentrations in water, sediment, and tissues of two fish species (Triplohysa pappenheimi, Gobio hwanghensis) from the Lanzhou section of the Yellow River, China. Environ Monit Assess 165: 97-102.
24. Weis P, Ashley JTF (2007) Contaminants in fish of the Hackensack meadowlands, New Jersey: Size, sex, and seasonal relationships as related to health risks. Arch Environ Con Tox 52:80-89.
25. Fabrik I, Svobodova Z, Adam V, Krizkova S, Trnkova L, et al. (2008) Metallothionein levels in sperm of various fish species. J Appl Ichthyol 24: 522-525.
26. Knapen D, Bervoets L, Verheyen E, Blust R (2004) Resistance to water pollution in natural gudgeon (*Gobio gobio*) populations may be due to genetic adaptation. Aquat Toxicol 67: 155-165.
27. Olsson PE, Kille P (1997) Functional comparison of the metal-regulated transcriptional control regions of metallothionein genes from cadmium-sensitive and tolerant fish species. Biochim Biophys Acta 1350: 325-334.
28. Bervoets L, Blust R (2003) Metal concentrations in water, sediment and gudgeon (*Gobio gobio*) from a pollution gradient: relationship with fish condition factor. Environ Pollut 126: 9-19.
29. Cooley HM, Evans RE, Klaverkamp JF (2002) Baseline measurements of indicators for sublethal effects of metals in lake whitefish (*Coregonus clupeaformis*). Arch Environ Con Tox 43: 418-423.
30. Fernandes C, Fontainhas-Fernandes A, Rocha E, Salgado MA (2008) Monitoring pollution in Esmoriz-Paramos lagoon, Portugal: Liver histological and biochemical effects in *Liza saliens*. Environ Monit Assess 145:315-322.
31. Knapen D, De Wolf H, Knaepkens G, Bervoets L, Eens M, et al. (2009) Historical metal pollution in natural gudgeon populations: inferences from allozyme, microsatellite and condition factor analysis. Aquat Toxicol 95: 17-26.
32. Bervoets L, Knaepkens G, Eens M, Blust R (2005) Fish community responses to metal pollution. Environ Pollut 138: 338-349.
33. De Cooman W, Seuntjens P, Bervoets L, Panis LI, De Wit M, et al. (1995) Research on the spatial variability of three sediment types in Flanders. Contaminated Soil 95:191-192.
34. Bervoets L, Solis D, Romero AM, Van Damme PA, Ollevier F. 1998. Trace metal levels in chironomid larvae (Chironomidae, Diptera) and sediments from a Bolivian River: impact of mining activities. Ecotox Environ Safe 41: 275-283.
35. Tessier A, Campbell PGC, Auclair JC, Bisson M (1984) Relationships between the partitioning of trace-metals in sediments and their accumulation in the tissues of the fresh-water mollusk *Elliptio complanata* in a mining area. Can J Fish Aquat Sci 41:1463-1472.
36. De Wit M, Blust R (1998) Determination of metals in saline and biological matrices by axial inductively coupled plasma atomic emission spectrometry using micrococentric nebulization. J Anal Atom Spectrom 13: 515-520.
37. Klein D, Sato S, Summer KH (1994) Quantification of oxidized metallothionein in biological material by a Cd saturation method. Anal Biochem 221:405-409.
38. Kito H, Ose Y, Mizuhira V, Sato T, Ishikawa T, et al. (1982) Separation and purification of (Cd, Cu, Zn)-metallothionein in carp hepato-pancreas. Comp Biochem Phys C 73: 121-127.
39. Bienengraber M, Forderkunz S, Klein D, Summer KH (1995) Determination of Cu-containing metallothionein: Comparison of Ag saturation assay, thiomolybdate assay, and enzyme-linked immunosorbent assay. Anal Biochem 228:69-73.
40. Bolger T, Connolly PL (1989) The selection of suitable indexes for the measurement and analysis of fish condition. J Fish Biol 34: 171–182.
41. Le Cren ED (1951) The length - weight relationship and seasonal cycle in gonad weight and condition in the perch (*Perca fluviatilis*). J Anim Ecol 20: 201-219.
42. MacDonald DD, Ingersoll CG, Berger TA (2000) Development and evaluation of consensus-based sediment quality guidelines for freshwater ecosystems. Arch Environ Con Tox 39: 20-31.
43. U.S. Environmental Protection Agency (1999) National recommended water quality criteria—Correction: EPA 822/Z-99-001. Washington, DC.
44. De Jonge M, Eyckmans M, Blust R, Bervoets L (2011) Are accumulated sulfide-bound metals metabolically available in the benthic oligochaete *Tubifex tubifex*? Environ Sci Technol 45: 3131-3137.
45. Maceda-Veiga A, Monroy M, de Sostoa A (2012) Metal bioaccumulation in the Mediterranean barbel (*Barbus meridionalis*) in a Mediterranean River receiving effluents from urban and industrial wastewater treatment plants. Ecotox Environl Safe 76:93-101.
46. Aleksander-Kwaterczak U, Helios-Rybicka E (2009) Contaminated sediments as a potential source of Zn, Pb, and Cd for a river system in the historical metalliferous ore mining and smelting industry area of South Poland. J Soil Sediment 9:13-22.

47. Altug G, Okgerman H (2008) Levels of some toxic elements in the surface sediment and some biota from the Sapanca Lake, Turkey. Fresen Environ Bull 17: 24-28.
48. Andres S, Ribeyre F, Tourencq JN, Boudou A (2000) Interspecific comparison of cadmium and zinc contamination in the organs of four fish species along a polymetallic pollution gradient (Lot River, France). Sci Total Environ 248: 11-25.
49. Blanchard M, Teil MJ, Carru AM, Garban B, Ollivon D, et al. (1999). Biota contamination by PCBs and trace metals in the freshwater estuary of the River Seine (France). Hydrobiologia 400:149-154.
50. Szarek-Gwiazda E, Amirowicz A (2006) Bioaccumulation of trace elements in roach, silver bream, rudd, and perch living in an inundated opencast sulphur mine. Aquat Ecol 40: 221-236.
51. Knapen D, Reynders H, Bervoets L, Verheyen E, Blust R (2007) Metallothionein gene and protein expression as a biomarker for metal pollution in natural gudgeon populations. Aquat Toxicol 82: 163-172.
52. Teil MJ, Blanchard M, Carru AM, Chesterikoff A, Chevreuil M (1996) Partition of metallic and organochlorinated pollutants and monoorthosubstituted PCB pattern in the trophic web from different areas of the river Seine. Sci Total Environ 181:111-123.
53. Gray JS (2002) Biomagnification in marine systems. The perspective of an ecologist. Mar Pollut Bull 45: 46-52.
54. Paris-Palacios S, Biagianti-Risbourg S, Fouley A, Vernet G (2000) Metallothioneins in liver of *Rutilus rutilus* exposed to Cu2+. Analysis by metal summation, SH determination and spectrofluorimetry. Comp Biochem Phys C 126: 113-122.
55. Paris-Palacios S, Biagianti-Risbourg S, Vernet G (2003) Metallothionein induction related to hepatic structural perturbations and antioxidative defences in roach (*Rutilus rutilus*) exposed to the fungicide procymidone. Biomarkers 8:128-141.
56. Hogstrand C, Lithner G, Haux C (1989) Relationship between metallothionein, copper and zinc in perch (Perca fluviatilis) environmentally exposed to heavy-metals. Mar Environ Res 28:179-182.
57. Clements WH, Rees DE (1997) Effects of heavy metals on prey abundance, feeding habits, and metal uptake of brown trout in the Arkansas River, Colorado. T Am Fish Soc 126: 774-785.
58. De la Torre FR, Ferrari L, Salibian A (2000) Long-term in situ toxicity bioassays of the Reconquista River (Argentina) water with *Cyprinus carpio* as sentinel organism. Water Air Soil Poll. 121: 205-215.
59. Dethloff GM, Bailey HC, Maier KJ (2001) Effects of dissolved copper on select haematological, biochemical, and immunological parameters of wild rainbow trout (*Oncorhynchus mykiss*). Arch Environ Con Tox 40, 371-380.
60. Ozmen M, Gungordu A, Kucukbay FZ, Guler RE (2006) Monitoring the effects of water pollution on *Cyprinus carpio* in Karakaya Dam Lake, Turkey. Ecotoxicology 15: 157-169.
61. Bervoets L, Van Campenhout K, Reynders H, Knapen D, Covaci A, et al. (2009) Bioaccumulation of micropollutants and biomarker responses in caged carp (*Cyprinus carpio*). Ecotox Environ Safe 72: 720-728.
62. Farag AM, Woodward DF, Goldstein JN, Brumbauch W, Meyer JS (1998) Concentrations of metals associated with mining waste in sediments, biofilm, benthic macroinvertebrates, and fish from the Coeur d'Alene River Basin, Idaho. Arch Environ Con Tox 34, 119-127.
63. Lobon-Cervia J, Montanes C, DeSostoa A (1991) Influence of environment upon the life history of gudgeon, Gobio gobio (L.) - a recent and successful colonizer of the Iberian Peninsula. J Fish Biol 39: 285-300.
64. Pereira JJ, Mercaldoallen R, Kuropat C, Luedke D, Sennefelder G (1993) Effect of cadmium accumulation on serum vitellogenin levels and hepatosomatic and gonadosomatic indexes of winter flounder (*Pleuronectes-americanus*). Arch Environ Con. Toxicol 4: 427-431.
65. Canadian Council of Ministers of the Environment (1999) Canadian environmental quality guidelines: Winnipeg, MB.
66. Flemish Government (2000) Besluit van de Vlaamse regering van 1 juni 1995 houdende vaststelling van het Vlaamse reglement betreffende de milieuvergunnung (Vlarem), zoals gewijzigd bij besluit van 17 juli 2000. *Belgisch Staatsblad*, 5 augustus 2000 (in Dutch).

An Integrated Approach Combining Chemical Analysis and an *In Vivo* Bioassay to Assess the Estrogenic Potency of a Municipal Solid Waste Landfill Leachate in Qingdao

Yufeng Gong[1], Hua Tian[1], Lijia Wang[2], Suping Yu[1], Shaoguo Ru[1]*

1 Marine Life Science College, Ocean University of China, Qingdao, China, **2** Qingdao Chengtou Xiaojianxi Leachate Treatment Corporation Limited, Qingdao, China

Abstract

Various adverse effects related to landfill leachate have made leachates an important issue in past decades, and it has been demonstrated that landfill leachate is an important source of environmental estrogens. In this study, we employed chemical analysis of some already evaluated estrogenic substances, in combination with a bioassay using several specific biomarkers (e.g., plasma vitellogenin and sex steroids, enzyme activity of gonad gamma-glutamyl transpeptidase, and gonadosomatic index) to evaluate the estrogenic activities in outlets from different stages of the leachate treatment process. The results indicated that 5 environmental estrogens (4-*t*-octylphenol, bisphenol A, di-ethyl phthalate, di-*n*-butyl phthalate, and diethylhexyl phthalate) were detected by a gas chromatography-mass spectrometry, and the concentrations in leachate samples were 6153 ng/L, 3642 ng/L, 2139 ng/L, 5900 ng/L, and 9422 ng/L, respectively. Leachate (1:200 diluted) induced the synthesis of plasma vitellogenin and led to decreased enzyme activity of gonad gamma-glutamyl transpeptidase and gonadosomatic index in male goldfish (*Carassius auratus*) after a 28-day exposure, while increased circulating 17β-estradiol level was also observed in males exposed to treated effluent. Although the target EEs were partially removed with removal rates varying from 87.2% to 99.77% by the "membrane bioreactor+reverse osmosis+aeration zeolite biofilter" treatment process, the treated effluent is still estrogenic to fish. The method combined chemical techniques with the responses of test organisms allowing us to identify the group of estrogen-like chemicals so that we were able to evaluate the overall estrogenic effects of a complex mixture, avoiding false negative assessments.

Editor: Toshi Shioda, Massachusetts General Hospital, United States of America

Funding: This study was supported by Natural Science Foundation of China (31170322), www.nsfc.gov.cn and Natural Science Foundation of Jiangsu Province (BK2011165), www.jsnsf.gov.cn. The funders had no role in study design, data collection and analysis, decision to publish, or preparation of the manuscript.

Competing Interests: The co-author Lijia Wang is affiliated with Qingdao Chengtou Xiaojianxi Leachate Treatment Corporation Limited.

* E-mail: rusg@ouc.edu.cn

Introduction

Currently, sanitary landfill has become the most practiced method for managing waste disposal due to its economic advantages. In 2011, approximately 6.36 billion tons of municipal solid waste (MSW) was deposited in landfills in China [1]. Landfill leachate is generated by the infiltration and percolation of rainfall, groundwater, or runoff water into and through the waste layers [2] and is a complex mixture of inorganic and organic compounds. It has already been established as a hazard to the environment [3]. Taking the risk of leachate leakage into account, leachate-induced biotoxicity has become of concern to societies throughout the world [4–6]. The proposed endocrine disrupting effects, or more specifically estrogenic effects, caused by landfill leachate were reported in Sweden [7–9], and landfill leachate has been considered an important source of environmental estrogens (EEs) in the aquatic habitat [10].

Various examinations regarding EEs in leachate are being conducted, and to our knowledge, the measuring methods used can be mainly classified into two categories: chemical analysis and biological based assays. Chemical analysis is a rapid and sensitive screen to measure a large number of compounds. A variety of chemical analysis methods, such as gas chromatography-mass spectrometry (GC-MS) [11,12], gas chromatography-flame ionization detector (GC-FID) [13], and high performance liquid chromatography (HPLC) [14], have been developed to identify estrogenic substances that potentially interfere with the reproductive system. Based on several combinations of instruments and detectors, the analytical results could provide excellent quantitative information of target compounds. However, it is limited in describing the overall estrogenic effects engendered by complex environmental samples containing multiple EEs because of the existence of additive, synergistic, or even antagonistic influences. Moreover, the pretreatment of chemical analysis may lead to the loss of some unknown EEs. These limitations can be overcome by utilizing biological responses of tested organisms, cellular, or even non-cellular materials in a biotest. Biological assays are typically classified as either *in vitro* or *in vivo* bioassays. Currently, several *in vitro* bioassays have been established and conducted for analyzing EEs, such as a simplified yeast two-hybrid assay [15], an E-screen test and enzyme-linked receptor assay [16], and a reporter-gene-based assay with a human breast-cancer-derived cell line [10]. There were also reports using chemical analysis and *in vitro* bioassays together to investigate the xenoestrogens in leachate [17–19], while *in vivo* bioassays have been rarely used in the examination of the biotoxicology of landfill leachate.

Compared to the *in vivo* test, the *in vitro* bioassay introduced in previous studies has obvious limitations because it does not take into account biological accumulation, biotransformation, or disruptions at other target sites in the reproductive axis, which could potentially result in a false negative evaluation or disappreciation of the EE concentrations [20,21]. For example, monocrotophos pesticide showed no estrogenic activity in the *in vitro* MCF-7 cell proliferation assay [22]; however, this pesticide's *in vivo* estrogenic potency was confirmed by a goldfish vitellogenin (Vtg) induction experiment in our previous study [23].

For the *in vivo* bioassay, the use of multiple biomarkers at different biological levels can be an effective tool in providing comprehensive and available information on EEs. As a female-specific protein, Vtg is a recognized indicator of exposure to estrogen-like substances [24–26]. However, Kime et al. [27] gave a suggestion that the result of evaluation using Vtg as an exclusive biomarker should be treated with caution, and some other specific indicators should be included. The concentrations of plasma sex steroids could help to reveal Vtg induction mechanism(s) [25,28], and enzyme activity of gonad gamma-glutamyl transpeptidase (γ-GTP) and gonadosomatic index (GSI) are always associated with the adverse effects on reproductive health of organisms, making them excellent complementary biomarkers for the indication of the presence of EEs [29,30].

In the past few years, the estrogenic contamination in certain lake, river, effluents of municipal wastewater treatment plant, and sediment has been evaluated by using chemical analysis combined with *in vivo* bioassays [31–35]. For example, Ribecco et al. [35] suggested the presence of estrogen-like substances in the sediment by revealing an increase in the expression of genes encoding estrogen receptor in *Solea solea*, though the levels of the primary pollutants present in the sediment did not exceed permissible Italian levels. In the present study, a valid method combining GC-MS and an *in vivo* bioassay was established to assess estrogenic activity in a real case study: a MSW landfill site in Qingdao. Frequently reported potential estrogenic compounds (natural and synthetic steroid estrogens, alkylphenols, phthalic acid esters, and BPA) were measured in outlets from different stages of the leachate treatment process, and then the removal ratios were calculated. Plasma Vtg and sex steroids, testis γ-GTP enzyme activity, and GSI were selected as specific bio-indicators to investigate the estrogenic effects in a semi-static exposure experiment.

Materials and Methods

Selection of the MSW Landfill Site

A MSW landfill site with a leachate treatment process in Qingdao was selected as a study model. It has been in operation since 2002 and mainly treats municipal solid waste and some bottom ash from MSW incinerators. The leachate treatment process used is an advanced membrane technology comprised of a membrane bioreactor treatment (MBR) with 4 bioreactors and an external ultrafiltration unit for sludge retention, a disctube reverse osmosis (DTRO) treatment, and a final aeration zeolite biofilter (see Figure 1). The leachate treatment facility is designed to collect and treat 900 m^3 of leachate per day and the treated effluent discharged into the nearby river meets Grade one A standard of the *Discharge Standard of Pollutants for Municipal Wastewater Treatment Plant* (GB18918-2002, China). The technical data of the leachate and the treated effluent samples employed in the experiment are shown in Table 1.

Chemical Analysis

In September 2011, composite aqueous samples were collected in solvent-clean brown glass bottles from storage tanks at different sampling points as shown in Figure 1 for the detection of target EEs, including natural and synthetic steroid estrogens [estrone (E_1), 17β-estradiol (E_2), estriol (E_3), and 17α-ethynilestradiol (EE_2)], bisphenol A (BPA), alkylphenols [nonylphenol (NP), 4-*n*-octylphenol (4-*n*-OP), and 4-*t*-octylphenol (4-*t*-OP)], and phthalic acid esters [di-ethyl phthalate (DEP), diethylhexyl phthalate (DEHP), di-*n*-butyl phthalate (D_nBP), and di-*n*-octylphthalate (DNOP)]. All samples were stored at 4°C, protected from light, and analyzed within 1 week.

Natural and synthetic steroids, alkylphenols, and BPA were extracted as previously described by Diniz et al. [33], with a few modifications. Briefly, samples (500 mL) were filtered through 0.45 µm glass fiber filters, mixed with 40 mL methanol and acidified with HCl to pH 3.0. Samples were then extracted through a 3 mL C18 solid phase extraction (SPE) cartridge which was previously conditioned with methanol and deionized water. EEs were eluted with 2×10 mL dichloromethane for steroids and BPA and with 2×10 mL methanol for alkylphenols. Elution fractions were evaporated to dryness under a gentle stream of nitrogen and reconstituted with 10% methanol to a final volume of 0.5 mL. The extraction of phthalic acid esters was adopted as follows: Samples (500 mL) were filtered through 0.45 µm glass fiber filters and extracted with 2×15 mL *n*-hexane for 10 min. The elution fractions were mixed and evaporated to a final volume of 0.5 mL.

The target compounds were analyzed using an Agilent 6890N/5973i GC-MS system (Agilent Technologies Inc., Palo Alto, USA). GC separations were carried out using an Agilent DB-17MS column (30 m×0.25 mm×0.25 µm). A sample volume of 1 µL was injected in splitless mode at an inlet temperature of 230°C. The mobile phase was He at a constant flow of 1 mL/min. For the detection of natural and synthetic steroids, alkylphenols, and BPA, the GC column temperature was programmed from 80°C (initial equilibrium time 2 min) to 230°C via a ramp of 15°C/min, 230°C–250°C via a ramp of 15°C/min, 250°C–280°C via a ramp of 5°C/min, and finally maintained at 280°C for 5 min. For the detection of phthalic acid esters, the GC column temperature was programmed from 60°C (initial equilibrium time 1 min) to 220°C via a ramp of 20°C/min, 220°C–280°C via a ramp of 5°C/min, and maintained at 280°C for 7 min. Identification of compounds was achieved by comparing the mass spectra and retention times of the chromatographic peaks with those of standard samples. Calibration curves were constructed by analyzing aliquots of a stock solution of authentic standards and all the correlation coefficients (r^2) were above 0.99. Table 2 shows the relevant parameters of the GC-MS. The removal ratios were calculated as following: the removal ratio = (1–concentration of a certain detected compound in treated effluent/the corresponding concentration in leachate)×100%.

In vivo Biological Assays

The fish were handled according to the National Institute of Health guidelines for the handling and care of experimental animals. The animal utilization protocol was approved by the Institutional Animal Care and Use Committee of the Ocean University of China. All surgery was performed under MS-222 anesthesia, and all efforts were made to minimize suffering.

Goldfish (*Carassius auratus*) are widespread throughout the world, and are a member of one of the largest vertebrate families, the *Cyprinidae*. Goldfish are omnivorous, feeding on plants, small crustaceans, and other food, which can reflect the food chain

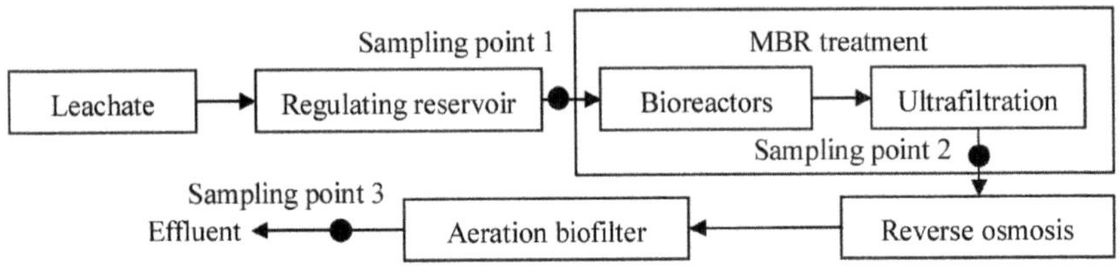

Figure 1. Schematic drawing of the treatment process for the leachate with sampling points (dots).

accumulation effect. In addition, many studies have been carried out to develop goldfish which could be handled easily and affordably in the laboratory, as a model animal for the evaluation of the estrogenic activities of various environmental chemicals [23,36–38]. Male goldfish (9.4±0.8 cm standard length; 27.1±4.2 g wet mass) were obtained from a local dealer in Qingdao, China, and were acclimated in the laboratory for 14 days. To estimate possible estrogenic activity on male fish, we simulated an accidental leachate release into groundwater resulting in a 1:200 dilution of the leachate; it was previously determined that a leachate dilution greater than 1:200 resulted in mortality (see Table S1 for data). Based on these data, male fish were exposed to 0.5% diluted leachate with dechlorinated tap water (sampling point 1), 0.5% diluted MBR outlet with dechlorinated tap water (sampling point 2), and 100% treated effluent (sampling point 3) for the *in vivo* bioassay. The semi-static toxicity test was conducted in 70 L aquaria containing 50 L solution water. Each group of fish (n = 24) were exposed in three aquaria (8 fish/tank). Additionally, 20 L exposure solutions were renewed daily and a male negative control group (dechlorinated tap water) was also set up. Male fish with intramuscular injection of E_2 dissolved in ethanol and peanut oil (1:1, v/v) at a concentration of 5 mg/mL was used as the positive control. The water temperature was 23±2°C and the dissolved oxygen was 7.0±0.1 mg/L. Fish were fed a non-estrogenic pelletized diet daily and no death was observed during exposure.

After a 28-day exposure in September 2011, male goldfish were anesthetized in 75 mg/L MS-222 (Sigma, St. Louis, MO, USA). Blood was taken from the caudal vein using chilled heparinized syringes and then goldfish were dissected. After centrifugation (3500 ×*g*, 10 min), plasma was frozen in liquid nitrogen for the detection of Vtg protein and sex steroids. Testis samples were removed, weighed, frozen in liquid nitrogen and stored at −80°C for enzyme activity analysis.

Vtg is usually measured by immunoassay and detected in western blot using specific antibodies. In this study, the plasma vitellogenin analysis was carried out by western blot as previously described by Pan & Tian et al. [39]. Discontinuous sodium dodecyl sulfate-polyacrylamide gel electrophoresis (SDS-PAGE) was performed using a resolving gel with 9% acrylamide and a stacking gel with 4% acrylamide. Rabbit anti-goldfish Vtg serum (prepared in previous studies) diluted 1:600 in blocking buffers and goat anti-rabbit IgG horseradish peroxidase conjugate (Beijing Solarbio Science & Technology Co., Ltd., Beijing, China) diluted 1:1600 in blocking buffers were used to detect the plasma Vtg protein.

Radioimmunoassay kits obtained from the Beijing North Institute of Biology Technology, China, were used to detect the plasma E_2 and testosterone (T) levels. The detection of plasma steroids was performed according to the manufacturer's instructions. The assay detection limits were 10 pg/mL for E_2 and 0.1 ng/mL for T. The inter- and intra-assay coefficients of variation for E_2 and T were controlled under 8% and 10%. The interaction ratio for E_2 and T detection with other hormones is < 0.03%.

γ-GTP testing kits (Nanjing Jiancheng Bioengineering Institute, Nanjing, China) were used for the γ-GTP enzyme activity test. Gonads were homogenized in 9 volumes (w/v) of homogenate buffer on ice. The supernatant was prepared using centrifugation at 4000 ×*g* for 10 min at 4°C. Analyses of γ-GTP enzyme activity were performed on ice according to the testing kit instructions, and protein contents were measured according to Bradford's method using bovine serum albumin (BSA) as the standard [40]. One unit (U) was defined as 0.01 OD produced by 1 mg protein in 1 h at 37°C. The assay detection limit was 0.08 U·mg/prot.

Statistics

The SPSS (version 16.0; USA) was employed to process all data. The data are expressed as the mean ± standard deviation (SD). Multiple comparisons were performed using one-way ANOVA analysis after checking for normality and equal variance. All differences were considered significant at $p<0.05$.

Table 1. Technical data of leachate and treated effluent.

Technical data	Leachate	Treated effluent
COD_{Cr} (mg/L)	15000	≤20
BOD_5 (mg/L)	5000	≤10
Ammonia nitrogen (mg/L)	2900	≤1
TN (mg/L)	3100	≤8
TP (mg/L)	15	≤0.5
TSS (mg/L)	2000	undetected

COD_{Cr}: chemical oxygen demand; BOD_5: five-day's biochemical oxygen demand; TN: total nitrogen; TP: total phosphorus; TSS: total suspended solids.

Table 2. Relevant parameters of the GC-MS.

EEs	linear equation	r^2	Qualification ions (m/z)	Quantitation ions (m/z)	Recovery	LODs (ng/L)
E_1	Y = 10.6X − 18.7	0.9972	342	257, 218, 327	ND	10
E_2	Y = 7.65X − 16.0	0.9961	285	416, 232, 401	82.3	5
E_3	Y = 2.05X − 10.6	0.9946	285	368, 232, 218	ND	5
EE_2	Y = 6.04X+40.9	0.9984	345	311, 504, 386	ND	3
4-*t*-OP	Y = 96.5X − 1060	0.9984	207	208, 278, 209	112	3
4-*n*-OP	Y = 37.4X − 292	0.9944	179	278, 180, 263	85.3	40
NP	Y = 23.3X+308	0.9994	179	180, 372, 193	190	40
BPA	Y = 86.3X − 751	0.9954	357	372, 207, 179	235	40
DEP	Y = 237X − 241	1.0000	121	149, 205, 223	ND	40
D_nBP	Y = 443X − 1620	1.0000	149	167, 205, 223	ND	10
DEHP	Y = 202X − 1820	1.0000	113	149, 167, 279	82.3	5
DNOP	Y = 159X − 2090	0.9996	149	167, 261, 279	ND	5

LODs: limits of detection; ND: not detected.

Results

Occurrence of EEs

The occurrence of EEs in outlets from different stages of the treatment process during this study is summarized in Table 3. 5 environmental estrogens (4-*t*-OP, BPA, DEP, D_nBP, and DEHP) were detected whereas natural and synthetic steroid estrogens, NP, and DNOP were under detection limits. The highest concentrations were found in leachate and reached 6153 ng/L, 3642 ng/L, 2139 ng/L, 5900 ng/L, and 9422 ng/L, respectively. The removal ratios of 4-*t*-OP, BPA, DEP, D_nBP, and DEHP were 99.7%, 99.04%, 99.69%, 87.2%, and 92.9%, respectively.

Induction of Plasma Vtg

Plasma samples were analyzed for Vtg by western blot (Figure 2). The results revealed the presence of one Vtg protein subunit in the plasma of 0.5% leachate exposed fish and 100% treated effluent exposed fish, which was also observed in the E_2 induced male but absent from the control male fish, indicating the induction of Vtg protein synthesis by 0.5% leachate and treated effluent exposure. In contrast, Vtg protein was absent from plasma of fish exposed to 0.5% MBR outlet.

Effects of Sex Steroids Levels in Plasma

As shown in Figure 3A, the level of E_2 in the plasma from treated effluent exposed fish (62.06±7.49 pg/mL) was significantly increased ($p<0.05$), while that of the other two groups was unchanged compared to the control male fish. Furthermore, the plasma T concentrations were not affected by any of the outlets from different stages of the treatment process (Figure 3B).

Effects on Gonad γ-GTP Activity and GSI

Enzymatic activities of γ-GTP in the gonad were down-regulated after a 28-day exposure of outlets from different stages of the treatment process (Figure 4A), among which the γ-GTP activity in the male fish exposed to 0.5% leachate was close to the detection limit (0.08 U·mg/prot).

The gonadosomatic index (GSI) was calculated by: GSI = (total gonad weight/wet mass) × 100%. As shown in Figure 4B, GSIs of male fish exposed to 0.5% leachate, outlet of MBR treatment, and treated effluent were significantly lower than control after a 28-day exposure.

Discussion

Chemical analysis was utilized to identify the chemical components of leachate samples. Yasuhara et al. [41,42] investigated approximately 190 types of chemical substances and more than 100 types of organic compounds including 1, 4-dioxane, phthalates, and bisphenol A in leachate samples in Japan. 41 sorts of phenols including low concentrations of 4-*t*-OP to high concentrations of BPA were detected in 38 municipal solid waste landfill sites [43]. Chemical analysis can serve as a useful tool to identify a group of EEs in leachate, by which the possible source of

Table 3. Mean concentrations of target EEs measured by GC-MS and EEQs of different samples.

Samples	Selected EEs (ng/L)					EEQ (ng/L)
	4-*t*-OP	BPA	DEP	D_nBP	DEHP	
Effluent	14.3	34.8	6.66	757	666	0.01
MBR outlet	13.41	5.88	2.87	198.5	530	0.008
Leachate	6153	3642	2139	5900	9422	3.78

The estrogen equivalent quantity (EEQ) was defined as the sum of the concentration for each individual compound after multiplying an estradiol equivalency factor derived from previous studies [17,73–76]. E_1, E_2, E_3, EE_2, and DNOP were under detection limits.

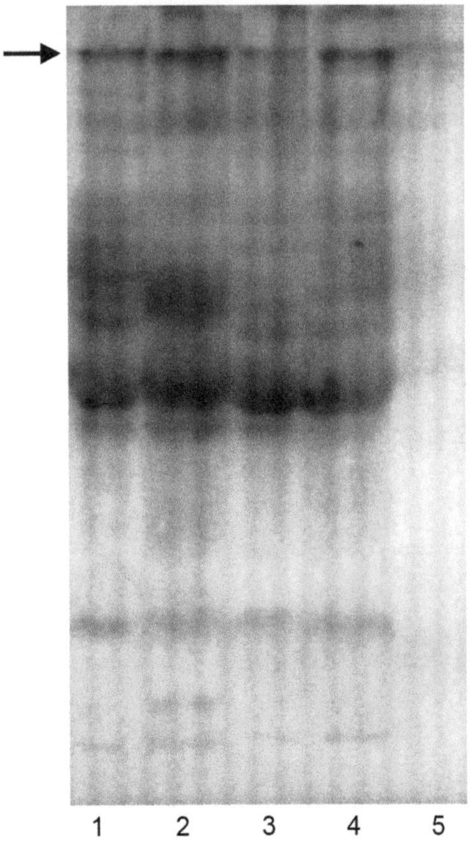

Figure 2. Western blot of plasma from treated and control male fish. Lane 1: plasma from E_2 injected fish (20×); Lane 2: plasma from effluent exposed fish; Lane 3: plasma from 0.5% diluted MBR outlet exposed fish; Lane 4: plasma from 0.5% diluted leachate exposed fish; Lane 5: plasma from negative control male. Primary antibody: rabbit anti-goldfish Vtg serum (diluted 1:600). Secondary antibody: goat anti-rabbit IgG horseradish peroxidase conjugate (diluted 1:1600).

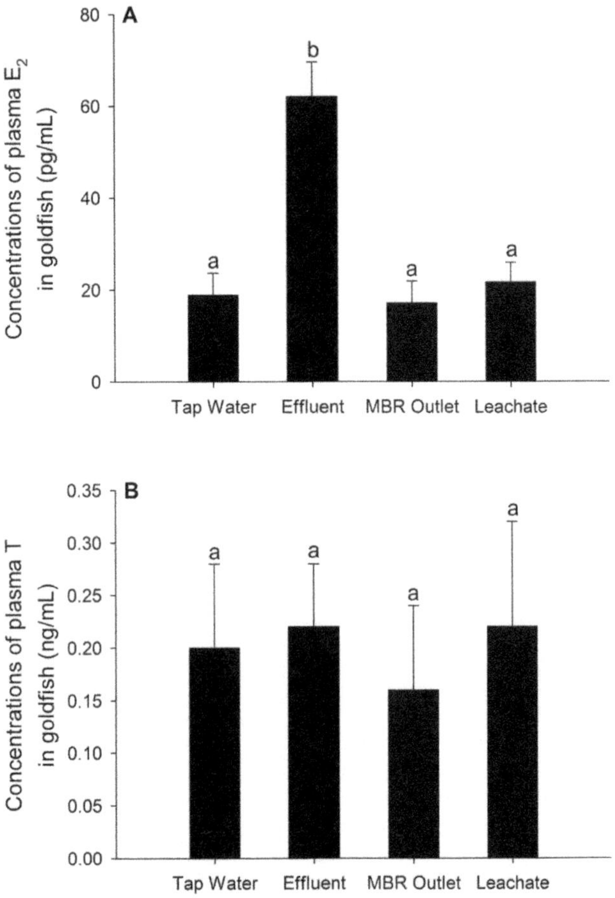

Figure 3. Quantification of plasma sex steroid levels in exposed male fish. Male goldfish were exposed to dechlorinated tap water, 100% treated effluent, 0.5% diluted MBR outlet, and 0.5% diluted landfill leachate (designated Tap Water, Effluent, MBR Outlet, and Leachate, respectively) for 28 days in a semi-static exposure system. Data are presented as the mean ± standard deviation (n = 24). Values with different letters denote statistically significant differences according to a one-way ANOVA analysis ($p<0.05$).

these compounds can be investigated. In this study, 5 identified EEs (4-*t*-OP, BPA, D_nBP, DEP, and DEHP) were detected in leachate samples collected from selected landfill sites, and high detection rates have been frequently reported for them in previous studies [11,12,17]. 4-*t*-OP is the most representative final breakdown product of alkylphenol polyethoxylates (APEOs), which are widely used as wetting and emulsion agents in industrial and household applications [33]. BPA is employed during the production of polycarbonate and epoxy resins, which are important raw materials of commodity products [44]. PAEs are mainly applied as plasticizers in the production of dope, binder, and a variety of plastics [45]. It has been suggested that the occurrence of these detected estrogen-like chemicals may be a result of the degradation of household appliances, more importantly household plastics. In addition, the bottom ash from MSW incinerators could be another source [43].

Of more importance, chemical analysis is recommended as a prevailing methodological approach for the quantification of known or suspected EEs. In this study, measured concentrations of 5 types of estrogenic compounds were found to be present in concentration ranges reported in the literature. Teuten et al. [46] reported that the highest concentration range of BPA was 0.18–4300 μg/L in leachates from tropical Asian countries. Jonsson et al. [47] found the occurrences of DEHP (5–460 μg/L), D_nBP (1–23 μg/L), and DEP (1–33 μg/L) in 17 European landfill leachates. However, the concentration of 4-*t*-OP found in this study was a little higher (0.03–3.4 μg/L by Teuten et al. [46]; 0.005–1.2 μg/L by Kurata et al. [43]; 0.04–0.31 μg/L by Asakura et al. [48]). The difference in 4-*t*-OP levels may be due to the composition of the landfill waste or the landfilling method.

Because of complex matrices of landfill leachate, the chemical analysis is limited in providing comprehensive knowledge of the overall estrogenic effects that might be engendered by leachate. This limitation can be overcome by the *in vivo* bioassay which utilized a series of specific biomarkers. In this study, a battery of bio-indicators including plasma Vtg and sex steroids, testis γ-GTP enzyme activity, and GSI were employed to evaluate the estrogenic activity at the molecular level, biochemical level, and tissue level.

Vtg is a female-specific phospholipoglycoprotein synthesized in the liver as nutrients for the developing embryo and is induced by endogenous estrogens [49,50]. Male and juvenile fish normally produce minute amounts of Vtg due to the lack of endogenous estrogens. Therefore, an exposure to estrogenic compounds would lead to an abnormal induction of plasma Vtg, which makes it an ideal biomarker to test for exogenous estrogen exposure [24,26,51]. Also as mentioned in the preliminary draft guideline

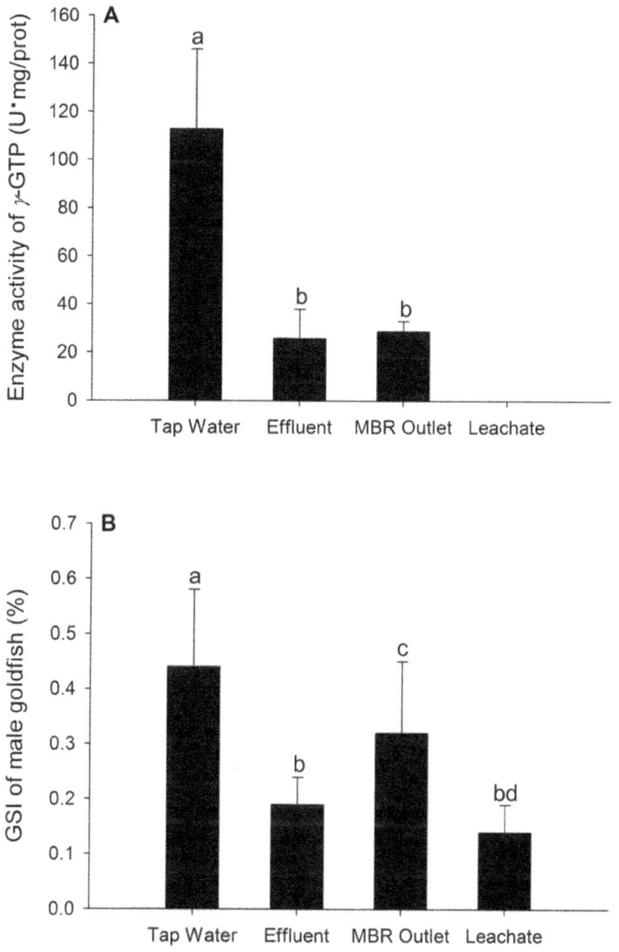

Figure 4. γ-GTP activity in gonad (A) and GSI (B) of exposed male goldfish. Male goldfish were exposed to dechlorinated tap water, 100% treated effluent, 0.5% diluted MBR outlet, and 0.5% diluted landfill leachate (designated Tap Water, Effluent, MBR Outlet, and Leachate, respectively) for 28 days in a semi-static exposure system. Data are presented as the mean ± standard deviation (n = 24). Values with different letters denote statistically significant differences according to a one-way ANOVA analysis ($p<0.05$).

"The Fish Screening Assay for Endocrine Active Substances" within Organisation for Economic Co-operation and Development (OECD), Vtg is employed as a core biomarker endpoint for detecting endocrine active chemicals (unpublished). Currently, the Vtg induction assay has been widely used in assessing the estrogenic activity of effluent from wastewater treatment plants [52–54]. Our current study demonstrated that vitellogenin secretion was significantly induced in male goldfish after exposure to 0.5% diluted leachate and 100% treated effluent, suggesting significant estrogenic properties of these outlets *in vivo*. It is likely that estrogenic substances present in leachate and treated effluent were taken up by the fish and thereby stimulated the biosynthesis of Vtg. Considering the weak Vtg induction found in plasma of male fish exposed to 0.5% MBR outlet, the estrogenic activity was thought to be relatively low. Because the normal concentrations of sex steroids were not affected in 0.5% leachate for exposed male fish, it appears that various components of landfill leachate may bind to the estrogen receptor directly to induce Vtg synthesis. Evidence from *in vitro* assays has indicated that growth hormone and prolactin can also induce Vtg [55,56].

Exposure to EEs is always associated with altered concentrations of plasma sex hormones, which play a decisive role in the regulation of sex differentiation, sexual characteristics, reproductive behavior, and reproductive capacity [8]. According to Lu et al. [32], male goldfish exposed to water from the Nanjing section of the Yangzi River had an elevated plasma E_2 concentration. Similarly, plasma and testicular T levels in prepubertal male rats were decreased after subcutaneous administration of BPA, a recognized environmental estrogen, as described by Nakamura et al. [57]. In the present study, exposure of goldfish to treated effluent resulted in significantly increased circulating E_2 levels compared to the negative control group, indicating a disturbance of sexual hormone homeostasis in male goldfish. The increased endogenous estradiol is one of the possible explanations of Vtg synthesis. In addition, this lower estrogenic potency of MBR outlet is further supported by the fact that the concentration of plasma E_2 in 0.5% MBR exposed males was not affected. Endogenous E_2 is mainly generated by the conversion of T [58] and the raised E_2 level is always accompanied by a decreased T level. In the current study, exposure to 100% treated effluent for 28-days caused significant increases in plasma E_2 levels but not in the T level, thus indicating the possible existence of androgenic activity. Studies in recent years have demonstrated that landfill leachate, municipal waste water, and industrial waste water may contain androgenic compounds [59–62]. For example, nine androgens and nine progestogens were detected in the influent and effluent of wastewater treatment plants in Beijing, China [63]. In addition, the biotransformation among androgens can be another reason for the unaffected T levels in plasma, as there are many types of androgens in male fish, such as androstenedione, 17α, 20β-algestone, and 11-ketotestosterone. Further detailed analysis is required to confirm this hypothesis.

The γ-GTP enzyme is mainly localized in the Sertoli cells of testis and the activity of the γ-GTP enzyme is a frequently used biomarker to indicate a lesion in the Sertoli cells and damage to the reproduction organs of male fish by xenobiotics at the biochemical level [30,64]. Previous studies have suggested that exposure to 4-*t*-OP or NP may lead to structural damage of Sertoli cells and decreased enzyme activity of γ-GTP in eelpout (*Zoarces viviparus*), the effect of which is similar to E_2 [65,66]. Because the γ-GTP activity in testis of male fish exposed to treated effluent was decreased, it is reasonable to conclude that the Sertoli cells might be damaged due to the remarkable estrogenic potency. The enzyme activity of γ-GTP of male fish exposed to 0.5% leachate was close to the detection limit (0.08 U·mg/prot), which indicates the high biotoxicity of leachate due to the existence of hazardous substances.

GSI is a highly comprehensive index to assess the damage of reproductive function at the tissue level. Generally speaking, the exposure to environmental estrogens always results in a pronounced decline in GSI of adult male fish [67–69]. The reproductive toxicity of leachate and treated effluent was certified by the testicular atrophy of males in this study. These significantly lower GSIs are likely to be a result of Vtg synthesis. Jobling et al. [70] reported the inverse correlation between plasma Vtg and GSI in male roaches (*Rutilus rutllus*). Considering that Vtg synthesis is an energy-wasting process, the reproduction and growth of exposed fish tends to be delayed.

The method using estrogen equivalent quantity (EEQ) is recommended for hazard assessment of different classes of estrogenic chemical mixtures. By the use of this approach, Muller et al. [71] assessed the removal efficiency of estrogens by a municipal wastewater treatment plant and found that the biological treatment process possessed the highest removal ratio.

The overall estrogenic equivalents of a mixture are defined by the concentration of individual compounds in a mixture times their relative potencies to E_2, as reported by Safe [72]. Thus, the overall estrogenic potency of a complex mixture, for example, leachate, can be calculated and compared. According to previous studies, the estrogen equivalent factors of the detected EEs were 5.6×10^{-4} for BPA, 6×10^{-5} for OP, 2.4×10^{-5} for D_nBP, 2.0×10^{-7} for DEHP, and 0.5×10^{-7} for DEP [17,73–76]. The EEQs of leachate, MBR outlet, and treated effluent when calculated by this method were 3.78 ng/L, 0.008 ng/L, and 0.01 ng/L, which seemed to be relatively low. However, remarkable estrogenic activities that led to the induction of Vtg, and depression of γ-GTP activity and GSI were observed in leachate and treated effluent exposed male fish. The biological responses in male goldfish did not agree either with the chemical data obtained by GC-MS. Similar results were also reported by Blankvoort et al. [60], and Gunnarsson et al. [77]. Leachate is supposed to compromise the validity of the EEQ approach based on chemical analysis because of the cross talk between multiple endocrine pathways [72], and hence it should be noted that an underestimation may be made if only using the results of chemical analysis. However, *in vivo* bioassays mostly provide an integrated result of the total potency of complex mixtures to induce some biological responses, just as mentioned above that due to the presence of some substances with androgenic activities, the plasma T levels were unexpectedly not changed. Thus batteries of biomarkers at the molecular level, biochemical level, and tissue level, may prove more useful than single biomarkers in characterizing certain endocrine disrupting effect. Accordingly such *in vivo* experimentation might be complicated to operate and costly in financial terms, thereby making it valuable in study rather than in routine controls. Taking these findings together, a combination of chemical analysis and an *in vivo* bioassay is necessary to provide comprehensive knowledge and valid assessments of estrogenicity of a complicated mixture, such as landfill leachate.

Based on the comprehensive assessment results, the efficiency of unit operations in removing the previously selected compounds can be well evaluated. Due to the heavily contaminated property and poor biodegradability of landfill leachate, conventional wastewater treatment cannot fulfill treatment requests. Advanced wastewater treatments, especially the membrane process, for example, reverse osmosis treatment, are widely applied as a landfill leachate treatment. Coors et al. [10] used a reporter-gene-based bioassay with a human breast-cancer-derived cell line to study the removal rate of two different membrane employing processes and found that the concentrated effluent of the reverse osmosis treatment still showed an estrogenic response. The selected leachate treatment facility employed a "membrane bioreactor+reverse osmosis+aeration zeolite biofilter" treatment process, with qualified removal ratios of conventional parameters (i.e., COD_{Cr}, BOD_5, TN, TP, and TSS), and the quality of the treated effluent meets Grade one A standard of the *Discharge Standard of Pollutants for Municipal Wastewater Treatment Plant* (GB18918-2002, China). A relatively high rejection ratio of detected EEs, which can reach 87.2% or above, was achieved by the landfill leachate treatment process. However, the treated effluent still showed estrogenic activity in bioassays, indicating the existence of environmental estrogens and the potential risk to the aquatic habitat.

Conclusions

To achieve both qualitative and quantitative analysis results for estrogenic potency of leachate and treated effluent, a combined approach incorporating two types of analyses was proposed and applied to a real case study in Qingdao. The following are the main findings: the method combined chemical techniques with the responses of test organisms can be effectively applied for analyzing estrogen compounds, and resulted in a more comprehensive characterization of the estrogenic activity of environmentally complex samples. In a concentration of 200 times diluted, the leachate still showed strong estrogenic activity and led to biotoxicity effects on male goldfish. The removal rate of "membrane bioreactor+reverse osmosis+aeration zeolite biofilter" treatment process for the target EEs can reach 87.2% or above; however, 5 types of EEs were still present in the effluent and contribute to an estrogen-like effect.

Supporting Information

Table S1 Biological responses of goldfish exposed to different dilutions of leachate. Male goldfish (12 fish each group) were exposed to different dilutions of the leachate (1:10, 1:20, 1:50, 1:100, and 1:200) in a semi-static exposure system after acclimated in the lab for 14 days. The leachate was diluted with dechlorinated tap water and a negative control group (dechlorinated tap water) was also set up. The mortality rate, survival time, and abnormal behavior were recorded.

Table S2 The ARRIVE guidelines checklist for animal research: reporting *in vivo* experiments.

Acknowledgments

The authors are grateful to all members in the lab for their help.

Author Contributions

Conceived and designed the experiments: YG SR HT. Performed the experiments: YG SY. Analyzed the data: YG SY LW. Contributed reagents/materials/analysis tools: SR HT LW. Wrote the paper: YG SR HT.

References

1. Ministry of Environmental Protection of The People's Republic of China (2011) Chinese environmental statistics annual book. Available: http://www.zhb.gov.cn.
2. Kjeldsen P, Barlaz MA, Rooker AP, Baun A, Ledin A, et al. (2002) Present and long-term composition of MSW landfill leachate: a review. Critical reviews in environmental science and technology 32: 297–336.
3. Lei L, Piao M, Nishizaki H, Nakamura H, Aoyama I (2006) Toxicity identification evaluation on industrial and municipal solid waste landfill leachates. Japanese Journal of Environmental Toxicology 9: 11–21.
4. Sisinno CL, Oliveira-Filho EC, Dufrayer MC, Moreira JC, Paumgartten FJ (2000) Toxicity evaluation of a municipal dump leachate using zebrafish acute tests. Bull Environ Contam Toxicol 64: 107–113.
5. Pivato A, Gaspari L (2006) Acute toxicity test of leachates from traditional and sustainable landfills using luminescent bacteria. Waste Manag 26: 1148–1155.
6. Klauck CR, Rodrigues MA, da Silva LB (2013) Toxicological evaluation of landfill leachate using plant (*Allium cepa*) and fish (*Leporinus obtusidens*) bioassays. Waste Manag Res 31: 1148–1153.
7. Noaksson E, Tjarnlund U, Bosveld AT, Balk L (2001) Evidence for endocrine disruption in perch (*Perca fluviatilis*) and roach (*Rutilus rutilus*) in a remote Swedish lake in the vicinity of a public refuse dump. Toxicol Appl Pharmacol 174: 160–176.
8. Noaksson E, Linderoth M, Bosveld AT, Norrgren L, Zebuhr Y, et al. (2003) Endocrine disruption in brook trout (*Salvelinus fontinalis*) exposed to leachate from a public refuse dump. Sci Total Environ 305: 87–103.
9. Noaksson E, Linderoth M, Tjarnlund U, Balk L (2005) Toxicological effects and reproductive impairments in female perch (*Perca fluviatilis*) exposed to leachate from Swedish refuse dumps. Aquat Toxicol 75: 162–177.

10. Coors A, Jones PD, Giesy JP, Ratte HT (2003) Removal of estrogenic activity from municipal waste landfill leachate assessed with a bioassay based on reporter gene expression. Environ Sci Technol 37: 3430–3434.
11. Asakura H, Matsuto T, Tanaka N (2007) Analytical study of endocrine-disrupting chemicals in leachate treatment process of municipal solid waste (MSW) landfill sites. Environ Sci 14: 79–87.
12. Asakura H, Matsuto T (2009) Experimental study of behavior of endocrine-disrupting chemicals in leachate treatment process and evaluation of removal efficiency. Waste Manag 29: 1852–1859.
13. He PJ, Zheng Z, Zhang H, Shao LM, Tang QY (2009) PAEs and BPA removal in landfill leachate with Fenton process and its relationship with leachate DOM composition. Sci Total Environ 407: 4928–4933.
14. Joseph L, Zaib Q, Khan IA, Berge ND, Park YG, et al. (2011) Removal of bisphenol A and 17alpha-ethinyl estradiol from landfill leachate using single-walled carbon nanotubes. Water Res 45: 4056–4068.
15. Kawagoshi Y, Tsukagoshi Y, Fukunaga I (2002) Determination of estrogenic activity in landfill leachate by simplified yeast two-hybrid assay. J Environ Monit 4: 1040–1046.
16. Dizer H, Fischer B, Sepulveda I, Loffredo E, Senesi N, et al. (2002) Estrogenic effect of leachates and soil extracts from lysimeters spiked with sewage sludge and reference endocrine disrupters. Environ Toxicol 17: 105–112.
17. Behnisch PA, Fujii K, Shiozaki K, Kawakami I, Sakai S (2001) Estrogenic and dioxin-like potency in each step of a controlled landfill leachate treatment plant in Japan. Chemosphere 43: 977–984.
18. Kawagoshi Y, Fujita Y, Kishi I, Fukunaga I (2003) Estrogenic chemicals and estrogenic activity in leachate from municipal waste landfill determined by yeast two-hybrid assay. J Environ Monit 5: 269–274.
19. Baderna D, Maggioni S, Boriani E, Gemma S, Molteni M, et al. (2011) A combined approach to investigate the toxicity of an industrial landfill's leachate: chemical analyses, risk assessment and *in vitro* assays. Environ Res 111: 603–613.
20. Folmar LC, Hemmer M, Denslow ND, Kroll K, Chen J, et al. (2002) A comparison of the estrogenic potencies of estradiol, ethynylestradiol, diethylstilbestrol, nonylphenol and methoxychlor *in vivo* and *in vitro*. Aquat Toxicol 60: 101–110.
21. Thorpe KL, Cummings RI, Hutchinson TH, Scholze M, Brighty G, et al. (2003) Relative potencies and combination effects of steroidal estrogens in fish. Environ Sci Technol 37: 1142–1149.
22. Zheng G, Zhou ZJ, Dai XF (2004) Detection of estrogenic effects of organphosphorus pesticides on MCF-7 human breast cancer cells. Chinese Journal of Industrial Medicine 17: 273–276.
23. Tian H, Ru S, Wang Z, Cai W, Wang W (2009) Estrogenic effects of monocrotophos evaluated by vitellogenin mRNA and protein induction in male goldfish (*Carassius auratus*). Comp Biochem Physiol C Toxicol Pharmacol 150: 231–236.
24. Tyler CR, van Aerle R, Hutchinson TH, Maddix S, Trip H (1999) An *in vivo* testing system for endocrine disruptors in fish early life stages using induction of vitellogenin. Environmental Toxicology and Chemistry 18: 337–347.
25. Sole M, Raldua D, Piferrer F, Barcelo D, Porte C (2003) Feminization of wild carp, *Cyprinus carpio*, in a polluted environment: plasma steroid hormones, gonadal morphology and xenobiotic metabolizing system. Comp Biochem Physiol C Toxicol Pharmacol 136: 145–156.
26. Marin MG, Matozzo V (2004) Vitellogenin induction as a biomarker of exposure to estrogenic compounds in aquatic environments. Marine pollution bulletin 48: 835–839.
27. Kime D, Nash J, Scott A (1999) Vitellogenesis as a biomarker of reproductive disruption by xenobiotics. Aquaculture 177: 345–352.
28. Diniz MS, Peres I, Castro L, Freitas AC, Rocha-Santos TA, et al. (2009) Effects of ECF-Kraft pulp mill effluent treated with fungi (*Rhizopus oryzae*) on reproductive steroids and liver CYP1A of exposed goldfish (*Carassius auratus*). Ecotoxicology 18: 1011–1017.
29. Komen J, Lodder P, Huskens F, Richter C, Huisman E (1989) Effects of oral administration of 17α-methyltestosterone and 17β-estradiol on gonadal development in common carp, (*Cyprinus carpio*). Aquaculture 78: 349–363.
30. Rasmussen TH, Teh SJ, Bjerregaard P, Korsgaard B (2005) Anti-estrogen prevents xenoestrogen-induced testicular pathology of eelpout (*Zoarces viviparus*). Aquat Toxicol 72: 177–194.
31. Lu G, Yan Z, Wang Y, Chen W (2011) Assessment of estrogenic contamination and biological effects in Lake Taihu. Ecotoxicology 20: 974–981.
32. Lu GH, Song WT, Wang C, Yan ZH (2010) Assessment of *in vivo* estrogenic response and the identification of environmental estrogens in the Yangtze River (Nanjing section). Chemosphere 80: 982–990.
33. Diniz MS, Mauricio R, Petrovic M, Lopez de Alda MJ, Amaral L, et al. (2010) Assessing the estrogenic potency in a Portuguese wastewater treatment plant using an integrated approach. J Environ Sci (China) 22: 1613–1622.
34. Ribecco C, Baker ME, Šášik R, Zuo Y, Hardiman G, et al. (2011) Biological effects of marine contaminated sediments on *Sparus aurata* juveniles. Aquat Toxicol 104: 308–316.
35. Ribecco C, Hardiman G, Šášik R, Vittori S, Carnevali O (2012) Teleost fish (*Solea solea*): A novel model for ecotoxicological assay of contaminated sediments. Aquat Toxicol 109: 133–142.
36. Tian H, Ru S, Bing X, Wang W (2010) Effects of monocrotophos on the reproductive axis in the male goldfish (*Carassius auratus*): Potential mechanisms underlying vitellogenin induction. Aquat Toxicol 98: 67–73.
37. Tian H, Ru S, Wang W, Bing X (2010) Effects of monocrotophos on the reproductive axis in the female goldfish (*Carassius auratus*). Comp Biochem Physiol C Toxicol Pharmacol 152: 107–113.
38. Zhang X, Tian H, Wang W, Ru S (2013) Exposure to monocrotophos pesticide causes disruption of the hypothalamic–pituitary–thyroid axis in adult male goldfish (*Carassius auratus*). Gen Comp Endocrinol 193: 158–166.
39. Pan Z, Tian H, Wang W, Wang J, Ru S (2012) Identification, purification, and immunoassay of stone flounder (*Kareius bicolouratus*) vitellogenin. Journal of the Korean Society for Applied Biological Chemistry 55: 219–227.
40. Bradford MM (1976) A rapid and sensitive method for the quantitation of microgram quantities of protein utilizing the principle of protein-dye binding. Anal Biochem 72: 248–254.
41. Yasuhara A, Shiraishi H, Nishikawa M, Yamamoto T, Uehiro T, et al. (1997) Determination of organic components in leachates from hazardous waste disposal sites in Japan by gas chromatography-mass spectrometry. Journal of Chromatography A 774: 321–332.
42. Yasuhara A, Shiraishi H, Nishikawa M, Yamamoto T, Nakasugi O, et al. (1999) Organic components in leachates from hazardous waste disposal sites. Waste Management and Research 17: 186–197.
43. Kurata Y, Ono Y, Ono Y (2008) Occurrence of phenols in leachates from municipal solid waste landfill sites in Japan. Journal of material cycles and waste management 10: 144–152.
44. Furhacker M, Scharf S, Weber H (2000) Bisphenol A: emissions from point sources. Chemosphere 41: 751–756.
45. Giam C, Atlas E, Powers M Jr, Leonard J (1984) Phthalic acid esters. Anthropogenic compounds: Springer. 67–142.
46. Teuten EL, Saquing JM, Knappe DR, Barlaz MA, Jonsson S, et al. (2009) Transport and release of chemicals from plastics to the environment and to wildlife. Philos Trans R Soc Lond B Biol Sci 364: 2027–2045.
47. Jonsson S, Ejlertsson J, Ledin A, Mersiowsky I, Svensson BH (2003) Mono- and diesters from o-phthalic acid in leachates from different European landfills. Water Res 37: 609–617.
48. Asakura H, Matsuto T, Tanaka N (2004) Behavior of endocrine-disrupting chemicals in leachate from MSW landfill sites in Japan. Waste Manag 24: 613–622.
49. Wallace RA, Begovac PC (1985) Phosvitins in Fundulus oocytes and eggs. Preliminary chromatographic and electrophoretic analyses together with biological considerations. J Biol Chem 260: 11268–11274.
50. Gerber-Huber S, Nardelli D, Haefliger JA, Cooper DN, Givel F, et al. (1987) Precursor-product relationship between vitellogenin and the yolk proteins as derived from the complete sequence of a *Xenopus* vitellogenin gene. Nucleic Acids Res 15: 4737–4760.
51. Sumpter JP, Jobling S (1995) Vitellogenesis as a biomarker for estrogenic contamination of the aquatic environment. Environ Health Perspect 103 Suppl 7: 173–178.
52. Purdom C, Hardiman P, Bye V, Eno N, Tyler C, et al. (1994) Estrogenic effects of effluents from sewage treatment works. Chemistry and Ecology 8: 275–285.
53. Li CR, Lee SH, Kim SS, Kim A, Lee KW, et al. (2009) Environmental estrogenic effects and gonadal development in wild goldfish (*Carassius auratus*). Environ Monit Assess 150: 397–404.
54. Wang R, Liu J, Yang X, Lin C, Huang B, et al. (2013) Biological response of high-back crucian carp (*Carassius auratus*) during different life stages to wastewater treatment plant effluent. Environ Sci Pollut Res Int 20: 8612–8620.
55. Carnevali O, Mosconi G, Yamamoto K, Kobayashi T, Kikuyama S, et al. (1992) Hormonal control of *in vitro* vitellogenin synthesis in *Rana esculenta* liver: effects of mammalian and amphibian growth hormone. Gen Comp Endocrinol 88: 406–414.
56. Carnevali O, Mosconi G, Yamamoto K, Kobayashi T, Kikuyama S, et al. (1993) *In-vitro* effects of mammalian and amphibian prolactins on hepatic vitellogenin synthesis in *Rana esculenta*. J Endocrinol 137: 383–389.
57. Nakamura D, Yanagiba Y, Duan Z, Ito Y, Okamura A, et al. (2010) Bisphenol A may cause testosterone reduction by adversely affecting both testis and pituitary systems similar to estradiol. Toxicol Lett 194: 16–25.
58. Simpson ER, Mahendroo MS, Means GD, Kilgore MW, Hinshelwood MM, et al. (1994) Aromatase cytochrome P450, the enzyme responsible for estrogen biosynthesis. Endocr Rev 15: 342–355.
59. Svenson A, Allard AS, Junestedt C, Cerne O, Ek M (2004) Assessment of androgenicity in leachates from Swedish landfills and treatments for its elimination. J Environ Sci Health A Tox Hazard Subst Environ Eng 39: 2817–2825.
60. Blankvoort BM, Rodenburg RJ, Murk AJ, Koeman JH, Schilt R, et al. (2005) Androgenic activity in surface water samples detected using the AR-LUX assay: indications for mixture effects. Environ Toxicol Pharmacol 19: 263–272.
61. Leusch FD, Chapman HF, van den Heuvel MR, Tan BL, Gooneratne SR, et al. (2006) Bioassay-derived androgenic and estrogenic activity in municipal sewage in Australia and New Zealand. Ecotoxicol Environ Saf 65: 403–411.
62. Wen R, Xie Y, Wan C, Fang Z (2013) Estrogenic and androgenic effects in mosquitofish (*Gambusia affinis*) from streams contaminated by municipal effluent in Guangzhou, China. Aquat Toxicol 132–133: 165–172.
63. Chang H, Wan Y, Wu S, Fan Z, Hu J (2011) Occurrence of androgens and progestogens in wastewater treatment plants and receiving river waters: comparison to estrogens. Water Res 45: 732–740.

64. Miura T, Yamauhi K, Nagahama Y, Takahashi H (1991) Induction of Spermatogenesis in Male Japanese Eel, *Anguilla japonica*, by a Single Injection of Human Chorionic Gonadotropin (Endocrinology). Zoological science 8: 63–73.
65. Christiansen L, Pedersen K, Korsgaard B, Bjerregaard P (1998) Estrogenicity of xenobiotics in rainbow trout (*Oncorhynchus mykiss*) using *in vivo* synthesis of vitellogenin as a biomarker. Marine Environmental Research 46: 137–140.
66. Rasmussen TH, Korsgaard B (2004) Estrogenic octylphenol affects seminal fluid production and its biochemical composition of eelpout (*Zoarces viviparus*). Comp Biochem Physiol C Toxicol Pharmacol 139: 1–10.
67. Jobling S, Sumpter JP, Sheahan D, Osborne JA, Matthiessen P (1996) Inhibition of testicular growth in rainbow trout (*Oncorhynchus mykiss*) exposed to estrogenic alkylphenolic chemicals. Environmental Toxicology and Chemistry 15: 194–202.
68. Toft G, Baatrup E (2001) Sexual characteristics are altered by 4-tert-octylphenol and 17beta-estradiol in the adult male guppy (*Poecilia reticulata*). Ecotoxicol Environ Saf 48: 76–84.
69. Cardinali M, Maradonna F, Olivotto I, Bortoluzzi G, Mosconi G, et al. (2004) Temporary impairment of reproduction in freshwater teleost exposed to nonylphenol. Reprod Toxicol 18: 597–604.
70. Jobling S, Tyler C, Nolan M, Sumpter J (1998) The identification of oestrogenic effects in wild fish. Research and development Technical Report W 119.
71. Muller M, Rabenoelina F, Balaguer P, Patureau D, Lemenach K, et al. (2008) Chemical and biological analysis of endocrine-disrupting hormones and estrogenic activity in an advanced sewage treatment plant. Environ Toxicol Chem 27: 1649–1658.
72. Safe SH (1998) Hazard and risk assessment of chemical mixtures using the toxic equivalency factor approach. Environ Health Perspect 106 Suppl 4: 1051–1058.
73. Api A (2001) Toxicological profile of diethyl phthalate: a vehicle for fragrance and cosmetic ingredients. Food and Chemical Toxicology 39: 97–108.
74. Körner W, Bolz U, Süβmuth W, Hiller G, Schuller W, et al. (2000) Input/output balance of estrogenic active compounds in a major municipal sewage plant in Germany. Chemosphere 40: 1131–1142.
75. ter Veld MG, Schouten B, Louisse J, van Es DS, van der Saag PT, et al. (2006) Estrogenic potency of food-packaging-associated plasticizers and antioxidants as detected in ERalpha and ERbeta reporter gene cell lines. J Agric Food Chem 54: 4407–4416.
76. Liu ZH, Kanjo Y, Mizutani S (2009) Removal mechanisms for endocrine disrupting compounds (EDCs) in wastewater treatment-physical means, biodegradation, and chemical advanced oxidation: a review. Sci Total Environ 407: 731–748.
77. Gunnarsson L, Adolfsson-Erici M, Bjorlenius B, Rutgersson C, Forlin L, et al. (2009) Comparison of six different sewage treatment processes-reduction of estrogenic substances and effects on gene expression in exposed male fish. Sci Total Environ 407: 5235–5242.

18

Long-Term Citizen-Collected Data Reveal Geographical Patterns and Temporal Trends in Lake Water Clarity

Noah R. Lottig[1]*, Tyler Wagner[2], Emily Norton Henry[3,4], Kendra Spence Cheruvelil[3,5], Katherine E. Webster[6], John A. Downing[7], Craig A. Stow[8]

1 Center for Limnology Trout Lake Station, University of Wisconsin Madison, Boulder Junction, Wisconsin, United States of America, **2** U.S. Geological Survey, Pennsylvania Cooperative Fish and Wildlife Research Unit, Pennsylvania State University, University Park, Pennsylvania, United States of America, **3** Department of Fisheries and Wildlife, Michigan State University, East Lansing, Michigan, United States of America, **4** Department of Outreach and Engagement, Oregon State University, Tillamook, Oregon, United States of America, **5** Lyman Briggs College, Michigan State University, East Lansing, Michigan, United States of America, **6** School of Natural Sciences, Trinity College, Dublin, Ireland, **7** Department of Ecology, Evolution, and Organismal Biology, Iowa State University, Ames, Iowa, United States of America, **8** Great Lakes Environmental Research Laboratory, National Oceanic and Atmospheric Administration, Ann Arbor, Michigan, United States of America

Abstract

We compiled a lake-water clarity database using publically available, citizen volunteer observations made between 1938 and 2012 across eight states in the Upper Midwest, USA. Our objectives were to determine (1) whether temporal trends in lake-water clarity existed across this large geographic area and (2) whether trends were related to the lake-specific characteristics of latitude, lake size, or time period the lake was monitored. Our database consisted of >140,000 individual Secchi observations from 3,251 lakes that we summarized per lake-year, resulting in 21,020 summer averages. Using Bayesian hierarchical modeling, we found approximately a 1% per year increase in water clarity (quantified as Secchi depth) for the **entire population** of lakes. On an **individual lake** basis, 7% of lakes showed increased water clarity and 4% showed decreased clarity. Trend direction and strength were related to latitude and median sample date. Lakes in the southern part of our study-region had lower average annual summer water clarity, more negative long-term trends, and greater inter-annual variability in water clarity compared to northern lakes. Increasing trends were strongest for lakes with median sample dates earlier in the period of record (1938–2012). Our ability to identify specific mechanisms for these trends is currently hampered by the lack of a large, multi-thematic database of variables that drive water clarity (e.g., climate, land use/cover). Our results demonstrate, however, that citizen science can provide the critical monitoring data needed to address environmental questions at large spatial and long temporal scales. Collaborations among citizens, research scientists, and government agencies may be important for developing the data sources and analytical tools necessary to move toward an understanding of the factors influencing macro-scale patterns such as those shown here for lake water clarity.

Editor: Peter S. Petraitis, University of Pennsylvania, United States of America

Funding: Funding for this research was provided by the National Science Foundation (MSB-1065786, EF-1065818, EF-1065649) and NTL-LTER (DEB-0822700). K.E.W. also thanks STRIVE grant 2011-W-FS-7 from the Environmental Protection Agency, Ireland for support. The funders had no role in study design, data collection and analysis, decision to publish, or preparation of the manuscript.

Competing Interests: The authors have declared that no competing interests exist.

* E-mail: nrlottig@wisc.edu

Introduction

Macrosystems ecology has emerged as a new ecology sub-discipline aimed at understanding ecosystem patterns and processes resulting from broad-scale environmental changes such as land use and climate change. This sub-discipline studies multi-scaled patterns and processes in biological, geophysical, and social components, and their interactions with each other and processes operating at finer and coarser scales [1]. One of the approaches that macrosystems ecologists are taking is integrating large datasets from a variety of sources. There may be many inherent challenges in assembling appropriate datasets for analysis in a macrosystems ecology context [2], however, we feel that well-established citizen monitoring programs can play an important role in macrosystems ecology, especially those that collect data across large spatial and temporal extents with standardized measurement, data entry, and quality control protocols [3]. The Christmas Bird Count is a well-known example of a citizen-based program (http://birds.audubon.org/about-christmas-bird-count), long recognized as a successful model of how to integrate citizens into meaningful science and conservation [4,5].

In freshwater ecosystems, the indicator of water quality with the longest history of standardized measurement by scientists and citizens alike is Secchi depth [6–8]. The first transparency measurements were made in the ocean during the early 1800s, followed by more careful investigations by Pietro Angelo Secchi in the mid 1800s, which led to his recognition as the method's founder (Fig. 1) [8]. Water clarity is easy and inexpensive to measure using a Secchi disk, and comparisons have shown that citizen-collected Secchi depth readings are nearly identical to those taken by professionals [9,10]. Consequently, the frequent, citizen-collected Secchi depth readings that are publically available online (e.g., Secchi Dip-In; http://www.secchidipin.org) often

Figure 1. Monument to Pietro Angelo Secchi in the Villa Borghese Park in Rome. Pietro Secchi, who lived in Italy during the 1800s, is recognized as the founder of the method of using a Secchi transparency disk to measure water clarity (Photo by K.E. Webster).

span multiple decades for individual lakes and can be used to complement data collected by state and federal agencies.

The U.S. Clean Water Act of 1972 regulates anthropogenic inputs of pollutants to receiving waters [11]. Therefore, Secchi depth records from citizen monitoring programs provide a potentially valuable resource for examining broad-scale responses of lakes to nutrient control efforts instituted in the subsequent 40 years. However, there have been few attempts at compiling and using these data to concurrently analyze spatial and temporal trends at broad scales. Previous studies generally have focused at the scale of a single U.S. state [7,12,13].

We compiled citizen-collected, publically available Secchi depth measurements to answer two questions: (1) what are the long-term trends in lake-water clarity across a broad geographic region?; and (2) how do trends differ as a function of spatial location, lake size, and when Secchi depth records were collected? To our knowledge, this is the first study to look at long-term water clarity trends using (1) a broad spatial extent (eight states with a total land area greater than 1.1 million km^2 in the Upper Midwest region of the United States), (2) a long temporal window (1938–2012), (3) a large number of lakes (>3,000), and (4) publically-available citizen-collected Secchi depth measurements (>140,000). We answer these questions using Bayesian hierarchical modeling, an approach that is useful because it can incorporate disparate data from multiple sources to examine relationships across multiple spatial and temporal extents [14,15]. Our results can be used as a basis for forming hypotheses to test the mechanisms driving large-scale spatial patterns and temporal trends in water clarity as well as informing local, regional, or federal policies and management actions.

Methods

Data Acquisition

We obtained 239,741 citizen-collected Secchi depth measurements from monitoring networks in eight states in the Upper Midwest region of the United States (Fig. 2; Table 1) [16]. Observations spanned the time period from 1938–2012 with the oldest records from the state of Minnesota; the other citizen monitoring networks began collecting data during the early 1970s, and the majority (99%) of measurements were taken post-1972. The number of waterbodies monitored varied widely among states, ranging from 5 in Iowa to 1,507 in Wisconsin (Table 1).

Prior to analysis, Secchi data were put through a series of QA/QC checks, limited to a consistent seasonal window, and summarized as annual averages. First, we converted readings to meters and removed 31 Secchi depth observations that exceeded 30 m. Given the distribution of all Secchi observations (data not shown), these observations appeared to be unrealistic for the region and were likely the result of transcription errors. Second, we limited our analysis to Secchi observations collected during June through August, which removed a source of inter-seasonal variation [17] and generally was the season when most of the monitoring networks collected data (e.g., Secchi Dip-In), providing consistency across sampling networks. Finally, the response variable used for all analyses was average summer Secchi depth calculated for each lake-year, estimated using raw-untransformed Secchi depth values. We also calculated the median sampling date of each lake for use as a predictor variable. For example, for a lake that was continuously monitored by citizens from 1980 through 2000, the median date of the lake record was 1990. Therefore, median sample date represents a surrogate for the time period a given lake was sampled over the entire 1938–2012 span of this study.

Using the latitude, longitude, and unique lake identifiers provided by the individual sampling networks, we merged the study lakes with the high-resolution National Hydrography Dataset (NHD; http://nhd.usgs.gov). We then verified the identity and location (latitude and longitude) of lakes in the dataset and obtained lake surface areas (ha) from the NHD. The final dataset included 148,991 Secchi observations, from which we calculated 21,020 summer lake specific average Secchi depth values for 3,251 lakes that spanned monitoring record lengths from 1 to 34 years (Fig. 2).

Statistical Analysis

To estimate the overall temporal trend (i.e., trend across all lakes) and assess the variability among lakes in temporal trends, we first fit a hierarchical model with $\log_e$-transformed Secchi depth as a linear function of year allowing the intercepts, slopes, and model error variances to differ among lakes. The form of the hierarchical model was as follows:

$$y_i \sim N\left(\alpha_{j(i)} + \beta_{j(i)} x_i, \sigma_j^2\right), \text{for } i = 1, \ldots n$$

$$\begin{pmatrix} \alpha_j \\ \beta_j \end{pmatrix} \sim N\left(\begin{pmatrix} \mu_\alpha \\ \mu_\beta \end{pmatrix}, \begin{pmatrix} \sigma_\alpha^2 & \rho\sigma_\alpha\sigma_\beta \\ \sigma_\alpha\sigma_\beta & \sigma_\beta^2 \end{pmatrix}\right), \text{for } j = 1, \ldots \text{J}$$

$$log(\sigma_j) \sim N\left(\mu_\sigma, \omega_\sigma^2\right), \text{for } j = 1, \ldots \text{J} \quad (1)$$

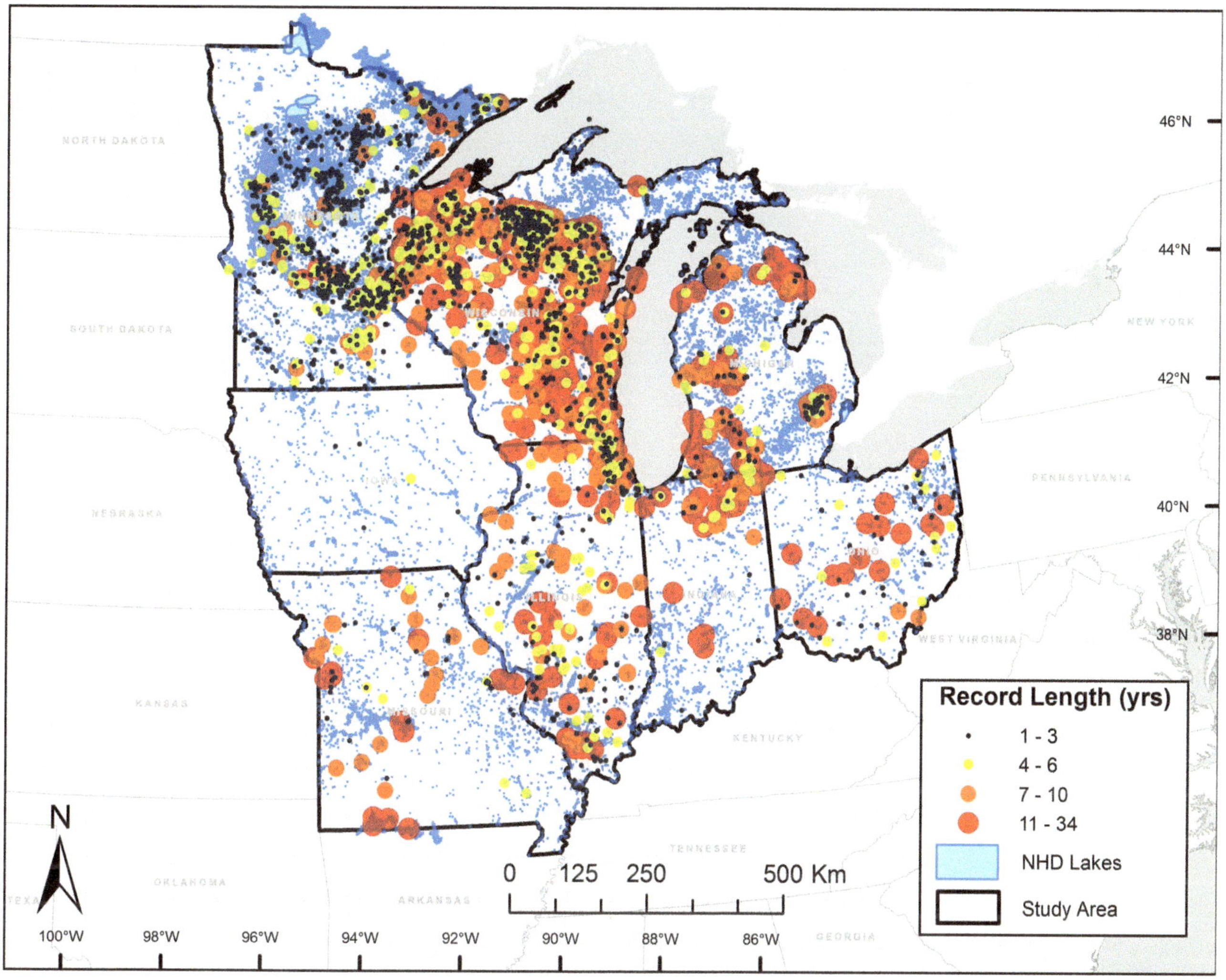

Figure 2. Upper Midwest states in the US and Secchi record lengths (years) for all citizen monitored lakes included in the study. Size and color of symbol designate record lengths. See Table 1 for information on citizen monitoring networks by state.

Where y_i is observation i of $\log_e$-transformed Secchi depth from lake j; α_j and β_j are the intercept and slope (temporal trend) for lake j, respectively; x is the year covariate (standardized; $(x_i - \bar{x})/SD(x_i)$) corresponding to the year Secchi observation i was measured; μ_α is the population-average intercept and μ_β is the population-average slope; σ^2_α and σ^2_β are the variances of the intercepts and slopes, respectively; and $\rho\sigma_\alpha\sigma_\beta$ describes the covariance between α_j and β_j, with ρ describing the correlation between α_j and β_j. We assumed a normal probability distribution for $\log_e$-transformed σ_j (lake-specific residual variances), with mean μ_σ and variance ω^2_σ. Non-informative normal priors were used for μ_α, μ_β, and μ_σ (i.e., $\mu_x \sim N(0,1000)$). Non-informative uniform priors were used for σ_α, σ_β, ω_σ (i.e., uniform distribution on $[0,100]$), and ρ (i.e., uniform distribution on $[-1,1]$).

Secondly, we fit models that were similar to Eq. 1, but that included predictor variables (i.e., latitude, lake surface area, median sample date) to model variation among lake-specific intercepts, slopes, and residual standard deviations. For example, using latitude (*lat*) as a lake-specific predictor, the model was as follows:

$$y_i \sim N\left(\alpha_{j(i)} + \beta_{j(i)} x_i, \sigma^2_j\right), \text{for } i = 1, \ldots n$$

$$\begin{pmatrix} \alpha_j \\ \beta_j \end{pmatrix} \sim N\left(\begin{pmatrix} \gamma^\alpha_0 + \gamma^\alpha_1 \times lat_j \\ \gamma^\beta_0 + \gamma^\beta_1 \times lat_j \end{pmatrix}, \begin{pmatrix} \sigma^2_\alpha & \rho\sigma_\alpha\sigma_\beta \\ \sigma_\alpha\sigma_\beta & \sigma^2_\beta \end{pmatrix}\right), \text{for } j = 1, \ldots \text{J}$$

$$log(\sigma_j) \sim N(\gamma^\sigma_0 + \gamma^\sigma_1 \times lat_j, \omega^2_\sigma), \text{for } j = 1, \ldots \text{J} \qquad (2)$$

Where γ^x_0 and γ^x_1 are intercepts and slopes, respectively. Non-informative normal priors were used for γ^x_0 and γ^x_1 and priors on other parameters were as described above. All covariates were standardized prior to analysis. Using an assessment method similar to classical significance testing, we examined 95% credible intervals (CRIs) for γ^x_1 to determine the importance of predictor variables (i.e., examined if 95% CRIs overlapped zero). The models were estimated using Bayesian estimation and the program JAGS was used for all analyses [18]. Three parallel chains were run with different initial values. After discarding the first 10,000

Table 1. Overview of the data sources and records available from the citizen monitoring networks included in this study.

State	Data Source	# Lakes[1]	# Lakes Monitored	% Lakes Monitored	# Secchi Observations[2]
Illinois	Volunteer Lake Monitoring Program[3]	2,857	303	10.6	40,681
Indiana	Indiana Clean Lakes Program[4]	1,879	95	5.06	8,610
Iowa	Secchi Dip-In Project[5]	1,081	5	0.46	47
Michigan	Michigan Clean Water Corps[6]	6,641	225	3.39	32,222
Missouri	Lakes of Missouri Volunteer Program[7]	1,886	79	4.19	7,581
Minnesota	Citizen Lake Monitoring Program[8]	14,043	961	6.84	13,392
Ohio	Citizen Lake Awareness and Monitoring[9]	1,257	75	5.97	7,819
Wisconsin	Citizen Lake Monitoring Network[10]	5,914	1507	25.5	129,389

Numbers represent totals through 2012.
[1]Number of lakes determined from 2007 National Hydrography Dataset (NHD) with surface area equal to or greater than 4 ha.
[2]Includes all citizen Secchi observations available from designated monitoring network.
[3]http://www.epa.state.il.us/water/vlmp.
[4]http://www.indiana.edu/~clp.
[5]http://www.Secchidipin.org.
[6]http://www.micorps.net/.
[7]http://www.lmvp.org.
[8]http://www.pca.state.mn.us/index.php/water/water-types-and-programs/surface-water/lakes/citizen-lake-monitoring-program/index.html.
[9]http://www.olms.org/citizen-lake-awareness-and-monitoring.
[10]http://dnr.wi.gov/lakes/CLMN.

samples, we retained every 3rd sample for a total of 20,000 samples. We examined the scale reduction factor $(\hat{R})$, a convergence statistic, for each parameter, trace plots, and plots of posterior distributions to assess convergence.

Results

Average lake specific summer Secchi depth for the 3,251 study lake population covered a wide range from 0.2 to 16 m, with a sample mean of 2.4 m (median 2.1 m; Fig 3A). For a majority of lakes (75%) the year-to-year variability of Secchi depth values within a specific lake was small (coefficient of variation <30%). However, at the population-level, the range was large (coefficient of variation between 6.5 and 184%).

For the **entire lake population**, temporal trends in Secchi depth from 1938 to 2012 were positive (average of all lake specific long-term trends), suggesting an overall increase in water clarity. The posterior estimate for the mean slope, μ_β, (Eq. 1) was 0.0092, indicating that Secchi depth, on average, has been increasing by 0.92% per year (95% CRI = 0.61–1.79%; Fig. 3B). On an **individual lake** basis, however, there was little evidence for long-term increases or decreases in water clarity for a majority of lakes (Fig. 3C). Of the 3,251 lakes examined, 128 (3.8%) and 224 (6.9%) had a 90% or greater probability of long-term declines or increases in Secchi depth, respectively (i.e., the number of the $\beta_j's$ estimated in Eq. 1 had a greater than 90% probability of being either <0 or >0; Fig. 3C).

Lake surface area was not related to average annual summer Secchi depth, long-term Secchi depth trend, or inter-annual variability in Secchi depth (posterior means and 95% CRIs for $\gamma_1^\alpha = -0.018$ [−0.034, −0.002]; $\gamma_1^\beta = 0.001$ [−0.005, 0.006]; $\gamma_1^\sigma = 0.004$ [−0.012, 0.020]). We found relationships between these response variables and the latitude of the monitored lakes (posterior mean and 95% CRI for $\gamma_1^\alpha = 0.249$ [0.224, 0.274]; $\gamma_1^\beta = 0.019$ [0.009, 0.029]; $\gamma_1^\sigma = -0.060$ [−0.086, −0.034]). Average annual summer Secchi depth (Fig. 4A) and long-term trend slope (Fig. 4B) increased and inter-annual variability in Secchi decreased (Fig. 4C) across the southern to northern latitudinal gradient (Fig. 2).

In addition to detecting spatial patterns in Secchi trends, we found that trends were related to when the data were collected within the multi-decadal time interval of 1938–2012 (i.e., the median year of sample collection). Median sample date was negatively related to both long-term Secchi depth trend and inter-annual variability (posterior mean and 95% CRI for $\gamma_1^\beta = -0.033$ [−0.043, −0.023]; $\gamma_1^\sigma = -0.137$ [−0.164, −0.108]; Fig. 5B, C). These results suggest that historical records were characterized by more positive annual Secchi depth trends relative to records collected more recently (Fig. 5B) and that inter-annual variability of Secchi depths was greater historically than what is observed from more current records (Fig. 5C). There was no evidence that relationships were due to differences in the water clarity of lakes being monitored by citizen volunteers over time, which has not changed (posterior mean and 95% CRI for $\gamma_1^\alpha = -0.012$ [−0.043, 0.017]; Fig. 5A).

Discussion

This study relied solely on citizen monitoring data to assess long-term water clarity (quantified as Secchi depth) trends and their relationships with spatial location, size of lake monitored, and time period monitoring took place. Overall, water clarity has been increasing slightly for the **entire population** of monitored lakes (average trend slope $[\mu_\beta] >$ zero). Eleven percent of lakes have strong trends in water clarity (90% probability slope is different than zero), with two-thirds corresponding to increased water clarity. We observed compelling spatial and temporal patterns with respect to the water clarity, long-term trends, and inter-annual variability of water clarity. Other studies have observed similar patterns in citizen monitored and remotely sensed Secchi depth [12,13] but over no more than two decades. We are unaware of any studies conducted with citizen-collected Secchi data at the spatial and temporal scales considered in this study.

Previous analyses of satellite-derived long-term Secchi depth patterns in the study region found that on average water clarity

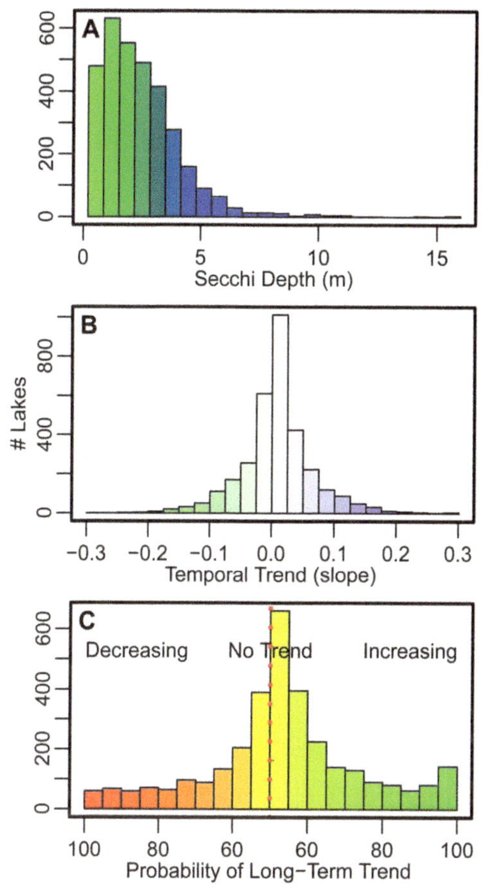

Figure 3. Distribution of lake-specific overall average Secchi depths (α_j from Eq. 1; A), long-term trends (β_j from Eq. 1; B), and probability of lake specific long-term trend (probability of $\beta_j's$ estimated in Eq. 1 of being either <0 or >0; C) based on a hierarchical Bayesian analysis. Long-term trend (B) positive values indicate long-term increases in water clarity (i.e., increased water clarity) and negative values indicate long-term declines in water clarity (i.e., decreased water clarity) for individual lakes. Probability of lake specific long-term trends (C) left of the dotted red line indicate the number of lakes with long-term annual declines in water clarity (Secchi depth becoming shallower) and associated probability of those long-term declines; values to the right indicate the number of lakes with long-term increases in annual water clarity (Secchi depth becoming deeper) and associated probability of those long-term increases.

trends have been relatively stable (Minnesota) [13,19] or increasing (Wisconsin) [12]. We observed that clarity is increasing on average by about 1% per year for the **entire population of lakes** included in this study. In large part, the long-term average increases in water clarity reported here were driven by data from lakes in Minnesota and Wisconsin, which comprise 76% of total number of lakes in the dataset (30% and 46% respectively). Excluding these two states, the average annual increase in clarity was negligible (0.17%). Interestingly, we found larger annual increases in Minnesota (1.59%) relative to Wisconsin (0.88%), which is contrary to the satellite-derived results that suggested relatively stable long-term patterns in Minnesota and more substantial increases in Wisconsin [12,13]. It is important to note, however, that the different approaches taken for these studies (e.g., citizen-collected Secchi data or satellite derived data, frequentist or Bayesian statistical approaches) to identify long-term trends are not suggesting drastically different patterns at regional scales (e.g., increasing versus decreasing), but rather subtle differences in the magnitudes of overall changes.

On an **individual lake** basis, results from our large multi-state study (e.g., Fig. 3) are generally consistent with small-scale lake-specific studies [7,20,21]. The analytical approaches employed in our study, however, provided an opportunity to combine space and time to analyze temporal trend patterns across a broad spatial extent. A majority of our lakes exhibited relatively stable (i.e., no significant long-term directional trend) water clarity with minimal inter-annual variability, whereas a small percentage of lakes (10.7%) exhibited strong evidence of long-term directional trends. These results are similar in magnitude to a recent analysis of satellite derived Secchi depths for >10,000 lakes, which found significant directional trends over 20 years in 10.8% of all Minnesota lakes [19]. Prior state-wide analyses of citizen volunteer data have also found similar patterns in water clarity over time. For example, over 50, 80, and 60% of the 50–122 lakes monitored by citizens in Minnesota, Wisconsin, and Michigan, respectively, showed no trend in water clarity over time [7,20,21]. Increasing trends in water clarity were found for 36, 31, and 10% of lakes in these three states, respectively, with a much smaller percentage of lakes in these studies showing decreasing water clarity over time [7,20,21]. Our study of a much larger number of lakes located across a very large spatial extent corroborates these previous studies and demonstrates that (1) most citizen monitored lakes in the Upper Midwest of the United States have had a relatively stable long-term water clarity record and (2) long-term increases in water clarity for individual lakes occur more frequently than do declines.

Some of the most compelling results emerging from our analyses are the relationships between long-term water clarity trends and the spatial location of lakes. Our results suggest that lakes situated at more southern latitudes tended to have trends of long-term decline in water clarity, whereas lakes situated at more northern latitudes had more of an overall shift towards long-term increases in water clarity (Fig. 5B). Although the water clarity patterns presented here were correlated with latitude, we suspect that these patterns were likely more directly influenced by a secondary, confounded, variable such as land use/cover [7,22] or climate [23]. For example, an analysis of satellite water clarity data from Minnesota over the last 20 years found similar patterns emerged at the state-level that were correlated with urban/agriculture (decreasing water clarity) and forested (increasing water clarity) land use types [19]. There is substantial spatial heterogeneity across our study extent, with extremes represented by large differences between the southern-most and northern-most regions in local and regional land use/cover patterns, as well as the underlying geology and soil patterns that are confounded with those human land uses. Consequently, understanding these regional water clarity patterns requires information about physical, chemical, and biological lake properties as well as data on watershed and regional landscape features at broad spatial and temporal scales [24]. However, a broad-scale, multi-thematic database containing such variables does not exist.

We also examined water clarity trends across the time period that encompassed passage of the U.S. Clean Water Act, leading to the expectation that an increase in water clarity would have occurred during the past 30–40 years. Our study did not find this pattern, at least not at the aggregate scale and for this population of lakes; in fact, the analyses provided evidence of stronger increases in the early part of the record. There are three possible reasons for this result. First, the data in our study are not continuous for all lakes pre- and post-Clean Water Act. Second, it is critical to acknowledge that lakes monitored by citizens are likely

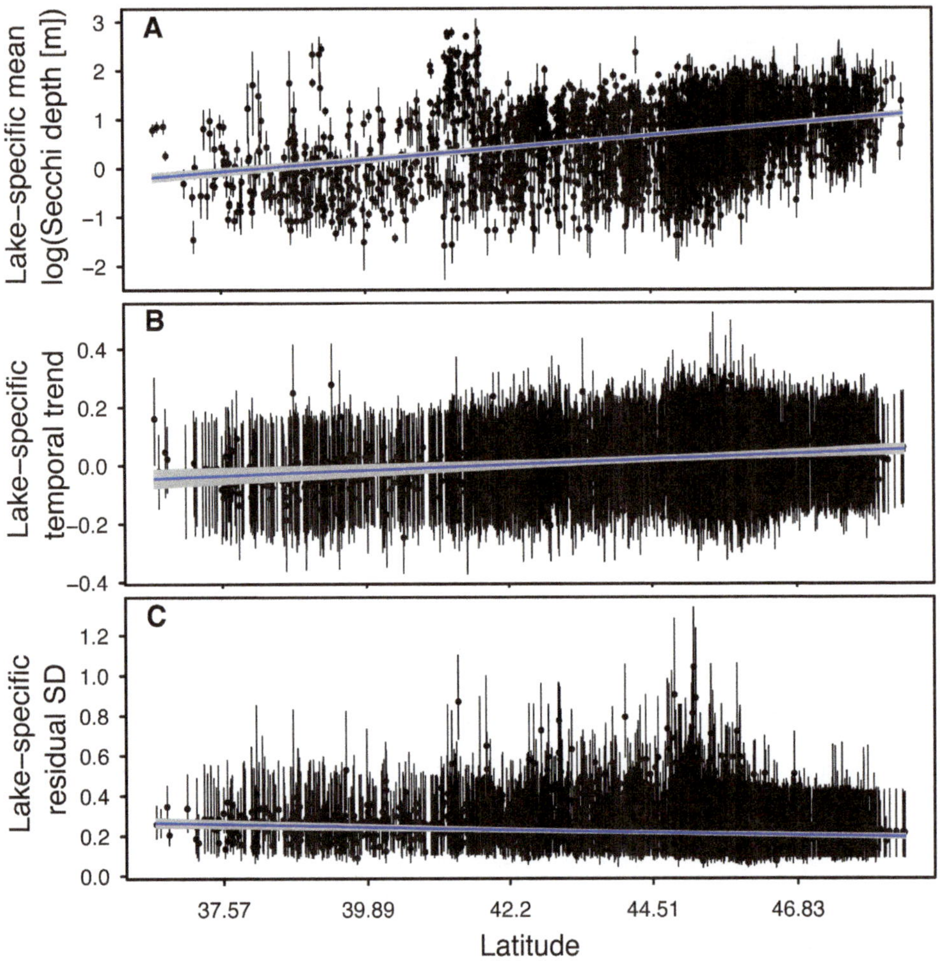

Figure 4. Relationships between average annual summer Secchi depth (A), long-term trend (B), and inter-annual variability (C) and spatial location (latitude) for 3,251 lakes in the Upper Midwest region of the United States. Solid points represent the parameter estimate for each individual lake and vertical lines represent the 95% credible interval of the estimates. Solid blue line represents the trend between each parameter and latitude. Grey shading around the solid blue lines represents the 95% credible interval of the latitudinal trend.

not representative of all lakes within the study extent. Although a substantial number of water clarity measurements have been made, only a fraction of all lakes in each state had water clarity actively monitored by citizen scientists (Table 1). Additionally, the distribution of surface areas of citizen-monitored lakes diverged from that of the entire distribution of lakes in our study extent (Fig. 6). The lakes that citizens monitor are disproportionately larger than would be expected based on the size distribution of lakes in general [25]. Similar biases towards overrepresentation of large lakes have been observed when examining the types of lakes monitored by state agencies [26]. We believe that this is the largest dataset of citizen-monitored Secchi data to be analyzed to date and suspect this bias may also exist for datasets from beyond our study region. The patterns presented here are characteristic of the suite of lakes included in our dataset and, thus, they may not be representative of all lakes in our study region or those beyond it. Consequently, careful consideration needs to be made should these results be extrapolated to lake populations other than the those considered in this study [26]. Third, water clarity is affected by a number of natural features that alter light attenuation including dissolved organic carbon, chlorophyll, and suspended sediments [17] and is altered by human-induced nutrient inputs [27], invasion by non-native species (e.g., zebra mussels) [28], acidification [29] and climate change (browning of waters) [23]. The complex set of factors affecting lake water clarity, many of which are correlated, makes it difficult to interpret the causes of trends in water clarity and has, perhaps, led to Secchi depth being considered a relatively coarse indicator of overall water quality. However, its use as an approximation of water quality status [30] and as a measure of the reliability of satellite-derived water clarity data [13], combined with ease of collection by citizen scientists, suggests this 150-year measure of water clarity will continue to provide valuable insights into the condition of lake ecosystems.

Conclusions

Although there has been a long history of examining water clarity trends and patterns augmented with citizen collected data, the results presented here incorporate over 140,000 citizen monitoring water clarity values, extending back to 1938, collected at greater spatial and temporal scales than any previous study we are aware of. It is impossible for aquatic management organizations to monitor all lakes all the time; citizen water clarity data have been critical in augmenting traditional agency monitoring efforts and provide vital data for identifying and understanding water quality patterns at broad spatial and temporal scales (e.g.,

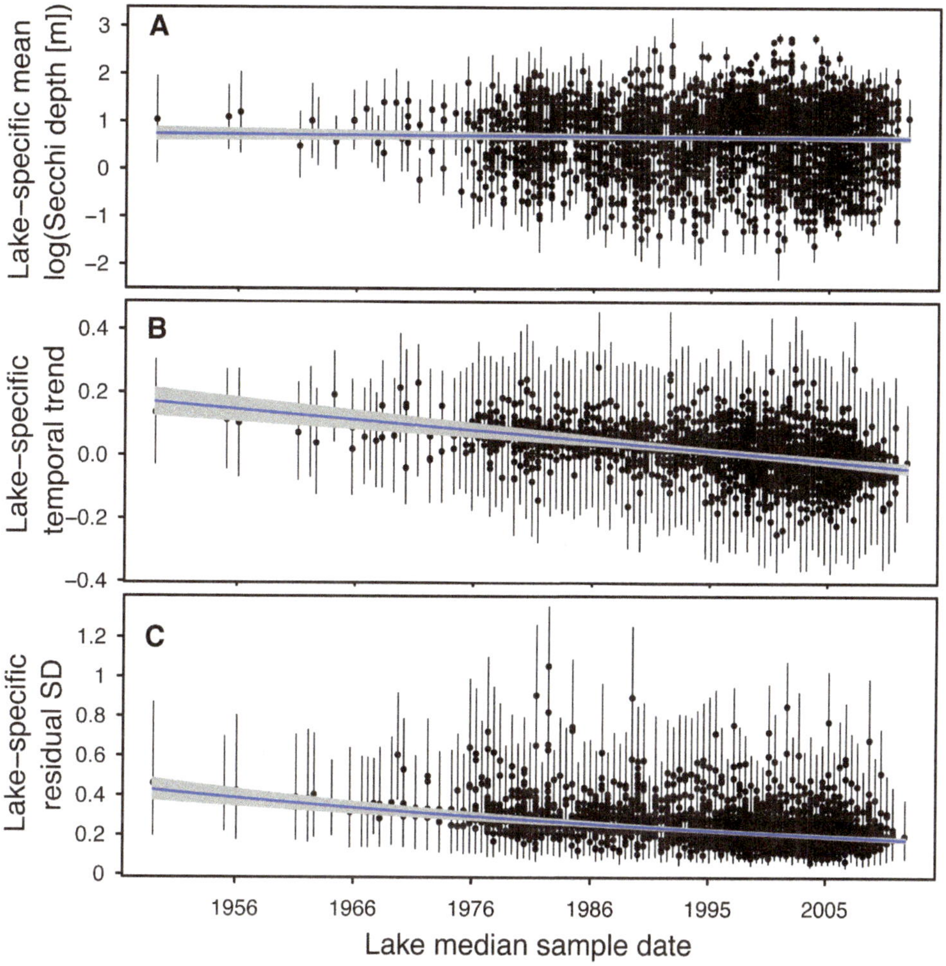

Figure 5. Relationships between average annual summer Secchi depth (A), long-term trend (B), and inter-annual variability (C) and median sample date (see text for details) for 3,251 lakes in the Upper Midwest region of the United States. Solid points represent the parameter estimate for each individual lake and vertical lines represent the 95% credible interval of the estimates. Solid blue line represents the trend between each parameter and latitude. Grey shading is the 95% credible interval of the latitudinal trend.

calibration of satellite-derived water clarity estimates). Here, we demonstrate the effectiveness of citizen data for quantifying long-term water clarity trends in individual lakes and identifying compelling population-level patterns in lake water clarity at the macro-scale. In an era of limited resources and during a push toward applying data-intensive science to environmental management and policy, the importance of citizen data such as these is likely to increase. Therefore, there is a clear need to continue to learn how to effectively leverage these types of data in the future.

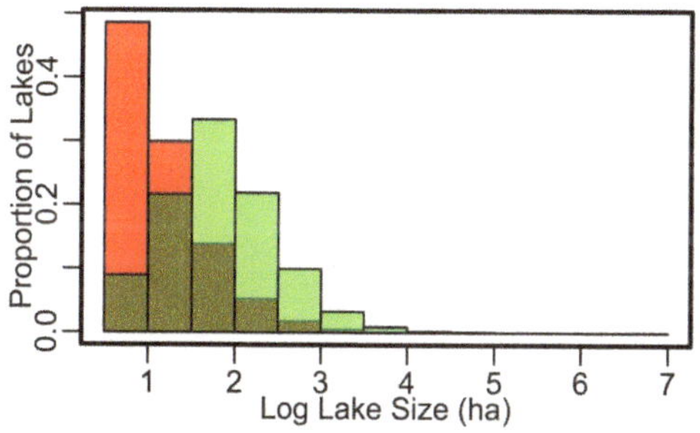

Figure 6. Distribution of lake sizes (surface areas) >4 hectare in the study region based on 2007 National Hydrography Dataset (red) relative to the distribution of lake sizes included in citizen science monitoring networks (green). Data are Log_{10} transformed.

These data needs coincide with the increasing recognition that stressors occurring at local, regional, and global scales interact to influence ecological patterns and processes from local to macro-scales [24]. At the scale of an **individual lake**, our data indicate that a majority of lakes in the Upper Midwest region have had relatively stable long-term water clarity patterns. However at the **regional scale**, we observe emerging macro-scale patterns in long-term water clarity trends and inter-annual variability. Understanding the linkages between the multi-scale water clarity patterns presented here and the factors like climate, land use/cover, and biotic interactions that have been shown to influence water clarity [7,23,31] have been elusive. Although citizen science affords a unique opportunity to obtain data such as water clarity at unprecedented spatial and temporal scales, we generally lack the large, multi-thematic data necessary to explain these patterns and fully apply a macrosystems ecology approach. Consequently, there is a need to compile multi-thematic data across multiple spatial and temporal scales in order to link patterns observed at local and regional scales to processes occurring across the landscape. We believe that collaborations among citizens, research scientists, and local, state and national agencies are important for developing the data sources and analytical tools necessary to move beyond identifying patterns toward an understanding of the critical factors influencing macro-scale patterns such as those shown here for lake water clarity.

Acknowledgments

We are grateful to the countless numbers of citizens who contributed Secchi data and the organizations that managed and provided us with data. It is only with the dedication of each citizen volunteer who diligently collected observations, state coordinators who trained volunteers and maintained databases, and high profile programs that have stimulated participation in lake monitoring at broad spatial extents that these invaluable and unique long-term data records have become available. We would also like to acknowledge the helpful contributions and suggestions from the individuals that reviewed this manuscript. We also thank the Macrosystems CSI Limnology research group for their contributions to the effort, especially Ed Bissell for all his assistance managing the data and GIS analyses. Any use of trade, firm, or product names is for descriptive purposes only and does not imply endorsement by the U.S. Government. GLERL contribution number 1703.

Author Contributions

Conceived and designed the experiments: NRL TW ENH KSC KEW JAD CAS. Analyzed the data: TW NRL. Wrote the paper: NRL TW ENH KSC KEW. Collect, compiled, and QAQCd data: NRL JAD ENH. Conducted Bayesian hierarchical modeling: TW. Created tables and figures: NRL TW. Contributed critical reviews and edited manuscript prior to submission: NRL TW ENH KSC KEW JAD CAS.

References

1. Heffernan J, Soranno PA, Angilletta MJ, Buckley LB, Gruner DS, et al. (2014) Macrosystems ecology: understanding ecological patterns and process at continental scales. Front Ecol Environ 12: 5–14 doi:10.1890/130017.
2. Ruegg J, Griese C, Bond-Lamberty B, Bowen GJ, Felzer BS, et al. (2014) Completing the data life cycle: using information management in macrosystems ecology research. Front Ecol Environ 12: 24–30 doi:10.1890/120375.
3. Hochachka WM, Fink D, Hutchinson RA, Sheldon D, Wong W-K, et al. (2011) Data-intensive science applied to broad-scale citizen science. Trends Ecol Evol 27: 130–137 doi: 10.1016/j.tree/2011.11.006.
4. Link WA, Sauer JR, Niven DK (2006) A hierarchical model for regional analysis of population change using Christmas Bird Count data, with application to the American Black Duck. Condor 108: 13–24.
5. Wilson S, Anderson EM, Wilson ASG, Bertram DF, Arcese P (2013) Citizen science reveals an extensive ehift in the winter distribution of migratory Western Grebes. PLoS ONE 8: e65408 doi:10.1371/journal.pone.0065408.s001.
6. Lathrop RC, Carpenter SR, Rudstam LG (1996) Water clarity in Lake Mendota since 1900: responses to differing levels of nutrients and herbivory. Can J Fish Aquat Sci 53: 2250–2261.
7. Bruhn LC, Soranno PA (2005) Long Term (1974–2001) Volunteer monitoring of water clarity trends in Michigan lakes and their relation to ecoregion and land use/cover. Lake Reserv Manag 21: 10–23.
8. Wernand MR (2010) On the history of the Secchi disc. JEOS:RP 5: 10013s doi:10.2971/jeos.2010.10013s.
9. Obrecht DV, Milanick M, Perkins BD, Ready D, Jones JR (1998) Evaluation of data generated from lake samples collected by volunteers. Lake Reserv Manag 14: 21–27. doi:10.1080/07438149809354106.
10. Canfield DE Jr., Brown CD, Bachmann RW, Hoyer MV (2002) Volunteer lake monitoring: testing the reliability of data collected by the Florida LAKEWATCH Program. Lake Reserv Manag 18: 1–9 doi:10.1080/07438140209353924.
11. Andreen WL (2013) Success and Backlash: The remarkable (continuing) story of the Clean Water Act. Geo Wash J Energy & Envtl L 4: 25–26.
12. Peckham SD, Lillesand TM (2006) Detection of spatial and temporal trends in Wisconsin lake water clarity using Landsat-derived estimates of Secchi depth. Lake Reserv Manag 22: 331–341 doi:10.1080/07438140609354367.
13. Olmanson LG, Bauer ME, Brezonik PL (2008) A 20-year Landsat water clarity census of Minnesota's 10,000 lakes. Remote Sense Environ 112: 4086–4097 doi:10.1016/j.rse.2007.12.013.
14. Stow CA, Lamon EC, Qian SS, Soranno PA, Reckhow KH (2009) Bayesian hierarchical/multilevel models for inference and prediction using cross-system lake data. In: Miao S, Carstenn S, Nungesser MK, editors. Real world ecology: large-scale and long-term case studies and methods. New York: Springer. pp. 111–136 doi:10.1007/978-0-387-77942-3_5.
15. Qian SS, Cuffney TF, Alameddine I, McMahon G, Reckhow KH (2010) On the application of multilevel modeling in environmental and ecological studies. Ecology 91: 355–361.
16. Lottig N, Wagner T, Henry E, Webster K, Downing J, et al. (2014) Upper Midwest Great Lake region citizen science Secchi data 1938–2012. Long Term Ecological Research Network doi:10.6073/pasta/5f94d3db0f3b7f1648579bf319aefacc.
17. Wetzel RG (2001) Limnology: lake and river ecosystems. 3rd ed. New York: Academic Press. 1006 p.
18. Plummer M (2011) JAGS Version 3.3.0 user manual. http://iweb.dl.sourceforge.net/project/mcmc-jags/Manuals/3.x/jags_user_manual.pdf.
19. Olmanson LG, Brezonik PL, Bauer ME (2013) Geospatial and temporal snalysis of a 20-year record of Landsat-based water clarity in Minnesota's 10,000 lakes. J Am Water Resour Assoc doi:10.1111/jawr.12138.
20. Webster KE (1998) Responses of lakes to drought: geomorphic and landscape controls. Ph.D. University of Wisconsin-Madison.
21. Heiskary S, Lindbloom J, Wilson CB (1994) Detecting water quality trends with citizen volunteer data. Lake Reserv Manag 9: 4–9.
22. Cross TK, Jacobson PC (2013) Landscape factors influencing lake phosphorus concentrations across Minnesota. Lake Reserv Manag 29: 1–12 doi:10.1080/10402381.2012.754808.
23. Gunn JM, Snucins E, Yan ND, Arts MT (2001) Use of water clarity to monitor the effects of climate change and other stressors on oligotrophic lakes. Environ Monit Assess 67: 69–88.
24. Soranno PA, Cheruvelil KS, Bissell EG, Bremigan MT, Downing JA, et al. (2014) Cross-scale interactions: quantifying multi-scaled cause-effect relationships in macrosystems. Front Ecol Environ 12: 65-73 doi:10.1890/120366.
25. Downing JA, Prairie YT, Cole JJ, Duarte CM, Tranvik LJ, et al. (2006) The global abundance and size distribution of lakes, ponds, and impoundments. Limnol Oceangr 51: 2388–2397.
26. Wagner T, Soranno PA, Cheruvelil KS, Renwick WH, Webster KE, et al. (2007) Quantifying sample biases of inland lake sampling programs in relation to lake surface area and land use/cover. Environ Monit Assess 141: 131–147 doi:10.1007/s10661-007-9883-z.
27. Jeppesen E, Sondergaard M, Jensen JP, Havens KE, Anneville O, et al. (2005) Lake responses to reduced nutrient loading - an analysis of contemporary long-term data from 35 case studies. Freshw Biol 50: 1747–1771 doi:10.1111/j.1365-2427.2005.01415.x.
28. Kissman CE, Knoll LB, Sarnelle O (2010) Dreissenid mussels (*Dreissena polymorpha* and *Dreissena bugensis*) reduce microzooplankton and macrozooplankton biomass in thermally stratified lakes. Limnol Oceangr 55: 1851–1859 doi:10.4319/lo.2010.55.5.1851.
29. Schindler DW, Curtis PJ, Parker BR, Stainton MP (1996) Consequences of climate warming and lake acidification for UV-B penetration in North American boreal lakes. Nature 379: 705–708.
30. Peeters ET, Franken RJ, Jeppesen E, MOSS B, Bécares E, et al. (2009) Assessing ecological quality of shallow lakes: Does knowledge of transparency suffice? Basic Appl Ecol 10: 89–96.
31. Carpenter SR, Kitchell JF, Hodgson JR (1985) Cascading trophic interactions and lake productivity. BioScience 35: 634–639.

19

Abundance, Composition and Activity of Ammonia Oxidizer and Denitrifier Communities in Metal Polluted Rice Paddies from South China

Yuan Liu[1], Yongzhuo Liu[1], Yuanjun Ding[1], Jinwei Zheng[1], Tong Zhou[1], Genxing Pan[1,2]*, David Crowley[3], Lianqing Li[1], Jufeng Zheng[1], Xuhui Zhang[1], Xinyan Yu[1], Jiafang Wang[1]

1 Institute of Resource, Ecosystem and Environment of Agriculture, Nanjing Agricultural University, Nanjing, China, **2** Center of Ecosystem Carbon Sink and Environment Remediation, Zhejiang Agricultural and Forestry University, Linan, Hangzhou, China, **3** Department of Environment Sciences, University of California Riverside, Riverside, California, United States of America

Abstract

While microbial nitrogen transformations in soils had been known to be affected by heavy metal pollution, changes in abundance and community structure of the mediating microbial populations had been not yet well characterized in polluted rice soils. Here, by using the prevailing molecular fingerprinting and enzyme activity assays and comparisons to adjacent non-polluted soils, we examined changes in the abundance and activity of ammonia oxidizing and denitrifying communities of rice paddies in two sites with different metal accumulation situation under long-term pollution from metal mining and smelter activities. Potential nitrifying activity was significantly reduced in polluted paddies in both sites while potential denitrifying activity reduced only in the soils with high Cu accumulation up to 1300 mg kg^{-1}. Copy numbers of *amoA* (AOA and AOB genes) were lower in both polluted paddies, following the trend with the enzyme assays, whereas that of *nirK* was not significantly affected. Analysis of the DGGE profiles revealed a shift in the community structure of AOA, and to a lesser extent, differences in the community structure of AOB and denitrifier between soils from the two sites with different pollution intensity and metal composition. All of the retrieved AOB sequences belonged to the genus *Nitrosospira*, among which species Cluster 4 appeared more sensitive to metal pollution. In contrast, *nirK* genes were widely distributed among different bacterial genera that were represented differentially between the polluted and unpolluted paddies. This could suggest either a possible non-specific target of the primers conventionally used in soil study or complex interactions between soil properties and metal contents on the observed community and activity changes, and thus on the N transformation in the polluted rice soils.

Editor: Manuel Reigosa, University of Vigo, Spain

Funding: The present research was funded by China National Science Foundation under grants number of 40671180 and of 40710019002. The funders had no role in study design, data collection and analysis, decision to publish, or preparation of the manuscript.

Competing Interests: The authors have declared that no competing interests exist.

* Email: gxpan@njau.edu.cn

Introduction

Nitrogen fertilizers are essential for rice production, but their overuse had also caused serious problems from runoff of nitrate and contributes to climate change from production of the greenhouse gas, nitrous oxide [1]. To addressing these problems, many of the environmental and management factors that affect biological transformations of nitrogen had been well studied to determine which soils are most prone to nitrogen losses [2,3]. However, there had been still open questions on the effects of heavy metal pollution on biological nitrogen transformations, mainly owing to the difficulty in determining the extent to which microbial communities adapted to the presence of heavy metals over time when metals were slowly introduced by atmospheric deposition, versus their response to sudden increases in metal concentrations when heavy metals were spiked into soils at high concentrations [4,5]. Another challenge in determining the effects of metals on nitrogen transformations had been the difficulty in relating changes in soil enzyme activities to changes in the abundance of molecular markers for specific microbial taxa and genes that could be involved in nitrogen transformations [6]. This is especially true for rice paddy soils where wet/dry cycles drive significant changes in redox that simultaneously affected metal bioavailability, microbial community structures, and rates for biological transformations of nitrogen.

Nitrous oxide (N_2O) had been widely accepted as the most radiative greenhouse gas increasing at a year rate of approximate 0.26% per year such that this greenhouse gas had reached with a concentration of 319 10^{-9} mol mol^{-1} in global air by IPCC [7]. However, it had been known also as a product resulted from uncompleted denitrification, in which reduction of nitrite was not completed to form N_2 in soils. Agriculture accounted for about 60% of the global total anthropogenic N_2O emission, of which rice paddies had been considered a major contributor [8–10]. Total N_2O emission from China's rice paddies was estimated at 29.0 Gg

N_2O per year, accounting for 7–11% of annual overall greenhouse gas emission from mainland China croplands [9]. Ammonia oxidization to nitrite had been well known as the initial and rate limiting step in nitrification, being mediated by microorganisms which carry genes encoding for the enzymes AOB [11,12] and/or AOA [13,14]. In contrast to the nitrifier which could be comprised by a few functional taxa, denitrifying bacteria responsible for denitrification could be broadly distributed among many different taxa using nitrate as an alternate electron acceptor for respiration [15].

China had been the largest rice producing country in the world with approximately 20% of global rice production [16]. In the last decade, metal pollution had been widely reported to occur in extensive rice production areas that included the lower Yangtze River delta [17,18], the Pearl River delta [19] and river valleys in the Jiangxi and Guangdong provinces [20]. Much attention had given to the potential health risk through food chain transfer of heavy metals and adverse effects on ecosystem health [21,22]. Recently, there had been observed in metal polluted rice paddies a decline in microbial biomass and fungal to bacterial ratio with an increase in the metabolic quotient that could lead to changes in C cycling [23]. Many laboratory studies had shown that heavy metal contamination in soil could affect the rates of microbial-mediated biogeochemical processes [24,25]. Soil nitrification rates had been known to be suppressed under metal pollution both in spiked soil samples in short term studies, and in polluted fields where metals typically accumulated at a slower rate [26]. In a study on surface wetland sediments [27], total denitrification activity was significantly decreased in multiple metal spiked wetland samples over a period of 7 days incubation with Cd, Cu or Zn ≥500 mg kg^{-1}, with Cd being the strongest inhibitor. Whereas, heavy metal pollution could exert effects on ammonia oxidizers [24,25,28,29] and denitrifying bacteria in field soils [30,31].

However, changes in composition and diversity of N-transforming microbial communities with metal pollution had not yet been studied in polluted rice fields. Among these, the potential effect of metal pollution on nitrification and denitrification processes in the rice fields should require primitive study for addressing changes in N cycling in polluted lands. Thus, understanding N loss via soil-air flux of N_2O emission and projecting future N_2O emissions from rice soils would depend on a better understanding of how nitrification and denitrification could contribute to the soil-atmosphere N_2O flux in rice agriculture and how these processes could be affected by heavy metals. In addition, nitrifier and denitrifier could also have different responses to stress of heavy metal pollution in soils owing to their different resource requirements, different metal contamination would exert different effects on the changes in the microbial communities and their biological functions on mediating N transformations in polluted rice fields.

Here we hypothesize that heavy metal contamination could have impact in the abundance, composition and activity of nitrifying and denitrifying communities, which could differ in soils with different soil properties as well as with different metal composition. An experiment was conducted to compare the community composition of nitrifier and denitrifier using two long-term metal polluted soils, which were compared to those in adjacent unpolluted fields at each location. We further studied the possible linkage between gene copy numbers that could serve as molecular markers for nitrogen transformation processes and process rates determined by direct enzyme assays measuring the potential activity of the nitrifying and denitrifying communities.

Ethics statement

The contaminated soils that we collected samples were under rice paddy used for rice production. No specific permission required nor any endangered or protected species involved. The polluted paddies both had a 30 year history of metal pollution due to discharge of mining waste water.

Materials and Methods

Site description

Two locations with soil pollution were selected for this study. Location DX (29°04′N, 117°43′E, Dexing County, Jiangxi) was situated in a copper mining area. The second location, designated DBS (24°26′N, 113°49′E, Wenyuan Municipality, Guangdong Province), was close to a multiple metal mine of zinc, cadmium, lead, and copper. The polluted paddy of at DX site was 1.5 km in distance from the top-hill ore mining and 1 km in distance to an adjacent Cu smelter, being affected by waste water discharge and atmospheric deposition. Whereas, the polluted paddy at DBS was polluted with irrigation by river water discharged from an upstream multiple metal mining ore and the associated smelter 3.5 km in distance. The mining and smelter activity had been taken place since late 1960's in both sites [23]. In each location, unpolluted rice paddies were selected that had no distinct access to polluted irrigation water. Soil samples from the polluted and non-polluted rice paddies at each location were designated as polluted (PS) and background soil (BGS) respectively. The climate at both locations are characterized by a subtropical monsoon climate with a mean annual temperature from 18°C to 25°C, and mean annual rainfall from 1200 mm to 1450 mm. In agriculture, double rice cropping has been traditionally practiced in the rice paddies in the area where these sites situated.

Soil sampling

Soil sampling was conducted at each location before rice planting in spring of 2009. Three composite topsoil samples at a depth of 0–15 cm were randomly collected respectively in 3 subplots from both PS and BGS fields. Each composite sample consisted of 5 sub-samples that were collected following a "Z" shaped sampling pattern with a distance of ~5 m in each subplot. The composite samples were mixed thoroughly and kept on ice until they were transported to the laboratory within two days after sampling. The samples were processed to remove gravels and plant detritus if any, and passed through a 2 mm sieve. Each composite sample was divided into three portions, one of which was used for DNA extraction. A second portion was stored at 4°C for measuring potential nitrifying and denitrifying activities. The third portion was air-dried at room temperature for analysis of soil chemical and physical properties as described below.

Analysis of soil properties and metal contents

Measurements of the basic properties and metal contents of the samples were conducted following the protocols described by Lu [32]. Briefly, soil pH was measured with a glass electrode using a 1/2.5 soil/water ratio. Soil textural class and particle size fractions were determined with a hydrometer method after dispersion with 0.5 mol L^{-1} NaOH solution. Soil organic carbon (SOC) was measured using wet digestion and oxidation with potassium dichromate. Total nitrogen was analyzed using the Kjeldahl method. For total heavy metal content determination, the samples were digested with a solution of $HF/HClO_4/HNO_3$ (10/2.5/2.5, v/v/v) and then extracted with 1 mol L^{-1} HCl. Content of Cd was determined with graphite furnace atomic absorption spectrometry (GFAAS, SpectrAA220Z, Varian, USA) while those of

Table 1. Soil physicochemical properties of the studied soil samples.

Site	Plot	pH (H_2O)	SOC ($g\ kg^{-1}$)	TN ($g\ kg^{-1}$)	Clay (%)	Silt (%)	Sand (%)
DX	BGS	4.87±0.05	22.79±1.58	1.58±0.05	21.0	31.2	47.8
	PS	4.10±0.06	22.25±0.35	1.96±0.07	27.0	32.2	40.8
DBS	BGS	5.58±0.13	15.23±0.60	1.03±0.03	22.2	23.2	54.6
	PS	5.45±0.16	19.11±0.68	1.49±0.08	27.0	28.8	44.2

BGS: Background soil; PG: Polluted soil. SOC: soil organic carbon; TN: total nitrogen.

Pb, Cu and Zn with flame atomic adsorption spectrophotometry (FAAS, TAS-986, China). Soil physic-chemical properties and metal contents determined are presented in Tables 1 and 2, respectively.

The Nemerow pollution index [33] was used to evaluate the overall extent of heavy metal pollution. It was calculated using the following equation:

$$P_n \sqrt{MaxP_i^2 + AveP_i^2)/2} \quad (1)$$

Where, P_n was the Nemerow pollution index value, and calculated as the sum of n metal elements analyzed for a soil sample; P_i was a single pollution intensity index of ith metal element with its measured concentration (C_i) divided by the guideline standard of environmental quality (RS_i). $MaxP_i$ and $AveP_i$ were the maximum and average pollution intensity respectively of the analyzed metals in a given soil. A relative metal accumulation degree using content of a single element in PS over in BGS was calculated and then averaged as the overall pollution intensity in a single site. All these calculated values are given in Table 2.

DNA extraction and real-time PCR assay

Total DNA was extracted from 0.25 g fresh soil with a PowerSoil DNA Isolation Kit (Mo Bio Laboratories Inc., CA) according to the manufacturer's protocol. The primers and thermal cycling procedures are listed in Table S1. The primer sets of *nirK*876 and *nirK*1040 were used in this study as they had been considered representing typical denitrifying bacterial community in multiple samples [36]. Each reaction was performed in a 25 µl reaction volume containing 15 ng of DNA, 1 µl of 10 µM of each primer and 12.5 µl of SYBR premix EX Taq TM (Takara Shuzo, Shinga, Japan). The amplified PCR products of *amoA (AOB and AOA)* and *nirK* genes were purified using PCR solution purification kit (Takara), ligated into pEASY-T3 cloning vector (Promega, Madison, WI) and cloned into *Escherichia coli* DH5α. Clones containing correct inserts were chosen as the standards for real-time PCR (qPCR). High amplification efficiencies of AOB-*amoA* (101%), AOA-*amoA (97%) and nirK* (102%) were obtained for gene quantification, with r^2 values of 0.996, 0.997 and 0.991 respectively.

PCR-DGGE of ammonia oxidizing and denitrifying communities

Profiles of the ammonia oxidizing and denitrifying communities were generated by PCR-DGGE using the primer sets *amoA*-1F-GC and *amoA*-2R, which had been considered specific for the AOB [34], Arch-*amoA*F-GC and Arch-*amoA*R which could be specific for AOA [35] and the *nirK*876-GC and *nirK*1040 set for the denitrifying bacterial communities [36]. The GC clamp described by Muyzer and Smalla [37] was incorporated into the 5′ end of primers. PCR reactions were performed with reaction mixtures that contained 12.5 µl Go TaqH Green Master Mix (Promega, Madison, WI), 1 µl of 10 µM of each primer, and 1 µl of DNA template. For DGGE analysis, PCR products were separated on an 8% (w/v) polyacrylamide gel of acrylamide/bisacrylamide (37.5/1, v/v) containing denaturing gradients of 45–60% for AOB, 25–50% for AOA and 50–65% for *nirK*. DGGE was performed in 1× TAE buffer at 60°C, 200 V for 5 min, then 140 V for 500 min. Gels were silver stained [38] and scanned using a gel document system (Bio-Rad, USA).

Table 2. Total heavy metal contents and Nemerow pollution index (Means ± S.D.) of the studied soils.

Sample	Cd (mg kg^{-1})	Pb (mg kg^{-1})	Cu (mg kg^{-1})	Zn (mg kg^{-1})	Nemerow index	Pollution intensity
DX-B	0.48±0.14b	58.95±0.90b	640.19±2.98b	94.57±17.21b	9.45±0.03b 19.74±1.86a	/
DX-P	1.55±0.14a	95.17±7.07a	1333.68±129.72a	163.90±15.41a		3.23±0.15B
DBS-B	0.29±0.00b	33.37±2.08b	21.87±1.02b	70.40±1.5b	0.76±0.001b 4.04±0.56a	/
DBS-P	1.49±0.24a	133.27±6.67a	224.83±5.68a	248.48±5.87a		5.32±0.23A

Different lowercase characters indicate significant different ($p<0.05$) between polluted soils (PS) and background soils (BGS) in a single site. Different capital letters for pollution intensity indicate significant different ($p<0.05$) between the two sites.

Sequencing and phylogenetic analysis

Prominent bands from the DGGE gels for the AOB and denitrifying communities were numbered and excised from the gels for sequence analysis. The retrieved sequences were compared with GenBank data base sequences using BLAST (Basic Local Alignment Search Tool) (http://www.ncbi.nlm.nih/gov/blast/) to search for best matches. Sequences of the DGGE bands had been deposited in GenBank under the accession numbers JF264803 to JF264813 (AOB) and JF264827 to JF264835 (*nirK*).

Measurement of potential nitrification and denitrification activities

Potential nitrification activity (PNA) was measured using the short-term nitrification assay described by Schmidt and Belser [39]. Briefly, 20 g moist soil was added to a 250 mL cotton-stopped flask with 100 ml of 0.5 mM phosphate buffer (pH 7.2), 0.5 mM $(NH_4)_2SO_4$ and 10 mM $KClO_3$ to stop further oxidation of NO_2^-. Triplicate flasks for each treatment were incubated for 24 h on a shaker at 175 rpm at 25°C. Net rate of NO_2^- production was calculated based on the linear correlation of NO_2^- concentration versus time.

Potential denitrification activity (PDA) assay was analyzed by the acetylene (C_2H_2) inhibition method following the procedure described by Tiedje et al. [40]. N_2O was measured using a gas chromatograph (Agilent 7890D, Santa Clara, CA, USA). Assays were conducted by addition of 20 g moist soil at 60% WHC to 250 mL glass bottles, in soil slurries produced by adding 20 mL of a substrate solution containing 1 mM glucose and 1 mM KNO_3. The linear rates of N_2O production over time were observed within 6 h of initiating the incubations.

Data processing and statistical analysis

The data were processed with Microsoft Excel 2010 and expressed in means plus/minus standard deviations. A paired t-test was used to evaluated differences between the polluted (PS) and background (BGS) samples from a single site with a significance defined at $p<0.05$. Digitized DGGE images were analyzed with Quantity One image analysis software (Version 4.0, Bio-Rad, USA). This software identified the band with the same position in the different lanes of the gel and also measures the intensity of identified bands. Cluster analysis and principal component analysis (PCA) of the DGGE profiles were performed to elucidate the microbial community structures based on relative band intensity and positions using the Minitab v.15 software. Redundancy analysis (RDA) was used to evaluate the relationship among DGGE fingerprints and soil properties (pH, SOC, TN, Nemerow index; total concentration of Cd, Pb, Cu and Zn) using the Canoco 4.5 software.

Results

Soil physicochemical properties and heavy metal pollution

As showed in Table 1, there were generally hardly differences in the basic properties between PS and BGS in a single site. For example, soil pH of the background soil at the DBS site (pH 5.58) was higher than at DX site (pH 4.87), without remarkable difference between PS and BGS in a single site. SOC and TN ranged from 15.23 g kg^{-1} to 22.79 g kg^{-1} and from 1.03 g kg^{-1} to 1.96 g kg^{-1} for the two sites, respectively. Here, N was generally higher in polluted soil than the background soil for less N consumed by rice with reduced production under metal pollution in both sites. However, C/N ratio seemed similar between the two sites both of polluted soils (14.4 and 14.8 respectively for DX and DBS) and of background soils (11.4 and 11.8 respectively for DX and DBS). As listed in Table 2, there were consistent differences in contents of total Cd, Pb, Cu and Zn between PS and BGS in a single site though the contents of a single heavy metal element varied with sites. While the total Cu content in PS was approximately 6 times higher from DX site than from DBS site, though a high Cu content up to 640 mg kg^{-1} owing to the lithological background [23]. As calculated based on environmental quality guideline value of China, Nemerow pollution index value was 4.0 for DBS and 20 for DX. While Cd, Pb and Zn contents of PS were similar between the two sites, the overall metal pollution intensity in PS estimated of the contents of all individual elements in PS over the BGS was significantly higher in DBS site than in DX site.

Abundances of ammonia oxidizer (AOB and AOA) and denitrifier (*nirK*)

The relative abundances of *amoA* and *nirK* gene copy numbers were assessed by *q*PCR (Table 3). The results showed that copy numbers of AOA *amoA* gene of the two soils ranged from 1.1×10^8 to 4.6×10^8 g^{-1} soil, and approximately 10–100 fold greater than those of AOB *amoA* genes, which ranged from 6.4×10^6 to 5.8×10^7 g^{-1} soil. Compared to background soil, copy numbers of AOB *amoA* gene were apparently reduced in PS by 19% at DX but 80% at DBS, corresponding to their metal pollution intensity difference. However, copy numbers of the AOA *amoA* gene were significantly reduced in polluted soil only at DX. Accordingly, the ratio of AOA to AOB in PS decreased at DX site but increased at DBS site while AOA genes were present in higher copy numbers than AOB genes at both sites, with ratios of AOA to AOB ranging from 2.5 to 59 across all samples.

Copy numbers of the *nirK* gene ranged from 2.32×10^8 to 8.33×10^8 g^{-1} soil in the studied paddies, while there was no significant change between PS and BGS in a single site. The ratios of *nirK*: *amoA* (sum of AOA and AOB gene copy numbers) ranged

Table 3. Ammonia oxidizer and denitrifier gene copy numbers and the relative ratios (Means ± S.D.) of the soils studied.

Sample	AOB ($\times 10^7$)	AOA ($\times 10^7$)	*nirK* ($\times 10^8$)	Ratio of AOA to AOB	Ratio of *nirK* to *amoA*
DX-B	0.79±0.07a	46.40±4.16a	6.25±1.86a	59.02±5.84a	1.32±0.35b
DX-P	0.64±0.05b	10.80±0.31b	8.33±0.25a	16.97±0.61b	7.92±0.20a
DBS-B	5.75±0.52a	14.80±5.78a	2.32±0.48a	2.55±0.89a	1.13±0.03a
DBS-P	1.13±0.06b	12.40±8.41a	3.38±0.68a	10.86±4.36a	3.34±1.03a

Different lowercase characters indicate significant different ($p<0.05$) between polluted soils (PS) and background soils (BGS) in a single site.

from 1.13 to 7.92 across all of the soil samples. Compare to GBSs, the *nirK*: *amoA* ratios increased in polluted soils from both locations (Table 3).

Community structures of ammonia oxidizer (*amoA*) and denitrifier (*nirK*)

Cluster analysis and principal component analysis (PCA) were used to group sampled soils based on similarity in relative band intensity and position of the DGGE profiles. PCA of DGGE profiles of AOB and AOA at the two sites yielded good summaries of data, as 86% (AOB) and 76% (AOA) of the total variability was explained by the first two components (Fig. S1). The AOB community did not show clear shifts between PS and BGS soils at DX site, but was separated between of PS and BGS of DBS site on the basis of PC2. Cluster analysis of AOB community from samples with or without heavy metal pollution revealed similarity of 59% at the DX site and 30% at the DBS site (Fig. 1a). In contrast, the AOA community profiles showed well separation between PS and BGS soils at each site, and separated on the basis of PC1 for DBS samples and on the basis of PC2 for DX samples (Fig. S1b). Cluster analysis clearly showed that AOA community under PS was distinguishable from BGS soils in both sites with an intragroup similarity of 40% for the DX site and 34% for the DBS site (Fig. 1b). Both PCA and cluster analysis of DGGE profiles of *nirK* gene revealed distinct communities between PS and BGS soils at the DX site with 26% similarity; but no clear differences in *nirK* community structure at the DBS site with 48% similarity.

Phylogenetic analysis

Selected DNA bands from the DGGE profiles were sequenced to identify the predominant taxa associated with these bands and the effects of metal pollution on specific taxa (Fig. S2). Some bands were present in the profiles from both BGS and PSs, but their intensity varied between locations. The results showed that all of the gene sequences represented in the DGGE profile of AOB *amoA* gene were associated with the genus *Nitrosospira*. The DGGE profile of AOB revealed several bands (B17, B18, B19 and B20) that were present in both PS and BGS from the DX site, and that had the same relative positions as bands present in BGS from the DBS site. However, sequences identified from the BGS at the DBS site belonged to Cluster 4, whereas bands B17–B20 from the DX site were affiliated with Cluster 12 (Fig. 2). None of the sequences of *nirK* from the different soils were affiliated with either *α-proteobacteria* or *β-proteobacteria* (Fig. 3). Whereas, most of the clones in the PS at the DBS site represented different taxa from those in BGS, sequences representing taxa from the DX site were similar for both BGS and PS.

Soil potential nitrifying and denitrifying activities

Potential nitrifying and denitrifying activities and differences associated with metal pollution varied with location. Transformation of ammonia to nitrate was approximately 3-fold lower in PSs compared to BGS at both locations (Fig. 4a). BGSs from both the DX and DBS sites released 0.008 and 0.023 mg NO_2^- -N kg^{-1} soil h^{-1}, respectively, which compared to 0.003 mg NO_2^- -N kg^{-1} soil h^{-1} at the DX site and 0.007 mg NO_2^- -N kg^{-1} soil h^{-1} at the DBS site in BSs. An 8-fold decrease in denitrification rates was found of PS (0.009 mg N_2O -N kg^{-1} soil h^{-1}) over that of BGS (0.001 mg N_2O -N kg^{-1} soil h^{-1}) at the DX site despite of no significant change between PS and BGS at the DX site. (Fig. 4b).

Correlation of microbial community structure with soil properties

Using redundancy analysis, we identified the factors which can best explained soil microbial community structure at each site (Fig. 5). In the redundancy analysis biplot, axis 1 and axis 2 values explained 80.2% and 79.7% of the variability, respectively, in AOB profiles and AOA profiles. Soil properties most important in explained AOB community composition were soil pH, SOC, TN, Nemerow index and Cu concentration, which were strongly related to the first axis, and soil pH, Nemerow index and Cu, Pb, Zn concentration, which were most important soil factors in determining AOA community composition. In the redundancy analysis biplot of *nirK* gene fragments, axis 1 explained 45.6% and axis 2 explained 29.3% of the variability. Soil pH, TN, Nemerow index and Cu concentration were strongly related to the first axis, and Cd concentration was strongly related to the second axis (Fig. 5c).

Discussion

Changes in ammonia oxidizer and denitrifier populations with metal pollution

Our results showed a strong impact of metal pollutants on ammonia oxidizer with the degree varying with the metal pollution situation in the two different locations. Copy numbers of *amoA* genes for eubacteria (10^6 to 10^7 per g dry soil) and archaea (10^8 per g dry soil) in the BGSs were both within the ranges that had been reported in previous studies of agricultural soils. In contrast to genes encoding enzymes for ammonia oxidation, the abundances of *nirK* genes representing the denitrifying bacteria seemed not affected by pollution in a single site. The *nirK* gene abundances ranged from 1.5×10^8 to 8.3×10^8 copies g^{-1} dry soil, which seemed within the range of those reported by Dandie et al. [41]. Decreases in copy numbers of *nirK* and *nosZ* genes were reported for estuary sediments with spiked copper, though recovered after prolonged incubation [31]. Likewise, in a study by Holtan-Hartwig et al. [42], a mixture of Cd, Cu and Zn caused a temporary reduction in N_2O production

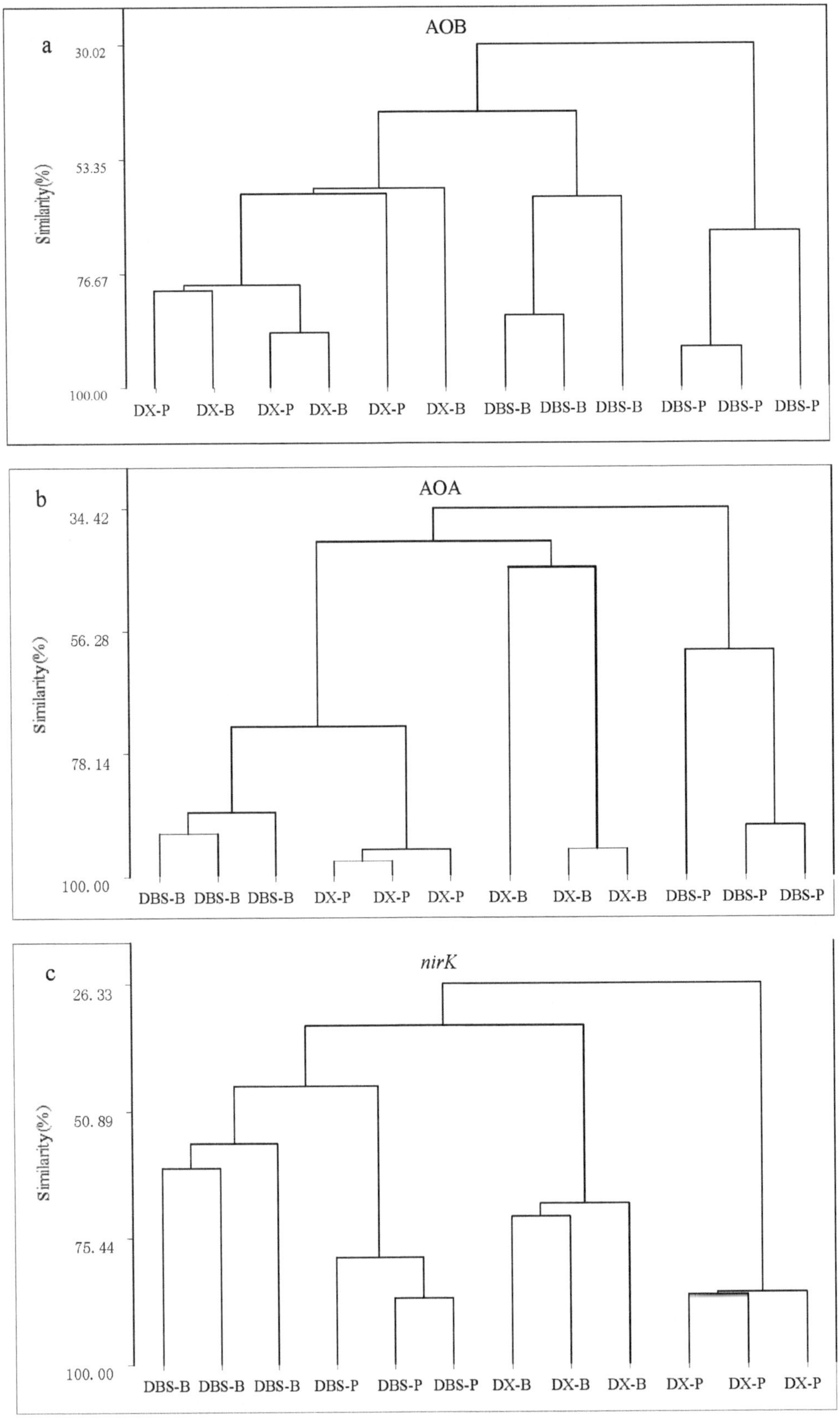

Figure 1. The cluster analysis of DGGE profiles of AOB (a), AOA (b) and *nirK* (c) gene fragments from the soil samples at the two sites. DX-B and DBS-B, background soil from site DX and DBS; DX-P and DBS-P, polluted soil from site DX and DBS. Euclidean distances were calculated from relative positions and intensities of bands, and the samples were clustered using Pearson's product-moment coefficient and a UPGMA (Unweighted Pair Group Methods with Arithmetic mean) algorithm.

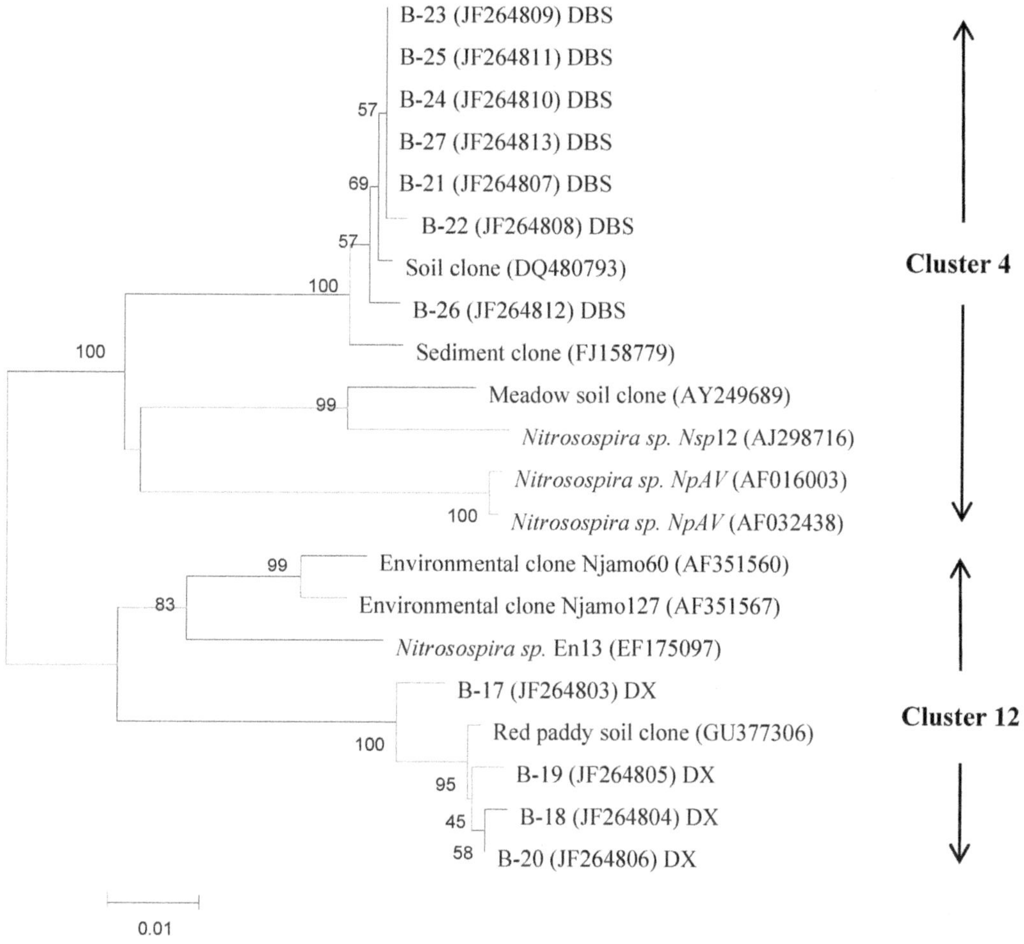

Figure 2. Neighbor-joining phylogenetic tree of AOB *amoA* sequences retrieved from the numbered DGGE bands of Fig. S2(A). Designation of the clones in the dashed line frames includes the following information: excised DGGE band number, accession number in the parentheses, followed by the sampling plot the clone retrieved from. Bootstrap values (>40%) with 1000 replicates are indicated at branch points. Scale bar indicates 1 change per 100 nucleotide positions.

in a sandy loam soil, which recovered within two months after spiking. Our results could be thus in general agreement with the hypothesis that denitrifier populations underwent adaptation to Cd, Pb, Cu and Zn pollution.

Sensitivity of ammonia oxidation to heavy metals had been known dependent on the specific metal or combination of metals. For example, a study by Liu et al. [29] showed no change in AOA abundance and community composition in a mercury-spiked vegetable soil. However, a work by Li et al. [28] showed a significant decrease in AOA abundance in four different soils (grassland and arable soils) with Cu amendments up to 1000 mg kg^{-1}. In a spiked study by Fait et al. [43], the ammonia oxidizer community was shown more vulnerable to Cu pollution than to Ni. The strong effects of copper on AOB populations was also shown in a study by van Beelen et al. [44], who showed that microbial communities more readily adapted to Zn and Ni contamination than to contamination from Cu and Cr in a sludge-amended soil. Therefore, AOA and AOB could be both significantly affected by metal pollution, but that AOA seemed more strongly reduced in Cu-polluted soil.

In this study, microorganisms carrying genes encoding for AOA were more abundant than those expressing AOB, with a ratio of AOA to AOB gene copy numbers ranging from 2.5 to 59. Moreover, the ratio of AOA/AOB gene copy numbers was 71% lower in PS at DX site but did not significantly changed in those at DBS site (Table 3). Ruyters et al. [25] reported an increase in AOA/AOB ratios in a grassland soil exposed to Zn up to 1300–4000 mg kg^{-1} soil for 12 months under pot experiment. Similarly, Li et al. [28] observed a sharp decrease in AOB populations in a Cu-spiked (1200 mg kg^{-1}) soil during a 21-day long microcosm study with grassland and cropland soils. In line with the potential impact on AOB population, the AOA/AOB ratio was clearly seen decreased in polluted soils while both AOA and AOB abundances were reduced significantly at DX site, where the Cu concentration was as high as 1333 mg kg^{-1}. In a similar fashion, the *nirK*/*amoA* ratios observed in our study ranged from 0.74 to 7.92 (Table 3), indicating a greater abundance of denitrifiers (*nirK*) relative to ammonia oxidizers (AOA and AOB) in all of the soils except for the BGS at DBS site. Similarly, Hai et al. [45] reported the predominance of *nirK* gene relative to *amoA* genes (AOA and AOB) in the rhizosphere of sorghum in an agricultural soil.

N availability to microbes of soil, usually assessed with C/N ratio, could affect the N removal in soils. Here, N was generally higher in polluted soil than the background soil in both sites for less N consumed by rice with reduced production under metal pollution. However, C/N ratio seemed similar between the two

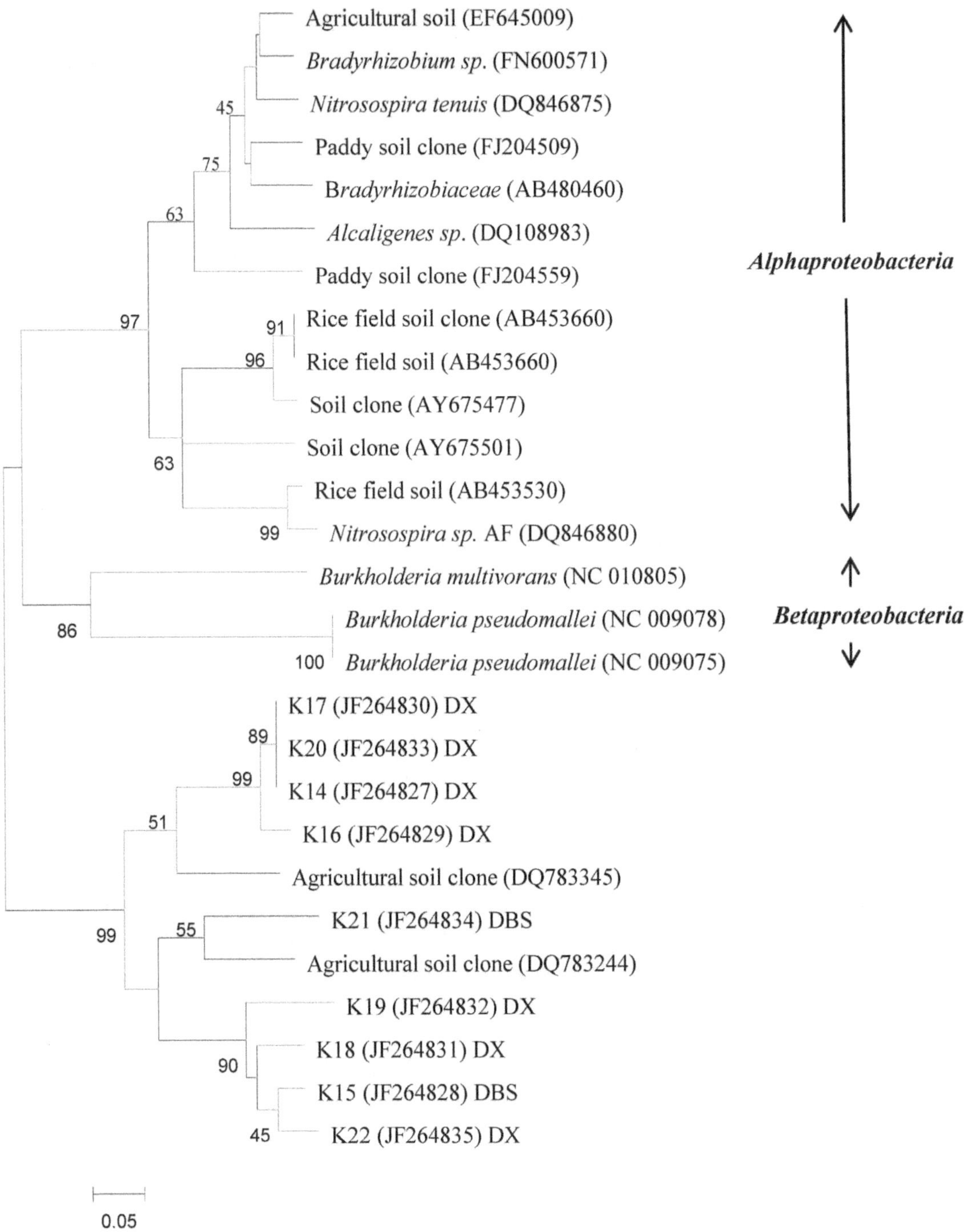

Figure 3. Neighbor-joining phylogenetic tree of *nirK* sequences retrieved from the numbered DGGE bands of Fig. S2(C). Designation of the clones in the dashed line frames includes the following information: excised DGGE band number, accession number in the parentheses, followed by the sampling plot the clone retrieved from. Bootstrap values (>40%) with 1000 replicates are indicated at branch points. Scale bar indicates 1 change per 100 nucleotide positions.

sites both of polluted soils (14.4 and 14.8 respectively for DX and DBS) and of background soils (11.4 and 11.8 respectively for DX and DBS). The finding that, over BGS, a decrease in AOB in PS was higher in DBS than in DX but AOA-to-AOB ratio in PS decreased in DX and increased in DBS suggested a role of metal pollution situation rather than N concentration on the community abundance, This was in line with the finding of a recent study of the authors group [46] that N_2O production was much higher (by~350%) in metal polluted soil (C/N ratio 11.6) than by 190% in unpolluted soil (C/N ratio 11.8) when amended with polluted straw amendment.

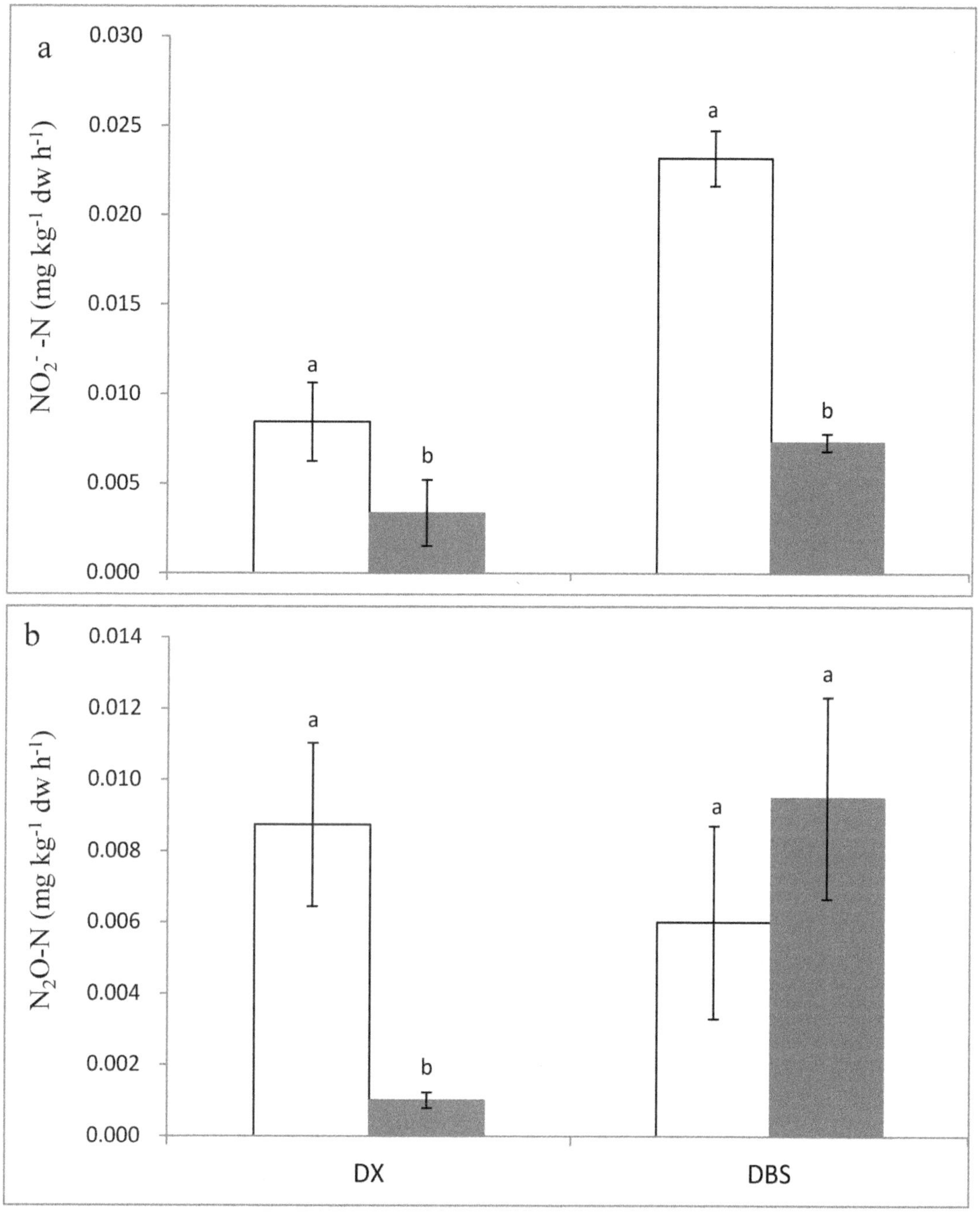

Figure 4. The potential nitrifying activity (a) and denitrifying activity (b) of the background (Blank) and polluted (Shaded) soils in DX and DBS sites. Different lowercase characters indicate significant difference ($p<0.05$) between the background and polluted soils in a single site.

Changes in community structure of ammonia oxidizer and denitrifier in metal polluted soil

Based on DGGE analysis between BGS and PS, both AOB and AOA communities were distinguishable at DBS site with a higher pollution degree by multiple metals but Whereas, only AOA community at DX site with a low pollution degree but predominated by Cu. Phylogenetic analysis revealed that all the AOB sequences in our study belonged to the genus *Nitrosospira* (Fig. 2). The dominance of *Nitrosospira spp.* both in BGSs and PSs here was already reported in earlier studies [47,48]. The clones from polluted soils in this study were affiliated with several clusters in the genus *Nitrosospira* as identified by Mertens et al. [24]. In the present work, sequences identified as belonging Cluster 4 were obtained only from the unpolluted soil at the DBS site, suggesting that multiple metal pollution of Cd, Pb, Cu and Zn had a strong impact on AOB community composition. However, it was unclear whether the metal tolerant populations in these heavy metal

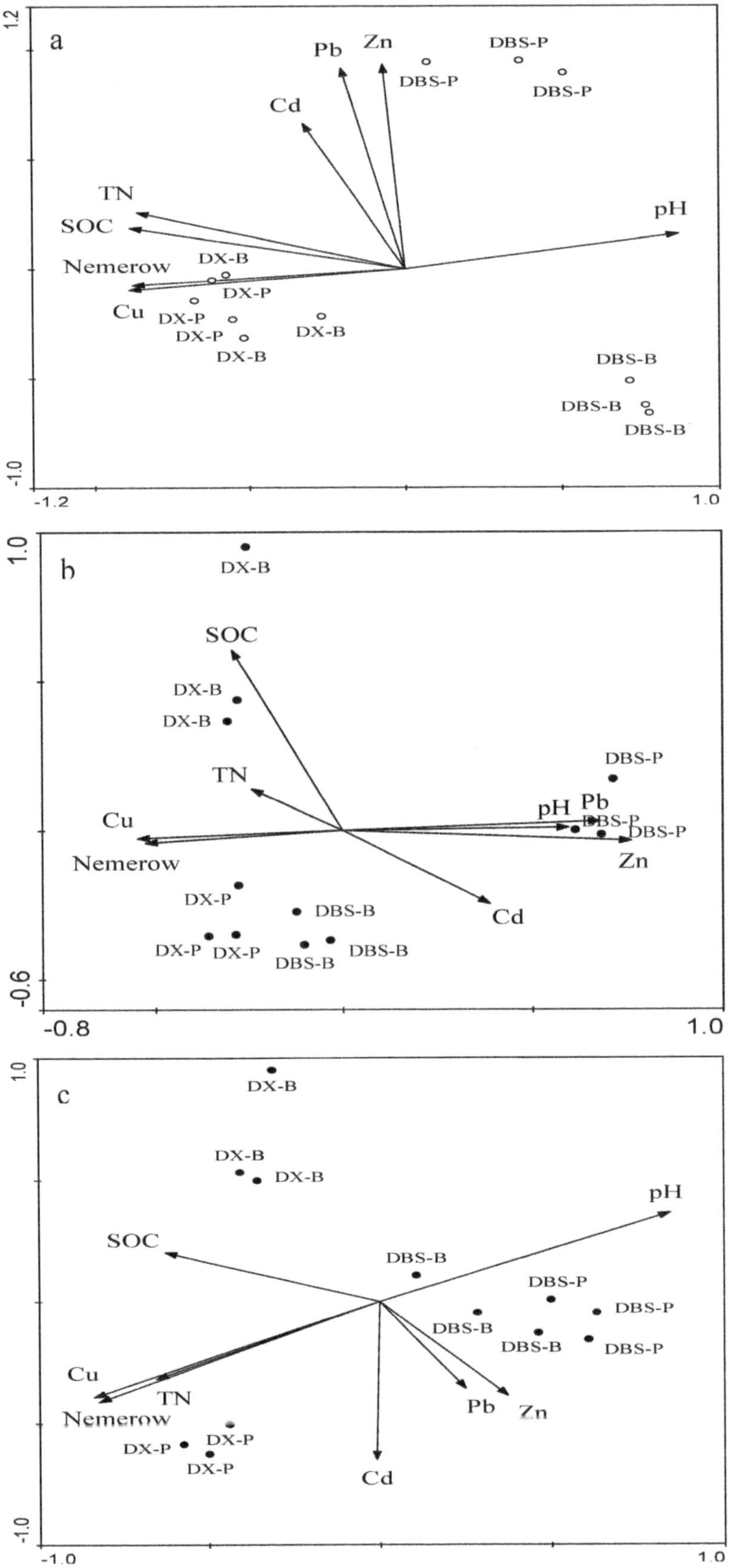

Figure 5. Redundancy analysis ordination plots of relationships between DGGE patterns of AOB (a), AOA (b) and *nirK* (c) gene and soil properties from the soil samples at the two sites. Soil properties included nemerow toxicity index, soil pH, SOC, TN, total concentration of Cd, Pb, Cu and Zn. DX-B and DBS-B, background soil from site DX and DBS; DX-P and DBS-P, polluted soil from site DX and DBS.

polluted soils were intrinsically tolerant to pollution or whether tolerance had been conferred by horizontal gene transfer. Some clones that were common to background soil were phylogenetically close to those that occurred in the polluted soil. As with the AOB community, the mechanistic details of AOA adaptations to heavy metal contamination remains unclear due to the limited ability to cultivate AOA species. Gene encoding AOA were also carried by members of the archaea that had been detected by molecular phylogenetic approaches, but that had not yet been cultured [49,50].

The communities represented by *nirK* were distinguishable between BGS and PS soils at DX site, but not at DBS site. Here, phylogenetic analysis showed that the clones from the polluted and background soils at a single site could be sorted into different groups. Furthermore, the clones in the polluted soils were widespread among samples from both sites. A study by Prasad et al. [51] showed that heavy metal tolerance varied widely among bacterial genera. These include several lineages belonging to the *Proteobacteria* that appeared to be highly tolerant to metal contaminated environments [52,53].

In addition to the influence of metal pollution, ammonia oxidizer and denitrifier community compositions were also affected by soil pH, SOC and TN. Soil organic carbon content is increased in polluted soils probably owing to inhibited decomposition rate [54]. On the other hand, soil pH had been found to affect microbial community structure directly by altering functional microbial groups [55], and indirectly by altering soil factors such as C and N substrate [56] and the availability of metals [57]. In this study, soil pH was lower but N higher in PS than in BGS at both sites, of course, both N and soil pH could be directly or indirectly affected by metal pollution [23]. There should be some interactions of soil chemical condition with metal pollution, for pH and organic matter could alter metal availability in soils to microorganisms.

Relationship between community changes and potential nitrifying and denitrifying activities

Nitrification had been known sensitive to heavy metal pollution [58–60], with heavy metal addition often decreasing soil nitrification rates [26,61]. Here, we observed that nitrifying activity was inhibited significantly in PSs compared to BGSs at both sites, which could be related to an altered soil microbial community composition with more resistant species. This could be partly explained by the remarkable decrease in both AOB and AOA populations with the metal pollution at both sites. AOB community composition was altered but its abundance unchanged in a long term Zn-polluted grassland, leading to an unchanged potential nitrification activity between polluted and unpolluted soils [62]. In contrast, a great decrease (up to 50%) in potential nitrification rates was found in copper spiked soils, with a simultaneous decrease in abundances of AOA and AOB plus a shift in AOB community structure and changes in the composition of the AOA community [63]. Of course, these community compositions could also subject to changes in soil pH and N level in soils under metal pollution [64], which could also be influenced with metal accumulation, as mentioned above. Nevertheless, metal impact on nitrification potential via metal toxicity or via alteration in abiotic conditions associated with C and N availability could be still a question.

Unlike nitrification activity, the potential denitrification activity was unchanged in the metal polluted soils at the DBS site with a soil pH of 5.5 but reduced significantly at DX site with Cu level up to 1333 mg kg^{-1} t and a low pH of 4.1. Apparently, this change was not related to difference in SOC and total N between the two sites. This seemed contrast to the reports that carbon additions lead to an increase in denitrifying activity but do not change the composition of the *nirK* community [65,66]. An inhibition of denitrification activity had been widely recognized under highly elevated Cu level in soil though soil pH could be an important factor for denitrifying activity. A decrease in denitrification activity in soils spiked or polluted with metals including high Cu was reported for wetland soil [27] and for pasture soil [67]. Attard et al. [68] argued that a change in denitrifying activity could be associated with a change in denitrifier abundance but not with a change in denitrifier community structure under land use changes. Here, a several fold decrease in denitrification activity was seen in DX but not in DBS, which corresponded with a shift in the composition of the *nirK* community in PS from DX but not from DBS (section 4.2). Notably, changes in potential denitrification activity could not be predicted from changes in the abundance of functional gene targeted with the *nirK* primers. As shown by Heylen et al. [69] in a culture-independent study through the *nir* gene sequence analysis of cultivated denitrifier, functional *nir* gene diversity did not match well the denitrifier diversity. There were likely uncertainties regarding the *nirK* gene data. Nitrite reductase was considered the key enzyme in denitrification, containing either cytochrome cd1 encoded gene (*nirS* denitrifier) or copper encoded gene (*nirK* denitrifier), could catalyze the reduction of NO_2^- to nitric oxide (NO). The *nirS* denitrifier appeared more abundant than *nirK* denitrifier, however, the latter could be more sensitive to soil environmental changes [70]. Recently, targeting both *nirK* and *nirS* genes in forest, grassland and agriculture systems had been proposed as an assay to elucidate the abundance and community structure of soil denitrifier [68,70,71]. However, it is still a question if this assay could better track the changes in denitrification activity with changes in denitrifying bacterial abundance in the polluted soils. Meanwhile, gene transcript numbers would be also a potential option to better predict the functional groups responsible for denitrification in these soils since they could reflect the active populations of the community.

Conclusion

Significant changes both in the activities and community structure of ammonia oxidizers and denitrifier existed with metal pollution in an interaction with soil abiotic factors in rice paddies. A consistent decrease in the AOB abundance and nitrifying activity in polluted soil was observed in two sites studied. However, a sharp decrease in AOA abundance and denitrifying activity were seen only in highly Cu-polluted soil though lower pH and higher N was seen in polluted soil compared to the background soil in both sites. By using molecular techniques employing DGGE, we observed a shift in the community structure of AOA, and to a lesser extent, of AOB and denitrifier populations that were associated with different metal composition of the polluted soils. The pollution effects on microbial abundance differed between populations of *amoA* and *nirK* genes in a single site but these changes were not seen correlated to changes in nitrification or denitrification activities. This could suggest either a possible non-specific target of the primers conventionally used in soil study or complex interactions between soil properties and metal contents on the observed community and activity changes. This study suggested that metal pollution could exert impacts on soil microbial communities responsible for N transformation and thus on potential N_2O production in rice paddies though the impacts on different communities could vary with metal composition and the associated changes in soil pH and N availability. However, future works would be required either with new molecular assays and/or on microbial responses to multiple metals under contrasting soil conditions in polluted agricultural soils.

Supporting Information

Figure S1 Principal component analysis (PCA) of DGGE profiles of AOB (a), AOA (b) and *nirK* (c) gene fragment from the soil samples at the two sites. DX-B and DBS-B, background soil from site DX and DBS; DX-P and DBS-P, polluted soil from site DX and DBS. Similar symbols with same color in PCA plot indicate the replicate samples.

Figure S2 DGGE profiles of AOB (A), AOA (B) and *nirK* (C) gene fragment from the soil samples at the two sites. M: 100 bp Marker. DX-B and DBS-B, background soil from site DX and DBS; DX-P and DBS-P, polluted soil from site DX and DBS. Arrows indicate the excised bands (B17–B27 and K14–K22) for sequencing.

Table S1 Primer sets and thermal profiles used for the absolute quantification of functional target genes.

Acknowledgments

The authors would like to thank the academic editor and the two anonymous reviewers for their helpful comments and suggestions.

Author Contributions

Conceived and designed the experiments: GXP YZL. Performed the experiments: YZL YL TZ. Analyzed the data: YL YZL GXP LQL JWZ XHZ JFZ. Contributed reagents/materials/analysis tools: YJD JFW XYY. Wrote the paper: YL GXP DC.

References

1. Kahrl F, Li Y, Su Y, Tennigkeit T, Wilkes A, et al. (2010) Greenhouse gas emissions from nitrogen fertilizer use in China. Environ Sci Policy 13: 688–694.
2. Luo J, De Klein C, Ledgard S, Saggar S (2010) Management options to reduce nitrous oxide emissions from intensively grazed pastures: a review. Agric Ecosyst Environ 136: 282–291.
3. Ju X, Xing G, Chen X, Zhang S, Zhang L, et al. (2009) Reducing environmental risk by improving N management in intensive Chinese agricultural systems. Proc Natl Acad Sci 106: 3041–3046.
4. Giller KE, Witter E, Mcgrath SP (1998) Toxicity of heavy metals to microorganisms and microbial processes in agricultural soils: a review. Soil Biol Biochem 30: 1389–1414.
5. Faulwetter JL, Gagnon V, Sundberg C, Chazarenc F, Burr MD, et al. (2009) Microbial processes influencing performance of treatment wetlands: a review. Ecol Engineering 35: 987–1004.
6. Wang Y, Shi J, Wang H, Lin Q, Chen X, et al. (2007) The influence of soil heavy metals pollution on soil microbial biomass, enzyme activity, and community composition near a copper smelter. Ecotoxicol Environ Saf 67: 75–81.
7. IPCC (2007) Climate change 2007: the physical science basis: Working group I contribution to the fourth assessment report of the IPCC. UK: Cambridge University Press.
8. Zou J, Huang Y, Jiang J, Zheng X, Sass RL (2005) A 3-year field measurement of methane and nitrous oxide emissions from rice paddies in China: Effects of water regime, crop residue, and fertilizer application. Global Biogeochemical Cycles 19: 10.1029.
9. Zou J, Huang Y, Zheng X, Wang Y (2007) Quantifying direct N2O emissions in paddy fields during rice growing season in mainland China: Dependence on water regime. Atmos Environ 41: 8030–8042.
10. Liu S, Qin Y, Zou J, Liu Q (2010) Effects of water regime during rice-growing season on annual direct N2O emission in a paddy rice-winter wheat rotation system in southeast China. Sci Total Environ 408: 906–913.
11. Jackson LE, Burger M, Cavagnaro TR (2008) Roots, nitrogen transformations, and ecosystem services. Plant Biol 59: 341.
12. Malchair S, De Boeck H, Lemmens C, Ceulemans R, Merckx R, et al. (2010) Diversity-function relationship of ammonia-oxidizing bacteria in soils among functional groups of grassland species under climate warming. Appl Soil Ecol 44: 15–23.
13. Nicol GW, Schleper C (2006) Ammonia-oxidising Crenarchaeota: important players in the nitrogen cycle? TRENDS in Microbiology 14: 207–212.
14. Wuchter C, Abbas B, Coolen MJ, Herfort L, van Bleijswijk J, et al. (2006) Archaeal nitrification in the ocean. Proc Natl Acad Sci 103: 12317–12322.
15. Philippot L, Hallin S, Schloter M (2007) Ecology of denitrifying prokaryotes in agricultural soil. Advan in Agrono 96: 249–305.
16. Frolking S, Qiu J, Boles S, Xiao X, Liu J, et al. (2002) Combining remote sensing and ground census data to develop new maps of the distribution of rice agriculture in China. Global Biogeochemical Cycles 16; 1091.
17. Wu X, Pan G, Li L (2006) Study on soil quality change in the Yangtze River Delta. Geogra Geo-Infor Sci 22: 88–91.
18. Hang X, Wang H, Zhou J, Ma C, Du C, et al. (2009) Risk assessment of potentially toxic element pollution in soils and rice (*Oryza sativa*) in a typical area of the Yangtze River Delta. Environ Pollut 157: 2542–2549.
19. Ma J, Pa G, Wang H, Xia Y (2004) Investigation on heavy metal pollution in a typical area of the Pearl River Delta. Chin J Soil Science 35: 636–638.
20. Xu C, Xia B, Qin J, He S, Li H, et al. (2007) Analysis and evaluation on heavy metal contamination in paddy soils in the lower stream of Dabaoshan Area, Guangdong Province. J Agro-Environ Sci 26: 549–553.
21. Arao T, Ae N (2003) Genotypic variations in cadmium levels of rice grain. Soil Sci and Plant Nutr 49: 473–479.
22. Liu J, Zhu Q, Zhang Z, Xu J, Yang J, et al. (2005) Variations in cadmium accumulation among rice cultivars and types and the selection of cultivars for reducing cadmium in the diet. J Sci Food Agric 85: 147–153.
23. Liu Y, Zhou T, Crowley D, Li L, Liu D, et al. (2012) Decline in topsoil microbial quotient, fungal abundance and C utilization efficiency of rice paddies under heavy metal pollution across South China. PloS One 7: 10.1371.
24. Mertens J, Broos K, Wakelin SA, Kowalchuk GA, Springael D, et al. (2009) Bacteria, not archaea, restore nitrification in a zinc-contaminated soil. The ISME journal 3: 916–923.
25. Ruyters S, Mertens J, Springael D, Smolders E (2010) Stimulated activity of the soil nitrifying community accelerates community adaptation to Zn stress. Soil Biol Biochem 42: 766–772.
26. Smolders E, Brans K, Coppens F, Merckx R (2001) Potential nitrification rate as a tool for screening toxicity in metal-contaminated soils. Environ Toxicol Chem 20: 2469–2474.
27. Sakadevan K, Zheng H, Bavor H (1999) Impact of heavy metals on denitrification in surface wetland sediments receiving wastewater. Water Sci Technol 40: 349–355.
28. Li X, Zhu Y, Cavagnaro TR, Chen M, Sun J, et al. (2009) Do ammonia-oxidizing archaea respond to soil Cu contamination similarly asammonia-oxidizing bacteria? Plant and soil 324: 209–217.
29. Liu Y, Zheng Y, Shen J, Zhang L, He J (2010) Effects of mercury on the activity and community composition of soil ammonia oxidizers. Environ Sci Pollut Res Int 17: 1237–1244.
30. Ruyters S, Mertens J, T'Seyen I, Springael D, Smolders E (2010) Dynamics of the nitrous oxide reducing community during adaptation to Zn stress in soil. Soil Biol Biochem 42: 1581–1587.
31. Magalhães CM, Machado A, Matos P, Bordalo AA (2011) Impact of copper on the diversity, abundance and transcription of nitrite and nitrous oxide reductase genes in an urban European estuary. FEMS Microbiol Ecol 77: 274–284.
32. Lu R (2000) Methods of Soil and Agro-chemical Aanalysis. Beijing (in Chinese): China Agricultural Science and Technology Press.
33. Nemerow NL (1991) Stream, lake, estuary, and ocean pollution. New York, USA. 0–472 p.
34. McTavish H, Fuchs J, Hooper A (1993) Sequence of the gene coding for ammonia monooxygenase in Nitrosomonas europaea. J Bacteriol 175: 2436–2444.
35. Francis CA, Roberts KJ, Beman JM, Santoro AE, Oakley BB (2005) Ubiquity and diversity of ammonia-oxidizing archaea in water columns and sediments of the ocean. Proc Natl Acad Sci USA 102: 14683–14688.
36. Henry S, Brue D, Stress B, Hallet S, Philippot L (2006) Quantitative detection of the nosZ gene, encoding nitrous oxide reductase, and comparison of the abundances of 16S rRNA, narG, nirK, and nosZ genes in soils. Applied and Environmental Microbiology 72(8): 5181–5189.
37. Muyzer G, Smalla K (1998) Application of denaturing gradient gel electrophoresis (DGGE) and temperature gradient gel electrophoresis (TGGE) in microbial ecology. Antonie van Leeuwenhoek 73: 127–141.
38. Sanguinetti C, Dias Neto E, Simpson A (1994) Rapid silver staining and recovery of PCR products separated on polyacrylamide gels. Biotechniques 17: 914.
39. Schmidt EL, Belser L (1994) Autotrophic nitrifying bacteria. Methods of Soil Analysis: Part 2-Microbiological and Biochemical Properties. 159–177 p.
40. Tiedje JM, Simkins S, Groffman PM (1989) Perspectives on measurement of denitrification in the field including recommended protocols for acetylene based methods: Springer. 217–240 p.
41. Dandie CE, Wertz S, Leclair CL, Goyer C, Burton DL, et al. (2011) Abundance, diversity and functional gene expression of denitrifier communities in adjacent riparian and agricultural zones. FEMS Microbiol Ecol 77: 69–82.

42. Holtan-Hartwig L, Bechmann M, Risnes Høyås T, Linjordet R, Reier Bakken L (2002) Heavy metals tolerance of soil denitrifying communities: N_2O dynamics. Soil Biol Biochem 34: 1181–1190.
43. Fait G, Broos K, Zrna S, Lombi E, Hamon R (2006) Tolerance of nitrifying bacteria to copper and nickel. Environ Toxicol Chem 25: 2000–2005.
44. van Beelen P, Wouterse M, Posthuma L, Rutgers M (2004) Location-specific ecotoxicological risk assessment of metal-polluted soils. Environ Toxicol Chem 23: 2769–2779.
45. Hai B, Diallo NH, Sall S, Haesler F, Schauss K, et al. (2009) Quantification of key genes steering the microbial nitrogen cycle in the rhizosphere of sorghum cultivars in tropical agroecosystems. Appl Environ Microbiol 75: 4993–5000.
46. Zhou T, Pan G, Li L Q, Zhang X H, Zheng J W, et al. (2014) Changes in greenhouse gas evolution in heavy metal polluted paddy soils with rice straw return: A laboratory incubation study. Eur J Soil Biol 63:1–6.
47. Stephen JR, Chang Y, Macnaughton SJ, Kowalchuk GA, Leung KT, et al. (1999) Effect of toxic metals on indigenous soil β-subgroup proteobacterium ammonia oxidizer community structure and protection against toxicity by inoculated metal-resistant bacteria. Appl Environ Microbiol 65: 95–101.
48. Koops H, Purkhold U, Pommerening-Röser A, Timmermann G, Wagner M (2006) The lithoautotrophic ammonia-oxidizing bacteria. The prokaryotes 5: 778–811.
49. Könneke M, Bernhard AE, José R, Walker CB, Waterbury JB, et al. (2005) Isolation of an autotrophic ammonia-oxidizing marine archaeon. Nature 437: 543–546.
50. De La Torre JR, Walker CB, Ingalls AE, Könneke M, Stahl DA (2008) Cultivation of a thermophilic ammonia oxidizing archaeon synthesizing crenarchaeol. Environ Microbiol 10: 810–818.
51. Prasad MNV (2001) In Metals in the Environment: Analysis by Biodiversity. CRC Press (New York). 285–321 p.
52. Feris K, Ramsey P, Frazar C, Moore JN, Gannon JE, et al. (2003) Differences in hyporheic-zone microbial community structure along a heavy-metal contamination gradient. Appl Environ Microbiol 69: 5563–5573.
53. Gillan DC, Danis B, Pernet P, Joly G, Dubois P (2005) Structure of sediment-associated microbial communities along a heavy-metal contamination gradient in the marine environment. Appl Environ Microbiol 71: 679–690.
54. Tyler G (1976) Heavy metal pollution, phosphatase activity, and mineralization of organic phosphorus in forest soils. Soil Biol Biochem 8: 327–332.
55. Rousk J, Brookes PC, Bååth E (2009). Contrasting soil pH effects on fungal and bacterial growth suggests functional redundancy in carbon mineralisation. Appl Environ Microbiol 75: 1589–1596.
56. Kemmitt SJ, Wright D, Goulding KWT, Jones DL (2006) pH regulation of carbon and nitrogen dynamics in two agricultural soils. Soil Biol Biochem 38: 898–911.
57. Flis SE, Glenn AR, Dilworth MJ (1993) Interaction between aluminium and root nodule bacteria. Soil Biol Biochem 25: 403–417.
58. Premi P, Cornfield A (1969) Effects of addition of copper, manganese, zinc and chromium compounds on ammonification and nitrification during incubation of soil. Plant and soil 31: 345–352.
59. Doelman P, Haanstra L (1984) Short-term and long-term effects of cadmium, chromium, copper, nickel, lead and zinc on soil microbial respiration in relation to abiotic soil factors. Plant and soil 79: 317–327.
60. Oorts K, Ghesquiere U, Swinnen K, Smolders E (2006) Soil properties affecting the toxicity of $CuCl_2$ and $NiCl_2$ for soil microbial processes in freshly spiked soils. Environ Toxicol Chem 25: 836–844.
61. Nies DH (1999) Microbial heavy-metal resistance. Appl Microbiol Biotechnol 51: 730–750.
62. Mertens J, Springael D, De Troyer I, Cheyns K, Wattiau P, et al. (2006) Long-term exposure to elevated zinc concentrations induced structural changes and zinc tolerance of the nitrifying community in soil. Environ Microbiol 8: 2170–2178.
63. Mertens J, Wakelin SA, Broos K, McLaughlin MJ, Smolders E (2010) Extent of copper tolerance and consequences for functional stability of the ammonia-oxidizing community in long-term copper-contaminated soils. Environ Toxicol Chem 29: 27–37.
64. Balabane M, Faivre D, van Oort F, Dahmani-Muller H (1999) Mutual effects of organic matter dynamics and heavy metals fate in a metallophyte grassland. Environ Pollut 105: 45–54.
65. Henry S, Texier S, Hallet S, Bru D, Dambreville C, et al. (2008) Disentangling the rhizosphere effect on nitrate reducers and denitrifiers: insight into the role of root exudates. Environ Microbiol 10: 3082–3092.
66. Miller M, Zebarth B, Dandie C, Burton D, Goyer C, et al. (2008) Crop residue influence on denitrification, N_2O emissions and denitrifier community abundance in soil. Soil Biol Biochem 40: 2553–2562.
67. Bardgett R, Speir T, Ross D, Yeates G, Kettles H (1994) Impact of pasture contamination by copper, chromium, and arsenic timber preservative on soil microbial properties and nematodes. Biol Fertil Soils 18: 71–79.
68. Attard E, Recous S, Chabbi A, De Berranger C, Guillaumaud N, et al. (2011) Soil environmental conditions rather than denitrifier abundance and diversity drive potential denitrification after changes in land uses. Global Change Biol 17: 1975–1989.
69. Heylen K, Gevers D, Vanparys B, Witteboll L, Geets J, et al. (2006) The incidence of *nirS* and *nirK* and their genetic heterogeneity in cultivated denitrifiers. Environ Microbiol 8(11), 2012–2021.
70. Bárta J, Melichová T, Vaněk D, Picek T, Šantrůčková H (2010) Effect of pH and dissolved organic matter on the abundance of *nirK* and *nirS* denitrifiers in spruce forest soil. Biogeochemistry 101(1–3): 123–132.
71. Petersen DG, Blazewicz SJ, Firestone M, Herman DJ, Turetsky M, et al. (2012) Abundance of microbial genes associated with nitrogen cycling as indices of biogeochemical process rates across a vegetation gradient in Alaska. Environ Microbiol 14(4): 993–1008.

20

Leaching Behavior of Heavy Metals and Transformation of Their Speciation in Polluted Soil Receiving Simulated Acid Rain

Shun-an Zheng[1,2,3]*, **Xiangqun Zheng**[1,2,3], **Chun Chen**[1,2,3]

1 Agro-Environmental Protection Institute, Ministry of Agriculture, Tianjin, People's Republic of China, **2** Key Laboratory of Production Environment and Agro-Product Safety, Ministry of Agriculture, Tianjin, People's Republic of China, **3** Tianjin Key Laboratory of Agro-Environment and Agro-Product Safety, Tianjin, People's Republic of China

Abstract

Heavy metals that leach from contaminated soils under acid rain are of increasing concern. In this study, simulated acid rain (SAR) was pumped through columns of artificially contaminated purple soil. Column leaching tests and sequential extraction were conducted for the heavy metals Cu, Pb, Cd, and Zn to determine the extent of their leaching as well as to examine the transformation of their speciation in the artificially contaminated soil columns. Results showed that the maximum leachate concentrations of Cu, Pb, Cd, and Zn were less than those specified in the Chinese Quality Standards for Groundwater (Grade IV), thereby suggesting that the heavy metals that leached from the polluted purple soil receiving acid rain may not pose as risks to water quality. Most of the Pb and Cd leachate concentrations were below their detection limits. By contrast, higher Cu and Zn leachate concentrations were found because they were released by the soil in larger amounts as compared with those of Pb and Cd. The differences in the Cu and Zn leachate concentrations between the controls (SAR at pH 5.6) and the treatments (SAR at pH 3.0 and 4.5) were significant. Similar trends were observed in the total leached amounts of Cu and Zn. The proportions of Cu, Pb, Cd, and Zn in the EXC and OX fractions were generally increased after the leaching experiment at three pH levels, whereas those of the RES, OM, and CAR fractions were slightly decreased. Acid rain favors the leaching of heavy metals from the contaminated purple soil and makes the heavy metal fractions become more labile. Moreover, a pH decrease from 5.6 to 3.0 significantly enhanced such effects.

Editor: Stephen J. Johnson, University of Kansas, United States of America

Funding: This work was financially supported by Central Public Research Institutes Basic Funds for Research and Development (Agro-Environmental Protection Institute, Ministry of Agriculture). The funders had no role in study design, data collection and analysis, decision to publish, or preparation of the manuscript.

Competing Interests: The authors have declared that no competing interests exist.

* E-mail: zhengshunan@gmail.com

Introduction

Acid rain has been a well-known environmental problem for decades and can lead to acidification of surface waters and soils. Acid deposition is formed from SO_2 and NO_x emitted to the atmosphere, largely because of fossil-fuel combustion. The most important sources are energy production, especially coal- and oil-fired power plants, and transportation sources, such as vehicles and ships. The air pollutants are transformed in the atmosphere to H_2SO_4 and HNO_3, transported across distances potentially as far as hundreds of kilometers, and deposited as precipitation (wet deposition) and as gas and particles (dry deposition) [1,2]. Acid deposition is also an environmental problem of increasing concern in China, where acid rain is mainly distributed in the areas of Yangtze River to the south, Qinghai-Tibet Plateau to the east, and in the Sichuan Basin. About 40% of the total territory of China is affected by the acid rain [3]. Sichuan Basin is one of the most severely hit-area of acid rain in southern China, where the rapid industrialization for last few decades has caused fast growth in sulfur emissions. Based on the monitoring data for 21 cities in Sichuan Province within the State-Controlled-Network of China, the number of cities with the annual average pH value of acid rain lower than 5.6 was 19. According to the environmental protection and monitoring agencies in Sichuan Province, the direct economic loss due to acid rain is estimated to be U. S. $ 3 billion for one year [4].

Sichuan basin is also known as the "Red Basin" because it is mainly covered by red or purple rock series of the Trias–Cretaceous system, from which the purple soils, one of the most important soils for agricultural production in subtropical areas of China, are developed and formed. Purple soil, classified as Eutric Regosols in FAO Taxonomy or Pup-Calric-Entisol in the Chinese soil taxonomy, is typically characterized by thin soil horizons and inherited many of the characteristics of parent materials or rocks, such as its color ranging from purple to red. But other changes in purple soil properties have taken place as a result of land use changes, agricultural practices, or eco-environment disturbances [5,6]. Due to the soil background and human activities, the soil has been severely contaminated by heavy metals (commonly including Cu, Pb, Cd and Zn) in many areas, resulting in potential risk to local human health and environment [7,8].

There is accumulating evidence that acid rain is able to enhance metal mobilization in soil ecosystems. Previous studies demonstrated that the H^+ ion in the acidic water displaces the cations from their binding sites, reduces the cation exchange capacity

(CEC), and increases the concentrations of these cations in the soil-water system. The negatively charged sulfate and nitrate ions in the acid rain can act as "counter-ions", which allow cations to be leached from the soil. Through a series of chemical reactions, cations such as K^+, Na^+, Ca^{2+} and Mg^{2+} are leached out and become unavailable to plants as nutrients [9,10]. Likewise, toxic ions, such as Cu, Pb and Cd, usually bound to the negatively charged surface of soil particles can be displaced by H^+ ion too [11].

Leaching is the process by which contaminants are transferred from a stabilized matrix to liquid medium, such as water or other solutions. However, the influence of acid rain on the leaching behavior of heavy metals and transformation of their speciation in polluted purple soil is not investigated in detail. The objectives of this study were to evaluate the leaching of heavy metals Cu, Pb, Cd and Zn in a contaminated purple soil affected by simulated acid rain (SAR) over a range of pH, and to identify how simulated acid rain influences the chemical speciation of these metals in purple soil. Column experiments were used in this study to provide information about element release, transport in soil, and chemistry of soil and leachates.

Materials and Methods

Soil Sample Collection and Analysis

The analyzed soil sample was collected from a 0 cm to 20 cm layer of agricultural purple soil in the Pengzhou Agro-ecological Station of the Chinese Academy of Agricultural Sciences in Sichuan Province, China. The study area has a subtropical humid monsoon climate with an average annual precipitation of 850 mm to 1000 mm and an average annual temperature of 15°C to 16°C. The soil samples were air-dried, ground, and passed through a 2 mm sieve. Selected soil characteristics determined by standard methods [12] were 6.27 for pH, 0.64 $g \cdot kg^{-1}$ for $CaCO_3$, 26.51 $g \cdot kg^{-1}$ for organic matter, 20.81 $g \cdot kg^{-1}$ for CEC, 265.41 $g \cdot kg^{-1}$ for clay (<0.002 mm), 51.84 $mg \cdot kg^{-1}$ for total copper, 43.92 $mg \cdot kg^{-1}$ for total lead, 0.36 $mg \cdot kg^{-1}$ for total cadmium, 121.23 $mg \cdot kg^{-1}$ for total zinc.

Contaminated Soil Preparation

Artificially contaminated soils were composed of the collected purple soil. The load quantities of Cu [$Cu(NO_3)_2 \cdot 3H_2O$], Pb [$Pb(NO_3)_2$], Cd [$Cd(NO_3)_2 \cdot 4H_2O$], and Zn [$Zn(NO_3)_2 \cdot 6H_2O$] were as follows: Cu+Pb+Cd+Zn = 400 $mg \cdot kg^{-1}$+500 $mg \cdot kg^{-1}$+1 $mg \cdot kg^{-1}$+500 $mg \cdot kg^{-1}$. Based on the levels of polluted soil defined by the Chinese Environmental Protection Agency, the selected concentrations in this study represented moderately contaminated soils (Grade III of the National Soil Heavy Metals Standards GB15618-1995). The thoroughly mixed soil samples were stored and incubated at a 75% water holding capacity for 12 mon. The chemical speciation of aged soil was then determined by the sequential extraction procedure of Tessier et al. [13]. The chemical reagents, extraction conditions, and their corresponding fractions are defined as follows:

1. Exchangeable fraction (EXC): 2 g of the soil sample (oven-dry weight), 16 mL 1.0 $mol \cdot L^{-1}$ $MgCl_2$, pH 7, shake vigorously in a reciprocating shaker for 1 h, 20°C.
2. Carbonate-bound fraction (CAR): 16 mL of pH 5, 1.0 $mol \cdot L^{-1}$ sodium acetate, shake vigorously in the reciprocating shaker for 5 h, 20°C.
3. The Fe/Mn oxide-bound fraction (OX): 40 mL of 0.04 $mol \cdot L^{-1}$ $NH_4OH \cdot HCl$ in 25% (v/v) acetic acid at pH 3 for 5 h at 96°C with occasional agitation.
4. The OM-bound fraction: 6 mL of 0.02 $mol \cdot L^{-1}$ HNO_3 and 10 mL of 30% H_2O_2 (pH adjusted to 2 with HNO_3), water bath, 85°C for 5 h with occasional agitation. 10 mL of 3.2 $mol \cdot L^{-1}$ NH_4OAc in 20% (v/v) HNO_3, shake vigorously in the reciprocating shaker for 30 min.
5. Residual fraction (RES): Dried in a force-air oven at 40°C, 24 h. Subsamples after sieving with 0.149 mm openings were used for determining Cu, Zn, Pb and Cd contents.

Extractions were performed in 100 mL polypropylene centrifuge tubes. Between each successive extraction, the supernatant was centrifuged at 1500 ×g for 30 min and then filtered using a membrane filter (0.45-μm nominal pore size).

Simulated Acid Rain Preparation

Simulated acid rain was designed according to the main ion composition and pH of the local rain water (the pH of rainfall varied from 3.0 to 5.6) [4]. Synthetic acid rain with pH values of 3.0, 4.5, and 5.6 was prepared from a stock H_2SO_4–HNO_3 solution (4:1, v/v). The concentrations of Ca^{2+}, NH_4^+, Mg^{2+}, SO_4^{2-}, CO_3^{2+}, Cl^-, and K^+ in SAR were 1.5, 2.62, 1.00, 10.00, 2.61, 11.17, and 1.78 $mg \cdot L^{-1}$, respectively.

Leaching Experiment

The column (Fig. 1A) was oriented vertically and slowly saturated from the bottom with deionized water until it reached the field-holding capacity. The soil column was allowed to stabilize for 24 h. The feed solutions were composed of SAR at pH 3.0, 4.5, and 5.6, with the last group as the control. The feed solutions were then introduced into the system using a peristaltic pump to percolate through the packed soil columns at a flow rate of (60±5) $mL \cdot h^{-1}$, which corresponds to the field infiltration velocity of 3.0 $cm \cdot h^{-1}$ [14] (Fig. 1B). The redox potential was measured in one-third of the columns to check the aeration of columns and to avoid waterlogging conditions. Each column was flushed with 3000 mL (1530 mm) of the incoming solution, which corresponds to 1.5 yr of precipitation (rain) in the study area. Experiments were conducted in triplicate at each pH treatment.

The leachate from the soil column was filtered through a 0.45 μm membrane filter, and 200 mL of the filtered leachate (equivalent to 100 mm of added SAR) was sampled using glass collectors to determine the heavy metal concentrations, pH, and electrical conductivity. After the tests, the columns were separated and extruded. All of the soils from the entire depth range were thoroughly mixed, dried, and ground to analyze the chemical speciation of the heavy metal residues using the sequential extraction procedure [13].

Metal Determination and Quality Control

The Cu, Zn, Pb, and Cd concentrations in soil were determined by digesting 0.5 g of the soil samples (oven-dry weight) with a HNO_3–HF–$HClO_4$ mixture followed by elemental analysis. The concentrations of these metals in all the solutions were analyzed by graphite furnace atomic absorption spectrometry (AA220Z; Varian, USA). The detection limits for Cu, Zn, Pb, and Cd were 2, 2, 1, and 0.5 $\mu g \cdot L^{-1}$, respectively. All the reagents used for analysis were of analytical grade or higher. All the containers were soaked in 10% HCl, rinsed thoroughly in deionized water, and dried before use. The standard substances such as the geochemical standard reference sample soil in China (GSS-15) were used to examine the precision and accuracy of determination. The relative errors (REs) between the sum of the metal concentration in individual fractions and the measured total metal concentration in

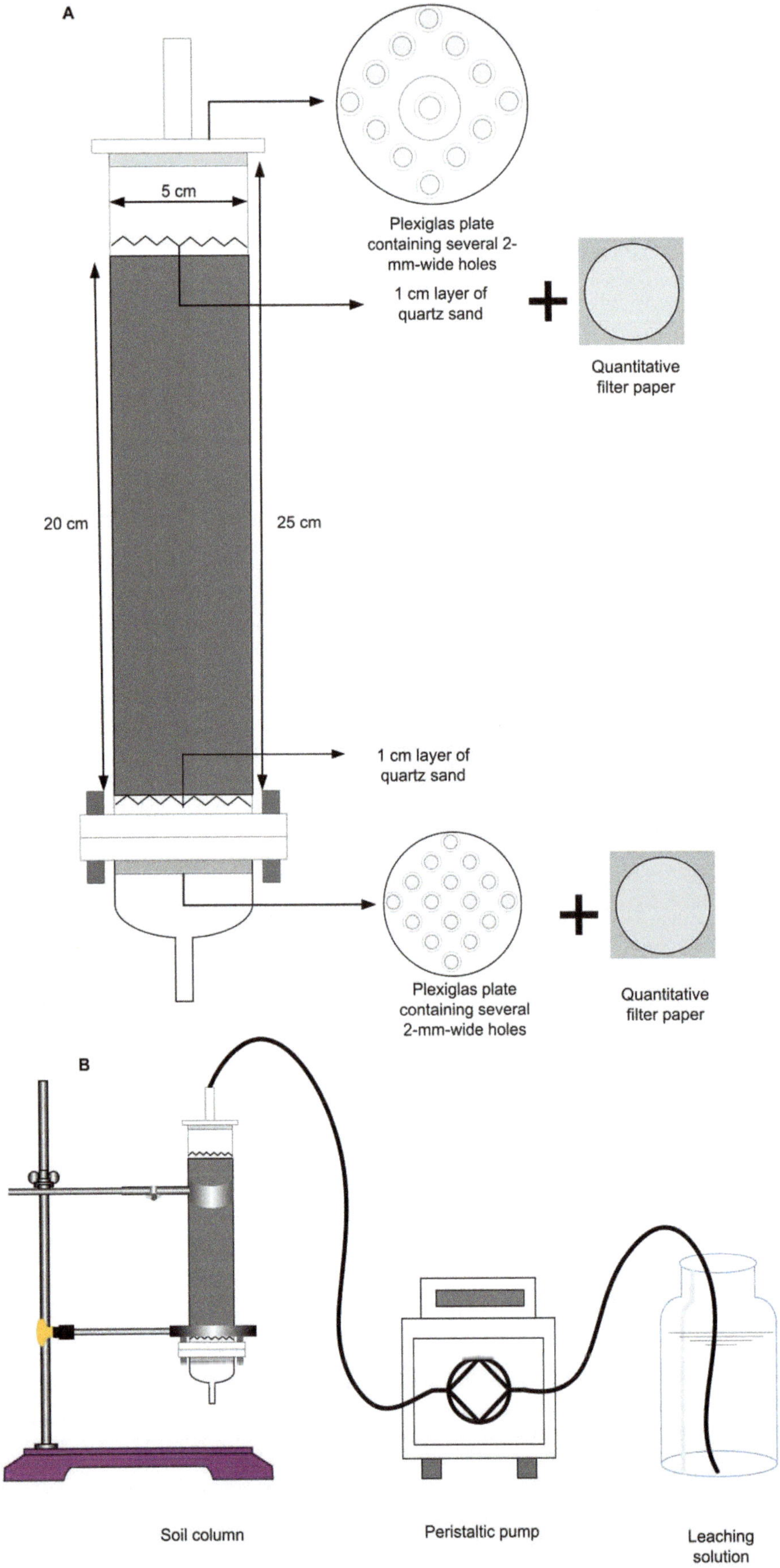
A
5 cm
Plexiglas plate containing several 2-mm-wide holes
1 cm layer of quartz sand
Quantitative filter paper
20 cm
25 cm
1 cm layer of quartz sand
Plexiglas plate containing several 2-mm-wide holes
Quantitative filter paper
B
Soil column
Peristaltic pump
Leaching solution

Figure 1. Leaching experimental design. (A) Schematic diagram of soil column. (B) Schematic analytical setup for the measurement of metal concentrations in the leaching experiment.

the soil samples, which ranged from −12.36% to 9.44%, were calculated to check the reliability of the sequential extraction procedure.

Statistical Analysis

Data were analyzed using the Origin 8.5.1 for Windows software at the 5% and 1% significance levels.

Results and Discussion

Heavy Metal Concentrations in the Leachate

The Pb and Cd concentrations in the leachate from SAR treatments at pH 3.0, 4.5, and 5.6 were generally very low throughout the leaching period, with most of them below their detection limits (data not shown). The maximum Pb and Cd concentrations in the leachate were only (17.0±2.16) and (2.8±0.42) $\mu g \cdot L^{-1}$, respectively, which are both less than the limits of the Chinese drinking water quality standards (50 and 10 $\mu g \cdot L^{-1}$, respectively). No significant differences were observed among the treatments of SAR at different pH levels. This result suggested that the polluted purple soil, which received SAR at a given pH in our study, may not cause groundwater contamination by Pb and Cd leaching. Low leachate concentrations of Pb and Cd can probably be attributed to the low mobility of Pb in soil as well as the comparatively low levels of Cd in artificially contaminated purple soil [15,16].

The Different results were observed with Cu and Zn. The leaching concentrations of Cu and Zn under SAR conditions are shown in Fig. 2. Higher Cu and Zn leachate concentrations were observed throughout the entire leaching period (ranging from 0.021 $mg \cdot L^{-1}$ to 1.49 $mg \cdot L^{-1}$ for Cu; 0.019 $mg \cdot L^{-1}$ to 2.34 $mg \cdot L^{-1}$ for Zn) as compared with those of Pb and Cd. However, the maximum Cu and Zn concentrations were still below the Chinese Quality Standards for Groundwater (Grade IV), thereby suggesting that leaching from the polluted purple soil under acid rain at the pH used in this study is unlikely to cause Cu and Zn contamination in water systems.

Fig. 2A shows the Cu concentrations in the leachates when SAR was used as the leaching reagent. Similar trends were observed for the changes in the Cu concentration of each treatment. The application of SAR produced a pulse of Cu in the leachates during the first stage of the experiment, with peaks of 1.49, 1.11, and 0.88 $mg \cdot L^{-1}$ at a pH of 3.0, 4.5, and 5.6, respectively. The concentrations decreased thereafter and reached <0.2 $mg \cdot L^{-1}$ at the end of the experiment. A small amount of Cu (<0.18 $mg \cdot L^{-1}$) was leached from the soil column in the first 200 mm of rainfall, which may be attributed to the required time for Cu to move from the upper surface layer to the bottom of the soil column and for the Cu-bound compounds to functionalize with the soil components. The SAR treatments at pH 3.0 and 4.5 yielded higher Cu concentrations than those in the control (SAR at pH 5.6), particularly before 1000 mm of SAR was added. The Cu leachate concentration increased as the pH values of SAR decreased. However, the Cu leachate concentrations for the treatments at pH 3.0 and 4.5 were approximately equal to those in the control as the leaching amounts were increased, thereby suggesting that the leachable Cu in all of the treatments was decreased. After 1200 mm of SAR was added, no significant differences were observed between the control and the treatments at pH 3.0 and 4.5.

The changes in the Zn leachate concentrations against the leaching amounts in the SAR treatments at different pH levels are shown in Fig. 2B. The variation curves of Zn with the three treatments were generally similar. Two peak values of the Zn concentration were found during the leaching experiment. The first peak appeared after approximately 300 mm to 500 mm of continuous leaching, and the second one appeared after 1200 mm of SAR was added. The peak values of the Zn concentration in the leachates were 2.34, 1.92, and 1.46 $mg \cdot L^{-1}$ for the treatments at pH 3.0, 4.5, and 5.6, respectively. These results suggested that the release of Zn from the contaminated purple soil can be observed in two stages. In the first leaching stage, leached Zn mainly existed as water soluble and exchangeable fractions. After being leached out, these mobile Zn began to adsorb and desorb in the soil column with the downward movement of leachate, and after 300–500 mm rainfall Zn concentration in the leachate reached the first peak value. During the second stage, Zn was released from different fractions such as the CAR-, OX-, and OM-bound Cu. A comparable amount of Zn was shifted from those fractions, which revealed the second peak values after 1200 mm of rainfall was added. Many studies have demonstrated that Zn in EXC and the water soluble fraction usually account for only 0.5% to 7% of the total Zn concentration [17–19], which could explain why the second peak values were higher than the first ones.

Total Leaching Amounts of Heavy Metals

An analysis of the total leaching amounts of heavy metals can directly reflect the leaching strength of the said heavy metals. The total leaching amounts of Pb and Cd were not calculated because most of their concentrations in the leachate were below the detection limits. The total leaching amounts of Cu and Zn in each treatment are shown in Table 1. Two-way ANOVA showed that the total amounts of the leached metals from the soil columns were significantly affected by the pH level and metal species as well as the interaction between these two factors (Table 2). Zn leached more easily from the soil column than Cu at each pH level (Tables 1 and 2). Similarly, Guo et al. [20] revealed that more Zn flowed out of the soil column than Cu in acidic soils under SAR treatments. The amount of leached metals increased at pH 3.0 and 4.5 as compared with that in the controls (pH 5.6), whereas the magnitude of this increase depended on the metal. Metallic elements become more soluble (leachable) under acidic conditions [21]. pH may likewise control the nature of the interactions between metals and soil surfaces [22]. Thus, higher Cu and Zn concentrations were induced by SAR and dissolved in columns at lower pH (Table 1). The H^+ ion in acid rain displaces the cations from their binding sites, which causes the increased amount of heavy metal desorption, as shown in the following reactions [Eq. (1–2)]:

$$SOH \cdots X^{2+} + H^+ \leftrightarrow SOH_2 + X^{2+} \quad (1)$$

$$SOX^{2+} + H^+ \leftrightarrow SOH + X^{2+} \quad (2)$$

The different affinities of the binding sites to metals may cause the different rates and amounts of the desorbed metal. For Zn, greater amounts were released (Table 1). H^+ ions can only replace

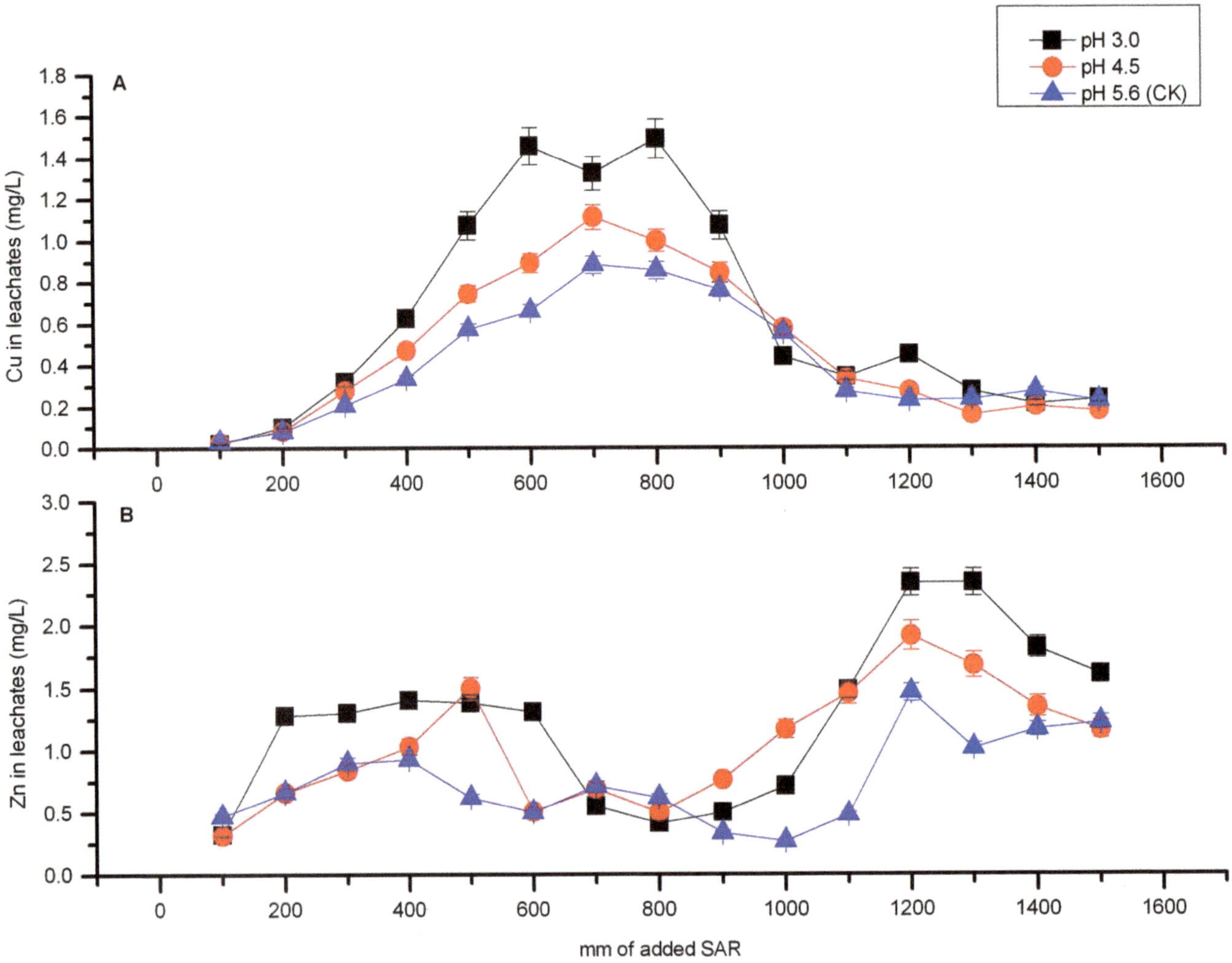

Figure 2. Metal concentrations in leachates as functions of addition of SAR. The concentrations of Cu and Zn in the leachates as functions of addition of SAR at different pHs. (A) Copper. (B) Zinc.

some metal cations, whereas the surface OH^- groups attached to the Fe or Al atoms can accept protons from the solution, thereby increasing the positive surface charges via the protonation of mineral surfaces [23,24]. Protonation is a very rapid reaction during the initial stage and depends on the amounts and properties of ferric oxides.

$$\text{Fe-OH} + H^+ \leftrightarrow \text{Fe-OH}_2^+ \leftrightarrow \text{dissolution}(Fe^{3+}) \quad (3)$$

The surface protonation in soil promotes the dissolution of ferric oxides in acidic solutions [Eq. (3)], which exhibit high adsorptive capacities for heavy metals and consequently increase the

Table 1. Total amounts of the metals leached from the soil columns in mg of metal per kg of soil.

Treatment	Total leaching amounts ($mg \cdot kg^{-1}$)	
	Cu	Zn
pH 3.0	(3.93±0.44) a	(7.81±0.61) a
pH 4.5	(2.98±0.36) b	(6.47±0.47) b
pH 5.6	(2.27±0.39) c	(4.72±0.33) c

Note: The values are means ± standard deviation. Different lower case letters show significant differences in the same treatment (ANOVA/LSD, $P<0.05$).

Table 2. Two-way ANOVA of pH levels and metal species effects for total amounts of the metals leached from the soil columns.

Source of variance	SS	MS	F
pH	48.22	48.22	245.75**
Metal species	16.93	8.46	43.14**
pH × metal species	1.64	0.82	4.18*
Error	2.35	0.20	
Total	69.14		

Note: **significant at 99% probability level, *significant at 95% probability level.

solubilization of these metals in their oxide fractions.

Chemical Speciation of Heavy Metals in Purple Soil Affected by SAR

The chemical fractions of Cu, Pb, Cd, and Zn in soils before and after the column leaching tests are shown in Fig. 3. For the original contaminated purple soil, Cu and Cd were dominantly associated with RES (37% to 41%), followed by OM (31% to 33%). Generally, EXC- and CAR-bound Cu and Cd accounted for <9% of the total amount of the respective metals. A significant fraction of Pb in the soils was bound in RES (66.31%), and four fractions accounted for <35% of the total Pb. Some studies have shown [15,25] that Pb is mostly present in the RES of soil at the surface or profile scale and is widely considered to exhibit very low geochemical mobility. The amount of Zn in soils was mainly associated with RES (42.69%), then with the OX (25.96%), followed by the CAR (14.44%), and OM (12.23%). The percent of exchangeable Zn in the soils was relatively low (4.69%).

Some variations in the proportions of the four metals in five fractions after the tests are shown by Fig. 3. The relative concentrations of the elements in the EXC and OX fractions were generally increased after the leaching experiment was conducted for the three treatments, whereas those of RES, OM, and CAR fractions were slightly decreased. The Pb and Cd concentrations in EXC ranged from 3.28% to 4.39% and 4.47% to 6.79%, respectively. By contrast, these metals increased by more than 3% in OX. The increase in the Cu concentrations in EXC was greater than that in the OX (EXC 3% to 7% vs. OX <2%). The relative concentrations of Zn in OM and OX significantly increased, particularly in the EXC fraction (average 4.32%).

The EXC fraction was usually the first to be brought into the solution and is readily available for plant uptake. The four trace elements in EXC, with high bioavailability and mobility, were increased from ~1.2% to 5.03%, which indicated the increased direct risk of Cu, Pb, Cd, and Zn contamination in the soil/groundwater system as caused by acid rain. The amounts of the non-residual fractions represent the amounts of potentially active trace elements [25]. Generally, the high proportion of trace elements in the non-residual fractions of soils may suggest the large contributions of anthropogenic elements. The non-residual fractions of Cu, Pb, Cd, and Zn in the soil zones after the leaching tests averaged at 62.55%, 37.43%, 65.95%, and 61.61%, respectively. The non-residual fractions of Cu, Pb, Cd, and Zn in the original soil samples averaged at 59.75%, 33.69%, 62.04%, and 57.31%, respectively. The increased non-residual fractions represented the increased potential risk of Cu, Pb, Cd, and Zn contamination in the soil/groundwater system. Simple correlation

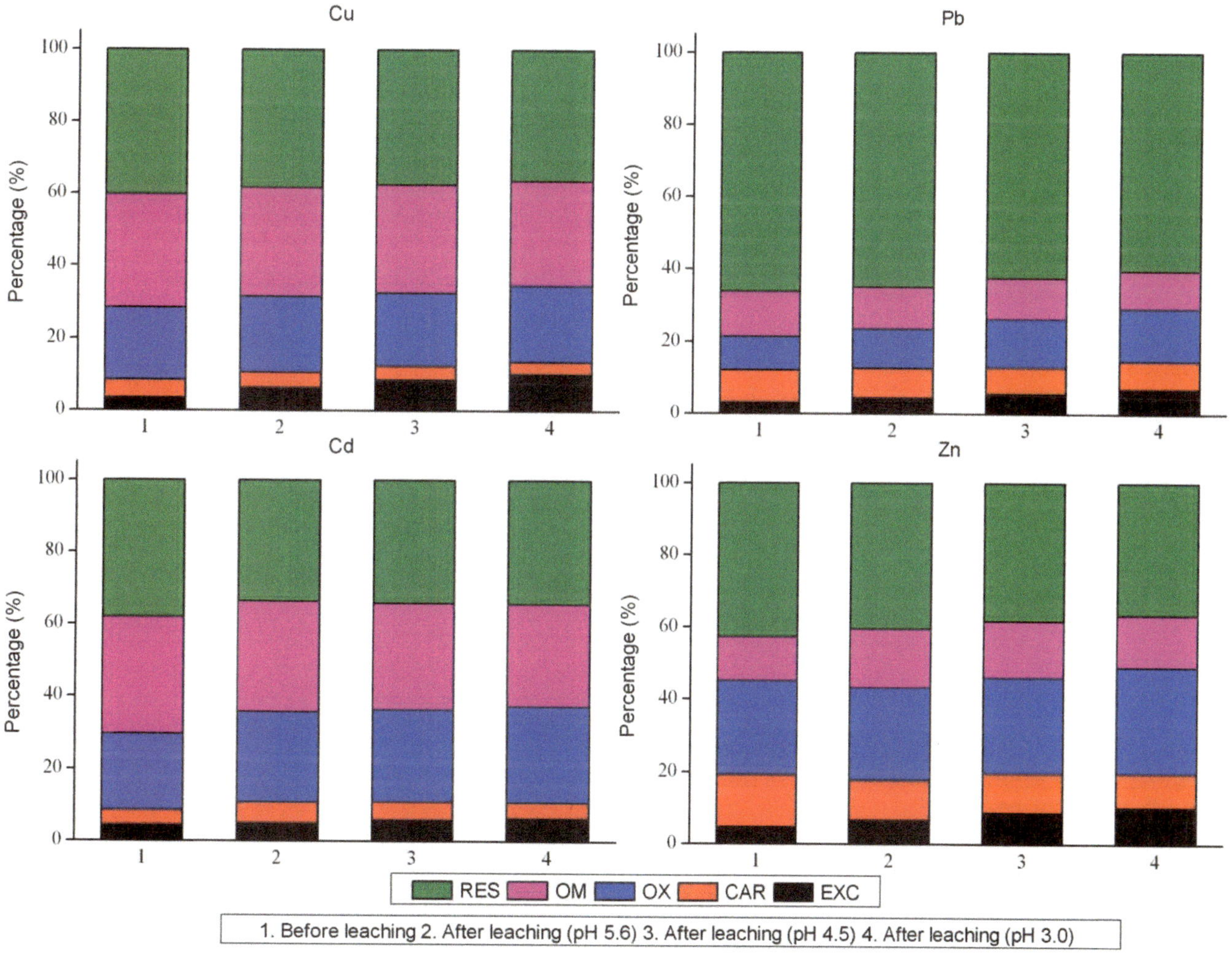

Figure 3. Variation of the percentage of metals in different fractions. Species distribution of Cu, Pb, Cd and Zn extracted with Tessier scheme before and after the column tests.

analysis indicated that the pH was significantly correlated with the exchangeable heavy metal content ($r = 0.974$, $P < 0.01$). Likewise, the pH was significantly correlated with the non-residual heavy metal content ($r = 0.968$, $P < 0.01$). These data suggested that acid rain favors the leaching of heavy metals from the contaminated purple soil. Consequently, acid rain can affect the mobilization potential of heavy metals. The decrease in the pH from 5.6 to 3.0 significantly enhanced such effects of acid rain, which are likely to cause more serious harmful effects on soil-vegetation systems.

Conclusions

The maximum leachate concentrations of Cu, Pb, Cd, and Zn from the soil that received SAR were less than those in the Chinese Quality Standard for Groundwater (Grade IV). Moreover, most of the Pb and Cd leachate concentrations were below their detection limits. By contrast, higher Cu and Zn leachate concentrations were observed because these metals were released in higher amounts as compared with Pb and Cd. The differences in the Cu and Zn leachate concentrations between the control (SAR at pH 5.6) and the treatments (SAR at pH 3.0 and 4.5) were significant. Similar trends were observed in the total leaching amounts of Cu and Zn. The proportions of Cu, Pb, Cd, and Zn in the EXC and OX fractions were generally increased after the leaching experiment at three pH levels, whereas those of the RES, OM, and CAR fractions were slightly decreased. Acid rain favors the leaching of heavy metals from the contaminated purple soil and causes the heavy metal fractions to become more labile. A decrease in the pH from 5.6 to 3.0 significantly enhanced such effects. However, the experiment is only an indoor simulation that used artificially contaminated purple soil. Thus, long-term field experiments on soils contaminated with acid rain should be conducted to study the cycling effect of heavy metal residues in the soil or those that leach into the groundwater.

Acknowledgments

The authors are especially grateful to Dr. Yirong Guo for critical review and perceptive comments on the manuscript. The comments and suggestions of the editor and two anonymous referees are appreciated.

Author Contributions

Conceived and designed the experiments: SZ. Performed the experiments: SZ CC. Analyzed the data: SZ XZ. Contributed reagents/materials/analysis tools: XZ. Wrote the paper: SZ.

References

1. Larssen T, Lydersen E, Tang D, He Y, Gao J, et al. (2006) Acid rain in China. Environ Sci Technol. 40: 418–425.
2. Singh A, Agrawal M (2008) Acid rain and its ecological consequences. J Environ. Biol. 29: 15–24.
3. Xie Z, Du Y, Zeng Y, Li Y, Yan M, et al. (2009) Effects of precipitation variation on severe acid rain in southern China. J Geogr Sci. 19: 489–501.
4. Ma L, Wang B, Yang J (2008) Spatial- Temporal Distribution of Acid Rain in Sichuan Province. Envion Sci Manage. 33: 26–29.
5. Li ZB, Li P, Han JG, Li M (2009) Sediment flow behavior in agro-watersheds of the purple soil region in China under different storm types and spatial scales. Soil Till Res. 105: 285–291.
6. Gao Y, Zhu B, Zhou P, Tang JL, Wang T, et al. (2009) Effects of vegetation cover on phosphorus loss from a hillslope cropland of purple soil under simulated rainfall: a case study in China. Nutr Cycl Agroecosys. 85: 263–273.
7. Wu F, Yang W, Zhang J, Zhou L (2010) Cadmium accumulation and growth responses of a poplar (Populus deltoids×Populus nigra) in cadmium contaminated purple soil and alluvial soil. J Hazard Mater. 177: 268–273.
8. Li Q, Wei C, Huang Y, Wang L, Wang D (2010) A study on the characteristics of heavy metals in orange ecosystem. Chin J Geochem. 29: 100–106.
9. Zhang JE, Ouyang Y, Ling DJ (2007) Impacts of simulated acid rain on cation leaching from the Latosol in south China. Chemosphere 67: 2131–2137.
10. Ling DJ, Huang QC, Ouyang Y (2010) Impacts of simulated acid rain on soil enzyme activities in a latosol. Ecotoxicol Environ Saf. 73: 1914–1918.
11. Wang M, Gu B, Ge Y, Liu Z, Jiang D, et al. (2009) Different responses of two Mosla species to potassium limitation in relation to acid rain deposition. J Zhejiang Univ-Sci B. 10: 563–571.
12. Lu RK (2000) Methods of soil and agrochemical analysis (in Chinese). Beijing: China Agricultural Science and Technology Press.
13. Tessier A, Campbell P, Bisson M (1979) Sequential extraction procedure for the speciation of particulate trace metals. Anal Chem. 51: 844–851.
14. Grolimund D, Borkovec M, Barmettler K, Sticher H (1996) Colloid-facilitated transport of strongly sorbing contaminants in natural porous media: a laboratory column study. Environ Sci Technol. 30: 3118–3123.
15. Martinez CE, Motto HL (2000) Solubility of lead, zinc and copper added to mineral soils. Environ Pollut. 107: 153–158.
16. Lee SZ, Chang L, Yang HH, Chen CM, Liu MC (1998) Adsorption characteristics of lead onto soils. J Hazard Mater. 63: 37–49.
17. Kabala C, Singh BR (2001) Fractionation and mobility of copper, lead, and zinc in soil profiles in the vicinity of a copper smelter. J Environ Qual. 30: 485–492.
18. Ma LQ, Rao GN (1997) Chemical fractionation of cadmium, copper, nickel, and zinc in contaminated soils. J Environ Qual. 26: 259–264.
19. Xiang HF, Tang HA, Ying QH (1995) Transformation and distribution of forms of zinc in acid, neutral and calcareous soils of China. Geoderma 66: 121–135.
20. Guo ZH, Liao BH, Huang CY (2005) Mobility and speciation of Cd, Cu, and Zn in two acidic soils affected by simulated acid rain. J Environ Sci. 17: 332–334.
21. Wilson MJ, Bell N (1996) Acid deposition and heavy metal mobilization. Appl Geochem. 11: 133–137.
22. Zhang H, Davison W, Tye AM, Crout NMJ, Young SD (2006) Kinetics of zinc and cadmium release in freshly contaminated soils. Environ Toxicol Chem. 25: 664–670.
23. Zhu MX, Jiang X, Ji GL (2005) Investigation of time-dependent reactions of H^+ ions with variable and constant charge soils: a comparative study. Appl Geochem. 20: 169–178.
24. Zhang FS, Zhang XN, Yu TR (1991) Reactions of hydrogen ions with variable charge soils: I. Mechanisms of reaction. Soil Sci. 151: 436–442.
25. Tack F, Verloo MG (1995) Chemical speciation and fractionation in soil and sediment heavy metal analysis: a review. Int J Environ An Ch. 59: 225–238.

Permissions

All chapters in this book were first published in PLOS ONE, by The Public Library of Science; hereby published with permission under the Creative Commons Attribution License or equivalent. Every chapter published in this book has been scrutinized by our experts. Their significance has been extensively debated. The topics covered herein carry significant findings which will fuel the growth of the discipline. They may even be implemented as practical applications or may be referred to as a beginning point for another development.

The contributors of this book come from diverse backgrounds, making this book a truly international effort. This book will bring forth new frontiers with its revolutionizing research information and detailed analysis of the nascent developments around the world.

We would like to thank all the contributing authors for lending their expertise to make the book truly unique. They have played a crucial role in the development of this book. Without their invaluable contributions this book wouldn't have been possible. They have made vital efforts to compile up to date information on the varied aspects of this subject to make this book a valuable addition to the collection of many professionals and students.

This book was conceptualized with the vision of imparting up-to-date information and advanced data in this field. To ensure the same, a matchless editorial board was set up. Every individual on the board went through rigorous rounds of assessment to prove their worth. After which they invested a large part of their time researching and compiling the most relevant data for our readers.

The editorial board has been involved in producing this book since its inception. They have spent rigorous hours researching and exploring the diverse topics which have resulted in the successful publishing of this book. They have passed on their knowledge of decades through this book. To expedite this challenging task, the publisher supported the team at every step. A small team of assistant editors was also appointed to further simplify the editing procedure and attain best results for the readers.

Apart from the editorial board, the designing team has also invested a significant amount of their time in understanding the subject and creating the most relevant covers. They scrutinized every image to scout for the most suitable representation of the subject and create an appropriate cover for the book.

The publishing team has been an ardent support to the editorial, designing and production team. Their endless efforts to recruit the best for this project, has resulted in the accomplishment of this book. They are a veteran in the field of academics and their pool of knowledge is as vast as their experience in printing. Their expertise and guidance has proved useful at every step. Their uncompromising quality standards have made this book an exceptional effort. Their encouragement from time to time has been an inspiration for everyone.

The publisher and the editorial board hope that this book will prove to be a valuable piece of knowledge for researchers, students, practitioners and scholars across the globe.

List of Contributors

Daniel Stalter
National Research Centre for Environmental Toxicology, The University of Queensland, Brisbane, Australia
Department Aquatic Ecotoxicology, Biological Sciences Division, Goethe University Frankfurt am Main, Frankfurt am Main, Germany

Kristin Quednow
Analytical Environmental Chemistry, Institute for Atmospheric and Environmental Sciences, Goethe University Frankfurt am Main, Frankfurt am Main, Germany

Alexandra Botzat
Department of Ecology - Conservation Ecology, Faculty of Biology, Philipps University of Marburg, Marburg, Germany

Jörg Oehlmann and Axel Magdeburg
Department Aquatic Ecotoxicology, Biological Sciences Division, Goethe University Frankfurt am Main, Frankfurt am Main, Germany

Seyedeh Belin Tavakoly Sany
Institute of Biological Sciences University of Malaya, Kuala Lumpur, Malaysia
Food Science and Technology Research Institute, ACECR Mashhad Branch, Mashhad, Iran

Majid Rezayi
Food Science and Technology Research Institute, ACECR Mashhad Branch, Mashhad, Iran
Chemistry Department, Faculty of Science, University Malaya, Kuala Lumpur, Malaysia

Rosli Hashim and Aishah Salleh
Institute of Biological Sciences University of Malaya, Kuala Lumpur, Malaysia

Ali Mehdinia
Department of Marine Science, Marine Living Group, Iranian National Institute for Oceanography, Tehran, Iran

Omid Safari
Faculty of Natural Resources and Environment, Ferdowsi University of Mashhad, Mashhad, Iran

Longxiang Li and Jieping Zhou
Institute of Remote Sensing and Digital Earth, Chinese Academy of Sciences, Olympic Science & Technology Park of CAS, Beijing, China

Jianhua Gong
Institute of Remote Sensing and Digital Earth, Chinese Academy of Sciences, Olympic Science & Technology Park of CAS, Beijing, China and Zhejiang-CAS Application Center for Geoinformatics, Jiashan, Zhejiang, China

Leslie M. Roche, Lea Kromschroeder and Kenneth W. Tate
Department of Plant Sciences, University of California, Davis, California, United States of America

Edward R. Atwill
School of Veterinary Medicine, University of California University of California, Davis, California, United States of America

Randy A. Dahlgren
Department of Land, Air, and Water Resources, University of California, Davis, California, United States of America

Saurabh Sudha Dhiman and Jung-Kul Lee
Department of Chemical Engineering, Konkuk University, Gwangjin-Gu, Seoul, Republic of Korea
Institute of SK-KU Biomaterials, Konkuk University, Gwangjin-Gu, Seoul, Republic of Korea

Gaurav Garg
Department of Biotechnology, Kurukshetra University, Kurukshetra, Haryana, India
Department of Biotechnology, Maharishi Markandeshwar University, Mullana-Ambala, Haryana, India

Jitender Sharma
Department of Biotechnology, Kurukshetra University, Kurukshetra, Haryana, India

Vipin C. Kalia
Microbial Biotechnology and Genomics, CSIR-Institute of Genomics and Integrative Biology, Delhi University Campus, Delhi, India

Yun Chan Kang
Department of Chemical Engineering, Konkuk University, Gwangjin-Gu, Seoul, Republic of Korea

Marisa Mazari-Hiriart
Laboratorio Nacional de Ciencias de la Sostenibilidad, Departamento de Ecología de la Biodiversidad, Instituto de Ecología, Universidad Nacional Autónoma de México, México, D.F., México

Gustavo Pérez-Ortiz
Laboratorio Nacional de Ciencias de la Sostenibilidad, Departamento de Ecología de la Biodiversidad, Instituto de Ecología, Universidad Nacional Autónoma de México, México, D.F., México
Departamento de Biología, Facultad de Ciencias, Universidad Nacional Autónoma de México, México, D.F., México

María Teresa Orta-Ledesma and Isaura Yañez-Noguez
Coordinación de Ingeniería Ambiental, Instituto de Ingeniería, Universidad Nacional Autónoma de México, México, D.F., México

Felipe Armas-Vargas
Posgrado en Ciencias de la Tierra, Instituto de Geología, Universidad Nacional Autónoma de México, México, D.F., México

Marco A. Tapia
Departamento de Biología, Facultad de Ciencias, Universidad Nacional Autónoma de México, México, D.F., México

Rosa Solano-Ortiz
Posgrado en Ciencias Biológicas, Facultad de Ciencias-Instituto de Ecología, Universidad Nacional Autónoma de México, México, D.F., México

Miguel A. Silva
Posgrado de Ciencias Bioquímicas, Departamento de Ingeniería Celular y Biocatálisis, Instituto de Biotecnología, Universidad Nacional Autónoma de México, Cuernavaca, Morelos, México

Yolanda López-Vidal
Programa de Inmunología Molecular Microbiana, Departamento de Microbiología y Parasitología, Facultad de Medicina, Universidad Nacional Autónoma de México, México, D.F., México

Carlos Díaz-Ávalos
Departamento de Probabilidad y Estadística, Instituto de Investigaciones en Matemáticas Aplicadas y en Sistemas, Universidad Nacional Autónoma de México, México, D.F., México

Joel T. Kidgell, Rocky de Nys, Nicholas A. Paul and David A. Roberts
MACRO - the Centre for Macroalgal Resources and Biotechnology, and School of Marine and Tropical Biology, James Cook University, Townsville, Queensland, Australia

Akiko Tomaru
Division of Microbiology, National Institute of Health Sciences, Setagaya-ku, Tokyo, Japan
Asian Natural Environmental Science Center, The University of Tokyo, Tokyo, Japan

Masanobu Kawachi and Mikihide Demura
Environmental Biology Division, National Institute for Environmental Studies, Ibaraki, Japan

Yasuwo Fukuyo
Asian Natural Environmental Science Center, The University of Tokyo, Tokyo, Japan

Jiabo Chen
Department of Natural Resources, College of Environment and Natural Resources, Zhejiang University, Hangzhou, Zhejiang Province, China

Jun Lu
Department of Natural Resources, College of Environment and Natural Resources, Zhejiang University, Hangzhou, Zhejiang Province, China
China Ministry of Education Key Lab of Environment Remediation and Ecological Health, Zhejiang University, Hangzhou, Zhejiang Province, China

Li Liu, Ling Chen, Lili Zhang and Lingling Wu
Key Laboratory of Yangtze Water environment, Ministry of Education, Tongji University, Shanghai, China

Ying Shao
Key Laboratory of Yangtze Water environment, Ministry of Education, Tongji University, Shanghai, China
Department of Ecosystem Analysis, Institute for Environmental Research (Biology V), Aachen Biology and Biotechnology, RWTH Aachen University, Aachen, Germany

Tilman Floehr, Hongxia Xiao, Yan Yan and Kathrin Eichbaum
Department of Ecosystem Analysis, Institute for Environmental Research (Biology V), Aachen Biology and Biotechnology, RWTH Aachen University, Aachen, Germany

Henner Hollert
Key Laboratory of Yangtze Water environment, Ministry of Education, Tongji University, Shanghai, China
Department of Ecosystem Analysis, Institute for Environmental Research (Biology V), Aachen Biology and Biotechnology, RWTH Aachen University, Aachen, Germany
College of Resources and Environmental Science, Chongqing University, Chongqing, China
School of Environment, Nanjing University, Nanjing, China

Xin Sui, Xiangrong Wang, Guotao Peng, Shoubing Wang and Zhengqiu Fan
Department of Environmental Science & Engineering, Fudan University, Shanghai, China

Honghui Huang
South China Sea Fisheries Research Institute, Chinese Academy of Fishery Sciences, Guangzhou, China

Katri Jalava
Department of Infectious Disease Surveillance and Control, National Institute for Health and Welfare, Helsinki, Finland
Environmental Health Department, City of Tampere, Finland

Hanna Rintala, Sirpa Hakkarainen, Juhani Merentie and Martti Pärnänen
Siilinjä rvi municipality, Finland

Jukka Ollgren, Marja Palander, Pia Räsänen, Sallamaari Siponen, Outi Nyholm, Aino Kyyhkynen, Markku Kuusi and Anja Siitonen
Department of Infectious Disease Surveillance and Control, National Institute for Health and Welfare, Helsinki, Finland

Leena Maunula, Joana Revez and Marja-Liisa Hänninen
Department of Food Hygiene and Environmental Health, University of Helsinki, Helskinki, Finland

Vicente Gomez-Alvarez, Hodon Ryu and Jorge W. Santo Domingo
Office of Research and Development, United States Environmental Protection Agency, Cincinnati, Ohio, United States of America

Jenni Antikainen and Raisa Loginov
Laboratory HUSLAB, Helsinki University Hospital, Helsinki, Finland

Ari Kauppinen, Ilkka Miettinen and Tarja Pitkänen
Department of Environmental Health, National Institute for Health and Welfare, Kuopio, Finland

Mailis Laht, Veiko Voolaid, Tanel Tenson and Veljo Kisand
Institute of Technology, University of Tartu, Tartu, Estonia

Antti Karkman and Marko Virta
Department of Food and Environmental Sciences, University of Helsinki, Helsinki, Finland

Christian Ritz
Department of Nutrition, Exercise and Sports, Faculty of Science, University of Copenhagen, Copenhagen, Denmark

Miklos Blaho, Tamas Herczeg, Adam Egri, Denes Szaz, Alexandra Farkas, Andras Barta and Gabor Horvath
Environmental Optics Laboratory, Department of Biological Physics, Physical Institute, Eötvö s University, Budapest, Hungary

Gyorgy Kriska
Group for Methodology in Biology Teaching, Biological Institute, Eötvö s University, Budapest, Hungary
Danube Research Institute, Centre for Ecological Research, Hungarian Academy of Sciences, Budapest, Hungary

Nikolett Tarjanyi
Danube Research Institute, Centre for Ecological Research, Hungarian Academy of Sciences, Budapest, Hungary

Laszlo Czinke
Environmental Optics Laboratory, Department of Biological Physics, Physical Institute, Eötvö s University, Budapest, Hungary
Group for Methodology in Biology Teaching, Biological Institute, Eötvö s University, Budapest, Hungary

Guangsheng Li, Yongmo Wang and Guoan Zhang
Hubei Insect Resources Utilization and Sustainable Pest Management Key Laboratory, College of Plant Science and Technology, Huazhong Agricultural University, Wuhan, Hubei, P.R. CHINA

Biao Liu
Nanjing Institute of Environmental Science, Ministry of Environmental Protection of the People's Republic of China, Nanjing, P.R. CHINA

Lieven Bervoets, Maarten De Jonge and Ronny Blust
Department of Biology, Systemic Physiological & Ecotoxicological Research, University of Antwerp, Antwerp, Belgium

Dries Knapen
Department of Biology, Systemic Physiological & Ecotoxicological Research, University of Antwerp, Antwerp, Belgium
Laboratory for Veterinary Physiology, Department of Veterinary Sciences, University of Antwerp, Antwerp, Belgium

Karen Van Campenhout
Department of Environment and Health, Flemish Government, Brussels, Belgium

Yufeng Gong, Hua Tian, Suping Yu and Shaoguo Ru
Marine Life Science College, Ocean University of China, Qingdao, China

Lijia Wang
Qingdao Chengtou Xiaojianxi Leachate Treatment Corporation Limited, Qingdao, China

Noah R. Lottig
Center for Limnology Trout Lake Station, University of Wisconsin Madison, Boulder Junction, Wisconsin, United States of America

Tyler Wagner
U.S. Geological Survey, Pennsylvania Cooperative Fish and Wildlife Research Unit, Pennsylvania State University, University Park, Pennsylvania, United States of America

Emily Norton Henry
Department of Fisheries and Wildlife, Michigan State University, East Lansing, Michigan, United States of America
Department of Outreach and Engagement, Oregon State University, Tillamook, Oregon, United States of America

Kendra Spence Cheruvelil
Department of Fisheries and Wildlife, Michigan State University, East Lansing, Michigan, United States of America
Lyman Briggs College, Michigan State University, East Lansing, Michigan, United States of America

Katherine E. Webster
School of Natural Sciences, Trinity College, Dublin, Ireland

John A. Downing
Department of Ecology, Evolution, and Organismal Biology, Iowa State University, Ames, Iowa, United States of America

Craig A. Stow
Great Lakes Environmental Research Laboratory, National Oceanic and Atmospheric Administration, Ann Arbor, Michigan, United States of America

Yuan Liu, Yongzhuo Liu, Yuanjun Ding, Jinwei Zheng, Tong Zhou, Lianqing Li, Jufeng Zheng, Xuhui Zhang, Xinyan Yu and Jiafang Wang
Institute of Resource, Ecosystem and Environment of Agriculture, Nanjing Agricultural University, Nanjing, China

Genxing Pan
Institute of Resource, Ecosystem and Environment of Agriculture, Nanjing Agricultural University, Nanjing, China
Center of Ecosystem Carbon Sink and Environment Remediation, Zhejiang Agricultural and Forestry University, Linan, Hangzhou, China

David Crowley
Department of Environment Sciences, University of California Riverside, Riverside, California, United States of America

Shun-an Zheng, Xiangqun Zheng and Chun Chen
Agro-Environmental Protection Institute, Ministry of Agriculture, Tianjin, People's Republic of China
Key Laboratory of Production Environment and Agro-Product Safety, Ministry of Agriculture, Tianjin, People's Republic of China
Tianjin Key Laboratory of Agro-Environment and Agro-Product Safety, Tianjin, People's Republic of China

Index